Tributes
Volume 7

Dialogues, Logics and Other Strange Things
Essays in Honour of
Shahid Rahman

Tributes Series Editor
Dov Gabbay

dov.gabbay@kcl.ac.u<

Dialogues, Logics and Other Strange Things

Essays in Honour of Shahid Rahman

edited by

Cédric Dégremont,
Laurent Keiff
and
Helge Rückert

ISBN 978-1-904987-13-0

College Publications
Scientific Director: Dov Gabbay
Managing Director: Jane Spurr
Department of Computer Science
King's College London, Strand, London WC2R 2LS, UK

http://www.collegepublications.co.uk

Original cover design by orchid creative   www.orchidcreative.co.uk
Printed by Lightning Source, Milton Keynes, UK

CONTENTS

Introduction

CÉDRIC DÉGREMONT, LAURENT KEIFF AND HELGE RÜCKERT

Shahid Rahman was born in 1956, in New Delhi, India. He holds Argentinean and German citizenship. His academic training took him through the Universidad Nacional del Sur (Bahia Blanca, Argentina), the Universität Erlangen-Nürnberg and the Universität des Saarlandes (Germany), and ranges from Mathematics to Philology, through Psychology and Philosophy. He holds a PhD in Philosophy (subsidiary subjects: Psychology and Philology) from the Universität des Saarlandes. His Dissertation (1993, supervisor: Kuno Lorenz) was an investigation of dialogical logic from the point of view of Category Theory. His Habilitation, in 1997 (Universität des Saarlandes), was devoted to Hugh MacColl, fictions and connexive logic. Since 2001, he has been Full Professor of Logic and Epistemology, University of Lille 3 (Human Sciences), France.

We would like to take the opportunity of this introduction to give a quick survey of the work of Shahid Rahman as a logician, an epistemologist and an editor. We think that doing so will also offer an interesting perspective on the wide diversity of topics covered in this volume. The main part of his work concerns dialogical logic. It has been Shahid Rahman's principal contribution to convert dialogical logic into a general framework for developing and combining logics. From the early days in the 50's on, in the view of its creators Paul Lorenzen and Kuno Lorenz, dialogical logic was thought to provide the "foundation" of intuitionistic logic. In 1987, Shahid Rahman suggested in a talk at the Universität des Saarlandes for the first time the link between dialogues and categorial *topoi*, though it was restricted to the foundations of intuitionistic dialogues. The second part of his PhD thesis, published in 1993, was dedicated to this topic. The first part of the thesis contains the first systematic study of the relations between dialogues and tableaux and their metatheory.

Shahid Rahman and his teams (first at the Universität des Saarlandes with Helge Rückert and later on with the group "Pragmatisme Dialogique" at Lille, founded in collaboration with Laurent Keiff), proposed a shift in the very notion of dialogical logic at a time when it was slowly disappearing from the philosophical research agenda. Under his influence, it converted into a general framework in which were studied, among others: IF-logic, free logic, propositional and first order normal modal logic, non-normal modal logic, hybrid logic, paraconsistent logic, linear logic, relevant logic and connexive logic. One of the points of the approach is that dialogical logic could be used as bridge between a model-theoretical and a proof-theoretical approach and even as a method of generating tableaux-systems for logics with no perspicuous model-theoretical semantics such as linear logic and connexive logic.

The dialogical perspective on non-classical logics has been the source of fruitful discussions. In 1995, in a talk at the Max Planck Institute of Computer Sciences in Saarbrücken, Dov Gabbay pointed out that the new developments concerning the interface between games and logic, which actually provided the technical basis of the dialogical approach, suggested the possibility to understand dialogues as a kind of pragmatic deductive systems. In 1999, Rahman and Rückert organized a workshop in Saarbrücken under the title *New Perspectives in Dialogical Logic*, that yielded a special volume of *Synthese* (vol. 127, 1-2, April/May 2001). The volume witnesses the new *élan* in the study of the links between games and logic coming mainly from computer science and artificial intelligence. Patrick Blackburn penned there a paper called "Modal Logic as Dialogical Logic" as a response to Rahman and Rückert's paper on dialogical modal logic (published earlier in *Logique et Analyse*, vol. 167-68, 1999) where connections are established between dialogues for modal logic and hybrid languages, in the general model-theoretic perspective on modal logic in which bisimulation is a central notion. In the same volume Gabriel Sandu pointed out one other important development in the interface between games and logic, namely linear logic. In fact linear logic and its associated generalization, the "ludics" project of Jean-Yves Girard, as well as the other game-theoretical approaches stemming from linear logic such as those of Andreas Blass or Samson Abramsky, are one of the research interests of the group in Lille.

In his Vision Statement for an entry on his work in the *International Directory of Logicians* (College Publications 2008, pp. 279-285) Shahid Rahman writes the following:

> Logic had a central role in the philosophy of science projects of the 20th century. The historical and sociological criticisms of the sixties banned logic from philosophy of science pointing out that science is more than a set of propositions linked by classical logic. This does not mean that the links between logic and philosophy and history of science must be abandoned. Rather that the reflection on sciences must be approached with more and new sophisticated tools of logic able to deal with propositions in a structure, the description of this

structure and the dynamics of this structure. The conjecture is that this will be possible by the confluence of game theoretical approaches to logic, belief revision and game, decision and argumentation theories.

Actually, this statement may be seen as a brief *manifesto* for the series *Logic, Epistemology and the Unity of Science* that he launched in 2001 with John Symons, the first volume of which appeared in 2004, edited by Rahman, Symons, Gabbay and Van Bendegem. Shahid Rahman, as an editor, has aimed to create adequate tools in order for the philosophical community to investigate the classical problems of the philosophy of science from the perspective of contemporary developments in mathematical logic, especially at the interface with game theory. Such is the driving inspiration for his other series at College Publications, namely the *Cahiers de Logique et d'Epistémologie* and its Spanish version: the *Cuadernos de Lógica, Epistemología y Lenguaje*, and the new *The Games of Logic, A Philosophical Perspective*, all of which being conceived as gathering places where philosophy should make possible the dialogue within the diversity of scientific expertise.

As a team leader, Shahid Rahman, together with his collaborators and his students work in the aforementioned perspective on a wide diversity of systems, including hybrid languages, first-order modal logic and dynamic free logic but also non-monotonic reasoning, belief revision, abduction and even rational choice.

The idea for this Festschrift was born on Shahid Rahman's 50th birthday. Given his liking for heterodoxy, we encouraged provocative contributions outside the scientific mainstream. This resulted in papers, very different in length and style, dealing with a wide variety of topics which, we think, make up for an interesting mix. The fact that many of the contributions have more than one author reflects Shahid Rahman's very communicative nature which has resulted in many scientific collaborations of his own. In one case (Schurz and Niiniluoto) the volume even includes a little dialogue with a paper, a comment on this paper and a rejoinder to the comment. The title for this volume *Dialogues, Logics and other Strange Things* is reminiscent of the title of Shahid Rahman's German PhD thesis *Dialoge, protologische Kategorien und andere Seltenheiten*.

We, the editors, wish to heartily thank all the contributors, as well as the further referees, who provided very valuable comments which led to an improvement of the overall scientific quality of this volume (in alphabetical order): Denis Bonnay (Paris), Robin Clark (Pennsylvania), David Corfield (Max Planck Institute, Tübingen), Frans H. van Eemeren (Amsterdam), Paul Egré (Paris), Marie-Hélène Gorisse (Lille), Siegfried Gottwald (Leipzig), Paul Harrenstein (München), Wilfrid Hodges (London), Terry Horgan (Arizona), Chris Meacham (University of Massachusetts, Amherst), Sieuwert van Otterloo (McKinsey & Company), Douglas Patterson (Kansas State), Robin Smith (Texas A&M).

Furthermore we would like to thank Benjamin Quarta, Jan Gogoll (both Manheim), Nicolas Troquard (Liverpool) and Jonathan A. Zvesper (Amsterdam) for editorial help, and especially Jane Spurr from College Publications.

Shahid, this one is for you! We hope you like it!

Coalition Games over Kripke Semantics: Expressiveness and Complexity

PHILIPPE BALBIANI, OLIVIER GASQUET, ANDREAS HERZIG,
FRANÇOIS SCHWARZENTRUBER AND NICOLAS TROQUARD

ABSTRACT. We show that Pauly's Coalition Logic can be embedded into a richer normal modal logic that we call Normal Simulation of Coalition Logic (NCL). We establish that the latter is strictly more expressive than the former by proving that it is NEXPTIME complete in the case of at least two agents.

1 Introduction

Recently Shahid Rahman, together with Cédric Dégremont [9], has proposed a dialogical proof system for deliberative STIT theories [6].

At the same time, two of us were investigating the properties of these theories, as well as their relationship with Coalition Logic, Alternating-time Temporal Logic (ATL) and epistemic extensions of these [2, 12, 4]. (See also [16].) The work of Cédric Dégremont shed new light on a subject that was studied only very rarely up to now viz. the proof-theory of deliberative STIT theories (a notable exception being [18]). Their contribution was one of the triggers of our interest in proof systems for logics of agency, paving the way for decidability and complexity results. The present paper provides such results for a fragment of the logic of Chellas's STIT, viz. Coalition Logic.

Coalition Logic (CL) was proposed by Pauly in [13] as a logic for reasoning about social procedures characterized by complex strategic interactions between agents, individuals or groups. Examples of such procedures are fair-division algorithms or voting processes. CL facilitates reasoning about abilities of coalitions in games by extending classical logic with operators $\langle\!\langle J \rangle\!\rangle \varphi$ for groups of agents J, reading: "the coalition J has a joint strategy to ensure that φ."[1]

In [3], we have shown that CL can be embedded into the logic of Chellas's STIT. STIT theory is the most prominent account of agency in philosophy of action. It is the logic of constructions of the form "agent i sees to it that φ holds". In

---

[1]Note that we use $\langle\!\langle J \rangle\!\rangle \varphi$ as an alternative notation for Pauly's non-normal operator $[J]\varphi$. We introduce this alternative syntax for two reasons: (1) the new syntax fits better with the quantifier combination $\exists - \forall$ underlying the semantics, and (2) we use Pauly's original syntax $[J]\varphi$ to denote Chellas's STIT operator, thereby emphasizing that this is a normal modal necessity operator.

the present paper we go beyond that and provide a proof theoretic analysis of the embedded fragment. In order to do that, we extend Xu's logic of Chellas's STIT with a 'next' operator, resulting in a logic we call NCL. We provide a complete and elegant axiomatization and prove that Xu's logic of Chellas's STIT and CL are embedded.

As designed by Pauly, semantics of Coalition Logic is in terms of neighborhood models, that is, models providing a *neighborhood function*, associating a world to a set of neighborhoods, or clusters. (See [8, Chap. 7] for details about those models.) Here, we present the normal logic NCL whose semantics is the well known *relational* or *Kripke semantics*. Our embedding is in itself an interesting result since it shows that Coalition Logic can be evaluated with respect to relational models.

Moreover, NCL extends CL with capabilities of reasoning about what a coalition is actually doing or about to do, as opposed to what it *could* do. We finally establish decidability and complexity results.

The remaining of this paper is along the following outline. Section 2 and 3 present respectively Coalition Logic and its normal simulation NCL. In Section 4, we check that NCL inherits all the principles of a version of deliberative STIT theories without tense operators. Analogously, Section 5 provides a translation from Coalition Logic to NCL. Section 6 is devoted to the studies of decidability and complexity. In Section 7, we devise about NCL expressiveness. In particular, we enlighten that NCL is more expressive than Coalition Logic, and informally discuss a possible application of the logic to the notion of 'power over' of agents and groups of agents. We finally conclude in Section 8.

2   Coalition Logic CL

2.1   Syntax of CL

Let AGT be a nonempty finite set of agents and *Prop* an infinite countable set of atomic formulas. The language $\mathcal{L}_{CL}$ (all formulas of Coalition Logic) is defined as follows:

$$\varphi ::= p \mid \neg\varphi \mid \varphi \vee \varphi \mid \langle\!\langle J \rangle\!\rangle \varphi$$

where p ranges over *Prop* and J ranges over subsets of AGT. The other Boolean operators are defined as usual.

2.2   Coalition model semantics

DEFINITION 1 (effectivity function). Given a nonempty set of states S, an *effectivity function* is a function $E : 2^{AGT} \rightarrow 2^{2^S}$. An effectivity function is said to be:

- *J-maximal* iff for all $X \subseteq S$, if $S \setminus X \notin E(AGT \setminus J)$ then $X \in E(J)$;

- *outcome-monotonic* iff for all $X, X' \subseteq S$ and for all $J \subseteq AGT$, if $X \in E(J)$ and $X \subseteq X'$ then $X' \in E(J)$;

- *superadditive* iff for all J_1, J_2, if $J_1 \cap J_2 = \emptyset$ then for all $X_1, X_2 \subseteq S$, if $X_1 \in E(J_1)$ and $X_2 \in E(J_2)$ then $X_1 \cap X_2 \in E(J_1 \cup J_2)$.

The function E intuitively associates every coalition J to a set of subsets of S (or set of outcomes) for which J is effective. That is, J can force the world to be in some state of X, for each $X \in E(J)$.

DEFINITION 2 (playable effectivity function). Given a nonempty set of states S, an effectivity function $E : 2^{AGT} \rightarrow 2^{2^S}$ is said to be *playable* iff the following conditions hold:

1. for all J, $\emptyset \notin E(J)$ (Liveness)

2. for all J, $S \in E(J)$ (Termination)

3. E is AGT-maximal

4. E is outcome-monotonic

5. E is superadditive

A coalition model is a pair $((S, E), V)$ where:

- S is a nonempty set of states;

- $E : S \rightarrow (2^{AGT} \rightarrow 2^{2^S})$ associates every state s with a playable effectivity function $E(s)$;

- $V : S \rightarrow 2^{Prop}$ is a valuation function.

We will write $E_s(J)$ instead of $E(s)(J)$ to denote the effectivity of J at the state s.

Truth conditions are standard for Boolean operators. We evaluate the coalitional operators against a coalition model M and a state s as follows:

$$M, s \models \langle\!\langle J \rangle\!\rangle \varphi \text{ iff } \{t \mid M, t \models \varphi\} \in E_s(J)$$

2.3   Game Semantics

In [14], Pauly investigates the link between coalition models and strategic games.

DEFINITION 3. A *strategic game* is a tuple $G = (S, \{\Sigma_i \mid i \in AGT\}, o)$ where S is a nonempty set, Σ_i is a nonempty set of choices for every agent $i \in AGT$, $o : \prod_{i \in AGT} \Sigma_i \rightarrow S$ is an outcome function which associates an outcome state in S with every combination of choices of agents (choice profile).

It appears that there is a strong link between a coalition model (whose effectivity structure is *playable* by definition) and a strategic game.

DEFINITION 4. Given a strategic game $G = (S, \{\Sigma_i | i \in AGT\}, o)$, the *effectivity function* $E_G : 2^{AGT} \to 2^{2^S}$ of G is defined as follows: for all J, let $E_G(J)$ be the set of all subsets X of S such that there exists a $Card(J)$-tuple σ in $\prod_{i \in J} \Sigma_i$ such that for all $Card(AGT \setminus J)$-tuples σ' in $\prod_{i \in AGT \setminus J} \Sigma_i$, $o(\sigma, \sigma')$ is in X.

Pauly then gives the following characterization:

THEOREM 5 ([14]). *An effectivity function E is playable iff it is the effectivity function of some strategic game.*

DEFINITION 6. Let $((S, E), V)$ be a coalition model. Let s be a state of S. A set $Y \subseteq S$ is called a *minimal effectivity outcome at s for J* iff (1) $Y \in E_s(J)$ and (2) for all $Y' \in E_s(J)$, if $Y' \subseteq Y$ then $Y' = Y$.

DEFINITION 7. The *non-monotonic core* of E is the mapping $\mu_E : 2^{AGT} \times S \to 2^{2^S}$ such that $\mu_E(J, s) = \{Y \mid Y$ is a minimal effectivity outcome at s for $J\}$.

The outcome of a strategic game is completely determined when every agent has made his choice.

PROPOSITION 8. *Let (S, E, V) be a coalition model. For all states $s \in S$, $\mu_E(AGT, s)$ is a nonempty set of singletons.*

PROOF. This is a corollary of Theorem 5.   ∎

2.4   Axiomatization of CL

The set of formulas that are valid in coalition models is completely axiomatized by the following principles [14].

(ProTau)   enough tautologies of the propositional calculus

(⊥) $\neg \langle\!\langle J \rangle\!\rangle \bot$

(⊤) $\langle\!\langle J \rangle\!\rangle \top$

(N) $\neg \langle\!\langle \emptyset \rangle\!\rangle \neg \varphi \to \langle\!\langle AGT \rangle\!\rangle \varphi$

(M) $\langle\!\langle J \rangle\!\rangle (\varphi \wedge \psi) \to \langle\!\langle J \rangle\!\rangle \varphi \wedge \langle\!\langle J \rangle\!\rangle \psi$

(S) $\langle\!\langle J_1 \rangle\!\rangle \varphi \wedge \langle\!\langle J_2 \rangle\!\rangle \psi \to \langle\!\langle J_1 \cup J_2 \rangle\!\rangle (\varphi \wedge \psi)$ if $J_1 \cap J_2 = \emptyset$

(MP)   from φ and from $\varphi \to \psi$ infer ψ

(RE)   from $\varphi \leftrightarrow \psi$ infer $\langle\!\langle J \rangle\!\rangle \varphi \leftrightarrow \langle\!\langle J \rangle\!\rangle \psi$

Note that the (N) axiom corresponds to the determinism of *choice profiles* (actions constituted by concurrent choices for every agent in the system): when every agent opts for a choice, the next state is fully determined, thus, if a formula is not settled true, the coalition of all agents (AGT) can always work together to make its negation true. The axiom (S) says that two disjoint coalitions can combine their efforts to ensure a conjunction of properties. Note that from (RE), (S) and $(\bot)$ it follows that if J_1 and J_2 are disjoint then $\langle\!\langle J_1 \rangle\!\rangle \varphi \wedge \langle\!\langle J_2 \rangle\!\rangle \neg \varphi$ is inconsistent. So, two disjoint coalitions cannot consistently ensure opposed facts.

3 Normal simulation of Coalition Logic NCL

3.1 Syntax of NCL

Let AGT be a nonempty finite set of agents and *Prop* an infinite countable set of atomic formulas. Without loss of generality, we assume that $AGT = \{0, \ldots, n-1\}$ where $n = Card(AGT)$. The language $\mathcal{L}_{NCL}$ of NCL has the following syntax, where p ranges over elements of *Prop* and J ranges over subsets of AGT:

$$\varphi ::= p \mid \neg\varphi \mid \varphi \vee \varphi \mid \mathbf{X}\varphi \mid [J]\varphi$$

The other Boolean operators are obtained as usual, and $\langle J \rangle \varphi =_{def} \neg[J]\neg\varphi$.

3.2 Semantics of NCL

The models of NCL are tuples $\mathcal{M} = (W, R, F_X, \pi)$ where:

- W is a nonempty set of worlds (alias contexts);

- R is a collection of equivalence relations R_J (one for every coalition $J \subseteq AGT$) such that:

    - $R_{J_1 \cup J_2} \subseteq R_{J_1} \cap R_{J_2}$
    - $R_\emptyset \subseteq R_J \circ R_{AGT \setminus J}$
    - $R_{AGT} = Id$

- $F_X : W \to W$ is a total function;

- $\pi : W \to 2^{Prop}$ is a valuation function.

The truth conditions of the operators are given by:

- $\mathcal{M}, w \models \mathbf{X}\varphi$ iff $\mathcal{M}, F_X(w) \models \varphi$

- $\mathcal{M}, w \models [J]\varphi$ iff $\forall u \in R_J(w)$, $\mathcal{M}, u \models \varphi$

3.3 Axiomatization of NCL

We give the following axiom schemas for NCL.

(ProTau)	enough tautologies of the propositional calculus
S5([J])	all S5-theorems, for every [J]
(Mon)	$[J_1]\varphi \vee [J_2]\varphi \rightarrow [J_1 \cup J_2]\varphi$
Elim($\emptyset$)	$\langle\emptyset\rangle\varphi \rightarrow \langle J\rangle\langle AGT \setminus J\rangle\varphi$
Triv(AGT)	$\varphi \rightarrow [AGT]\varphi$
K(**X**)	all K-theorems for **X**
D(**X**)	$\mathbf{X}\varphi \rightarrow \neg\mathbf{X}\neg\varphi$
Det(**X**)	$\neg\mathbf{X}\neg\varphi \rightarrow \mathbf{X}\varphi$

We admit the standard inference rules of modus ponens and necessitation for [$\emptyset$] and **X**. From the former, necessitation for every [J] follows by the inclusion axiom (Mon). A formula φ is a theorem of NCL, in symbols $\vdash_{\mathsf{NCL}} \varphi$, iff it can be derived from the above axioms and inference rules within a finite number of steps.

LEMMA 9. $\vdash_{\mathsf{NCL}} \langle\emptyset\rangle\varphi \rightarrow \langle J_1\rangle\langle J_2\rangle\varphi$ if $J_1 \cap J_2 = \emptyset$.

PROOF. By Elim($\emptyset$) we have $\vdash_{\mathsf{NCL}} \langle\emptyset\rangle\varphi \rightarrow \langle J_1\rangle\langle AGT \setminus J_1\rangle\varphi$. Now by hypothesis $J_1 \cap J_2 = \emptyset$, or equivalently $J_2 \subseteq AGT \setminus J_1$. Thus by (Mon) $\vdash_{\mathsf{NCL}} \langle AGT \setminus J_1\rangle\varphi \rightarrow \langle J_2\rangle\varphi$. We obtain $\vdash_{\mathsf{NCL}} \langle J_1\rangle\langle AGT \setminus J_1\rangle\varphi \rightarrow \langle J_1\rangle\langle J_2\rangle\varphi$ by [J_1]-necessitation and K([J_1]). We conclude that $\vdash_{\mathsf{NCL}} \langle\emptyset\rangle\varphi \rightarrow \langle J_1\rangle\langle J_2\rangle\varphi$. ∎

THEOREM 10. *Our axiomatization of* NCL *is both sound and complete with respect to the class of all models of* NCL, *i.e.,* $\vdash_{\mathsf{NCL}} \varphi$ *iff for all models* $\mathcal{M} = (W, R, F_X, \pi)$ *of* NCL *and for all worlds* $w \in W$, $\mathcal{M}, w \models \varphi$.

PROOF. Soundness is obtained by a routine argument while completeness is immediate from Sahlqvist's theorem. Cf. [1]. ∎

4 Translating Chellas's STIT logic into NCL

We call here Chellas's STIT logic a possible presentation of so-called *deliberative STIT theories* [11] without tense operators. An axiomatics was provided by Xu and named *Ldm* in [6, Chap. 17]. It is the logic of Chellas's STIT operators for individual agents plus an operator for historical necessity $\square$. Recently in [5], three of us have shown that *Ldm* axiomatics could be simplified and in particular that historical necessity was superfluous in presence of at least two agents. This work resulted in an alternative axiomatics of *Ldm* noted *ALdm*.

Via *ALdm*, we show that Chellas's STIT logic embeds in NCL.

4.1   Syntax of *ALdm*

Let *AGT* be a nonempty finite set of agents and *Prop* an infinite countable set of atomic formulas. Without loss of generality, we assume that $AGT = \{0, \ldots, k\}$ where k is a non-negative integer. The language $\mathcal{L}_{ALdm}$ of *ALdm* has the following syntax, where p ranges over elements of *Prop* and i ranges elements of *AGT*:

$$\varphi ::= p \mid \neg\varphi \mid \varphi \vee \varphi \mid [i]\varphi \mid \Box\varphi$$

The other classical connectives are obtained as usual. Let $\langle i \rangle \varphi =_{def} \neg[i]\neg\varphi$ and $\Diamond\varphi =_{def} \neg\Box\neg\varphi$.

4.2   Semantics for *ALdm*

DEFINITION 11.   An *ALdm*-model is a tuple $M = (W, R, V)$ where:

- W is a nonempty set of contexts;

- R is a collection of equivalence relations R_i (one for every agent $i \in AGT$) such that for all $w, v \in W$ and for all $l, m, n \in AGT$, if $(w, v) \in R_l \circ R_m$, then $\exists u \in W, (w, u) \in R_n$ and $(u, v) \in \bigcap_{i \in AGT \setminus \{n\}} R_i$;

- $V : W \to 2^{Prop}$ is a valuation function.

Truth conditions are as follows:

- $M, w \models [i]\varphi$ iff for all $u \in R_i(w)$, $\mathcal{M}, u \models \varphi$

- $M, w \models \Box\varphi$ iff for all $u \in (R_1 \circ R_0)(w)$, $\mathcal{M}, u \models \varphi$

and as usual for Boolean operators.

4.3   *ALdm* axiomatics

Axiom schemas for *ALdm* are given as follows:

S5(i)	all S5-theorems, for every $[i]$
Def($\Box$)	$\Box\varphi \leftrightarrow [1][0]\varphi$
(GPerm$_k$)	$\langle l \rangle \langle m \rangle \varphi \to \langle n \rangle \bigwedge_{i \in AGT \setminus \{n\}} \langle i \rangle \varphi$

In the axiom (GPerm$_k$), note that $l, m, n \in AGT$. As proved in [5], from (GPerm$_k$) we can derive Xu's "axiom scheme for independence of agents" (AIA$_k$) [6, Chap. 17]. For instance, (AIA$_1$) corresponds to:[2]

$$\Diamond[0]\varphi_0 \wedge \Diamond[1]\varphi_1 \to \Diamond([0]\varphi_0 \wedge [1]\varphi_1)$$

---

[2](AIA$_1$) corresponds to the case of two agents 0 and 1.

(AIA_k) is central in STIT since it captures the notion of independence of agents. We show later in Lemma 15 that a group version of (AIA_1) is a theorem of NCL too.

THEOREM 12 ([5]). *Our axiomatization of ALdm is both sound and complete with respect to the class of all ALdm-models.*

4.4   Embedding *ALdm* in NCL

NCL is easily proved to be a conservative extension of *ALdm*. To illustrate that, we give the following translation from formulas of *ALdm* to formulas of NCL.

$$
\begin{aligned}
tr_0(p) &= p \\
tr_0(\Box\varphi) &= [\emptyset]\varphi \\
tr_0([i]\varphi) &= [\{i\}]tr_0(\varphi)
\end{aligned}
$$

and homomorphic for the Boolean operators.

LEMMA 13. *The translation of $(GPerm_k)$ by tr_0 is a theorem of NCL.*

PROOF. By applying (Mon) to $\langle\{l\}\rangle$ and $\langle\{m\}\rangle$ in the right part of the tautology $\langle\{l\}\rangle\langle\{m\}\rangle\varphi \rightarrow \langle\{l\}\rangle\langle\{m\}\rangle\varphi$ and next S5($[\emptyset]$) we have $\vdash_{\mathsf{NCL}} \langle\{l\}\rangle\langle\{m\}\rangle\varphi \rightarrow \langle\emptyset\rangle\varphi$. Then by Elim($\emptyset$) we obtain $\vdash_{\mathsf{NCL}} \langle\{l\}\rangle\langle\{m\}\rangle\varphi \rightarrow \langle\{n\}\rangle\langle AGT \setminus \{n\}\rangle\varphi$.

Now, by classical principles on instances of (Mon) $\langle AGT \setminus \{n\}\rangle\varphi \rightarrow \langle\{i\}\rangle\varphi$ for every $i \in AGT \setminus \{n\}$, we have $\vdash_{\mathsf{NCL}} \langle AGT \setminus \{n\}\rangle\varphi \rightarrow \bigwedge_{i \in AGT\setminus\{n\}}\langle\{i\}\rangle\varphi$. We conclude that $\vdash_{\mathsf{NCL}} \langle\{l\}\rangle\langle\{m\}\rangle\varphi \rightarrow \langle\{n\}\rangle\bigwedge_{i \in AGT\setminus\{m\}}\langle\{i\}\rangle\varphi$. ∎

We prove that NCL is a conservative extension of *ALdm* in presence of at least two agents.

THEOREM 14. *φ is a theorem of ALdm iff $tr_0(\varphi)$ is a theorem of NCL.*

PROOF.

$\boxed{\Rightarrow}$ Remind that besides $(GPerm_k)$, the only other axioms of *ALdm* are S5 axioms for $[i]$. From S5($[J]$) and Lemma 13, we have that every translated axiom of *ALdm* is a theorem of NCL. Moreover, translated inference rules preserve validity.

$\boxed{\Leftarrow}$ Let $M = \langle W, R, V\rangle$ be an *ALdm*-model, $x \in W$ be a world and φ a *ALdm*-formula.

Assume $M, x \models \varphi$. We transform M into an NCL-model $\mathcal{M} = (W', R', F'_X, \pi)$ as follows:

- $W' = W$;

- $R'_\emptyset = R_1 \circ R_0$;

- $R'_{AGT} = id$;

- $R'_J = \bigcap_{j \in J} R_j$, if $J \neq \emptyset$ and $J \neq AGT$;

- $F'_X = id$;

- $\pi(w) = V(w)$, for every $w \in W'$.

It is easy to check that the constructed model $\mathcal{M}$ satisfies every constraint on NCL-models and $\mathcal{M}, x \models tr_0(\varphi)$.

∎

5   Translating Coalition Logic into NCL

First, we show a theorem in NCL which generalizes (AIA$_1$) from individuals to coalitions, and that will be instrumental later in the proof of superadditivity in Theorem 16.

LEMMA 15. $\vdash_{NCL} \langle \emptyset \rangle [J_0] \varphi_0 \wedge \langle \emptyset \rangle [J_1] \varphi_1 \rightarrow \langle \emptyset \rangle ([J_0] \varphi_0 \wedge [J_1] \varphi_1)$ for $J_0 \cap J_1 = \emptyset$.

PROOF.   Suppose $J_0 \cap J_1 = \emptyset$. We establish the following deduction:

1. $\langle \emptyset \rangle [J_0] \varphi_0 \rightarrow \langle J_1 \rangle \langle J_0 \rangle [J_0] \varphi_0$

   by Lemma 9

2. $\langle \emptyset \rangle [J_0] \varphi_0 \rightarrow \langle J_1 \rangle [J_0] \varphi_0$

   from previous line by S5($[J_0]$)

3. $\langle \emptyset \rangle [J_0] \varphi_0 \wedge [J_1] \varphi_1 \rightarrow \langle J_1 \rangle [J_0] \varphi_0 \wedge [J_1] [J_1] \varphi_1$

   from previous line by S5($[J_1]$)

4. $\langle \emptyset \rangle [J_0] \varphi_0 \wedge [J_1] \varphi_1 \rightarrow \langle J_1 \rangle ([J_0] \varphi_0 \wedge [J_1] \varphi_1)$

   from previous line by S5($[J_1]$)

5. $\langle \emptyset \rangle (\langle \emptyset \rangle [J_0] \varphi_0 \wedge [J_1] \varphi_1) \rightarrow \langle \emptyset \rangle \langle J_1 \rangle ([J_0] \varphi_0 \wedge [J_1] \varphi_1)$

   from previous line by $[\emptyset]$-necessitation and K($[\emptyset]$)

6. $\langle \emptyset \rangle [J_0] \varphi_0 \wedge \langle \emptyset \rangle [J_1] \varphi_1 \rightarrow \langle \emptyset \rangle \langle J_1 \rangle ([J_0] \varphi_0 \wedge [J_1] \varphi_1)$

   from previous line by S5($[\emptyset]$)

7. $\langle \emptyset \rangle [J_0] \varphi_0 \wedge \langle \emptyset \rangle [J_1] \varphi_1 \rightarrow \langle \emptyset \rangle ([J_0] \varphi_0 \wedge [J_1] \varphi_1)$

   from previous line by (Mon) and S5($[\emptyset]$)

■

Now we give the following translation from Coalition Logic to NCL.

$$tr(p) \quad = \quad p$$
$$tr(\langle\!\langle J\rangle\!\rangle\varphi) \quad = \quad \langle\emptyset\rangle[J]\mathbf{X}tr(\varphi)$$

and homomorphic for the Boolean operators.

THEOREM 16. *If φ is a theorem of CL then $tr(\varphi)$ is a theorem of NCL.*

PROOF. First, the translations of the CL axiom schemas are valid:

- $tr(\neg\langle\!\langle J\rangle\!\rangle\bot) = \neg\langle\emptyset\rangle[J]\mathbf{X}\bot$

  By D($\mathbf{X}$), $\vdash_{NCL} \mathbf{X}\bot \leftrightarrow \bot$. By S5([$J$]), $\vdash_{NCL} [J]\bot \leftrightarrow \bot$. It remains to prove that $\vdash_{NCL} \neg\langle\emptyset\rangle\bot$, which follows from S5([$\emptyset$]).

- $tr(\langle\!\langle J\rangle\!\rangle\top) = \langle\emptyset\rangle[J]\mathbf{X}\top$

  By K($\mathbf{X}$), $\vdash_{NCL} \mathbf{X}\top \leftrightarrow \top$. By S5([$J$]), $\vdash_{NCL} [J]\top \leftrightarrow \top$. Finally, by S5([$\emptyset$]), $\vdash_{NCL} \langle\emptyset\rangle\top$.

- $tr(\neg\langle\!\langle\emptyset\rangle\!\rangle\neg\varphi \rightarrow \langle\!\langle AGT\rangle\!\rangle\varphi) = \neg\langle\emptyset\rangle[\emptyset]\mathbf{X}\neg tr(\varphi) \rightarrow \langle\emptyset\rangle[AGT]\mathbf{X}tr(\varphi)$.

  As $\vdash_{NCL} [AGT]\psi \leftrightarrow \psi$ by Triv(AGT), and as $\vdash_{NCL} \langle\emptyset\rangle[\emptyset]\psi \leftrightarrow [\emptyset]\psi$ by S5([$\emptyset$]), the translation of (N) is equivalent to $\neg[\emptyset]\mathbf{X}\neg tr(\varphi) \rightarrow \langle\emptyset\rangle\mathbf{X}tr(\varphi)$. This is again equivalent to $\langle\emptyset\rangle\neg\mathbf{X}\neg tr(\varphi) \rightarrow \langle\emptyset\rangle\mathbf{X}tr(\varphi)$ which is proved a theorem from Det($\mathbf{X}$).

- $tr(\langle\!\langle J\rangle\!\rangle(\varphi \wedge \psi) \rightarrow \langle\!\langle\varphi\rangle\!\rangle \wedge \langle\!\langle J\rangle\!\rangle\psi) = \langle\emptyset\rangle[J]\mathbf{X}(tr(\varphi) \wedge tr(\psi)) \rightarrow \langle\emptyset\rangle[J]\mathbf{X}tr(\varphi) \wedge \langle\emptyset\rangle[J]\mathbf{X}tr(\psi)$

  First, $\vdash_{NCL} \mathbf{X}(tr(\varphi) \wedge tr(\psi)) \rightarrow \mathbf{X}tr(\varphi) \wedge \mathbf{X}tr(\psi)$ by K($\mathbf{X}$). We have $\vdash_{NCL} [J]\mathbf{X}(tr(\varphi) \wedge tr(\psi)) \rightarrow [J]\mathbf{X}tr(\varphi) \wedge [J]\mathbf{X}tr(\psi)$ by [J]-necessitation and we conclude by [$\emptyset$]-necessitation.

- $tr(\langle\!\langle J_1\rangle\!\rangle\varphi \wedge \langle\!\langle J_2\rangle\!\rangle\psi \rightarrow \langle\!\langle J_1 \cup J_2\rangle\!\rangle(\varphi \wedge \psi)) = \langle\emptyset\rangle[J_1]\mathbf{X}tr(\varphi) \wedge \langle\emptyset\rangle[J_2]\mathbf{X}tr(\psi) \rightarrow \langle\emptyset\rangle[J_1 \cup J_2]\mathbf{X}(tr(\varphi) \wedge tr(\psi))$

  Assume $J_1 \cap J_2 = \emptyset$. The proof that $\vdash_{NCL} \langle\emptyset\rangle[J_1]\mathbf{X}tr(\varphi) \wedge \langle\emptyset\rangle[J_2]\mathbf{X}tr(\psi) \rightarrow \langle\emptyset\rangle[J_1 \cup J_2]\mathbf{X}(tr(\varphi) \wedge tr(\psi))$ is done by the following steps:

  1. $\langle\emptyset\rangle[J_1]\mathbf{X}tr(\varphi) \wedge \langle\emptyset\rangle[J_2]\mathbf{X}tr(\psi) \rightarrow \langle\emptyset\rangle([J_1]\mathbf{X}tr(\varphi) \wedge [J_2]\mathbf{X}tr(\psi))$
     by Lemma 15

  2. $[J_1]\mathbf{X}tr(\varphi) \wedge [J_2]\mathbf{X}tr(\psi) \rightarrow [J_1 \cup J_2]\mathbf{X}tr(\varphi) \wedge [J_1 \cup J_2]\mathbf{X}tr(\psi)$
     by (Mon)

3. $\langle\emptyset\rangle([J_1]\mathbf{X}tr(\varphi) \wedge [J_2]\mathbf{X}tr(\psi)) \rightarrow \langle\emptyset\rangle([J_1 \cup J_2](\mathbf{X}tr(\varphi) \wedge \mathbf{X}tr(\psi))$

from previous line and $[\emptyset]$-necessitation

4. $\langle\emptyset\rangle[J_1]\mathbf{X}tr(\varphi) \wedge \langle\emptyset\rangle[J_2]\mathbf{X}tr(\psi) \rightarrow \langle\emptyset\rangle[J_1 \cup J_2]\mathbf{X}(tr(\varphi) \wedge tr(\psi))$

from line 1 and 3 by standard modal principles

Clearly the translation of modus ponens preserves validity. To prove that the translation of CL's (RE) preserves validity suppose $tr(\varphi \leftrightarrow \psi) = tr(\varphi) \leftrightarrow tr(\psi)$ is a theorem of NCL. We have to prove that $tr(\langle\!\langle J\rangle\!\rangle\varphi \leftrightarrow \langle\!\langle J\rangle\!\rangle\psi) = \langle\emptyset\rangle[J]\mathbf{X}tr(\varphi) \leftrightarrow \langle\emptyset\rangle[J]\mathbf{X}tr(\psi)$ is a theorem of NCL. This follows from the theoremhood of $tr(\varphi) \leftrightarrow tr(\psi)$ by standard modal principles. ∎

LEMMA 17. *Let* $M = ((S, E), V)$ *be a coalition model. Let selec be a function associating to each state s in S a choice profile selec(s) in* $\mu_E(AGT, s)$.[3] *Let the tuple* (W, R, F_X, π) *be constructed as follows:*

- $W = \{\langle s, Y\rangle \mid s \in S, Y \in \mu_E(AGT, s)\}$

- $R_J = \{(\langle s, Y\rangle, \langle s, Y'\rangle) \mid$ *there is* $Z \in \mu_E(J, s)$ *such that* $Y \cup Y' \subseteq Z\}$

- $F_X(\langle s, Y\rangle) = \langle s', Z\rangle$, *where* $Y = \{s'\}$ *and* $Z = selec(s')$

- $\pi(\langle s, X\rangle) = V(s)$

Then (W, R, F_X, π) *is a model of NCL.*

PROOF. The proof consists in checking that the constructed model satisfies every constraint on NCL models. Note that we are permitted to define F_X this way because of Proposition 8. ∎

THEOREM 18. *If* φ *is CL-satisfiable then* $tr(\varphi)$ *is satisfiable in NCL.*

PROOF. Given a coalition model $M = ((S, E), V)$ we construct a model $\mathcal{M}_{NCL} = (W, R, F_X, \pi)$ of NCL as in Lemma 17. We prove by structural induction that $M, s \models \varphi$ iff $\exists Y \in \mu_E(AGT, s)$ such that $\mathcal{M}_{NCL}, \langle s, Y\rangle \models tr(\varphi)$.

The cases of atoms and classical connectives are straightforward. We just consider the case of $\varphi = \langle\!\langle J\rangle\!\rangle\psi$.

1. Suppose, $M, s \models \langle\!\langle J\rangle\!\rangle\psi$. Then, there is $Z' \in E_s(J)$ such that for all $t \in Z', M, t \models \psi$. Thus, there is $Z \in \mu_E(J, s)$ such that for all $t \in Z, M, t \models \psi$. By induction hypothesis, for all $t \in Z, \mathcal{M}_{NCL}, \langle t, selec(t)\rangle \models tr(\psi)$.

2. By construction, for all $t \in Z, \forall Y \in \mu_E(AGT, s)$ such that $Y \subseteq Z, F_X(\langle s, Y\rangle) = \langle t, selec(t)\rangle$.

---

[3] Such a function exists by the axiom of choice.

3. By (1) and (2) it follows that $\forall Y \in \mu_E(AGT, s)$ such that $Y \subseteq Z$, $\mathcal{M}_{NCL}, \langle s, Y \rangle \models$ $\mathbf{X}tr(\psi)$, and thus, since $Z \in \mu_E(J, s)$, it follows that $\exists Y \subseteq Z$ such that $\mathcal{M}_{NCL}, \langle s, Y \rangle \models [J]\mathbf{X}tr(\psi)$.

4. And then, there is $Y' \in \mu_E(AGT, s)$ such that $\mathcal{M}_{NCL}, \langle s, Y' \rangle \models \langle \emptyset \rangle[J]\mathbf{X}tr(\psi)$.

The other direction of the induction step is verified by reverse arguments. ∎

COROLLARY 19. *φ is a theorem of CL iff $tr(\varphi)$ is a theorem of NCL.*

PROOF. The left-to-right direction is Theorem 16. The right-to-left direction follows from Pauly's completeness result for Coalition Logic and Theorem 18. ∎

6   Decidability and complexity of NCL

In this section, we study the satisfiability problem of an NCL-formula.[4] We first study the fragment of NCL without time.

6.1   NCL without time

In the remaining, we call NCLwt(n) the particular instance of NCL with n agents and without the temporal operator $\mathbf{X}$. NCLwt(n) is the fragment of NCL(n) defined by the BNF:

$$\varphi ::= p \mid \neg \varphi \mid \varphi \vee \varphi \mid [J]\varphi$$

where J ranges over the set of subsets of $\{0, ..., n-1\}$.

DEFINITION 20. A *NCLwt-model* is a tuple $\mathcal{M} = (W, R, \pi)$ where:

- W is a set of contexts;

- R is a collection of equivalence relations R_J (one for every coalition $J \subseteq AGT$) such that:

  - $R_{J_1 \cup J_2} \subseteq R_{J_1} \cap R_{J_2}$
  - $R_\emptyset \subseteq R_J \circ R_{AGT \setminus J}$
  - $R_{AGT} = Id$

- $\pi : W \to 2^{Prop}$ is a valuation function.

PROPOSITION 21. *NCLwt(n) is both sound and complete w.r.t. to the class of NCLwt-models where $R_\emptyset$ is the universal relation.*

PROOF. Soundness is obtained by a routine argument while completeness is immediate from Sahlqvist's theorem. ∎

---

[4]It was the subject of [15] (in French).

We now define filtration.

DEFINITION 22. Let $SF(\Phi)$ be the set of all subformulas of an NCLwt-formula Φ.

The set of formulas we use to filter is defined as follows:

DEFINITION 23. Given an NCLwt-formula Φ, let $Cl(\Phi)$ be the set $SF(\Phi) \cup \{[J]\varphi \mid J \subseteq AGT, \varphi \in SF(\Phi)\}$.

PROPOSITION 24. $card(Cl(\Phi)) \leq 2^{card(AGT)} \times |\Phi|$, where $|\Phi|$ is the length of the formula Φ.

The filtered model is defined as follows:

DEFINITION 25. Let $M = (W, \{R_J, J \subseteq AGT\}, \pi)$ be an NCLwt-model. We define the equivalence relation $\leftrightsquigarrow^{Cl(\Phi)}$ on W as follows:

$$\forall x, y \in W, x \leftrightsquigarrow^{Cl(\Phi)} y \text{ iff } (\forall \varphi \in Cl(\Phi), M, x \models \varphi \Leftrightarrow M, y \models \varphi)$$

DEFINITION 26. Given $M = (W, \{R_J, J \subseteq AGT\}, \pi)$ an NCLwt-model and Φ an NCLwt-formula, we define the *filtered model* $M' = (W', \{R'_J, J \subseteq AGT\}, \pi')$ as follows:

- The worlds in W' are the equivalence classes of worlds in W under the relation $\leftrightsquigarrow^{Cl(\Phi)}$;

- For all $J \subseteq AGT$, we define R'_J as $|x|R'_J|y|$ iff $\forall \varphi \in SF(\Phi), \forall J' \subseteq J, M, x \models [J']\varphi$ iff $M, y \models [J']\varphi$;

- For all $|x|_{\leftrightsquigarrow} \in W'$ and for all atomic formulas p in $SF(\Phi)$, $p \in \pi'(|x|_{\leftrightsquigarrow})$ iff $p \in \pi(x)$.

We now study the properties of the filtered model.

PROPOSITION 27. M' contains at most $2^{2^{card(AGT)} \times |\Phi|}$ worlds.

PROOF. We use Proposition 24 and the fact that

$$
\begin{array}{rcl}
W' & \to & 2^{Cl(\Phi)} \\
|x|_{\leftrightsquigarrow} & \mapsto & \{\varphi \in Cl(\Phi) \mid M, x \models \varphi\}
\end{array}
$$

is a well-defined injective application.   ∎

PROPOSITION 28. *If M is an NCLwt-model such that $R_\emptyset = W \times W$ and Φ any formula, then the model M' filtered by Φ is an NCLwt-model.*

PROPOSITION 29. *Given a formula Φ, for all φ in $SF(\Phi)$ and for all $x \in W$, we have $M', |x|_{\leftrightsquigarrow} \models \varphi \Leftrightarrow M, x \models \varphi$.*

PROOF. By induction on φ.   ∎

THEOREM 30. *An NCLwt(n)-formula Φ is satisfiable iff it is satisfiable in a model with $2^{2^{card(AGT)} \times |\Phi|}$ worlds.*

PROOF. By Propositions 21, 27, 28 and 29.   ∎

THEOREM 31. *The problem of deciding satisfiability of an NCLwt(1)-formula is NP-complete.*

PROOF. NCLwt(1) is nothing but S5 because:

- $\langle \emptyset \rangle$ is S5-operator;

- $\langle AGT \rangle$ is a trivial operator since $\langle AGT \rangle \varphi \leftrightarrow \varphi$, and can thus be eliminated.

    ∎

PROPOSITION 32. *If $n \geq 2$ then the problem of deciding satisfiability of an NCLwt(n)-formula is in NEXPTIME.*

PROOF. According to Theorem 30, we can test the satisfiability of a formula by examining all models with $2^{2^{card(AGT)} \times |\varphi|}$ worlds.   ∎

We are going to compare NCLwt(n) and [S5; S5] defined as follows:

DEFINITION 33. [S5; S5] is the modal logic defined by:

- a language with two modal operators $\langle 1 \rangle$ and $\langle 2 \rangle$;

- an axiomatics with $S5(\langle 1 \rangle)$, $S5(\langle 2 \rangle)$ and the axiom of permutation $[1][2]\varphi \leftrightarrow [2][1]\varphi$.

Note that the Church-Rosser axiom $\langle 1 \rangle [2]\varphi \rightarrow [2]\langle 1 \rangle \varphi$ is a [S5; S5]-theorem.

PROPOSITION 34. *[S5; S5] is characterized by the class of frames $F = (W, R_1, R_2)$ such that:*

- R_1 *and* R_2 *are equivalence relations ;*

- $R_1 \circ R_2 = R_2 \circ R_1 = W \times W$.

THEOREM 35 ([10]). *The problem of deciding satisfiability of an [S5; S5]-formula is NEXPTIME-hard.*

PROPOSITION 36. *If $n \geq 2$ then NCLwt(n) is a conservative extension of [S5; S5].*

PROOF. We define $tr : \mathcal{L}_{S5^2} \rightarrow \mathcal{L}_{NCLwt}$ which replaces the two operators $\langle 1 \rangle$ and $\langle 2 \rangle$ of [S5; S5] by $\langle \{1\} \rangle$ and $\langle \{2\} \rangle$ respectively.

First, the reader may easily verify that $\vdash_{[S5;S5]} \varphi$ implies $\vdash_{NCLwt} tr(\varphi)$. Second, if φ is a satisfiable $[S5; S5]$ formula, there is a $[S5; S5]$-model (M, x), where we suppose that $R_1 \circ R_2 = W \times W$, such that $M, x \models \varphi$. We extend M to an NCLwt-model M' by stipulating:

- If $1 \in J$ and $2 \in J$ then $R_J = Id_W$;

- If $1 \in J$ and $2 \notin J$ then $R_J = R_1$;

- If $1 \notin J$ and $2 \in J$ then $R_J = R_2$;

- If $1 \notin J$ and $2 \notin J$ then $R_J = W \times W$.

It is straightforward to check that M' is an NCLwt-model and that $M', x \models tr(\varphi)$.
∎

COROLLARY 37. *If $n \geq 2$ then the problem of deciding satisfiability of an NCLwt(n)-formula is NEXPTIME-hard.*

PROOF. From Theorem 35 and Proposition 36. ∎

THEOREM 38. *If $n \geq 2$ then the problem of deciding satisfiability of an NCLwt(n)-formula is NEXPTIME-complete.*

PROOF. From Proposition 32 and Corollary 37. ∎

6.2 NCL(n)

For NCL(n), i.e. with the temporal operator **X**, it is difficult to apply filtration directly because we cannot assume that $R_\emptyset$ is the universal relation anymore. First let us introduce the notion of a frozen formula.

DEFINITION 39. Let $\mathbf{X}at(\varphi)$ be the set of all the subformulas of φ of the form $\mathbf{X}\psi$ that are not proper subformulas of some other subformula $\mathbf{X}\varphi_1$.

DEFINITION 40. Let $Freeze(\varphi)$ be the formula φ where all subformulas $\mathbf{X}\psi \in \mathbf{X}at(\varphi)$ are replaced by a new atomic formula $p_{\mathbf{X}\psi}$.

EXAMPLE 41. For $\varphi = \mathbf{X}p \vee \langle\{1\}\rangle(p \wedge \mathbf{X}(\langle\{2\}\rangle p \vee \mathbf{X}q) \wedge \langle\{2,4\}\rangle\mathbf{X}\mathbf{X}q)$, we have:

- $\mathbf{X}at(\varphi) = \{\mathbf{X}p, \mathbf{X}(\langle\{2\}\rangle p \vee \mathbf{X}q), \mathbf{X}\mathbf{X}q\}$;

- $Freeze(\varphi) = p_{\mathbf{X}p} \vee \langle\{1\}\rangle(p \wedge p_{\mathbf{X}(\langle\{2\}\rangle p \vee \mathbf{X}q)} \wedge \langle\{2,4\}\rangle p_{\mathbf{X}\mathbf{X}q})$.

Figure 1 shows a non-deterministic algorithm to decide the satisfiability problem of an NCL(n)-formula φ. The procedure NCL(n)-SAT-ND uses two subroutines:

- NCLwt-SAT-ND is a non-deterministic decision procedure for NCLwt-SAT;

---

function recursive $\text{NCL(n)-SAT-ND}(\varphi)$
    if φ does not contain any **X** **then**
        return $\text{NCLwt-SAT-ND}(\varphi)$
   else
        $\varphi' := \mathit{Freeze}(\varphi)$
        $\mathbf{J} := \mathbf{X}\text{at}(\varphi')$
        if $\text{NCLwt-SAT-ND}(\varphi') = \text{UNSATISFIABLE}$ **then**
            return UNSATISFIABLE
        $(M = (W, \{R_J\}, \pi), r) := \text{NCLwt-GiveModel-ND}(\varphi')$
        for $y \in W$,
            $\psi_y := \bigwedge_{\mathbf{X}\psi \in J / p_{\mathbf{X}\psi} \in \pi(y)} \psi \wedge \bigwedge_{\mathbf{X}\psi \in J / p_{\mathbf{X}\psi} \notin \pi(y)} \neg\psi$
            if $\text{NCL(n)-SAT-ND}(\psi_y) = \text{UNSATISFIABLE}$ **then**
                return UNSATISFIABLE
        endFor
        return SATISFIABLE
    endIf
endFunction

Figure 1. NCL(n)-SAT-ND

- $\text{NCLwt-GiveModel-ND}(\varphi')$ nondeterministically chooses an $\text{NCLwt}(n)$-pointed-model (M, r) such that $M, r \models \varphi'$.

As usual, an $\text{NCLwt}(n)$-pointed-model is a pair (M, r) where M is an $\text{NCLwt}(n)$-model and r is a context of M.

Both NCLwt-SAT-ND and $\text{NCLwt-GiveModel-ND}(\varphi')$ are optimal, that is to say:

- in the mono-agent case, they run in a polynomial space. In particular, $\text{NCLwt-GiveModel-ND}(\varphi')$ returns a model of polynomial size;

- if there are more than two agents ($n \geq 2$), they run in exponential time. $\text{NCLwt-GiveModel-ND}(\varphi')$ returns a model with $2^{2^n \times |\varphi'|}$ worlds.

The procedure goes along the following idea: if an NCL-formula φ does not contain any **X** symbol, then we can immediately use the first NCLwt-SAT-ND procedure. Else we begin by treating the satisfiability of the NCLwt-formula $\mathit{Freeze}(\varphi)$. If it is satisfiable, we choose an NCLwt-model M, r such that $M, r \models \mathit{Freeze}(\varphi)$. Then we try to know if the valuation of the propositions in $\mathbf{X}\text{at}(\varphi)$ is consistent in every world y of the model M. This is why we test whether ψ_y is NCL-satisfiable for every world y of the model M by recursive calls to NCL(n)-SAT-ND.

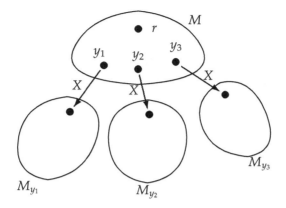

Figure 2. Construction of the NCL-model ($\boxed{\Rightarrow}$-sense proof of Theorem 43).

THEOREM 42. *NCL-SAT-ND terminates for all φ.*

PROOF. By induction on the modal degree w.r.t. the **X** operator. ∎

THEOREM 43. *NCL-SAT-ND(φ) returns SATISFIABLE iff φ is satisfiable.*

PROOF.

The proof is done by induction.

$\boxed{\Rightarrow}$ We can construct an NCL-model for φ by gluing together the model M that is built in NCLwt-GiveModel-ND with an NCL-model M_y for each ψ_y (as exemplified in Figure 2).

$\boxed{\Leftarrow}$ If φ is satisfiable then there is an NCL-pointed-model (N, x) which satisfies φ. Then we proceed as follows (see Figure 3).

1. We extract from (N, x) an NCLwt(n) model G for *Freeze(φ)*.

2. We filter G: we obtain G'.

3. One execution of NCLwt-SAT-ND is such that $M = G'$. We then take into account that any ψ_y is satisfiable (in (N, t) where t is a world of N).

Finally, NCL-SAT-ND(φ) returns SATISFIABLE. ∎

We now state the complexity of NCL(1).

PROPOSITION 44. *In the mono-agent setting, the problem of deciding the satisfiability of an NCL-formula is in PSPACE.*

PROOF. By induction we can prove that an execution of NCL-SAT-ND uses polynomial space. We take into account that:

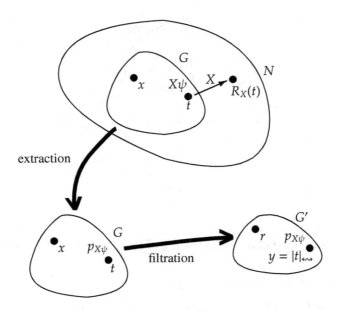

Figure 3. Picture explaining the $\boxed{\Leftarrow}$ part of Theorem 43.

- NCLwt-SAT-ND and NCLwt-GiveModel-ND run in polynomial space;

- M contains a polynomial number of worlds.

Then we use that NPSPACE = PSPACE (Savitch's theorem).   ∎

PROPOSITION 45. *In the mono-agent setting, we have a polynomial reduction from the satisfiability problem of a K-formula to the satisfiability problem of an NCL-formula by the following translation* $tr : \mathcal{L}_K \to \mathcal{L}_{NCL}$ *defined by:* $tr(\Box\varphi) = X[\emptyset]tr(\varphi)$.

COROLLARY 46. *In the mono-agent setting, the problem of deciding the satisfiability of an NCL-formula is PSPACE-hard.*

THEOREM 47. *In the mono-agent setting, the problem of deciding the satisfiability of an NCL-formula is PSPACE-complete.*

PROOF. From Proposition 44 and Corollary 46.   ∎

We finally give the complexity of the satisfiability problem of an NCL(n)-formula for $n \geq 2$.

PROPOSITION 48. *The problem of deciding the satisfiability of an* $NCL(n)$-*formula is in NEXPTIME.*

PROOF. By induction, we can prove that an execution of NCL-SAT-ND uses exponential time. We take into account that:

- NCLwt-SAT-ND and NCLwt-GiveModel-ND run in exponential time;

- M contains an exponential number of worlds.

■

PROPOSITION 49. *NCL is a conservative extension of NCLwt.*

PROOF. First, an NCLwt proof is an NCL proof. Second, if φ is a satisfiable NCLwt-formula in an NCLwt-model (M, x), we extend M to M' by adding $F_X = Id_W$. We have $M', x \models \varphi$. ■

COROLLARY 50. *In presence of at least two agents, the problem of deciding the satisfiability of an* NCL-*formula is NEXPTIME-hard.*

PROOF. From Corollary 37 and Proposition 49. ■

THEOREM 51. *In presence of at least two agents, the problem of deciding the satisfiability of an* NCL-*formula is NEXPTIME-complete.*

PROOF. From Proposition 48 and Corollary 50. ■

7 Discussion

Coalition Logic is basically a logic of ability, in the sense that its main operator formalizes sentences of the form "agent i is able to ensure φ". As we have seen, NCL embeds CL and is of course suitable for such kind of reasoning about abilities of agents and coalitions. However, the introduction of a STIT-style operator is a move to more expressivity.

Authors in logics of action have often been interested in the notion of 'making do'. It can be linked to the idea of an agent having the *power over* another agent [7]. On the model of Figure 4, it is easy to check that the formula $\langle \emptyset \rangle [\{i\}] X [\{j\}] \varphi$ is satisfied at w_1 and w_2. A direct reading of this formula is "agent i sees to it that next, agent j sees to it that φ".

For example, in an organizational or normative setting, it in fact reflects adequately the agentive component of a *delegation*. As an illustration of our logic, we see here how NCL can grasp tighter notions of ability than Coalition Logic.

Chellas's STIT logic has some annoying properties: if we try to model influence of an agent on an other, we are inclined to state it via the formula $[i][j]\varphi$. It is

Figure 4. Representation of an NCL-model with two moments and two agents: i chooses the columns ($R_{\{i\}}$) and j chooses the rows ($R_{\{j\}}$). The *grand* coalition can determine a unique outcome: $R_{\{i,j\}} = Id$ is represented by the 'small squares'. Nature ($\emptyset$) cannot distinguish outcomes of a same moment: $R_\emptyset = R_{\{i\}} \circ R_{\{j\}}$ is represented by the 'big boxes'. Arrows are F_X transitions.

nevertheless equivalent to $\Box\varphi$. (Recall the historical necessity operator in Section 4.) Hence, in this logic, an agent can force another agent to do something if and only if this something is settled. We must admit this is a poor notion of influence.

In previous attempts to extend straightforwardly the logic of Chellas's STIT with a 'next' operator ([3]), the formula $[\{i\}]X[\{j\}]\varphi \to X\Box\varphi$ was valid. It means that if i forces that next j ensures φ then next, φ is inevitable. Inserting an X operator between the agent's actions gives us a refined notion of influence. Still, it is not completely satisfying, since it suggests that an agent influences another agent j to do φ by forcing the world to be at a moment where φ is settled. Since an agent at a moment sees to everything being historically necessary (in formula: for every i, $\Box\varphi \to [i]\varphi$), it means that an agent i influences an agent to do φ if and only if it influences every agent to do φ, i included.

On the contrary, the following formula is not a theorem of NCL:

$$[\{i\}]X[\{j\}]\varphi \to X[\emptyset]\varphi.$$

In particular in the model of Figure 4, the following formulas are true at w_1 and w_2:

- $\langle\emptyset\rangle[\{i\}]X[\{j\}]\varphi$

- $\langle\emptyset\rangle[\{i\}]\mathbf{X}[\{j\}]\neg\varphi$

It somewhat grasps the fact that agent i controls the truth value of φ by exerting influence on j. An interesting account of similar concepts but focused on propositional control is given by [17]. Of course, the CL 'fused' operator is not designed for those issues, and Coalition Logic is not suitable for modeling the notion of *power over*.

Even though our quick study does not permit to prove that NCL is indeed a good logic to reason about influence, we think that the consistency of $\langle\emptyset\rangle[\{i\}]\mathbf{X}([\{j\}]\varphi \wedge \langle\emptyset\rangle[\{j\}]\neg\varphi)$ which is at first sight a drawback, is in fact an interesting property: an agent can force an agent j to ensure φ even if j would also be able to ensure $\neg\varphi$. It somewhat leaves some place to indeterminism and unsuccessful delegations. What should constrain a delegated agent is not physics but norms. If one wants to rule that property out, one could simply release Det($\mathbf{X}$) and add the axiom schema $\mathbf{X}\varphi \rightarrow \mathbf{X}[\emptyset]\varphi$. The nature of time in NCL is simply a very convenient one for embedding CL and is amenable at will. We believe it particularly deserves a work effort in the future.

8 Conclusion

To conclude, we have investigated the properties of a normal modal logic version of Pauly's Coalition Logic that we call NCL. We have shown that due to its richer language it is strictly more complex than the latter: satisfiability checking is NEXPTIME-complete (for more than two agents).

We think that the versatility of NCL models allows for smoothness in modeling. The information 'contained' in a context, viz. the physical description of the world *and* the actual strategy profile of agents may permit to capture fine-grained notions relevant for multi-agent systems via Kripke models in the realm of normal modal logics.

We believe that the present framework opens new perspectives on a proof theoretic investigation of logics of agency.

BIBLIOGRAPHY

[1] Patrick Blackburn, Maarten de Rijke, and Yde Venema. *Modal Logic*. Cambridge Tracts in Theoretical Computer Science. Cambridge University Press, 2001.

[2] Jan Broersen, Andreas Herzig, and Nicolas Troquard. Embedding Alternating-time Temporal Logic in strategic STIT logic of agency. *Journal of Logic and Computation*, 16(5):559–578, 2006.

[3] Jan Broersen, Andreas Herzig, and Nicolas Troquard. From Coalition Logic to STIT. In Wiebe van der Hoek, Alessio Lomuscio, Erik de Vink, and Mike Wooldridge, editors, *Third International Workshop on Logic and Communication in Multi-Agent Systems (LCMAS 2005)*, *Edinburgh, Scotland, UK*, volume 157 of *Electronic Notes in Theoretical Computer Science*, pages 23–35. Elsevier, 2006.

[4] Jan Broersen, Andreas Herzig, and Nicolas Troquard. Normal simulation of coalition logic and an epistemic extension. In Dov Samet, editor, *Proceedings of TARK 2007*, pages 92–101, Brussels, Belgium, 2007. Presses Universitaires de Louvain.

[5] Philippe Balbiani, Andreas Herzig, and Nicolas Troquard. Alternative axiomatics and complexity of deliberative stit theories. *Journal of Philosophical Logic*, 37(4): 387–406, 2008.

[6] Nuel Belnap, Michael Perloff, and Ming Xu. *Facing the future: agents and choices in our indeterminist world*. Oxford, 2001.

[7] Cristiano Castelfranchi. The micro-macro constitution of power. In *Protosociology*, number 18-19. 2003.

[8] Brian Chellas. *Modal Logic: an introduction*. Cambridge University Press, 1980.

[9] Cédric Dégremont. Dialogical Deliberative Stit. Master's thesis, University of Lille 3, 2006.

[10] Dov Gabbay, Agnes Kurucz, Frank Wolter, and Michael Zakharyaschev. *Many-Dimensional Modal Logics: Theory and Applications*. Number 148 in Studies in Logic and the Foundations of Mathematics. Elsevier, North-Holland, 2003.

[11] John Horty and Nuel Belnap. The deliberative STIT: A study of action, omission, and obligation. *Journal of Philosophical Logic*, 24(6):583–644, 1995.

[12] Andreas Herzig and Nicolas Troquard. Knowing How to Play: Uniform Choices in Logics of Agency. In Gerhard Weiss and Peter Stone, editors, *5th International Joint Conference on Autonomous Agents & Multi-Agent Systems (AAMAS-06)*, pages 209–216, Hakodate, Japan, 8-12 mai 2006. ACM Press.

[13] Marc Pauly. *Logic for Social Software*. PhD thesis, University of Amsterdam, 2001. ILLC Dissertation Series 2001-10.

[14] Marc Pauly. A modal logic for coalitional power in games. *Journal of Logic and Computation*, 12(1):149–166, 2002.

[15] François Schwarzentruber. Décidabilité et complexité de la logique normale des coalitions. Master's thesis, Univ. Paul Sabatier Toulouse III, 2007.

[16] Nicolas Troquard. *Independent agents in branching time*. PhD thesis, Univ. Paul Sabatier Toulouse III & Univ. degli studi di Trento, 2007.

[17] Wiebe van der Hoek and Michael Wooldridge. On the dynamics of delegation, cooperation, and control: A logical account. In *Proc. of the Fourth International Joint Conference on Autonomous Agents and Multi-Agent Systems (AAMAS'05)*, 2005.

[18] Heinrich Wansing. Tableaux for multi-agent deliberative-STIT logic. In Guido Governatori, Ian Hodkinson, and Yde Venema, editors, *Advances in Modal Logic, Volume 6*, pages 503–520. King's College Publications, 2006.

Towards a Dialogic Interpretation of Dynamic Proofs

DIDERIK BATENS

ABSTRACT. Two matters are cleared up in this paper. First dynamic proofs are defined in a precise way and an approach to them is elaborated. The approach is in a sense Hilbertian and anti-Tarskian: logical inference is characterized in terms of types of proofs, rather than in terms of properties of the consequence relation.

The second point concerns a dialogic or game-theoretical interpretation of dynamic proofs. Dynamic proofs in themselves do not form a demonstration of the derivability of a formula from a premise set. Apart from the proof, such a demonstration requires a specific metalevel argument. It will be shown that the metalevel argument may be phrased, in a natural and appealing way, in terms of the existence of a winning strategy for the proponent.[1]

1 Introduction

Dialogic and game theoretical s are fascinating enterprises. Whether a formula A is a consequence of a premise set Γ may be understood and sometimes settled in terms of a dialogue between a *proponent*, who claims that $\Gamma \vdash_L A$, and an *opponent*, who denies it. The underlying idea is not who wins the dialogue or game, but whether there exists a winning strategy for one of the players. Obviously, the rules of the dialogue have to be spelled out in a precise way. One has to delineate the moves allowed to each player in view of a certain state of the dialogue, the conditions under which the proponent (respectively opponent) wins, and the conditions under which the proponent (possibly the opponent) has a winning strategy.

One of the fascinating aspects of dialogic is that it offers an interpretation of the idea of logical consequence: it is impossible that the conclusion is false if the premises are true. This is matched to: whatever the choices made by the opponent, the proponent has a winning strategy.

Of course computational aspects come into play. If the logic is decidable and $\Gamma \nvdash_L A$, the opponent will have a winning strategy. Sometimes there is only a

[1] Research for this paper was supported by subventions from Ghent University and from the Fund for Scientific Research – Flanders. For comments on a previous draft, I am grateful to Joke Meheus, Peter Verdée, Kristof De Clercq, the two referees, and especially Christian Straßer.

positive test for the consequence relation (it is semi-decidable). In that case $\Gamma \not\vdash_L A$ has to be connected to the absence of a winning strategy for the proponent.

While dialogic is fascinating in the case of usual logics, there are other logics for which there is not even a positive test. The proofs of such logics are typically dynamic: a formula derived at some point in a proof from a premise set Γ, may be considered not to be a consequence of Γ at a later point in the proof in view of the insights in the premises that were gained by continuing the proof. The dynamics need not stop there: at a still later point, the proof may provide further insights in Γ and, as a result of this, it is possible that the formula has to be considered again as a consequence of Γ. Incidentally, there are many such logics, for example (predicative) logics that characterize defeasible reasoning processes. It is well-known that these occur frequently in scientific as well as in everyday contexts.

It will be shown in this paper that it is natural to apply a dialogic or game theoretical approach to consequence relations defined by dynamic proofs. I shall do so by discussing the dynamic proofs I am most familiar with, viz. those of adaptive logics.

Static proofs will be introduced in Section 2 and dynamic proofs in Section 3. Although everyone can recite Hilbert's definition of a proof, the extent to which logicians rely on implicit presuppositions concerning proofs is striking. I try to repair this in these sections. Next, I shall briefly introduce adaptive logics in Section 4. This will enable me to spell out the dialogic approach for a very large family of logics that is well-studied and of which concrete examples are readily available. The dialogues themselves are presented and discussed in Section 5. Some open problems are mentioned in Section 6.

2   Static proofs

Let $\mathcal{L}$ be a language with a denumerable alphabet and $\mathcal{W}$ its set of closed formulas. A *logic* L is a function that maps every set of closed formulas to a set of closed formulas, L: $\wp(\mathcal{W}) \to \wp(\mathcal{W})$. In other words, a logic assigns a consequence set to every set of formulas. The L-consequence set of Γ will be denoted by $Cn_L(\Gamma)$. Alternatively, that A is a L-consequence of Γ is denoted by $\Gamma \vdash_L A$. Obviously, $Cn_L(\Gamma) = \{A \mid \Gamma \vdash_L A\}$.

By a *rule* I shall mean a metalinguistic expression of the form "to derive A from Σ",[2] henceforth Σ/A, in which A is a metalinguistic formula and Σ is a recursive (or decidable) set of metalinguistic formulas. A condition may be attached to the rules, provided it can be decided whether the condition is fulfilled by inspecting the proofs are defined below.

The rule Σ/A is *finitary* iff Σ is a finite set. If Σ is empty, A is usually called an axiom schema. Every set $\mathcal{R}$ of rules will be taken to contain the *Premise* rule, which allows one to introduce members of the premise set.

---

[2] $A, B, \ldots$ and Σ, Π will be used as metametalinguistic variables.

Dynamic proofs are most easily described in annotated format. For the sake of comparison, I shall do the same for static proofs. Lines of static annotated proofs are composed of a line number, a formula, and a justification. The justification of a line l consists of a (possibly empty) set of line numbers N_l and the name of a rule R_l. Given a set $\mathcal{R}$ of rules and a list[3] of lines, a line l in the list is $\mathcal{R}$-*correct* iff (i) all members of N_l precede l in the list, (ii) $R_l \in \mathcal{R}$, and (iii) the formula of l is obtained by application of R_l to the formulas of the lines N_l.

As many other results of human action, proofs may be approached as results or as processes. On the process approach, proofs are seen as chains of proof stages.[4]

DEFINITION 1. A $\mathcal{R}$-*stage* from Γ is a list of $\mathcal{R}$-*correct* lines in which only members of Γ are introduced by the Premise rule.

DEFINITION 2. Where L and L' are $\mathcal{R}$-stages from Γ, L' is an *extension* of L iff all elements of L occur in the same order in L' and some elements of L' do not occur in L.

The new lines may be appended to the former stage or inserted in it, provided that the resulting list is a $\mathcal{R}$-stage from Γ. To facilitate terminology, I shall consider the empty list as the first stage of every $\mathcal{R}$-proof from every premise set.

DEFINITION 3. A static $\mathcal{R}$-*proof* from Γ is a chain of $\mathcal{R}$-stages from Γ, the first element of which is the empty list and all other elements of which are extensions of their predecessors.

DEFINITION 4. A static $\mathcal{R}$-*proof of A from* Γ is a $\mathcal{R}$-proof from Γ containing a stage of which a line has A as its formula.

DEFINITION 5. $\Gamma \vdash_{\mathcal{R}} A$, A is $\mathcal{R}$-*derivable from* Γ, iff there is a static $\mathcal{R}$-proof of A from Γ.

The *syntactic characterization of a logic* **L** is often identified with the logic itself. In this case proofs are named after the logic rather than after a set of rules characterizing the logic.

DEFINITION 6. A logic **L** *has static proofs* iff there is a recursive set $\mathcal{R}$ such that $\Gamma \vdash_{\mathbf{L}} A$ iff $\Gamma \vdash_{\mathcal{R}} A$.

That a logic **L** has static proofs has a number of interesting and easily provable consequences. Let $\mathcal{R}_{\mathbf{L}}$ be a recursive set of rules such that $\Gamma \vdash_{\mathbf{L}} A$ iff $\Gamma \vdash_{\mathcal{R}_L} A$. Note that there is a set $\mathcal{R}_{\mathbf{L}}$ whenever **L** has static proofs.

Let us start with some properties related to the cardinality of entities introduced earlier.

---

[3] A list is an enumeration of a (possibly infinite) set in which each member of the set is associated with a positive integer, which indicates its place in the list - see [10, Ch. 1].

[4] I am indebted to Andrzej Wiśniewski who convinced me, several years ago, to see (dynamic) proofs as a chain of stages.

DEFINITION 7. A set of rules $\mathcal{R}$ is *redundant* iff there is a $(\Sigma/A) \in \mathcal{R}$ and a $\Sigma' \subset \Sigma$ such that, for all Γ and A, $\Gamma \vdash_{\mathcal{R}} A$ iff $\Gamma \vdash_{\mathcal{R}-\{\Sigma/A\} \cup \{\Sigma'/A\}} A$.

DEFINITION 8. A set of rules $\mathcal{R}$ is *minimal* iff, for every $\mathcal{R}' \subset \mathcal{R}$, there are Γ and A for which $\Gamma \vdash_{\mathcal{R}} A$ and $\Gamma \nvdash_{\mathcal{R}'} A$.

THEOREM 9. *If $\mathcal{R}$ is a non-redundant and minimal set of rules, then all members of $\mathcal{R}$ are finitary.*

Proof. Suppose that the antecedent is true and consider a $\mathcal{R}$-proof from some premise set. In view of Definition 3, every stage of the proof is a list of lines. So, if a $R \in \mathcal{R}$ would be non-finitary, the result of the application of R cannot occur in any stage of the proof. As $\mathcal{R}$ is non-redundant and minimal, $R \notin \mathcal{R}$. ∎

THEOREM 10. *If $\mathcal{R}$ is a set of rules, there is a non-redundant and minimal set of rules $\mathcal{R}'$ for which $\Gamma \vdash_{\mathcal{R}'} A$ iff $\Gamma \vdash_{\mathcal{R}} A$.*

Proof. Immediate from Definitions 7 and 8. ∎

THEOREM 11. *If $\mathbf{L}$ has static proofs and $\Gamma \vdash_{\mathbf{L}} A$, then there is a static $\mathcal{R}_{\mathbf{L}}$-proof of A from Γ in which A is the formula of the last line of the last stage.*

Proof. Suppose that the antecedent is true. In view of Definitions 4, 5, and 6, A is the formula of a line l of a stage of a $\mathcal{R}_{\mathbf{L}}$-proof from Γ. Let L be the list obtained by truncating the stage after line l. As $\langle \emptyset, L \rangle$ is a chain of stages,[5] the consequence of the theorem holds true. ∎

Next we turn to some traditional properties of consequence relations.

THEOREM 12. *If $\mathbf{L}$ has static proofs, then $\mathbf{L}$ is Compact (if $A \in Cn_{\mathbf{L}}(\Gamma)$ then $A \in Cn_{\mathbf{L}}(\Gamma')$ for some finite $\Gamma' \subseteq \Gamma$).*

Proof. Immediate in view of Theorem 11. ∎

THEOREM 13. *If $\mathbf{L}$ has static proofs, then $\mathbf{L}$ is Reflexive ($\Gamma \subseteq Cn_{\mathbf{L}}(\Gamma)$).*

Proof. Immediate in view of Definitions 3 and 6 and of the presence of the Premise rule in every set of rules. ∎

THEOREM 14. *If $\mathbf{L}$ has static proofs, then $\mathbf{L}$ is Transitive (if $A \in Cn_{\mathbf{L}}(\Delta)$ and $\Delta \subseteq Cn_{\mathbf{L}}(\Gamma)$, then $A \in Cn_{\mathbf{L}}(\Gamma)$).*

---

[5]There are obviously (finitely) many suitable chains, for example the one that ends with L and in which every list is obtained from its predecessor by appending one line.

Proof. Suppose that the antecedent is true and that $A \in Cn_L(\Delta)$ and $\Delta \subseteq Cn_L(\Gamma)$. In view of Theorem 12, there is a finite $\Delta' \subseteq \Delta$ such that $\Delta' \subseteq Cn_L(\Gamma)$ and $A \in Cn_L(\Delta')$. Let $\Delta' = \{B_1, \ldots, B_n\}$. In view of Definitions 3 and 6, there is, for each B_i ($1 \leq i \leq n$), a $\mathcal{R}_L$-proof p_i of B_i from Γ and there is a $\mathcal{R}_L$-proof of A from Δ'. Let L be the result of deleting from the last stage of the latter proof all lines on which a member of Δ' is introduced by the premise rule. The list obtained by concatenating the last stages of $\mathsf{p}_1, \ldots, \mathsf{p}_n$ and finally L is easily seen to be the last stage of a $\mathcal{R}_L$-proof of A from Γ. So $A \in Cn_L(\Gamma)$ by Definition 6. ∎

THEOREM 15. *If* **L** *has static proofs, then* **L** *is Monotonic* ($Cn_L(\Gamma) \subseteq Cn_L(\Gamma \cup \Gamma')$ *for all* Γ').

Proof. In view of Definition 4, every $\mathcal{R}_L$-proof of A from Γ is a $\mathcal{R}_L$-proof of A from $\Gamma \cup \Gamma'$. So the theorem follows by Definition 6. ∎

LEMMA 16. *If* **L** *has static proofs, every line that occurs in a stage of a* $\mathcal{R}_L$-*proof can be written as a finite string of a denumerable alphabet.*

Proof. This is obvious for the line number and the formulas. The justification of a line contains a finite set of line numbers (in view of Theorem 9) and the name of a rule. So all line numbers involved can be written as a finite string of a finite alphabet and, given that $\mathcal{R}_L$ is a denumerable set, finite strings of a finite alphabet are sufficient to name all rules. The three elements of a line and the elements of the justification can obviously be separated by finitely many symbols. ∎

So we use a denumerable alphabet to write proof lines as finite strings. Actually, if the lemma would not hold, humans would not be able to write proofs.

There is a *positive test* for a logic **L** (**L** is *semi-decidable*) iff there is a mechanical procedure that, for every decidable Γ and A, leads after finitely many steps to the answer YES iff $\Gamma \vdash_L A$ (but may not provide an answer at any finite point if $\Gamma \nvdash_L A$).

THEOREM 17. *If* **L** *has static proofs, then there is a positive test for* **L**.

Proof. Suppose that Γ is a decidable set of formulas and that $\Gamma \vdash_L A$. In view of Definitions 4 and 6 and Theorem 11, there is a finite $\mathcal{R}_L$-proof of A from Γ and the last stage of this proof is a finite list of lines.

All finite lists of finite strings of the alphabet in which proofs are written can be ordered into a list **L**. It is well-known (and easily seen) to be decidable whether a member of **L** is a $\mathcal{R}_L$-correct list of lines, and hence the last stage of a $\mathcal{R}_L$-proof of A from Γ. As some member of **L** is bound to be the last stage of a $\mathcal{R}_L$-proof of A from Γ, it can be identified after finitely many steps. ∎

The upshot is that logics that have static proofs are logics of the usual kind. Incidentally, most usual logics have static proofs. However, not all known logics have static proofs. A well-known exception is second order logic, which is not compact. An obvious example, taken from [10, p. 283] concerns the premise set comprising second order axioms for arithmetic (roughly Peano arithmetic plus the second order axiom of mathematical induction) together with all formulas of the form $\sim c = i$ (for $i \in \{0, 0', 0'', \ldots\}$ and c a constant that is added to the language of arithmetic). This set is inconsistent, but (on the supposition that arithmetic is consistent) every finite subset of the set is consistent. So, for example, $0 = 0'$ is a consequence of the set, but not of any finite subset of it. However, there clearly is no static proof of $0 = 0'$ from the set. All positive integers are used up by the premises and none is left to be associated with $0 = 0'$. So, even if some static proofs are infinite chains comprising infinite stages, static proofs require compactness.

The materials of this section may be adjusted to non-annotated proofs as well as to the result-directed approach to proofs, which identifies a proof with its last stage. This is left as an easy exercise for the interested reader.

3   Dynamic proofs

Dynamic proofs may be realized in several ways. I shall present the way they grew out of the work of my research group during the last twenty years. The proofs are those of adaptive logics. It is a long term aim of the research group to characterize all forms of dynamic reasoning. Perhaps it is possible to do so in terms of adaptive logics. Otherwise new types of logics will have to be devised.

Dynamic proofs are meant to characterize a logic, in other words a consequence relation. So the dynamics should be handled in a controlled way. It should not depend on decisions of the person constructing the proof, but on something 'objective'.

The peculiar form of dynamic proofs considered here relies on the following intuitive idea. Certain formulas, called *abnormalities*, are considered as false unless the premises 'prevent one' from doing so. As we shall see in Section 4, abnormalities are formulas that have a specific logical form - the form may differ from one logic to the other.

The role of abnormalities is double. First, the rules allow one to derive some formulas on a *condition*, which is a set of abnormalities. Next, a *Marking definition* determines whether a line is *marked* or *unmarked* at a stage. The (classical) disjunctions of abnormalities that are derived unconditionally at a stage, express the insight which the stage provides about the abnormality of the premises. In view of this insight a condition is or is not considered acceptable at that stage. If the condition of a line is considered unacceptable at a stage, the line is marked at that stage. The formula of a line is considered as derived at a stage iff the line is unmarked at that stage.

To make the annotated proofs transparent, their lines are quadruples consisting of a line number, a formula, a justification, and a condition (which may be the empty set). A logic **L** is syntactically characterized by a couple comprising a set of rules and a Marking definition. In the next section, I shall present specific rules as well as specific marking definitions. In the present section, I use these notions in a more abstract way.

The rules and the Marking definition have strictly different functions. The rules determine which lines may be added to a proof and do not in any way interfere with the marks. The Marking definition determines which lines are marked at a stage and which unmarked. It does not interfere with the rules which may be applied to marked as well as to unmarked lines.

So the dynamics of the proof is reduced to the marks which may come and go as the proof proceeds, i.e. from one stage to the other. As the marks are governed by a definition which is independent of the rules, it is possible to describe the effect of the rules separately: static proofs of formulas on a condition, which is a very simple generalization of the static proofs described in Section 2. I now set out to describe this in a more precise way.

The change to *lines* was mentioned before. *Rules* have the form $\Sigma/\mathsf{A}{:}\Pi$, which abbreviates "to derive A on the condition Π from Σ". The *Premise* rule may introduce premises on an empty or non-empty condition. The *application* of rules requires some attention: if each member of Σ occurs as the formula of a line of stage L, and Ξ is the union of the conditions of those lines, then the result of an application of $\Sigma/\mathsf{A}{:}\Pi$ is the formula A on the condition $\Xi \cup \Pi$. The underlying idea is simple: the derivability of the members of Σ depends on Ξ and the rule enables one to derive A on the condition Π from Σ. Next, one trivially adjusts the further concepts: *finitary* rule - no restriction is imposed on Π, - $\mathcal{R}$-*correct* line of L, $\mathcal{R}$-*stage* from Γ, *extension* of a stage, and *static* $\mathcal{R}$-*proof* from Γ. From here on I spell out the new stuff because the changes are (obvious but) more significant.

DEFINITION 18. A static $\mathcal{R}$-*proof of* $A{:}\Delta$ *from* Γ is a static proof from Γ containing a stage of which a line has A as its formula and Δ as its condition.

DEFINITION 19. $\Gamma \vdash_{\mathcal{R}} A{:}\Delta$, A on the condition Δ is $\mathcal{R}$-*derivable from* Γ, iff there is a static $\mathcal{R}$-proof of $A{:}\Delta$ from Γ.

DEFINITION 20. Where **L** is defined by a recursive set $\mathcal{R}$ of rules and a Marking definition, $\Gamma \vdash_{\mathbf{L}} A{:}\Delta$ iff $\Gamma \vdash_{\mathcal{R}} A{:}\Delta$.

DEFINITION 21. $\mathcal{R}$ be *redundant* iff, for some $(\Sigma/\mathsf{A}{:}\Pi) \in \mathcal{R}$, (i) there is a $\Sigma' \subseteq \Sigma$ and a $\Pi' \subseteq \Pi$ for which $\Sigma' \subset \Sigma$ or $\Pi' \subset \Pi$ and (ii) $\Gamma \vdash_{\mathcal{R}} A{:}\Delta$ iff $\Gamma \vdash_{\mathcal{R}-\{\Sigma/\mathsf{A}{:}\Pi\}\cup\{\Sigma'/\mathsf{A}{:}\Pi'\}} A{:}\Delta$.

DEFINITION 22. $\mathcal{R}$ be *minimal* iff, for every $\mathcal{R}' \subset \mathcal{R}$, there are Γ, A, and Δ for which $\Gamma \vdash_R A:\Delta$ and $\Gamma \nvdash_{R'} A:\Delta$.

The proofs of the following theorems are similar to those of the corresponding theorems from Section 2.

THEOREM 23. *If R is a non-redundant and minimal set of rules, then all members of R are* finitary.

THEOREM 24. *If R is a set of rules, then* there is *a non-redundant and minimal set of rules R' such that $\Gamma \vdash_{R'} A:\Delta$ iff $\Gamma \vdash_R A:\Delta$.*

THEOREM 25. *Where L is defined by a recursive R of rules and a Marking definition, if $\Gamma \vdash_L A:\Delta$, then there is a static R-proof of $A:\Delta$ from Γ in which A is the formula and Δ the condition of the* last *line of the* last *stage. (Finiteness of Conditional Derivation)*

A R-proof that has s as its last stage will also be called a proof *at stage s*. Where no confusion arises, I shall sometimes call finite stages by the number of the line that was last added to them. This will also enable me to refer to stages that are predecessors of the present stage.

We shall see in the next section that, in some cases, there are reasons to consider extensions of infinite stages (stages comprising infinitely many lines). Clearly, appending a line to such a stage does not result in a list of formulas. Recall, however, that we have only to consider finitary rules. So if all formulas required to apply some rule occur in the proof, they all occur at or before a finite point in s; let the last formula occur at the nth line of s. So the result of the application can be inserted between the nth and $(n + 1)$th line of s. It is unimportant whether one renumbers the lines from $n + 1$ on, or gives the inserted line an unusual number, say $n.1$. All that matters is that the extension of s is a list of formulas.

People who might have principled objections against the insertion of lines in a proof should realize that the result of the insertion may obviously also be obtained by appending only. The resulting objects, viz. the stages, are identical; only their history, viz. the preceding chain of stages, is different.

Having fixed the derivability of a formula on a condition, let us now move on to the derivability of a formula, i.e. bring in the Marking definitions.

DEFINITION 26. *A is derived at stage s* of a R-proof from Γ iff A is the formula of a line of s and this line is unmarked at stage s.

DEFINITION 27. *A is R-derivable at a stage* from Γ iff A is derived at a stage of a R-proof from Γ.

A formula may be derived at one stage and non-derived at the next, or vice versa. This is typical of dynamic proofs and there is nothing wrong with it. Yet we also want to define a more stable notion of derivability to express where the dynamics leads to in the end - whether we are able to find out where the dynamics leads to is a different matter. This stable notion will be called final derivability.

DEFINITION 28. Where p is a $\mathcal{R}$-proof from Γ at stage s, p is *stable with respect to* line i iff (i) line i occurs in s and (ii) if line i is marked, respectively unmarked, at stage s, then it is marked, respectively unmarked, in all extensions of s.

The intuitive idea behind final derivability is that A is derived from Γ on an unmarked line of a stage of a $\mathcal{R}$-proof from Γ and that the proof is stable with respect to that line. Note that, in order to decide that the definition applies to a proof, one needs a *metalevel argument* about all possible extensions of this proof. This should be taken into account when forging a definition of final derivability.

For some **L**, A, and Γ, only infinite **L**-proofs from Γ contain an unmarked line on which A is derived *and* are stable with respect to this line. That is obviously inconvenient. Such a proof can be reasoned about, at the metalevel, but cannot be produced. This is a good reason to look for a different approach. However, it is difficult to do so at the abstract level of the present section. So let us postpone this to the next section, in which a specific family of logics is presented. If you wonder what concrete dynamic proofs look like, you will obtain an answer there.

Note that the static proofs from Section 2 are a special case of dynamic proofs. Those static proofs are just like dynamic proofs except that all lines have an empty condition, whence no line is ever marked. This is why derivability at a stage coincides with final derivability for static proofs.

4   Adaptive logics

The motivation for studying adaptive logics cannot be presented here. I refer the reader for example to [4].

Nearly all known adaptive logics have been phrased in *standard format*,[6] which has major advantages as will become clear below. The format was first introduced in [3] and most extensively studied in [5], which contains details and metatheoretic proofs. Not too long from now, the best reference should be [6]. From here on, I disregard adaptive logics that are not in standard format. So all claims on adaptive logics should be read as claims on adaptive logics in standard format (even if some claims hold for all adaptive logics).

While describing the standard format, I shall illustrate it with two related logics, which I shall use for examples in Section 5. The information on these logics is presented in separate paragraphs.

An adaptive logic **AL** in standard format is defined by a triple:

   1. A *lower limit logic* **LLL**: a logic that has static proofs, has a characteristic

---

[6] An exception is Graham Priest's LP^m from [11], which at the predicative level defines abnormalities with respect to models rather than with respect to the formulas verified by models. See [2] for a discussion of some odd effects.

semantics, and contains **CL** (Classical Logic).[7]

2. A *set of abnormalities* Ω: a set of **LLL**-contingent formulas, characterized
   by a (possibly restricted) logical form **F**.

3. An *adaptive strategy*: Reliability or Minimal Abnormality.[8]

Examples Two related adaptive logics are $\mathbf{CLuN}^r$ and $\mathbf{CLuN}^m$. Their lower limit
logic is **CLuN** (*C*lassical *L*ogic allowing for gl*u*ts with respect to *N*egation),
viz. full positive **CL** with $(A \supset {\sim}A) \supset {\sim}A$ added as the only axiom for the
standard negation , and extended with the classical connectives - see note
7.[9] While $A \vee {\sim}A$ is a **CLuN**-theorem, $A \wedge {\sim}A$ is **CLuN**-contingent. The
set of abnormalities Ω comprises all formulas of the form $\exists(A \wedge {\sim}A)$ (the
existential closure of $A \wedge {\sim}A$). The strategies are respectively Reliability
and Minimal Abnormality (as the superscripts reveal).

Extending **LLL** with an axiom that declares all abnormalities logically false
results in the *upper limit logic* **ULL**. If a premise set Γ does not require that
any abnormalities are true, the **AL**-consequences of Γ are identical to its **ULL**-
consequences. If the premise set requires some abnormalities to be true, the
AL-consequence set is stronger than the **LLL**-consequence set (except for border
cases) and is weaker than the **ULL**-consequence set.

Examples The upper limit logic of $\mathbf{CLuN}^r$ and of $\mathbf{CLuN}^m$ is **CL**. It is obtained
by extending **CLuN** with the axiom $(A \wedge {\sim}A) \supset B$.

In the expression $Dab(\Delta)$, Δ will always be a finite subset of Ω, and $Dab(\Delta)$
will denote the *classical* disjunction (see note 7) of the members of Δ. $Dab(\Delta)$ is
called a Dab-formula (a disjunction of abnormalities). Even where the standard
disjunction has the same meaning as the classical disjunction, I shall consider only
classical disjunctions of abnormalities as Dab-formulas. While the matter is of
little importance in most contexts, it is extremely convenient for the dialogues
presented in Section 5.

---

[7]That **LLL** contains **CL** is realized by adding classical logical symbols (those having the same
meaning as in **CL**) to the language. These will be written as $\tilde{\ }$, $\check{\vee}$, $\check{\exists}$, etc. The classical symbols
have mainly a technical use and are not meant to occur in the premises or conclusions of standard
applications.

[8]Strategies are ways to cope with derivable disjunctions of abnormalities. The effects of the Reliability strategy and the Minimal Abnormality strategy become clear below in the text. Other strategies
were mainly devised in order to characterize consequence relations from the literature in terms of an
adaptive logic. All those strategies can be reduced to Reliability or Minimal Abnormality under a
translation.

[9]Suitable axioms for classical negation are $(A \supset {\tilde{\ }}A) \supset {\tilde{\ }}A$ and $A \supset ({\tilde{\ }}A \supset B)$. The other classical
connectives have the same meaning as the corresponding standard connectives in **CLuN**.

Adaptive logics have dynamic proofs in the sense of the previous section. The rules of adaptive logics in standard format are defined in terms of the lower limit logic. Where

$$A \qquad \Delta$$

abbreviates that A occurs in the proof on the condition Δ, the (generic) rules are:[10]

PREM   If $A \in \Gamma$:

$$\frac{\cdots \quad \cdots}{A \quad \emptyset}$$

RU   If $A_1, \ldots, A_n \vdash_{\text{LLL}} B$:

$$\begin{array}{cc} A_1 & \Delta_1 \\ \cdots & \cdots \\ A_n & \Delta_n \\ \hline B & \Delta_1 \cup \ldots \cup \Delta_n \end{array}$$

RC   If $A_1, \ldots, A_n \vdash_{\text{LLL}} B \,\check{\vee}\, Dab(\Theta)$:

$$\begin{array}{cc} A_1 & \Delta_1 \\ \cdots & \cdots \\ A_n & \Delta_n \\ \hline B & \Delta_1 \cup \ldots \cup \Delta_n \cup \Theta \end{array}$$

We need some technicalities in preparation of the marking definitions. $Dab(\Delta)$ is a *minimal Dab-formula* at stage s of an **AL**-proof iff $Dab(\Delta)$ has been derived at that stage on the condition $\emptyset$ whereas there is no $\Delta' \subset \Delta$ for which $Dab(\Delta')$ has been derived on the condition $\emptyset$. A *choice set* of $\Sigma = \{\Delta_1, \Delta_2, \ldots\}$ is a set that contains an element out of each member of Σ. A *minimal choice set* of Σ is a choice set of Σ of which no proper subset is a choice set of Σ. Consider a proof from Γ at stage s and let $Dab(\Delta_1), \ldots, Dab(\Delta_n)$ be the minimal Dab-formulas at that stage. $U_s(\Gamma) = \Delta_1 \cup \ldots \cup \Delta_n$ and $\Phi_s(\Gamma)$ is the set of minimal choice sets of $\{\Delta_1, \ldots, \Delta_n\}$.

DEFINITION 29. Marking for Reliability: Line i is marked at stage s iff, where Δ is its condition, $\Delta \cap U_s(\Gamma) \neq \emptyset$.

Example Let $\Gamma_1 = \{\sim p, \sim q, p \vee r, q \vee s, p \vee q, p\}$. Consider the following stage 9 of a proof from Γ_1.

1	$\sim p$	Premise	$\emptyset$	
2	$\sim q$	Premise	$\emptyset$	
3	$p \vee r$	Premise	$\emptyset$	
4	$q \vee s$	Premise	$\emptyset$	
5	$p \vee q$	Premise	$\emptyset$	
6	p	Premise	$\emptyset$	
7	r	1, 3; RC	$\{p \wedge \sim p\}$	$\checkmark^9$

---

[10]The set of rules generated by the three generic rules is redundant and not minimal. This is irrelevant for the point I am making below in the text.

8	s	2, 4; RC	$\{q \wedge \sim q\}$	$\checkmark^9$
9	$(p \wedge \sim p) \check{\vee} (q \wedge \sim q)$	1, 2, 5; RU	$\emptyset$	

Note that $U_9(\Gamma_1) = \{p \wedge \sim p, q \wedge \sim q\}$. So, at stage 9 of the proof, neither r nor s is derived from Γ_1. However, the proof may be continued as follows

7	r	1, 3; RC	$\{p \wedge \sim p\}$	$\checkmark^{10}$
8	s	2, 4; RC	$\{q \wedge \sim q\}$	
9	$(p \wedge \sim p) \check{\vee} (q \wedge \sim q)$	1, 2, 5; RU	$\emptyset$	
10	$p \wedge \sim p$	1, 6; RU	$\emptyset$	

Only line 7 is marked at stage 10 of the proof because $U_{10}(\Gamma_1) = \{p \wedge \sim p\}$. Extending the proof will not change anything to the marks of lines 1–10 because $U_{10+i}(\Gamma_1) = U_{10}(\Gamma_1)$ for all i.

DEFINITION 30. Marking for Minimal Abnormality: Line i is marked at stage s iff, where A is derived on the condition Δ at line i, (i) there is no $\varphi \in \Phi_s(\Gamma)$ such that $\varphi \cap \Delta = \emptyset$, or (ii) for some $\varphi \in \Phi_s(\Gamma)$, there is no line at which A is derived on a condition Θ for which $\varphi \cap \Theta = \emptyset$.

This reads more easily: where A is derived on the condition Δ on line i, line i is *unmarked* at stage s iff (i) there is a $\varphi \in \Phi_s(\Gamma)$ for which $\varphi \cap \Delta = \emptyset$ and (ii) for every $\varphi \in \Phi_s(\Gamma)$, there is a line at which A is derived on a condition Θ for which $\varphi \cap \Theta = \emptyset$.

Example Let $\Gamma_2 = \{\sim p, \sim q, p \vee r, q \vee s, p \vee q\}$. Consider the following proof from Γ_2.

1	$\sim p$	Premise	$\emptyset$	
2	$\sim q$	Premise	$\emptyset$	
3	$p \vee r$	Premise	$\emptyset$	
4	$q \vee s$	Premise	$\emptyset$	
5	$p \vee q$	Premise	$\emptyset$	
6	r	1, 3; RC	$\{p \wedge \sim p\}$	$\checkmark^{10}$
7	s	2, 4; RC	$\{q \wedge \sim q\}$	$\checkmark^{10}$
8	$r \vee s$	6; RU	$\{p \wedge \sim p\}$	
9	$r \vee s$	7; RU	$\{q \wedge \sim q\}$	
10	$(p \wedge \sim p) \check{\vee} (q \wedge \sim q)$	1, 2, 5; RU	$\emptyset$	

Line 8 may also be justified by "1, 3; RC" and line 9 by "2, 4; RC". Note that $\Phi_{10}(\Gamma_2) = \{\{p \wedge \sim p\}, \{q \wedge \sim q\}\}$. So, at stage 10 of the proof, neither r nor s is derived from Γ_2. However, $r \vee s$ is derived because the condition of line 9 does not overlap with $\{p \wedge \sim p\}$ and the condition of line 8 does not overlap with $\{q \wedge \sim q\}$. Both lines 8 and 9 would be marked on the Reliability strategy.

The sense of the marking definitions (and their relation to the semantics) is studied in other papers, for example [5], and cannot be discussed here.

As was stated in the previous section, final derivability is established by a proof and a metalevel reasoning. The existence of the proof should of course not be established at the metalevel. So the proof should be finite.

Here is the promised definition of final derivability.

DEFINITION 31. *A* is *finally derived* on line i of an **AL**-proof from Γ at stage s iff (i) *A* is the formula of line i, (ii) line i is not marked at stage s, and (iii) every extension of the proof in which line i is marked may be further extended in such a way that line i is unmarked.

DEFINITION 32. $\Gamma \vdash_{\mathbf{AL}} A$ (*A* is *finally* **AL**-*derivable* from Γ) iff *A* is finally derived on a line of a proof from Γ.

Let me first show that this definition is adequate with respect to the intuitive understanding of final derivability - see the previous section. To do so we need some preparation. $Dab(\Delta)$ is a *minimal Dab-consequence of* Γ iff $\Gamma \vdash_{\mathbf{LLL}} Dab(\Delta)$ and there is no $\Delta' \subset \Delta$ for which $\Gamma \vdash_{\mathbf{LLL}} Dab(\Delta')$. Where $Dab(\Delta_1), Dab(\Delta_2), \ldots$ are the minimal *Dab*-consequences of Γ, $U(\Gamma) = \Delta_1 \cup \Delta_2 \cup \ldots$ and $\Phi(\Gamma)$ is the set of minimal choice sets of $\{\Delta_1, \Delta_2, \ldots\}$.[11]

Let **AL**^{r} and **AL**^{m} be adaptive logics the third element of which is Reliability, respectively Minimal Abnormality. Theorems 33 and 34 are proved as Theorems 6 and 8 in [5].

THEOREM 33. $\Gamma \vdash_{\mathbf{AL}^r} A$ *iff there is a finite* $\Delta \subset \Omega$ *for which* $\Gamma \vdash_{\mathbf{LLL}} A \check{\vee} Dab(\Delta)$ *and* $\Delta \cap U(\Gamma) = \emptyset$.

THEOREM 34. $\Gamma \vdash_{\mathbf{AL}^m} A$ *iff, for every* $\varphi \in \Phi(\Gamma)$, *there is a finite* $\Delta \subset \Omega$ *such that* $\Delta \cap \varphi = \emptyset$ *and* $\Gamma \vdash_{\mathbf{LLL}} A \check{\vee} Dab(\Delta)$.

LEMMA 35. *If* $\Gamma \vdash_{\mathbf{AL}^r} A$, *then there is an* **AL**^{r}-*proof from* Γ *in which A is derived on an unmarked line and that is stable with respect to that line.*

Proof. Suppose that $\Gamma \vdash_{\mathbf{AL}^r} A$. By Theorem 33 there is a (finite) $\Delta \subset \Omega$ for which $\Gamma \vdash_{\mathbf{LLL}} A \check{\vee} Dab(\Delta)$ and $\Delta \cap U(\Gamma) = \emptyset$. As **LLL** has static proofs, there is a finite **AL**^{r}-proof in which $A \check{\vee} Dab(\Delta)$ is derived on the condition $\emptyset$. From this, *A* is derived on the condition $Dab(\Delta)$ (in one step by RC), say on line i. Let this be an **AL**^{r}-proof at the finite stage s and call this proof p_0.

There are only countably many minimal *Dab*-consequences of Γ, say $Dab(\Delta_1), Dab(\Delta_2), \ldots$. For each of these, there is a finite **AL**-proof, call it p_i, in which $Dab(\Delta_i)$ is derived on the condition $\emptyset$.

---

[11]It is useful to compare the definition of $U(\Gamma)$ with that of $U_s(\Gamma)$ and to compare the definition of $\Phi(\Gamma)$ with that of $\Phi_s(\Gamma)$. In each case, the latter set is an estimate of the former depending on the insights provided by the proof at a stage.

Consider the proof p' of which the last stage, call it s', is the concatenation $\langle p_0, p_1, p_2, \ldots \rangle$. As all minimal *Dab*-consequences of Γ have been derived on the condition $\emptyset$ in s', $U_{s'}(\Gamma) = U(\Gamma)$. As $\Delta \cap U(\Gamma) = \emptyset$, line i is unmarked. Moreover, as all minimal *Dab*-consequences of Γ have been derived on the condition $\emptyset$ in s', $U_{s''}(\Gamma) = U_{s'}(\Gamma) = U(\Gamma)$ for every extension s'' of s'. So line i is unmarked in every extension s'' of s', which means that p' is stable with respect to line i. ∎

For some Γ, $\Phi(\Gamma)$ is uncountable. However, the set of Δ such that, for some $\varphi \in \Phi(\Gamma)$, $\Delta \cap \varphi = \emptyset$ and $\Gamma \vdash_{\mathbf{LLL}} A \check{\vee} Dab(\Delta)$, is a countable set - each of *these* Δ is a finite set of formulas. Moreover, for each such Δ, there is a finite proof of $A \check{\vee} Dab(\Delta)$. Let $\{p'_1, p'_2, \ldots\}$ be the countable set of these proofs. The proof of Lemma 36 proceeds exactly as that of Lemma 35, except that we now define p' as a proof that has as stage s' the concatenation $\langle p_1, p'_1, p_2, p'_2, \ldots \rangle$, which warrants that $\Phi_{s'}(\Gamma) = \Phi(\Gamma)$ and that, for every extension s'' of s', $\Phi_{s''}(\Gamma) = \Phi_{s'}(\Gamma) = \Phi(\Gamma)$.

LEMMA 36. *If $\Gamma \vdash_{\mathbf{AL}^m} A$, then there is an $\mathbf{AL}^m$-proof from Γ in which A is derived on an unmarked line and that is stable with respect to that line.*

Whether the third element of an adaptive logic is Reliability or Minimal Abnormality, the following lemma holds.

LEMMA 37. *If A is derived on an unmarked line of an $\mathbf{AL}$-proof from Γ that is stable with respect to that line, then $\Gamma \vdash_{\mathbf{AL}} A$*

Proof. Suppose that the antecedent is true. As the unmarked line on which A is derived will not be marked in any extension of the proof, A is finally $\mathbf{AL}$-derived in this proof. ∎

THEOREM 38. *$\Gamma \vdash_{\mathbf{AL}} A$ iff A is derived on an unmarked line of an $\mathbf{AL}$-proof from Γ that is stable with respect to that line.*

Proof. Immediate in view of Lemmas 35, 36, and 37. ∎

Having established that Definition 32 is adequate, I show two theorems concerning the finiteness of the proof and extensions mentioned in Definition 31. Obviously, an $\mathbf{AL}$-proof is *finite* iff each stage of the proof is a finite list of formulas.

THEOREM 39. *If $\Gamma \vdash_{\mathbf{AL}} A$, then A is finally derived on a line of a finite $\mathbf{AL}$-proof from Γ.*

Proof. Suppose that the antecedent is true. If the strategy is Reliability, there is a finite $\Delta \subset \Omega$ such that $\Gamma \vdash_{\mathbf{LLL}} A \check{\vee} Dab(\Delta)$ and $\Delta \cap U(\Gamma) = \emptyset$ (by Theorem 33). If the strategy is Minimal Abnormality, there is a $\Delta \subset \Omega$ such that $\Gamma \vdash_{\mathbf{LLL}} A \check{\vee} Dab(\Delta)$ and $\Delta \cap \varphi = \emptyset$ for some $\varphi \in \Phi(\Gamma)$ (by Theorem 34).

As $\mathbf{LLL}$ is compact, there is a finite $\Gamma' \subseteq \Gamma$ for which $\Gamma' \vdash_{\mathbf{LLL}} A \check{\vee} Dab(\Delta)$. So there is a finite $\mathbf{AL}$-stage from Γ' in which occur all members of Γ' followed by a line in which A is derived on the condition Δ by application of RC. As $\Gamma \vdash_{\mathbf{AL}} A$, A is finally derived in this proof in view of definitions 29, 30, 31, and 32. ∎

THEOREM 40. *If the strategy is Reliability, Definitions 31 and 32 are still adequate if the extensions mentioned in Definition 31 are finite.*

Proof. Case 1: $\Gamma \vdash_{\mathbf{AL}^r} A$. Let A be finally derived on line i in an $\mathbf{AL}^r$-proof from Γ, let Δ be the condition of line i, and let s be the last stage of this proof. Consider a finite extension s' of s in which line i is marked. Stage s' counts at most finitely many minimal Dab-formulas, say $Dab(\Theta_1), \ldots, Dab(\Theta_n)$, for which $\Theta_i \cap \Delta \neq \emptyset$ $(1 \leq i \leq n)$. In view of Definitions 29, 31, and 32, there is, for each of these Θ_i, a $\Theta_i' \subset \Theta_i$ such that $\Gamma \vdash_{\mathbf{AL}} Dab(\Theta_i'){:}\emptyset$ and $\Theta_i' \cap \Delta = \emptyset$. Append the last stage of the proof of each of these $Dab(\Theta_i){:}\emptyset$ to s' and let the result be s''. Stage s'' counts finitely many lines and $\Delta \cap U_{s''}(\Gamma) = \emptyset$.

Case 2: $\Gamma \nvdash_{\mathbf{AL}^r} A$. In view of Theorem 33 it holds for all $\Delta \subset \Omega$ that $\Delta \cap U(\Gamma) \neq \emptyset$ if $\Gamma \vdash_{\mathbf{LLL}} A \check{\vee} Dab(\Delta)$. Suppose that A has been derived on the condition Δ on a line, say i, of a finite $\mathbf{AL}^r$-proof from Γ and that the last stage of this proof is s. It follows that there is a minimal Dab-consequence $Dab(\Theta)$ of Γ for which $\Theta \cap \Delta \neq \emptyset$. As $\Gamma \vdash_{\mathbf{LLL}} Dab(\Theta)$, Θ can be derived on the condition $\emptyset$ in a finite extension s' of s and there is no extension of s' in which line i is unmarked. ∎

5   The dialogues

As the dialogues I want to propose are somewhat unusual, let me first say a few words about usual dialogues. It is not difficult to define these, first for **CLuN**, and next for **CLuN**^{r} and **CLuN**^{m}. Tableau methods presented in [8] and [9] form a good start. The tableau methods may even be simplified by extending the language with classical negation , whence there is no need for signed formulas.

Adaptive logics have no theorems of their own. If theorems are defined by $\emptyset \vdash A$, then the theorems of the adaptive logic, for example **CLuN**^{r}, are identical to those of its upper limit logic, in the example **CL**. If theorems are defined by "for all Γ, $\Gamma \vdash A$", then the theorems of the adaptive logic are identical to those of its lower limit logic, in the example **CLuN** - obviously all theorems of the lower limit logic are theorems of the upper limit logic. This means that one cannot define the adaptive consequence relation in terms of theorems, but that dialogues for the consequence relation should be devised. So one will have to adjust the description of a dialogue from, for example, [12] or [13], and there will be a few peculiarities, to which I return briefly in Section 6.

Let us now turn to the unusual dialogues I announced. The proponent claims that $\Gamma \vdash_{\mathbf{AL}} A$ and the opponent denies this. We let the proponent and opponent construct a proof together, giving each a specific task. The proponent starts. If, at the end of the dialogue, A is derived in the proof, the proponent wins; otherwise the opponent wins. This kind of dialogue is completely silly if the logic has static proofs. Indeed, the opponent has no specific role to play: no contribution to the proof forms a means to attack the derivability of the conclusion from the premises.

The situation is dramatically different for logics that have dynamic proofs. If the conclusion A is not derivable from the premises Γ by the lower limit logic, then the proponent can only derive it on a non-empty condition. We have seen that the resulting proof does not constitute a demonstration of $\Gamma \vdash_{AL} A$. Actually, no proof forms such a demonstration. So it seems natural to construct a demonstration of $\Gamma \vdash_{AL} A$ as a dialogue between a proponent, who tries to show that A is finally derivable but has to defend herself against moves of the opponent. Let me first comment on the natural character of the approach.

First a comparison. Every logician is acquainted with the situation in which he or she tries to find out whether a formal system has a certain property. If one is convinced that the property holds, one will attempt to prove so. If one does not find the proof, this very fact will undermine the conviction. At some point one will become convinced that the property does not hold and one will try to find a counterexample - often insights from the failing proof will indicate in which direction to look for a counterexample. If, in turn, one fails to produce a counterexample, this may induce one to look again for a proof, etc. The alternating phases may be seen as a dialogue between a proponent and an opponent.

Let us now look more closely at adaptive logics. The idea is that abnormalities are presupposed to be false, unless and until proven otherwise.[12] So two different aims should be realized in a well-directed proof: to establish the conclusion on some condition and to establish that the condition is safe - in the case of Reliability, this means that no member of the condition is unreliable; in the case of Minimal Abnormality, it means that the condition does not overlap with a minimal choice set of all Dab-consequences of the premises *and* that, for each such minimal choice set φ, the conclusion can be derived on a condition that does not overlap with φ. So it is indeed natural to see this as a dialogue in which the proponent first establishes the conclusion on some condition, next the opponent tries to show that the condition is unsafe, next the proponent tries to reestablish the safety of the condition, and so on. Several variant dialogues are possible, even for the same strategy. They will be considered in some detail below.

Although no dynamic proof will establish that a conclusion is finally derived from a premise set, the metalevel reasoning that is required next to the proof can be seen in dialogic terms: the conclusion is finally derivable iff the proponent can uphold it against every possible attack.

It seems to me that this is at the heart of all forms of defeasible reasoning:

---

[12]The expression is taken from the oldest paper on adaptive logics, [1]. It is obviously vague if disjunctions of abnormalities (Dab-formulas) are derivable on the condition $\emptyset$. The strategy removes the vagueness.

that one establishes a conclusion on some condition and that the condition can be maintained in the face of every possible attack.

Given the differences between the two strategies, I shall consider the variant dialogues for one strategy, and next for the other. Let us start with *Reliability*.

Stability with respect to a line

In this type of dialogue, the proponent first establishes the conclusion on some condition on an unmarked line, say line l, of a (finite or infinite) proof. Next, the opponent may extend the proof. The opponent wins if he produces an extension in which line l is marked; otherwise the proponent wins. The proponent has a winning strategy iff she can produce a proof that warrants her winning.

This approach is all right, but requires that the proponent sometimes starts off by producing an infinite proof. Consider the premise set $\Gamma_3 = \{p \vee q, \sim q, (q \wedge \sim q) \vee (r_i \wedge \sim r_i), (q \wedge \sim q) \supset (r_i \wedge \sim r_i) \mid i \in \mathbb{N}\}$ and let the proponent aim at establishing $\Gamma_3 \vdash_{\mathbf{CLuN}^r} p$. Consider a finite proof, produced by the proponent, that starts off with

1	$p \vee q$	Premise	$\emptyset$
2	$\sim q$	Premise	$\emptyset$
3	p	1, 2; RC	$\{q \wedge \sim q\}$

and moreover contains a (forcibly finite) number of triples of lines of the following form

j	$(q \wedge \sim q) \vee (r_i \wedge \sim r_i)$	Premise	$\emptyset$
$j+1$	$(q \wedge \sim q) \supset (r_i \wedge \sim r_i)$	Premise	$\emptyset$
$j+2$	$r_i \wedge \sim r_i$	$j, j+1$; RU	$\emptyset$

Clearly, line 3 is unmarked in this proof. However, if the opponent extends the proof with the lines

k	$(q \wedge \sim q) \vee (r_l \wedge \sim r_l)$	Premise	$\emptyset$
$k+1$	$(q \wedge \sim q) \check{\vee} (r_l \wedge \sim r_l)$	k; RU	$\emptyset$

for a r_l that does not yet occur in the proof, then line 3 is marked.[13] So the proponent looses. Of course, she should have a winning strategy, because $\Gamma_3 \vdash_{\mathbf{CLuN}^r} p$. And indeed there is one, but it requires that the proponent introduces all premises and all connected lines $j+1$ and $j+2$, which means that she should produce an infinite proof in her first move. This is not handy. Infinite proofs cannot be produced, but should be handled by a metalevel reasoning. It would be more attractive if at least the first move in the dialogue would be a proof that can actually be produced.

---

[13]Line 3 is marked at stage $k + 1$ of the proof, not at stage k. This follows from the convention that "*Dab*-formula" strictly refers to a *classical* disjunction of abnormalities. Showing that the formula of line k is equivalent to a *Dab*-formula requires a deductive step, which here is taken by the opponent.

Also, not much dialogue is involved in this kind of game. The outcome fully depends on the first move of the proponent. She has a winning strategy iff she is able to produce, as her first step, a proof that is stable with respect to an unmarked line at which the conclusion is derived.

Incidentally, some readers might balk at the artificiality of the premise set Γ_3. It is indeed hard to imagine real life applications in which the depicted complication would arise. Nevertheless, in describing logics, one should consider all possible complications, whether they are artificial or not.

Many turns

This kind of dialogue is definitely more fascinating than the previous one. In her first move, the proponent produces a finite proof that contains an unmarked line, say line l, in which the conclusion is derived on a condition Δ. Next, the opponent tries to show that Δ is unreliable by producing a finite extension of the proof. If the opponent's move is successful, line l is marked at the last stage of the extended proof. The proponent reacts by trying to finitely extend the proof in such a way that line l is unmarked. And so on. The proponent has a winning strategy iff she is able to answer every move of the opponent, viz. iff she is able to extend every new extension in such a way that line l is unmarked.

To illustrate the matter, consider again Γ_3 from the previous dialogue and let the logic be **CLuN**^{r}. Suppose that the proponent starts as follows:

1	$p \vee q$	Premise	$\emptyset$
2	$\sim q$	Premise	$\emptyset$
3	p	1, 2; RC	$\{q \wedge \sim q\}$

The opponent may reply, for example, by the following extension:

4	$(q \wedge \sim q) \vee (r_1 \wedge \sim r_1)$	Premise	$\emptyset$
5	$(q \wedge \sim q) \check{\vee} (r_1 \wedge \sim r_1)$	4; RU	$\emptyset$

Note that line 3 is marked at stage 5 of the proof. The proponent will answer by extending the proof as follows:

6	$(q \wedge \sim q) \supset (r_1 \wedge \sim r_1)$	Premise	$\emptyset$
7	$r_1 \wedge \sim r_1$	6, 4; RU	$\emptyset$

As $U_7(\Gamma_3) = \{r_1 \wedge \sim r_1\}$, line 3 is unmarked at stage 7 of the proof. Of course, the opponent may still attack, and will perhaps attack more forcibly:

8	$(q \wedge \sim q) \vee (r_2 \wedge \sim r_2)$	Premise	$\emptyset$
9	$(q \wedge \sim q) \check{\vee} (r_2 \wedge \sim r_2)$	8; RU	$\emptyset$
$\vdots$			
24	$(q \wedge \sim q) \vee (r_{10} \wedge \sim r_{10})$	Premise	$\emptyset$
25	$(q \wedge \sim q) \check{\vee} (r_{10} \wedge \sim r_{10})$	24; RU	$\emptyset$

Line 3 is now again marked, but the proponent will reply by

26 $(q \land \sim q) \supset (r_2 \land \sim r_2)$ Premise $\emptyset$
27 $r_2 \land \sim r_2$ 26, 8; RU $\emptyset$
$\vdots$
42 $(q \land \sim q) \supset (r_{10} \land \sim r_{10})$ Premise $\emptyset$
43 $r_{10} \land \sim r_{10}$ 42, 24; RU $\emptyset$

after which line 3 is unmarked. So the proponent defended herself adequately.

The premise set is a dull one, but it candidly illustrates that, in some cases, every attack can be answered successfully, whereas a successful attack is possible after any finite number of defenses - actually infinitely many different attacks are possible at any finite point.

Several comments are appropriate. First, the proponent is able to answer every move of the opponent. So the proponent has a winning strategy in the dialogue for $\Gamma_3 \vdash_{\mathbf{CLuN}^r} p$. Next, only after an infinite sequence of attacks and defenses will all means to attack have been used up. Finally, one may introduce a convention to terminate the dialogue after finitely many turns. The proponent and opponent may make a deal about this either beforehand or during the dialogue, the choice may be with one of them, or whatever one likes. Especially if the premise set is less orderly and hence more interesting, it seems attractive to let the dialogue go on until the opponent gives up. This will enable the opponent to try out different lines of attack in view of the premises. All that is essential is that the proponent is given a defense after every attack.

POP

This is the optimized simplification of the previous dialogue type. The proponent starts by producing a finite proof in which the conclusion is derived on a condition Δ on an unmarked line, say line l. Next, the opponent tries to show that Δ is unreliable by producing a finite extension. If the opponent is successful, line l is marked at the last stage of the extended proof. The proponent reacts by trying to finitely extend the proof in such a way that line l is unmarked. The proponent wins the dialogue if line l is unmarked after her reaction. The proponent has a winning strategy iff she is able to win, whatever be the opponent's attack.

Calling premises

The proponent starts by producing a finite proof that contains an unmarked line, say l, in which the conclusion is derived from the premises on a condition Δ. At this point, the opponent does not extend the proof, but delineates a finite set Γ' of premises, which he will be allowed to use in his attack. The proponent finitely extends her proof, introducing whatever premises she wants. Next, the opponent attacks, viz. extends the extension, introducing as premises only members of Γ' and deriving only formulas from the premises he himself introduced. The proponent wins the dialogue if line l is unmarked after the opponent's attack; otherwise the opponent wins. The proponent has a winning strategy iff she can proceed in such a way that she wins the dialogue. This means that she is able to produce a proof in which the conclusion is derived, say at line l, and that, whatever finite $\Gamma' \subseteq \Gamma$ the opponent chooses, she can produce an extension of her proof that the opponent cannot extend in such a way that line l is marked, provided that the opponent only derives formulas that follow from Γ'. This type of dialogue illustrates that the proponent can defend herself by finitely many moves against all Dab-formulas that are derivable from any finite $\Gamma' \subseteq \Gamma$ *just in case* the conclusion is finally derivable from Γ.

More dialogue types may be possible, but those described before are sufficient to make the point I was trying to make. I still have to prove that the dialogues are adequate. Let a dialogue of each of these types be called a dialogue for $\Gamma \vdash_{AL^r} A$, in which Γ is the premise set and A is the conclusion. The proof of the following theorem is rather simple because the dialogues 'interpret' definitions and theorems from Section 4.

THEOREM 41. *For the four dialogue types described holds: the proponent has a winning strategy in the dialogue for $\Gamma \vdash_{AL^r} A$ iff $\Gamma \vdash_{AL^r} A$.*

Proof. Let us start with *POP*. $\Rightarrow$ Suppose that the proponent derives A on the condition Δ, say at line l, and that $\Gamma \nvdash_{AL^r} A$. In view of Theorem 33, it follows that there is a minimal Dab-consequence of Γ, say $Dab(\Theta)$, for which $\Delta \cap \Theta \neq \emptyset$. By the compactness of **LLL**, $Dab(\Theta)$ is **LLL**-derivable from a finite $\Gamma' \subseteq \Gamma$. So the opponent will introduce the members of Γ' and derive $Dab(\Theta)$, at which point line l is marked. As there is no way to extend the proof in such a way that line l is unmarked, the proponent has no winning strategy. $\Leftarrow$ Obvious in view of Definitions 31 and 32 and Theorem 40.

Many turns. $\Rightarrow$ As for POP, except that it is sufficient for the opponent to derive $Dab(\Theta)$ in one of his attacks. $\Leftarrow$ In view of Definitions 31 and 32 and Theorem 40, the proponent has a successful reply after every attack.

Stability with respect to a line. This is an obvious consequence of the proof for POP in view of Theorem 38.

Calling premises. $\Rightarrow$ As for POP, except that the opponent calls the members of Γ'. $\Leftarrow$ Suppose that $\Gamma \vdash_{AL^r} A$. In view of Theorem 33, there is a finite $\Delta \subset \Omega$ for which $\Gamma \vdash_{LLL} A \check{\vee} Dab(\Delta)$ and $\Delta \cap U(\Gamma) = \emptyset$. So the proponent derives A on the condition Δ, say on line l. Where $\Gamma' \subseteq \Gamma$ is a finite set, let $Dab(\Theta_1), \ldots, Dab(\Theta_n)$ be the minimal Dab-consequences of Γ' - every finite Γ' has finitely many minimal Dab-consequences. As $\Delta \cap U(\Gamma) = \emptyset$, there is, for every Θ_i ($1 \leq i \leq n$) for which $\Theta_i \cap \Delta \neq \emptyset$, a $\Theta_i' \subset \Theta_i$ for which $Dab(\Theta_i')$ is a minimal Dab-consequence of Γ and $\Delta \cap \Theta_i' = \emptyset$. So it is sufficient that the proponent derives these finitely many $Dab(\Theta_i')$ - this requires only a finite proof in view of the compactness of

LLL - in order to warrant that line l will not be marked if the proof is extended by consequences of Γ'. ∎

To prove the adequacy of the Calling Premises dialogue, it is essential that the opponent cannot rely in his extension on premises introduced by the proponent. This is related to the fact that most adaptive logics are not compact. The following example illustrates the lack of compactness of **CLuN^r**. Let $\Gamma_4 = \{((p \vee q) \wedge \sim q) \wedge ((q \wedge \sim q) \vee (r_1 \wedge \sim r_1))\} \cup \{((q \wedge \sim q) \supset (r_i \wedge \sim r_i)) \wedge ((q \wedge \sim q) \vee (r_{i+1} \wedge \sim r_{i+1})) \mid i \in \mathbb{N}\}$. All **CLuN**-models of Γ_4 verify $r_i \wedge \sim r_i$ for all $i \in \mathbb{N}$, and some verify no other abnormality. So $U(\Gamma_4) = \{r_i \wedge \sim r_i \mid i \in \mathbb{N}\}$, whence $\Gamma_4 \vdash_{\mathbf{CLuN}^r} p$. However, there is no finite $\Gamma' \subset \Gamma_4$ for which $\Gamma' \vdash_{\mathbf{CLuN}^r} p$. The example also illustrates that, if the opponent were allowed to rely on premises introduced by the proponent, then he would be able to win the dialogue for $\Gamma_4 \vdash_{\mathbf{CLuN}^r} p$, even if he chose $\Gamma' = \emptyset$. But of course the proponent should have a winning strategy because $\Gamma_4 \vdash_{\mathbf{CLuN}^r} p$.

The last comment on the dialogues for Reliability concerns decidability. Although the proponent has a winning strategy or does not have one, for each specific Γ and A, it is very well possible that we are unable to find out which of the two is the case. This is related to the undecidability at the predicative level. The absence of a positive test for (predicative) final derivability has the effect that a concrete dialogue will not constitute a demonstration that A is derivable from Γ. Only establishing that the proponent has a winning strategy will do so and it is only possible to establish this by a reasoning at the metalevel. All this is unavoidable in the case of defeasible reasoning, unless one artificially restricts it to decidable fragments.[14]

Let us now move on to the *Minimal Abnormality* strategy. In general, Minimal Abnormality requires more complex proofs than Reliability. For some Γ and A, A can only be derived on an unmarked line if A is derived on several conditions (and hence on several lines). So the proponent has not only to derive *Dab*-formulas in order to show that some of the *Dab*-formulas in the opponent's attack are not minimal. Often, the proponent should also derive the intended conclusion on several conditions. The proof from Γ_2 (p. 38) illustrates this. If line 9 were absent from the proof, line 8 would be marked.

Stability with respect to a line

This dialogue is identical to its namesake for Reliability. And so is the inconvenience: in some cases the only winning strategy for the proponent requires that she produces an infinite proof in her first move. The dialogue for $\Gamma_3 \vdash_{\mathbf{CLuN}^m} p$ illustrates this.

---

[14] If the premises and conclusion belong to a **CL**-decidable fragment of the language and the premise set is finite, then it is decidable whether the proponent has a winning strategy. This follows from a forthcoming result on the embedding of (full predicative) **CLuN** into **CL** - for the result on the propositional case see [7].

Many turns

This dialogue is identical to its namesake for Reliability, except that not all re-strictions on the finiteness of the proof and its extensions can be upheld. Actually, several complications should be considered.

Let $\Gamma_5 = \{(p_i \wedge {\sim}p_i) \vee (p_j \wedge {\sim}p_j) \mid i \neq j; i, j \in \mathbb{N}\} \cup \{q \vee (p_i \wedge {\sim}p_i) \mid i \in \mathbb{N}\}$. As $\Phi(\Gamma_5) = \{\{p_i \wedge {\sim}p_i \mid i \in \mathbb{N}\} - \{p_j \wedge {\sim}p_j\} \mid j \in \mathbb{N}\}$, it is easily seen (in view of Theorem 34) that $\Gamma_5 \vdash_{\mathbf{CLuN}^m} q$ (because q can be derived on the condition $\{p_j \wedge {\sim}p_j\}$ for every $j \in \mathbb{N}$). This seems to work fine with a finite proof and finite extensions. The proponent starts off with, for example, the proof

1	$q \vee (p_1 \wedge {\sim}p_1)$	Premise	$\emptyset$
2	q	1; RC	$\{p_1 \wedge {\sim}p_1\}$

after which the opponent offers a finite reply, an extension of 1–2 in which line 2 is marked. There are infinitely many such extensions. All that is required for line 2 to be marked is that, where s is the last stage of the extension, there is a $\varphi \in \Phi_s(\Gamma_5)$ for which $p_1 \wedge {\sim}p_1 \in \varphi$. A simple example is the extension:

3	$(p_0 \wedge {\sim}p_0) \vee (p_1 \wedge {\sim}p_1)$	Premise	$\emptyset$
4	$(p_0 \wedge {\sim}p_0) \,\check{\vee}\, (p_1 \wedge {\sim}p_1)$	Premise	$\emptyset$

Consider such an extension and let it count l lines. As the extension is finite, at most finitely many letters p_i occur in it. So the proponent can simply pick a p_i that does not occur in the extension and add the lines:

$l + 1$	$q \vee (p_i \wedge {\sim}p_i)$	Premise	$\emptyset$
$l + 2$	q	$l + 1$; RC	$\{p_i \wedge {\sim}p_i\}$

As p_i does not occur up to line l, $p_i \wedge {\sim}p_i$ is not a member of any $\varphi \in \Phi_{l+2}(\Gamma_5)$ and hence line $l + 2$ is unmarked. Moreover, as some $\varphi \in \Phi_{l+2}(\Gamma_5)$ are bound not to contain $p_1 \wedge {\sim}p_1$, line 2 is unmarked. So all seems well: $\Gamma_5 \vdash_{\mathbf{CLuN}^m} q$ and the proponent has a reply to every attack of the opponent on 1–2.

However, consider $\Gamma_6 = \{(p_i \wedge {\sim}p_i) \vee (p_j \wedge {\sim}p_j) \mid i \neq j; i, j \in \mathbb{N}\} \cup \{q \vee (p_i \wedge {\sim}p_i) \mid i \in \mathbb{N} - \{0\}\}$ - so $\Gamma_6 = \Gamma_5 - \{q \vee (p_0 \wedge {\sim}p_0)\}$. As $\Phi(\Gamma_6) = \{\{p_i \wedge {\sim}p_i \mid i \in \mathbb{N}\} - \{p_j \wedge {\sim}p_j\} \mid j \in \mathbb{N}\}$, we now have (in view of Theorem 34) that $\Gamma_6 \nvdash_{\mathbf{CLuN}^m} q$ (because q cannot be derived on the condition $\{p_0 \wedge {\sim}p_0\}$). The only point at which the proponent turns out to loose the game is after all *Dab*-formulas of the form $(p_i \wedge {\sim}p_i) \,\check{\vee}\, (p_j \wedge {\sim}p_j)$ have been derived. As there is no line on which q is derived on the condition $\{p_0 \wedge {\sim}p_0\}$, all lines on which q is derived are marked at this stage, call it s. Indeed, every condition on which q has been derived, overlaps with $\{\{p_i \wedge {\sim}p_i \mid i \in \mathbb{N}\} - \{p_0 \wedge {\sim}p_0\}\} \in \Phi_s(\Gamma_6) = \Phi(\Gamma_6)$.

As was shown in the previous paragraphs, this dialogue type requires infinity in one way or other. Either we have to allow that the opponent attacks (finitely many times) by an infinite extension and that the proponent defends by infinite extensions, or we have to allow the game to go on infinitely. The need to refer to infinite proofs at a stage will return in the other types of dialogue. The first move of the proponent, to the contrary, may be required to be finite.

Some people will not like dialogues that require infinite lists of formulas. Yet, the requirement is unavoidable for characterizing final derivability on the Minimal Abnormality strategy. Note that this is not too bad. Even for usual dialogues, the question is not who wins the game, but whether the proponent has a winning strategy. In order to show this, one may need to refer to infinitely many possible dialogues even in the case of **CL**. Each of these is finite, whereas the dialogues considered in this paper are infinite if Minimal Abnormality is the strategy. Of course, if it can be demonstrated that the proponent has a winning strategy, then this metalinguistic demonstration is finite.

POP

The long discussion of the previous dialogue type gives us at once the insights required for describing this type. The dialogue is identical to its namesake for Reliability, except that the extension of the proof and the extension of the extension should be allowed to be infinite. As was remarked before, the existence or absence of a winning strategy for the proponent has to be established at the metalevel anyway.

Calling premises

This dialogue type is identical to that for Reliability, except that the opponent is allowed to delineate an infinite set of premises and that, after this, the proponent is allowed to produce an infinite extension of her proof.

The proof of the following theorem proceeds as the proof of Theorem 41. There is one difference. The task of the proponent in a defense is double. First, for some minimal Dab-formulas $Dab(\Theta)$ derived by the opponent, she should derive a $Dab(\Theta')$ with $\Theta' \subset \Theta$.[15] Next, she should derive the conclusion on a set of conditions. By doing so, she should produce a stage s in which the following situation holds: for every $\varphi \in \Phi_s(\Gamma)$, the conclusion should be derived on a condition Δ for which $\Delta \cap \varphi = \emptyset$.[16] Note that she is able to do so, for each dialogue type, just in case $\Gamma \vdash_{\mathbf{AL}^m} A$.

THEOREM 42. *For the four dialogue types described holds: the proponent has a winning strategy in the dialogue for $\Gamma \vdash_{\mathbf{AL}^m} A$ iff $\Gamma \vdash_{\mathbf{AL}^m} A$.*

Given the absence of a positive test (in general), the computational complexity of adaptive logics is greater than that of classical (predicative) logic and it is greater

---

[15] In the case of the Calling Premises dialogue, the proponent should do this for all Dab-formulas $Dab(\Theta)$ that are **LLL**-derivable from Γ'.

[16] In the case of the Calling Premises dialogue, no Dab-formulas derivable from Γ' should make this condition false.

for Minimal Abnormality than for Reliability. This does not prevent one, however, from describing dialogue types and to show them adequate.

6   In conclusion

The main point I tried to make was that it is natural to understand final derivability in dynamic proofs in terms of dialogues: roughly, that a formula is finally derivable from a premise set iff there is a derivation of the formula that can be defended against every attack. Of course the allowed moves had to be made precise. I presented different types of dialogue for adaptive logics. This shows that some variation is possible. Note also that the different dialogue types clarify different aspects of final derivability and illustrate different possible definitions of final derivability. These are more or less attractive with respect to the philosophical interpretation or from a computational point of view.

An interesting open question concerns the combination of the types of dialogues described in the previous section with more usual dialogues. Put differently, it would be interesting to know what remains of the different moves described above if the proponent and opponent are given the usual dialogic means, viz. not proofs and extensions of proofs, but attacks on and defenses of formulas. Clearly the attacks and defenses can be most easily defined by studying the semantics of adaptive logics (which I had to skip in the present paper). Such dialogues seem to have some interesting aspects. Consider for example a usual dialogue corresponding to the POP dialogue described in the previous section. In a first phase the proponent tries to establish the conclusion on some condition. In a second phase, the opponent tries to establish that there is a selected (reliable, respectively minimally abnormal) model in which the conclusion is false (and hence a member of the condition is true). In the third phase, the proponent tries to show that the constructed model is not a selected one. Another interesting aspect are the restrictions on the introduction of atomic formulas. It seems natural to keep the restriction that the proponent cannot introduce literals in the phase in which she attempts to derive the conclusion. In the phase in which the opponent is attempting to establish abnormalities (that jointly correspond to abnormalities in a selected model), I surmise that the restriction should be adjusted in such a way that only the proponent can introduce abnormalities.

A very different open problem concerns dynamic proofs. It seems unproblematic to define final derivability in terms of a proof that is stable with respect to a certain line (on which the conclusion has been derived). We have seen that Definitions 31 and 32 present a more attractive way to characterize final derivability. It is unclear, however, whether this characterization is adequate, in the sense of Theorem 38, for all logics that have dynamic proofs. It should not be too difficult to delineate the set or sets of conditions on the set of rules $\mathcal{R}$ that warrant that the characterization in terms of Definitions 31 and 32 are adequate. Such a result would solve a problem which is now approached in a piecemeal way, namely whether all logics having dynamic proofs can be characterized, possibly under a translation, by an adaptive logic.

BIBLIOGRAPHY

[1] Diderik Batens. Dynamic dialectical logics. In Graham Priest, Richard Routley, and Jean Norman, editors, *Paraconsistent Logic. Essays on the Inconsistent*, pages 187–217. Philosophia Verlag, München, 1989.

[2] Diderik Batens. Linguistic and ontological measures for comparing the inconsistent parts of models. *Logique et Analyse*, 165–166:5–33, 1999. Appeared 2002.

[3] Diderik Batens. A general characterization of adaptive logics. *Logique et Analyse*, 173–175:45–68, 2001. Appeared 2003.

[4] Diderik Batens. The need for adaptive logics in epistemology. In Dov Gabbay, S. Rahman, J. Symons, and J. P. Van Bendegem, editors, *Logic, Epistemology and the Unity of Science*, pages 459–485. Kluwer Academic Publishers, Dordrecht, 2004.

[5] Diderik Batens. A universal logic approach to adaptive logics. *Logica Universalis*, 1:221–242, 2007.

[6] Diderik Batens. *Adaptive Logics and Dynamic Proofs. A Study in the Dynamics of Reasoning.* 200x. Forthcoming.

[7] Diderik Batens, Kristof De Clercq, and Natasha Kurtonina. Embedding and interpolation for some paralogics. The propositional case. *Reports on Mathematical Logic*, 33:29–44, 1999.

[8] Diderik Batens and Joke Meheus. A tableau method for inconsistency-adaptive logics. In Roy Dyckhoff, editor, *Automated Reasoning with Analytic Tableaux and Related Methods*, volume 1847 of *Lecture Notes in Artificial Intelligence*, pages 127–142. Springer, 2000.

[9] Diderik Batens and Joke Meheus. Shortcuts and dynamic marking in the tableau method for adaptive logics. *Studia Logica*, 69:221–248, 2001.

[10] George S. Boolos, John P. Burgess, and Richard J. Jeffrey. *Computability and Logic*. Cambridge University Press, 2002. (Fourth edition).

[11] Graham Priest. Minimally inconsistent **LP**. *Studia Logica*, 50:321–331, 1991.

[12] Shahid Rahman and Tero Tulenheimo. From games to dialogues and back: Towards a general frame for valitity. In Ondrej Majer, Ahti-Veikko Pietarinen and Tero Tulenheimo, editors, *Games: Unifying Logic, Language and Philosophy*, Springer. To appear.

[13] Shahid Rahman and Laurent Keiff. On how to be a dialogician. In D. Vanderveken, editor, *Logic, Thought and Action*, volume 2 of *Logic, Epistemology and Unity of Science*, pages 359–408. Springer, Dordrecht, 2005.

Man Muss Immer Umkehren!

JOHAN VAN BENTHEM

1 Introduction

The 19th century geometrist Jacobi famously said that one should always try to invert every geometrical theorem. But his advice applies much more widely! Choose any class of relational frames, and you can study its valid modal axioms. But now turn the perspective around, and fix some modal axiom beforehand. You can then find the class of frames where the axiom is guaranteed to hold by 'modal correspondence' analysis - and we all know the famous examples of that. It may look as if this style of analysis is tied to one particular semantics, say relational frames: but it is not. Correspondence analysis also works on neighbourhood models, telling us, e.g., just which modal axioms collapse these to binary relational frames. We will show how this same style of inverse thinking also applies to modern dynamic logics of information change. Basic axioms for knowledge after information update $!A$ tell us what sort of operation must be used for updating a given model $\mathcal{M}$ to a new one incorporating A. Likewise, we will show how modal axioms for (conditional) beliefs that hold after revision actions $*A$ actually fix one particular operation of changing the relative plausibility orderings which agents have on the universe of possible worlds. And finally, going back to the traditional heartland of logic, we show how we can read standard predicate-logical axioms as constraints on the sort of abstract 'process models' that lie at the heart of first-order semantics, properly understood. In all these cases, in order for the inversion to work and illuminate a given subject, we need to step back and reconsider our standard modeling. But that, I think, is what Shahid Rahman is all about.

2 Standard modal frame correspondences

One of the most attractive features of the semantics of modal logic is the match between modal axioms and corresponding patterns in the accessibility relation between worlds. This can be seen by giving a class of models, say temporal or epistemic, and then axiomatizing its set of modal validities. On top of the minimal modal logic which holds under all circumstances, one gets additional axioms reflecting more specific structure. For general background to modal completeness theory, as well as the rest of this paper, we refer to the *Handbook of Modal Logic*

[14] (P. Blackburn, J. van Benthem & F. Wolter, eds.) which has come out with Elsevier Science Publishers, Amsterdam, 2006.

Now, as a counterpoint to the completeness, the point of modal correspondence theory [4] is that one can also invert this line of thought. One takes some appealing modal axiom whose validity is to be guaranteed, and then finds out which accessibility patterns must now be assumed. Just to get into the spirit, consider the perennial modal **K4**-axiom $\Box p \to \Box\Box p$. Let us call a modal formula true at a point s in a semantic frame $\mathcal{F} = (W, R)$ if it is true at s under all atomic valuations V on $\mathcal{F}$. Here is perhaps the mother of all correspondences:

FACT 1. $\mathcal{F}, s \models \Box p \to \Box\Box p$ iff $\mathcal{F}$'s accessibility relation R is transitive at the world s: i.e., $\mathcal{F}, s \models \forall y(Rxy \to \forall z(Ryz \to Rxz))$.

Proof. If the relation is transitive, $\Box p \to \Box\Box p$ clearly holds under every valuation. Conversely, let $\mathcal{F}, s \models \Box p \to \Box\Box p$. In particular, the **K4**-axiom will hold if we take $V(p)$ to be $\{y | Rsy\}$. But then, the antecedent $\Box p$ holds at s, and hence so does the consequent $\Box\Box p$. And the latter states the transitivity, by the definition of $V(p)$.
∎

The theory behind this example involves the Sahlqvist Theorem [20, 4]: all modal axioms of a certain syntactic shape generalizing the usual modal laws of **T, S4** and **S5** allow for systematic first-order translation. As a beneficial side-effect, the inversion in perspective also makes us look differently at familiar modal axioms, and see patterns unnoticed before! One famous very non-Sahlqvist principle is Löb's axiom $\Box(\Box p \to p) \to \Box p$ in arithmetical provability logic. This expresses an interesting higher-order feature of accessibility patterns:

FACT 2. Löb's Axiom is true at the point s in a frame $\mathcal{F} = (W, R)$ iff

(a) $\mathcal{F}$ is upward R-well-founded at s, and

(b) $\mathcal{F}$ is transitive at s

Here, a relation R is upward well-founded at s if it allows no upward sequences $sRw_1Rw_2R\ldots$: i.e., s starts no infinite sequences or cycles. Correspondence theory is still alive and expanding today. [10, 11] show how the syntactic details of Fact 2 for Löb's Axiom lead to a systematic analysis of structural properties of accessibility definable in **LFP(FO)**, first-order logic with added fixed-point operators. As a consequence, one can also analyze well-known modal fixed-point languages like the μ-calculus in new ways. But further modal axioms define accessibility patterns still beyond this level, corresponding to truly monadic second-order frame properties - with the McKinsey Axiom $\Box\Diamond p \to \Diamond\Box p$ as a prime example.

3 Modal distribution and neighbourhood models

Some people think correspondence analysis is tied up exclusively with one particular view of what modal models must be like, viz. directed graphs. But it will work on any sort of structure, even ones that look 'higher-order'. E.g., [19] showed how one can do correspondence analysis of intuitionistic axioms on Beth models, taking points and branches as primitive objects. Here is an example closer to modal logic itself [5], [7, Chapter 11]. *Neighbourhood models* generalize directed graphs by having accessibility relations R relating single points x to sets of points Y. These structures have concrete motivations in scenarios of 'deductive support' in logic programs, topological semantics and modal logics of space, or modal logics of players' powers of reaching outcomes in games. One can then interpret the key modality via the following generalization of the usual truth condition:

$$\mathcal{M}, s \models \Diamond\varphi \text{ iff there is a set of points } Y \text{ with } RsY$$
$$\text{and for all } y \in Y, \mathcal{M}, y \models \varphi$$

The resulting minimal logic loses distributivity of the modality over both conjunction and disjunction, though it retains upward monotonicity. (The technical reason is that the $\exists\forall$ quantifier combination in the preceding truth condition suppresses both forms of distribution.) Moreover, the complexity of satisfiability drops from the PSPACE-complete for modal **K** to NP-complete: i.e., 'from worse to bad'. But this move to a more general semantics also means that formerly minimally valid principles now acquire substantial content. Particularly, we have this

FACT 3. The distribution axiom $\Diamond(p \vee q) \leftrightarrow (\Diamond p \vee \Diamond q)$ is valid on a neighbourhood frame iff that frame is generated by a binary world-to-world relation Rxy with RxY iff $\{y : Rxy\} \subseteq Y$.

[5] investigates correspondences over neighbourhood frames in more details, and finds generalizations for the correspondence-theoretic content of major modal principles. Consider its existential neighbourhood version of the **K4**-axiom:

FACT 4. $\Diamond\Diamond p \to \Diamond p$ corresponds to the following modified rule of Cut (i.e., 'Generalized Transitivity'):

$$\forall x \forall Y \forall \{Z_y \mid y \in Y\}: RxY \wedge RyZ_y \text{ (for all } y \in Y) \to Rx \bigcup\{Z_y \mid y \in Y\}$$

Just as over directed graphs, such correspondences can be computed automatically by a substitution algorithm (cf. [15]), this time, producing relational conditions in a weak sub-language of second-order logic.

4 Geometry: two-sorted modal logic

Another source of correspondence thinking which goes 'out of the box' is in geometry. [6, 9] makes a plea for a many-sorted view of space, with points and lines, or points and arrows, on a par. Matching modal languages will now be two-sorted, with one kind of formulas referring to points, and another to lines or arrows. Of course, relations can be of many kinds in this setting, beyond the original idea of accessibility for worlds. The resulting modal geometry has appealing concrete correspondences between modal axioms and spatial patterns. We give two different examples here.

Consider modal *Arrow Logic*, a two-sorted language describing both points and transitions between them as first-class semantic citizens. We focus on the latter here. Basic arrow models are of the form $\mathcal{M} = (A, C^3, R^2, I^1, V)$ with A a set of 'arrows' with three predicates: C^3x, yz (x is a composition of y and z), R^2x, y (y is a reversal of x), I^1x (x is an identity arrow). The modal language is interpreted with the following key clauses:

$$\mathcal{M}, x \models \varphi \bullet \psi \quad \text{iff} \quad \text{there are } y, z \text{ with } Cx, yz \text{ and } \mathcal{M}, y \models \varphi, \mathcal{M}, z \models \psi$$
$$\mathcal{M}, x \models \breve{\varphi} \qquad \text{iff} \quad \text{there exists } y \text{ with } Rx, y \text{ and } \mathcal{M}, y \models \varphi$$

Here is the content of two famous principles for converse and composition from Tarski's Relational Algebra, re-stated as modal axioms of Arrow Logic:

FACT 5.

$(\varphi \bullet \psi)\breve{} \rightarrow \breve{\psi} \bullet \breve{\varphi}$ corresponds to $\forall xyz : Cx, yz \rightarrow Cr(x), r(z)r(y)$

$\varphi \bullet \neg(\breve{\varphi} \bullet \psi) \rightarrow \neg\psi$ corresponds to $\forall xyz : Cx, yz \rightarrow Cz, r(y)x$

Given these properties of our relations, we can view composition triangles like the one depicted here from any arrow we please, by taking reversals:

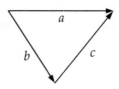

Figure 1. Arrow Composition

Basic arrow logics are decidable. But an ominous threshold is the existential principle of *associativity*, which makes these logics undecidable, just as happens with relational set algebra:

FACT 6. The associativity axiom $((\varphi \bullet \psi) \bullet \chi) \rightarrow (\varphi \bullet (\psi \bullet \chi))$ corresponds to $\forall xyzuv : (Cx, yz \ \& \ Cy, uv) \rightarrow \exists w : (Cx, uw \ \& \ Cw, vz)$

This says that abstract composition structures have to be rich enough to admit of 'recombination'. Complex or not, this same axiom is highly appealing from a geometrical standpoint.

Our second example comes from more standard modal logics of space [1]. Here is what Associativity says when we shift our correspondence analysis to modal logics of geometry where the ternary relation now rather stands for affine betweenness Bx, yz: x *lies on the segment $y - z$*.

We use the following modality:

$$\mathcal{M}, s \models C\varphi\psi \text{ iff } \exists t, u : Bs, tu \ \& \ \mathcal{M}, t \models \varphi \ \& \ \mathcal{M}, u \models \psi$$

FACT 7. The associativity law $C(C\varphi\psi)\chi \rightarrow C\varphi(C\psi\chi)$ corresponds to Pasch's Axiom $\forall xyzuv : (Bu, xy \ \& \ Bv, uz) \rightarrow \exists s(Bv, xs \ \& \ Bs, yz)$.

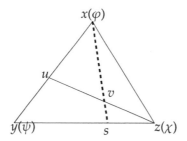

Figure 2. Pasch Triangle Axiom

Thus, depending on the semantic environment, correspondence analysis may reveal very different content for well-known modal principles, linking them up in surprising ways with mathematical structure known from other sources. Using these insights, let's now move elsewhere.

5 Knowledge and information update

Another major paradigm for modal logic is the analysis of knowledge and other information-related attitudes. Here, models stand for information patterns describing current states of one or more agents in interaction. The modal language is as usual: $p \mid \neg\varphi \mid \varphi \lor \psi \mid K_i\varphi$ and perhaps common knowledge C_G, while models $\mathcal{M} = (W, \{\sim_i \mid i \in G\}, V)$ have worlds W, accessibility relations $\sim_i$, and a valuation V. The standard epistemic truth condition reads: $\mathcal{M}, s \models K_i\varphi$ iff for all t with $s \sim_i t$: $\mathcal{M}, t \models \varphi$. The usual modal frame correspondences apply here, both for knowledge modalities and for the common knowledge, treated as a fixed-point operator in the sense of Section 2. So far, so good.

But now consider a modern trend, the analysis of informational actions which change a current epistemic model [12]. For instance, an event of public announcement !P works as follows: learning that P is true eliminates the worlds where P is false. In a picture:

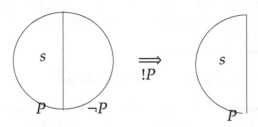

Figure 3. Eliminative Update Step

To describe this, we need a dynamic-epistemic logic, with a key operator

$$\mathcal{M}, s \models [!P]\varphi \quad \text{iff} \quad \text{if } \mathcal{M}, s \models P, \text{ then } \mathcal{M}|P, s \models \varphi$$

The logic of this can be axiomatized completely. In particular, one compositional reduction principle explains when agents acquire knowledge after an announcement of some 'hard fact' P:

FACT 8. The following modal axiom is sound for public announcement:

$$[!P]K_i\varphi \; \leftrightarrow \; (P \rightarrow K_i[!P]\varphi)$$

Well-understood, this axiom expresses non-trivial assumptions about epistemic agents. In particular, the interchange of modalities for knowledge and observed events expresses their capacity of *perfect memory*. This is indeed presupposed in the following soundness argument:

Proof. Compare two models: $(\mathcal{M}, s)$ and $(\mathcal{M}|P, s)$ before and after the update [it helps to draw pictures]. The formula $[P!]K_i\varphi$ says that, in $\mathcal{M}|P$, all worlds $\sim_i$-accessible from s satisfy φ. The matching worlds in $\mathcal{M}$ are those $\sim_i$-accessible from s and which satisfy P. Moreover, truth values of formulas may change in an update step, say from ignorance to knowledge. Hence the correct description of these worlds in $\mathcal{M}$ is not that that they satisfy φ (which they do in $\mathcal{M}|P$), but rather $[P!]\varphi$: they become φ after the update. Finally, a small detail. !P is a partial operation, as P has to be true for its truthful public announcement. Thus, we make our assertion on the right only if !P is executable, i.e., P is true. Altogether then,

$[P!]K_i\varphi$ says the same as $P \rightarrow K_i(P \rightarrow [P!]\varphi)$ - which can still be simplified to the above form $P \rightarrow K_i[P!]\varphi$. ∎

The use of pictures here is not just a convenience. They also reflect genuine visual intuitions about what might be called the *geometry of knowledge* and update.

But now, Jacobi's advice once more. We have seen that treating public announcement of hard facts as world elimination validates the above axiom. What about a converse? Suppose that this axiom looks independently plausible for information update, *which operations on models would validate it?* The answer is again a correspondence argument [13]. We consider abstract model-changing operations ♡p (here the 'heart' symbol is just our ad-hoc notation for an arbitrary model change) taking epistemic models $\mathcal{M}$ with sets of p-worlds inside to new models $\mathcal{M}♡p$ - with some mild conditions on available worlds for their domains. A simple proof then shows

FACT 9. Eliminative update is the only model-changing operation which satisfies the equivalence $[♡p]Kq \leftrightarrow (p \rightarrow K(p \rightarrow [♡p]q))$.

Proof. We assume that the world s is in p: the operation ♡p would not be defined at $\mathcal{M}, s$ otherwise. First, from left to right, the axiom implies the following. Take q equal to the set of worlds which are $\sim$-accessible from the current world s in the new model $\mathcal{M}♡p$. This validates the left-hand side of the axiom. Then the right-hand side must be true, and it says that all worlds $\sim$-accessible from s before the operation ♡p are in q. This shows that the relation change respects all already existing links from p-worlds to p-worlds. By a similar argument, using the axiom in the converse direction, we see that indeed *only* such links are preserved into the new model after the operation ♡p: no new ones are created. Taken together, this is precisely the natural 'link-cutting' version of the above public epistemic update. ∎

This argument can be sharpened up, defining the universe of relevant epistemic frames and transition relations explicitly, and stipulating how individual worlds can be related across frames. In such a setting, three axioms capture eliminative update for public announcement:

First, the equivalence (a) $\langle !p \rangle \top \leftrightarrow p$ makes sure that inside a given model $\mathcal{M}$, the only worlds surviving into $\mathcal{M}♡p$ are those in the set denoted by p. Next, a reduction axiom (b) $\langle !p \rangle Eq \leftrightarrow p \wedge E\langle !p \rangle q$ for the existential modality Eq ("q is true in some world") says that the domain of $\mathcal{M}♡p$ contains no objects beyond the set p in $\mathcal{M}$. Finally, the knowledge axiom in Fact 8 ensures that the epistemic relations are the same in $\mathcal{M}$ and $\mathcal{M}♡p$, so that our abstract update operation really takes a submodel.

This example suggests taking modal correspondence analysis to the current world of dynamic epistemic logics for larger families of informative events, including partial observation and hiding. In particular, one would want to show that the basic product update mechanism of [3] for such more sophisticated scenarios is essentially the only model construction in some suitable abstract space validating the general **DEL** reduction axiom

$$[\mathbf{E}, e]K_i\varphi \;\leftrightarrow\; (PRE_e \;\rightarrow\; \bigwedge\{K_i[\mathbf{E}, f]\varphi \mid f \sim_i e \text{ in } \mathbf{E}\})$$

6   Geometry of belief revision

We do not just receive information which smoothly updates our current knowledge. There are also more dramatic episodes of facts which challenge our current beliefs, and lead to dynamic processes of belief revision. Here, too, the preceding considerations can be brought to bear. Beliefs can be interpreted over modal models with a comparison relation of relative plausibility between worlds. The key modality then becomes:

$$\mathcal{M}, s \models B_i\varphi \;\text{ iff }\; \mathcal{M}, t \models \varphi \text{ for all worlds } t$$
$$\text{which are } \textit{minimal} \text{ in the ordering } \lambda xy. \leq_{i,s} xy.$$

But soon, this turns out less than what one needs - and a more general notion of conditional belief helps 'pre-encode' beliefs we would have if we were to learn certain new things:

$$\mathcal{M}, s \models B_i^\psi\varphi \;\text{ iff }\; \mathcal{M}, t \models \varphi \text{ for all worlds } t \text{ which are minimal}$$
$$\text{for } \lambda xy. \leq_{i,s} xy \text{ in the set } \{u \mid \mathcal{M}, u \models \psi\}$$

The resulting logic consists of the standard principles of the minimal conditional logic: that of Lewis minus the connectivity axiom; cf. [16, 21]. Now, for the 'hard facts' of Section 5, it is easy to check the following reduction axiom:

$$[!P]B_i^\psi\varphi \;\leftrightarrow\; (P \rightarrow B_i^{P \wedge [!P]\psi}[!P]\varphi)$$

But more interesting is the response of agents to 'soft triggers', events which make a proposition more plausible, though not definitively ruling out that it might be false. Such triggers will not eliminate worlds in the current model, but they will *change the plausibility pattern*.

One typical response of this sort is *lexicographic upgrade* ⇑ P, described variously as what a trusting, or a radical agent might do. This changes the current model $\mathcal{M}$ to $\mathcal{M} \Uparrow P$:

P-worlds become better than all $\neg P$-worlds;
within these two zones, the old order remains.

The complete dynamic logic of this operation of model change can be axiomatized - bringing it into the language through a matching modality:

$$\mathcal{M}, s \models [\Uparrow P]\varphi \quad \text{iff} \quad \mathcal{M} \Uparrow P, s \models \varphi$$

Then the following key recursion principle emerges for the conditional beliefs which agents will have after a lexicographic plausibility change occurred for some soft trigger with P:

$$[\Uparrow P]B^{\psi}\varphi \quad \leftrightarrow \quad (E(P \wedge [\Uparrow P]\psi) \wedge B^{P \wedge [\Uparrow P]\psi}[\Uparrow P]\varphi)$$
$$\vee \, (\neg E(P \wedge [\Uparrow P]\psi) \wedge B^{[\Uparrow P]\psi}[\Uparrow P]\psi)$$

Here E is again the global existential modality - or a similar existential epistemic modality. This time, we will not go into the details of the soundness argument. But we do note that, extending the analysis for information update, a nice modal correspondence result can be proved, showing that we have captured the essence here. Again, we are working in a universe of frames connected by abstract relation changing operations $\heartsuit p$:

FACT 10. The law $[\heartsuit p]B^r q \quad \leftrightarrow \quad (E(p \wedge r) \wedge B^{p \wedge r}q) \vee (\neg E(p \wedge r) \wedge B^r q))$ holds in a universe of frames iff the operation $\heartsuit p$ is lexicographic upgrade.

Proof. Let $\leq_s xy$ in $\mathcal{M}\heartsuit p$. We show that $\leq_s$ is the relation produced by lexicographic upgrade. Let r be the set $\{x, y\}$ and set $q = \{x\}$. Then the left-hand side of our axiom is true. There are two cases for truth on the right-hand side then. *Case 1*: one of x, y is in p, and hence $p \wedge r = \{x, y\}$ (1.1) or $\{y\}$ (1.2) or $\{x\}$ (1.3). Moreover, $B^{p \wedge r}q$ holds in $\mathcal{M}$ at s. If (1.1), we have $\leq_s xy$ in $\mathcal{M}$. If (1.2), we must have $y = x$, and again $\leq_s xy$ in $\mathcal{M}$. Case (1.3) can only occur when $x \in p$ and $y \notin p$. Thus, all new relational pairs in $\mathcal{M}\heartsuit p$ satisfy the above description of the lexicographic reordering. *Case 2* is when we have $\neg(E(p \wedge r))$ and none of x, y are in p. This can be analyzed analogously, now using the truth of the disjunct $B^r q$. Thus, each pair in the new ordering comes from an old one by the given description.

Conversely, it can be shown in the same style that all pairs $\leq_s xy$ satisfying the condition for lexicographic upgrade do make it into the new order. Here is one case, the others are easier. Suppose that $x \in p$ while $y \notin p$. Again, set $r = \{x, y\}$ and $q = \{x\}$. Then $p \wedge r = \{x\}$, and moreover, we have $B^q p \wedge r$ by reflexivity. Thus, the left disjunct of the right-hand side of our reduction axiom is true. It follows that the left-hand side of the axiom must be true for this same choice of r and q. Thus, the formula $[\heartsuit p]B^r p \wedge r$ is true in the model $(M), s$ − and therefore, in the new model $M\heartsuit p, s$, the best worlds in $\{x, y\}$ are in $\{x\}$: i.e., $\leq_s xy$. ∎

Again, [13] actually analyzes the technicalities here a bit more carefully. Indeed, the preceding arguments obviously turn around very simple predicate substitutions, like those used for the Sahlqvist Theorem.

In the area of belief revision, this correspondence analysis has further attractions, since no single action of plausibility change works once and for all. E.g., more conservative, or less trusting, agents, might respond to a soft trigger by the operation $\uparrow P$, which only puts *the best P-worlds* on top, and leaves everything else as it was in $\mathcal{M}$. With a matching modality, one finds the corresponding reduction axiom for this new policy:

$$[\uparrow P]B^\psi\varphi \quad \leftrightarrow \quad (B^P\neg[\uparrow P]\psi \ \wedge \ B^{[\uparrow P]\psi}[\uparrow P]\varphi)$$
$$\vee \ (\neg B^P\neg[\uparrow P]\psi) \ \wedge \ (B^{P\wedge[\uparrow P]\psi}[\uparrow P]\varphi)$$

Again a correspondence argument may be used to show that this exactly determines the belief change policy $\uparrow P$.

Correspondence theory tells us the exact correlation between principles describing changes of (conditional) beliefs and suitably definable semantic changes in plausibility patterns. Thus, as with knowledge, it reveals an appealing geometry of belief and belief changes.

7   Modal foundations for first-order logic

My final example of the power of correspondence analysis and inversion goes back to the heartland where it all started. Modal logic started as an extension of, or maybe a fine-structure fragment of, standard first-order predicate logic. But well-understood, the latter system itself is very 'modal'! Consider Tarski's clause for the existential quantifier

$$\mathcal{M}, \alpha \models \exists x\varphi \ \textit{iff for some } d \in |\mathcal{M}| : \mathcal{M}, \alpha^x_d \models \varphi$$

Here, the variable assignments are essential in decomposing quantified statements. But much less than this is needed to give a compositional semantics for first-order quantification, viz. merely the abstract pattern

$$\mathcal{M}, \alpha \models \exists x\varphi \ \textit{iff for some } \beta : R_x\alpha\beta \text{ and } \mathcal{M}, \beta \models \varphi$$

Here, the assignments become abstract states, and the concrete relation $\alpha =_x \beta$ which holds between α and α^x_d has become just a binary relation R_x. Evidently, this is the semantics of a minimal poly-modal language. This state semantics has an independent appeal. First-order evaluation is an informational process that changes computational states [7], and formulas are compound procedures over basic atomic actions of testing for a fact, and shifting the value of some variable.

Accordingly, the usual validities of first-order logic, which one can look up in a good textbook like [17], split into two groups. One group consists of the minimal modal logic: (a) all classical Boolean propositional laws, (b) Modal Distribution: $\forall x(\varphi \rightarrow \psi) \rightarrow (\forall x\varphi \rightarrow \forall x\psi)$, (c) Modal Necessitation: *if* $\vdash \varphi$, *then* $\vdash \forall x\varphi$, and (d) a definition of $\exists x\varphi$ as $\neg\forall x\neg\varphi$. Much first-order inference can be described this way. But now, we can also look at further first-order axioms, and see what these say on top of this. What we expect is that they reflect additional properties of the evaluation process.

Again, this may be brought out using modal frame correspondences. The full story is in [8], but we cite a few high-lights here. First, there are some universal properties of the specific relations $\alpha =_x \beta$ among assignments. The fact that these are equivalence relations is reflected in valid **S5**-style axioms such as $\exists x\exists x\varphi \rightarrow \exists x\varphi$. As usual, the latter corresponds to the transitivity of state shifting. Indeed, the total system corresponding to making these general assumptions, but without any existential ones on 'fullness' of the set of available assignments leads us to an interesting, and still *decidable*, version of first-order logic called **CRS** [18], related to generalized relational algebras.

But now about the sources of the undecidability of our usual first-order logic! Modal frame correspondences help us discover these. 'Deconstructing' the further first-order axioms in [17], one is quickly left to focus on the valid *interchange laws* for quantifiers. These turn out to correspond (just by virtue of their syntactic Sahlqvist forms) to well-known significant existential geometrical properties of the evaluation process:

FACT 11.

    a $\exists y\exists x\varphi \rightarrow \exists x\exists y\varphi$ expresses 'Path Reversal':

       $\forall\alpha\beta\gamma((R_x\alpha\beta \;\&\; R_y\beta\gamma) \rightarrow \exists\delta(R_y\alpha\delta \;\&\; R_x\delta\beta))$

    b $\exists y\forall x\varphi \rightarrow \forall x\exists y\varphi$ expresses 'Confluence':

       $\forall\alpha\beta\gamma((R_y\alpha\beta \;\&\; R_x\alpha\gamma) \rightarrow \exists\delta(R_x\beta\delta \;\&\; R_y\gamma\delta))$

Thus, by stepping into a broader class of semantic structures, we give predicate-logical validities different voices. Some remain universally valid - but others express various specific properties of the space of available computational states. And when that space becomes full enough, with grid structures associated with known undecidable Tiling Problems, we get undecidability of its modal theory.

But such a larger universe also brings further rewards. E.g., more distinctions can be made on abstract state models for first-order logic than on standard Tarski models. In particular, there are now separate denotations for substitution operators $[t/x]$. Also one can naturally interpret *polyadic quantifiers* $\exists \mathbf{x}$ for tuples of variables $\mathbf{x}$ to the first-order language, in terms of *simultaneous change* of values in

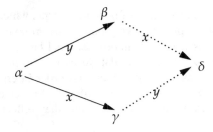

Figure 4. Diamond for Confluence

their registers. This genuinely enriches the first-order vocabulary, while still retaining decidability over abstract state models [2]. Of course, modal correspondence analysis still applies to axiomatic principles in these richer languages, too.

8 Conclusion

Modal correspondence analysis arises when we invert an established perspective, look back at the models we have chosen, and ask what sort of semantic content attaches to proposed syntactic axioms. This style of thinking may look like an abstract and somewhat curious interest at times. But we hope to have shown its benefits across a wide range of cases. We can check whether proposed axioms really capture what they are supposed to do, we can find surprising new content to familiar principles, and sometimes, we are led to the construction of new semantic domains, and new languages over these. In the Old Testament, Lot's wife was punished for looking back along her path. There was great unfairness in that – but even more, in logic, looking 'the other way' can only benefit us!

BIBLIOGRAPHY

[1] Marco Aiello and Johan van Benthem. A modal walk through space. *Journal of Applied Non-Classical Logics*, 12(3-4):319–364, 2002.

[2] Hajnal Andréka, Johan van Benthem, and Istvan Németi. Modal logics and bounded fragments of predicate logic. *Journal of Philosophical Logic*, 27(3):217–274, 1998.

[3] Alexandru Baltag, Lawrence S. Moss, and Slawomir Solecki. The logic of public announcements, common knowledge, and private suspicions. In *TARK '98: Proceedings of the 7th conference on theoretical aspects of rationality and knowledge*, pages 43–56, San Francisco, CA, USA, 1998. Morgan Kaufmann Publishers Inc.

[4] Johan van Benthem. *Modal Logic and Classical Logic*. Bibliopolis, Napoli, 1983.

[5] Johan van Benthem. Logic as programming. *Fundam. Inform.*, 17(4):285–317, 1992.

[6] Johan van Benthem. Complexity of contents versus complexity of wrappings. In M. Marx, M. Masuch, and L. Pólos, editors, *Arrow Logic and Multimodal Logic*, pages 203–219, Stanford, 1996. CSLI Publications.

[7] Johan van Benthem. *Exploring Logical Dynamics*. CSLI Publications, Stanford, 1996.

[8] Johan van Benthem. Modal foundations for predicate logic. *Logic Journal of the IGPL*, 5(2):259–286, 1997.

[9] Johan van Benthem. Temporal patterns and modal structure. *Logic Journal of the IGPL*, 7(1):7–26, 1999.

[10] Johan van Benthem. Minimal predicates, fixed-points, and definability. *Journal of Symbolic Logic*, 70(3):696–712, 2005.

[11] Johan van Benthem. Modal frame correspondences and fixed-points. *Studia Logica*, 83(1):133–155, 2006.

[12] Johan van Benthem. One is a lonely number: on the logic of communication. In Z. Chatzidakis, P. Koepke, and W. Pohlers, editors, *Logic Colloquium '02*, pages 96–129, Wellesley MA, 2006. ASL and A.K. Peters.

[13] Johan van Benthem. Dynamic Logic for Belief Revision. *Journal of Applied Non-Classical Logics*, 17(2):129–155, 2007.

[14] P. Blackburn, F. Wolter, and J. van Benthem, editors. *Handbook of Modal Logic*. Elsevier, Amsterdam, 2006.

[15] Patrick Blackburn, Maarten de Rijke, and Yde Venema. *Modal Logic*. Cambridge University Press, 2001.

[16] John P. Burgess. Quick completeness proofs for some logics of conditionals. *Notre Dame Journal of Formal Logic*, 22(1):76–84, 1981.

[17] Herbert Enderton. *A Mathematical Introduction to Logic*. Academic Press, 1972.

[18] István Németi. The equational theory of cylindric relativized set algebras is decidable. Technical report, Preprint No 63/85, Mathematical Institute, Hungarian Academy of Sciences, Budapest, 1985.

[19] Pieter Hendrik Rodenburg. *Intuitionistic correspondence theory*. Ph. d. dissertation, Mathematical Institute, Universiteit van Amsterdam, 1986.

[20] Henrik Sahlqvist. Correspondence and completeness in the first- and second-order semantics for modal logic. In S. Kanger, editor, *Proceedings of the Third Scandinavian Logic Symposium*, pages 110–143, Amsterdam, 1975. North-Holland.

[21] Frank Veltman. *Logics for Conditionals*. Dissertation, Philosophical Institute, University of Amsterdam, 1985.

A Sound and Complete Temporal Logic for Belief Revision

GIACOMO BONANNO

ABSTRACT. Branching-time temporal logic offers a natural setting for a theory of belief change, since belief revision deals with the interaction of belief and information over time. We propose a temporal logic that, besides the next-time operator, contains a belief operator and an information operator. It is shown that this logic is sound and complete with respect to the class of branching-time frames augmented, for each instant t, with a set of states and two binary relations on it, representing beliefs and information, respectively.

1 Introduction

Belief revision deals with the interaction between initial beliefs and new evidence. As new information is acquired over time, beliefs are correspondingly changed to accommodate that information. Temporal logic provides a natural framework for a theory of belief revision. We propose a basic logic for belief revision which, besides the next-time operator $\bigcirc$, contains a belief operator B and an information operator I. The information operator is not a normal operator and is formally similar to the "all I know" operator introduced by Levesque [9]. On the semantic side we consider branching-time frames to represent different possible evolutions of beliefs. For every date t, beliefs and information are represented by binary relations $\mathcal{B}_t$ and $\mathcal{I}_t$ on a set of states Ω_t. As usual, the link between syntax and semantics is provided by the notion of valuation and model. The truth of a formula in a model is defined at a state-instant pair (ω, t). We prove soundness and completeness of this basic logic with respect to the class of frames considered. Extensions of this basic logic are studied elsewhere (Bonanno [3], [4]). In particular, it is shown in [3] that a suitable extension of the basic logic considered in this paper provides an axiomatic characterization of the AGM theory of belief revision (Alchourrón et al. [1]).

2 Syntax

We consider a propositional language with five modal operators: the next-time operator $\bigcirc$ and its inverse $\bigcirc^{-1}$, the belief operator B, the information operator I and the "all state" operator A. The intended interpretation is as follows:

$\bigcirc \varphi$:   "at every next instant it will be the case that φ"

$\bigcirc^{-1}\varphi$:   "at every previous instant it was the case that φ"

$B\varphi$:   "the agent believes that φ"

$I\varphi$:   "the agent is informed that φ"

$A\varphi$:   "it is true at every state that φ".

The "all state" operator A is needed in order to capture the non-normality of the information operator I (see below). For a thorough discussion of the "all state" operator see Goranko and Passy [7].

The formal language is built in the usual way (see Blackburn et al. [2]) from a countable set of atomic propositions, the connectives $\neg$ and $\vee$ (from which the connectives $\wedge$, $\rightarrow$ and $\leftrightarrow$ are defined as usual) and the modal operators $\bigcirc$, $\bigcirc^{-1}$, B, I and A. Let $\Diamond\varphi \overset{def}{=} \neg \bigcirc \neg\varphi$, and $\Diamond^{-1}\varphi \overset{def}{=} \neg \bigcirc^{-1} \neg\varphi$. Thus the interpretation of $\Diamond\varphi$ is "at *some* next instant it will be the case that φ" while the interpretation of $\Diamond^{-1}\varphi$ is "at some previous instant it was the case that φ".

We denote by $\mathbb{L}_0$ the basic logic of belief revision defined by the following axioms and rules of inference.

AXIOMS:

1. All propositional tautologies.

2. Axiom K for $\bigcirc$, $\bigcirc^{-1}$, B and A:

$$(\Box\varphi \wedge \Box(\varphi \rightarrow \psi)) \rightarrow \Box\psi \quad \text{for } \Box \in \{\bigcirc, \bigcirc^{-1}, B, A\} \quad \text{(K)}$$

3. Temporal axioms relating $\bigcirc$ and $\bigcirc^{-1}$:

$$\varphi \rightarrow \bigcirc\Diamond^{-1}\varphi \quad \text{(O}_1\text{)}$$
$$\varphi \rightarrow \bigcirc^{-1}\Diamond\varphi \quad \text{(O}_2\text{)}$$

4. Backward Uniqueness axiom:

$$\Diamond^{-1}\varphi \rightarrow \bigcirc^{-1}\varphi \quad \text{(BU)}$$

5. S5 axioms for A:

$$A\varphi \rightarrow \varphi \quad \text{(T}_A\text{)}$$
$$\neg A\varphi \rightarrow A\neg A\varphi \quad \text{(5}_A\text{)}$$

6. Inclusion axiom for B (note the absence of an analogous axiom for I):

$$A\varphi \rightarrow B\varphi \quad \text{(Incl}_B\text{)}$$

7. Axioms to capture the non-standard semantics for I:

$$(I\varphi \wedge I\psi) \rightarrow A(\varphi \leftrightarrow \psi) \quad (I_1)$$
$$A(\varphi \leftrightarrow \psi) \rightarrow (I\varphi \leftrightarrow I\psi) \quad (I_2)$$

RULES OF INFERENCE:

1. Modus Ponens: $\frac{\varphi, \; \varphi \rightarrow \psi}{\psi}$ (MP)

2. Necessitation for A, $\bigcirc$ and $\bigcirc^{-1}$: $\frac{\varphi}{\Box\varphi}$ for $\Box \in \{\bigcirc, \bigcirc^{-1}, A\}$ (Nec).

Note that from MP, $Incl_B$ and Necessitation for A one can derive necessitation for B ($\frac{\varphi}{B\varphi}$). On the other hand, necessitation for I is *not* a rule of inference of this logic (indeed it is not validity preserving).

3 Semantics

On the semantic side we consider branching-time structures with the addition of a set of states, a belief relation and an information relation for every instant t.

DEFINITION 1. A *next-time branching frame* is a pair $\langle T, \rightarrowtail \rangle$ where T is a (possibly infinite) set of instants or dates and $\rightarrowtail$ is a binary relation on T satisfying the following properties: $\forall t_1, t_2, t_3 \in T$,

(1)	uniqueness	if $t_1 \rightarrowtail t_3$ and $t_2 \rightarrowtail t_3$ then $t_1 = t_2$,
(2)	acyclicity	if $\langle t_1, ..., t_n \rangle$ is a sequence with $t_i \rightarrowtail t_{i+1}$ for every $i = 1, ..., n-1$, then $t_n \neq t_1$.

The interpretation of $t_1 \rightarrowtail t_2$ is that t_2 is an *immediate successor* of t_1 or t_1 is the *immediate predecessor* of t_2: every instant has at most one unique immediate predecessor but can have several immediate successors.

Given a next-time branching frame $\langle T, \rightarrowtail \rangle$, we denote by $\prec$ the transitive closure of $\rightarrowtail$. Thus, for $t, t' \in T$, $t \prec t'$ if and only if there is a sequence $\langle t_1, ..., t_n \rangle$ in T such that $t_1 = t$, $t_n = t'$ and $t_i \rightarrowtail t_{i+1}$ for all $i = 1, ..., n-1$. The interpretation of $t \prec t'$ is that t is a *predecessor of* t' or t' is a *successor* of t.

REMARK 2. (Backward linearity of $\prec$). It is straightforward to show that if $t_0, t_1, t_2 \in T$ are such that $t_0 \prec t_2$ and $t_1 \prec t_2$ then either $t_0 = t_1$ or $t_0 \prec t_1$ or $t_1 \prec t_0$.

DEFINITION 3. A *general temporal belief revision frame* is a tuple $\langle T, \rightarrowtail, \Omega, \{\Omega_t, \mathcal{B}_t, \mathcal{I}_t\}_{t \in T} \rangle$, where $\langle T, \rightarrowtail \rangle$ is a next-time branching frame; Ω is a set of states; for every $t \in T$, $\emptyset \neq \Omega_t \subseteq \Omega$; and $\mathcal{B}_t$ and $\mathcal{I}_t$ are binary relations on Ω_t.

The interpretation of $\omega \mathcal{I}_t \omega'$ is that at state ω and time t according to the information received it is possible that the true state is ω'. On the other hand, the interpretation of $\omega \mathcal{B}_t \omega'$ is that at state ω and time t in light of the information received (if any) the individual considers state ω' possible (an alternative expression is "ω' is a doxastic alternative to ω at time t"). We shall use the following notation:

$$\mathcal{B}_t(\omega) = \{\omega' \in \Omega_t : \omega \mathcal{B}_t \omega'\} \text{ and, similarly, } \mathcal{I}_t(\omega) = \{\omega' \in \Omega_t : \omega \mathcal{I}_t \omega'\}.$$

Thus $\mathcal{B}_t(\omega)$ is the set of states that are reachable from ω according to the relation $\mathcal{B}_t$ and similarly for $\mathcal{I}_t(\omega)$.

General temporal belief revision frames can be used to describe either a situation where the objective facts describing the world do not change – so that only the beliefs of the agent change over time – or a situation where both the facts and the doxastic state of the agent change. In the computer science literature the first situation is called belief revision, while the latter is called belief update (Katsuno and Mendelzon [8]). In this paper we restrict attention to belief revision.

DEFINITION 4. Given a general temporal belief revision frame, define the binary relation $\hookrightarrow$ on $\Omega \times T$ as follows: $(\omega, t) \hookrightarrow (\omega', t')$ if and only if (1) $\omega = \omega'$, (2) $\omega \in \Omega_t \cap \Omega_{t'}$ and either (3a) $t \rightarrowtail t'$ or (3b) $t \prec t'$ and, for every $x \in T$ if $t \prec x$ and $x \prec t'$ then $\omega \notin \Omega_x$.

The interpretation of $(\omega, t) \hookrightarrow (\omega, t')$ is that, from the point of view of state ω, instant t is the immediate predecessor of t'. Thus the immediate predecessor of an instant can be different at different states.[1]

Given a general temporal belief revision frame $\langle T, \rightarrowtail, \Omega, \{\Omega_t, \mathcal{B}_t, \mathcal{I}_t\}_{t \in T} \rangle$ one obtains a *model based on it* by adding a function $V : S \rightarrow 2^\Omega$ (where S is the set of atomic propositions and 2^Ω denotes the set of subsets of Ω) that associates with every atomic proposition q the set of states at which q is true. Note that defining a valuation this way is what frames the problem as one of belief revision, since the truth value of an atomic proposition depends only on the state and not on the date.[2] Given a model, a formula φ, an instant t and a state ω such that $\omega \in \Omega_t$, we write $(\omega, t) \models \varphi$ to denote that φ is true at state ω and time t. Let $\|\varphi\|$ denote the truth set of φ, that is, $\|\varphi\| = \{(\omega, t) \in \Omega \times T : \omega \in \Omega_t \text{ and } (\omega, t) \models \varphi\}$ and let $\lceil\varphi\rceil_t \subseteq \Omega_t$ denote the set of states at which φ is true *at time* t, that is, $\lceil\varphi\rceil_t = \{\omega \in \Omega_t : (\omega, t) \models \varphi\}$. Truth at a pair (ω, t) is defined recursively as follows:

---

[1] A special class of general temporal belief revision frames is the class that satisfies the resctriction that, for every $t \in T$, $\Omega_t = \Omega$. It is straightforward to show that, within this class, $(\omega, t) \hookrightarrow (\omega', t')$ if and only if $\omega = \omega'$ and $t \rightarrowtail t'$, so that the immediate predecessor of an instant t is the same at every state. This is the class of frames called 'temporal belief revision frames' in [3]. Hence the addition of the adjective 'general' in Definition 3.

[2] Belief update would require a valuation to be defined as a function $V : S \rightarrow 2^X$ where $X = \{(\omega, t) \in \Omega \times T : \omega \in \Omega_t\}$.

if $q \in S$, $(\omega, t) \models q$ if and only if $\omega \in \Omega_t$ and $\omega \in V(q)$.

$(\omega, t) \models \neg \varphi$ if and only if $\omega \in \Omega_t$ and $(\omega, t) \not\models \varphi$.

$(\omega, t) \models \varphi \vee \psi$ if and only if either $(\omega, t) \models \varphi$ or $(\omega, t) \models \psi$ (or both).

$(\omega, t) \models \bigcirc \varphi$ if and only if, for all $t' \in T$, if $(\omega, t) \hookrightarrow (\omega, t')$ then $(\omega, t') \models \varphi$.

$(\omega, t) \models \bigcirc^{-1} \varphi$ if and only if, for all $t'' \in T$, if $(\omega, t'') \hookrightarrow (\omega, t)$ then $(\omega, t'') \models \varphi$.

$(\omega, t) \models B\varphi$ if and only if $\mathcal{B}_t(\omega) \subseteq \lceil \varphi \rceil_t$, that is,
            if $(\omega', t) \models \varphi$ for all $\omega' \in \mathcal{B}_t(\omega)$.

$(\omega, t) \models I\varphi$ if and only if $\mathcal{I}_t(\omega) = \lceil \varphi \rceil_t$, that is, if (1) $(\omega', t) \models \varphi$
            for all $\omega' \in \mathcal{I}_t(\omega)$, and (2) if $(\omega', t) \models \varphi$ then $\omega' \in \mathcal{I}_t(\omega)$.

$(\omega, t) \models A\varphi$ if and only if $\lceil \varphi \rceil_t = \Omega_t$, that is, if $(\omega', t) \models \varphi$ for all $\omega' \in \Omega_t$.

Note that, while the truth condition for the operator B is the standard one, the truth condition for the operator I is non-standard: instead of simply requiring that $\mathcal{I}_t(\omega) \subseteq \lceil \varphi \rceil_t$ we require equality: $\mathcal{I}_t(\omega) = \lceil \varphi \rceil_t$. Thus our information operator is formally similar to the "all I know" operator introduced by Levesque [9], although the interpretation is different.

A formula φ is *valid in a model* if $(\omega, t) \models \varphi$ for every $(\omega, t) \in \Omega \times T$ with $\omega \in \Omega_t$. A formula φ is *valid in a frame* if it is valid in every model based on it.

4   Soundness and completeness

PROPOSITION 5. *Logic $\mathbb{L}_0$ is sound with respect to the class of general temporal belief revision frames, that is, every theorem of $\mathbb{L}_0$ is valid in every general temporal belief revision frame.*

Proof. We need to show that (1) the rules of inference are validity preserving and (2) the axioms of $\mathbb{L}_0$ are valid in an arbitrary general temporal belief revision frame.

The proof of (1) is entirely standard and is omitted. The proof of validity of axiom K for $\bigcirc$, $\bigcirc^{-1}$, B and A and for the temporal axioms (O_1) and (O_2) is also standard and is omitted.

Validity of the backward uniqueness axiom $\diamondsuit^{-1} \varphi \rightarrow \bigcirc^{-1} \varphi$. Let (ω, t) be such that $(\omega, t) \models \diamondsuit^{-1} \varphi$. Then there exists a $t' \in T$ such that $(\omega, t') \hookrightarrow (\omega, t)$ and $(\omega, t') \models \varphi$. By Definition 4,

(1) $\omega \in \Omega_{t'}$, $t' \prec t$ and either $t' \rightarrowtail t$ or, for every $x \in T$
            such that $t' \prec x$ and $x \prec t$, $\omega \notin \Omega_x$.

Fix an arbitrary $t_0 \in T$ and suppose that $(\omega, t_0) \hookrightarrow (\omega, t)$. Then, by Definition 4,

(2) $\omega \in \Omega_{t_0}$, $t_0 \prec t$ and either $t_0 \rightarrowtail t$ or, for every $x \in T$
            such that $t_0 \prec x$ and $x \prec t$, $\omega \notin \Omega_x$.

We want to show that $t_0 = t'$, so that $(\omega, t_0) \models \varphi$ and, therefore, $(\omega, t) \models \bigcirc^{-1}\varphi$. Since $t' \prec t$ and $t_0 \prec t$, by backward linearity of $\prec$ (see Remark 2), either $t_0 = t'$ or $t_0 \prec t'$ or $t' \prec t_0$. The case $t_0 \prec t'$ contradicts (2) since, by (1), $t' \prec t$ and $\omega \in \Omega_{t'}$ (note that by definition of branching-time frame - see Definition 1 - if $t_0 \rightarrowtail t$ then there is no x such that $t_0 \prec x$ and $x \prec t$). Similarly, the case $t' \prec t_0$ contradicts (1), since, by (2), $t_0 \prec t$ and $\omega \in \Omega_{t_0}$. Thus $t_0 = t'$.

Validity of the S5 axioms for A. Suppose that $(\omega, t) \models A\varphi$. Then $(\omega', t) \models \varphi$ for every $\omega' \in \Omega_t$, thus in particular for $\omega' = \omega$. Similarly, if $(\omega, t) \models \neg A\varphi$ then there exists an $\omega' \in \Omega_t$ such that $(\omega', t) \models \neg\varphi$. Hence $(\omega'', t) \models \neg A\varphi$ for every $\omega'' \in \Omega_t$ and, therefore, $(\omega, t) \models A\neg A\varphi$.

The proof of validity of the inclusion axiom for B (Incl_B) is straightforward and is omitted.

Validity of axiom I_1: $I\varphi \wedge I\psi \rightarrow A(\varphi \leftrightarrow \psi)$. Suppose that $(\omega, t) \models I\varphi \wedge I\psi$. Then $\mathcal{I}_t(\omega) = \lceil\varphi\rceil_t$ and $\mathcal{I}_t(\omega) = \lceil\psi\rceil_t$. Thus $\lceil\varphi\rceil_t = \lceil\psi\rceil_t$, so that $\lceil\varphi \leftrightarrow \psi\rceil_t = \Omega_t$, yielding $(\omega, t) \models A(\varphi \leftrightarrow \psi)$.

Validity of axiom I_2: $A(\varphi \leftrightarrow \psi) \rightarrow (I\varphi \leftrightarrow I\psi)$. Suppose that $(\omega, t) \models A(\varphi \leftrightarrow \psi)$. Then $\lceil\varphi \leftrightarrow \psi\rceil_t = \Omega_t$ and, therefore, $\lceil\varphi\rceil_t = \lceil\psi\rceil_t$. Thus, $(\omega, t) \models I\varphi$ if and only if $\mathcal{I}_t(\omega) = \lceil\varphi\rceil_t$, if and only if $\mathcal{I}_t(\omega) = \lceil\psi\rceil_t$, if and only if $(\omega, t) \models I\psi$. Hence $(\omega, t) \models I\varphi \leftrightarrow I\psi$.   ∎

PROPOSITION 6. *Logic $\mathbb{L}_0$ is complete with respect to the class of general temporal belief revision frames, that is, if φ is a formula which is valid in every general temporal belief revision frame then φ is a theorem of $\mathbb{L}_0$.*

To prove Proposition 6 we need to show that, for every consistent formula φ, there is a state-instant pair (ω, t) in a model based on a general temporal belief revision frame such that $(\omega, t) \models \varphi$. We follow the constructive approach of Burgess [5]: given a consistent formula φ_0, we construct a chronicle (see Definition 11 below) where φ_0 is true at some state-instant pair and then extend it to a perfect chronicle. First some preliminary definitions and lemmas.

Let $\mathbb{M}_{\mathbb{L}_0}$ denote the set of maximally consistent sets of formulas of logic $\mathbb{L}_0$.

DEFINITION 7. Define the binary relations $\mathcal{A}^c$, $\mathcal{B}^c$ and $\hookrightarrow^c$ on $\mathbb{M}_{\mathbb{L}_0}$ as follows:

1. $m\mathcal{A}^c m'$ if and only if $\{\varphi : A\varphi \in m\} \subseteq m'$, that is, if $A\varphi \in m$ implies $\varphi \in m'$;

2. $m\mathcal{B}^c m'$ if and only if $\{\varphi : B\varphi \in m\} \subseteq m'$, that is, if $B\varphi \in m$ implies $\varphi \in m'$;

3. $m \hookrightarrow^c m'$ if and only if $\{\varphi : \bigcirc\varphi \in m\} \subseteq m'$, that is, if $\bigcirc\varphi \in m$ implies $\varphi \in m'$.

REMARK 8. For every $\square \in \{A, B, \bigcirc\}$ and for every $m, m' \in \mathbb{M}_{\mathbb{L}_0}$, $\{\varphi : \square\varphi \in m\} \subseteq m'$ if and only if $\{\neg\square\neg\varphi : \varphi \in m'\} \subseteq m$ (see Chellas [6] Theorem 4.30(1),

p. 158) . Furthermore, $\mathcal{A}^c$ is an equivalence relation because of axioms T_A and 5_A (Chellas [6] Theorem 5.13 (2) and (5), p. 175), $\mathcal{B}^c$ is a subrelation of $\mathcal{A}^c$ because of axiom $Incl_B$, and the relation $\hookrightarrow^c$ satisfies the following properties: (1) because of the temporal axioms O_1 and O_2, $m \hookrightarrow^c m'$ if and only if $\{\varphi : \bigcirc^{-1}\varphi \in m'\} \subseteq m$,[3] and (2) because of axiom BU, if $m_1 \hookrightarrow^c m$ and $m_2 \hookrightarrow^c m$ then $m_1 = m_2$.[4]

The following lemma is well-known (see Blackburn et al. [2], Lemma 4.20, p. 198).

LEMMA 9. *Let $m \in \mathbb{M}_{\mathbb{L}_0}$. Then: (1) if $\neg A \neg \varphi \in m$ then there exists an $m' \in \mathbb{M}_{\mathbb{L}_0}$ such that $m \mathcal{A}^c m'$ and $\varphi \in m'$, (2) if $\neg B \neg \varphi \in m$ then there exists an $m' \in \mathbb{M}_{\mathbb{L}_0}$ such that $m \mathcal{B}^c m'$ and $\varphi \in m'$, (3) if $\Diamond \varphi \in m$ then there exists an $m' \in \mathbb{M}_{\mathbb{L}_0}$ such that $m \hookrightarrow^c m'$ and $\varphi \in m'$, (4) if $\Diamond^{-1}\varphi \in m$ then there exists an $m' \in \mathbb{M}_{\mathbb{L}_0}$ such that $m' \hookrightarrow^c m$ and $\varphi \in m'$.*

LEMMA 10. *Let $m, m' \in \mathbb{M}_{\mathbb{L}_0}$ be such that $m \mathcal{A}^c m'$ and let φ be a formula such that $I\varphi \in m$ and $\varphi \in m'$. Then, for every formula ψ, if $I\psi \in m$ then $\psi \in m'$, that is, $\{\psi : I\psi \in m\} \subseteq m'$.*

Proof. Suppose that $m \mathcal{A}^c m'$, $I\varphi \in m$ and $\varphi \in m'$. Fix an arbitrary ψ such that $I\psi \in m$. Then $I\varphi \wedge I\psi \in m$. Since $(I\varphi \wedge I\psi) \rightarrow A(\varphi \leftrightarrow \psi)$ is a theorem, it belongs to every MCS, in particular to m. Hence $A(\varphi \leftrightarrow \psi) \in m$. Then, since $m \mathcal{A}^c m'$, $\varphi \leftrightarrow \psi \in m'$. Since $\varphi \in m'$, it follows that $\psi \in m'$. ∎

DEFINITION 11. A *chronicle* is a general temporal belief revision frame together with a function $\mu : \{(\omega, t) \in \Omega \times T : \omega \in \Omega_t\} \rightarrow \mathbb{M}_{\mathbb{L}_0}$ that associates with every state-instant pair an MCS. A chronicle μ is *coherent* if it satisfies the following properties:

(1) if $\bigcirc\varphi \in \mu(\omega, t)$ and $(\omega, t) \hookrightarrow (\omega, t')$ then $\varphi \in \mu(\omega, t')$, that is, $\mu(\omega, t) \hookrightarrow^c \mu(\omega, t')$;

(2) if $A\varphi \in \mu(\omega, t)$ then, for all $\omega' \in \Omega_t$, $\varphi \in \mu(\omega', t)$, that is, if $\omega' \in \Omega_t$ implies $\mu(\omega, t) \mathcal{A}^c \mu(\omega', t)$;

---

[3]Proof. Suppose that $m \hookrightarrow^c m'$ and $\bigcirc^{-1}\varphi \in m'$. Then $\Diamond \bigcirc^{-1} \varphi \in m$. Since the following is an instance of axiom O_1: $\neg\varphi \rightarrow \bigcirc\neg\bigcirc^{-1}\varphi$ and is propositionally equivalent to $\Diamond\bigcirc^{-1}\varphi \rightarrow \varphi$, it belongs to m. Thus $\varphi \in m$. Conversely, suppose that $\{\varphi : \bigcirc^{-1}\varphi \in m'\} \subseteq m$. Then (see Chellas [6] Theorem 4.30(1), p. 158) $\{\Diamond^{-1}\varphi : \varphi \in m\} \subseteq m'$. We want to show that $m \hookrightarrow^c m'$, that is, that if $\bigcirc\varphi \in m$ then $\varphi \in m'$. Fix an arbitrary φ such that $\bigcirc\varphi \in m$. Then $\Diamond^{-1} \bigcirc \varphi \in m'$. Since axiom O_2 is equivalent to $\Diamond^{-1}\bigcirc\varphi \hookrightarrow^c \varphi$, the latter belongs to m'. Thus $\varphi \in m'$.

[4]Proof. Suppose that $m_1 \hookrightarrow^c m$ and $m_2 \hookrightarrow^c m$ and $m_1 \neq m_2$. Then, by definition of maximally consistent set (MCS), there exists a formula φ such that $\varphi \in m_1$ and $\neg\varphi \in m_2$. It follows that $\Diamond^{-1}\varphi \in m$ and $\Diamond^{-1}\neg\varphi \in m$. By axiom BU, $(\Diamond^{-1}\varphi \rightarrow \bigcirc^{-1}\varphi) \in m$ and $(\Diamond^{-1}\neg\varphi \rightarrow \bigcirc^{-1}\neg\varphi) \in m$. Thus $(\bigcirc^{-1}\varphi \wedge \bigcirc^{-1}\neg\varphi) \in m$, which implies, since $m_1 \hookrightarrow^c m$, that $\varphi \wedge \neg\varphi \in m_1$, contradicting the definition of MCS.

(3) if $B\varphi \in \mu(\omega, t)$ and $\omega' \in \mathcal{B}_t(\omega)$ then $\varphi \in \mu(\omega', t)$, that is, if $\omega' \in \mathcal{B}_t(\omega)$ implies $\mu(\omega, t)\mathcal{B}^c\mu(\omega', t)$;

(4a) if $I\varphi \in \mu(\omega, t)$ and $\omega' \in \mathcal{I}_t(\omega)$ then $\varphi \in \mu(\omega', t)$;

(4b) if $I\varphi \in \mu(\omega, t)$ and $\omega' \in \Omega_t$ and $\varphi \in \mu(\omega', t)$ then $\omega' \in \mathcal{I}_t(\omega)$.

Let φ_0 be a consistent formula. Then by Lindenbaum's lemma there is an $m_0 \in \mathbb{M}_{\mathbb{L}_0}$ such that $\varphi_0 \in m_0$. Construct the following chronicle: $T = \{t\}$, $\rightarrowtail\ = \varnothing$, $\Omega = \Omega_t = \{\alpha\}$, $\mathcal{B}_t(\alpha) = \varnothing$,

$$\mathcal{I}_t(\alpha) = \begin{cases} \varnothing & \text{if, for every } \varphi,\ \varphi \notin m_0 \text{ whenever } I\varphi \in m_0 \\ \{\alpha\} & \text{if, for some } \varphi,\ I\varphi \in m_0 \text{ and } \varphi \in m_0 \end{cases}$$

and $\mu(\alpha, t) = m_0$.

LEMMA 12. *The above chronicle is coherent.*

Proof. Condition (1) of Definition 11 is satisfied trivially since $\rightarrowtail\ = \varnothing$. Condition (2) is satisfied because the relation $\mathcal{A}^c$ is reflexive. Condition (3) is satisfied trivially since $\mathcal{B}_t(\alpha) = \varnothing$. Now we turn to conditions (4a) and (4b). If there is no φ such that $I\varphi \in m_0$ and $\varphi \in m_0$, then (4a) is satisfied trivially because, by construction, $\mathcal{I}_t(\alpha) = \varnothing$, and (4b) is satisfied trivially because if $I\varphi \in m_0$ then $\varphi \notin m_0 = \mu(\alpha, t)$. Suppose therefore that, for some φ, $I\varphi \in m_0$ and $\varphi \in m_0$. Fix an arbitrary formula ψ and suppose that $I\psi \in m_0$. It follows from Lemma 10, using the fact that $m_0\mathcal{A}^c m_0$, that $\psi \in m_0$. Thus (4a) and (4b) are satisfied since, by construction, $\mathcal{I}_t(\alpha) = \{\alpha\}$.   ∎

DEFINITION 13. Fix a chronicle $\langle \mathcal{R}, \mu \rangle$ where $\mathcal{R} = \langle T, \rightarrowtail, \Omega, \{\Omega_t, \mathcal{B}_t, \mathcal{I}_t\}_{t\in T} \rangle$. We say that the chronicle $\langle \mathcal{R}', \mu' \rangle$, with $\mathcal{R}' = \langle T', \rightarrowtail', \Omega', \{\Omega'_t, \mathcal{B}'_t, \mathcal{I}'_t\}_{t\in T'} \rangle$ is an *extension* of $\langle \mathcal{R}, \mu \rangle$ if

(1) $T \subseteq T'$,

(2) $\Omega \subseteq \Omega'$, and, for every $t \in T$, $\Omega_t \subseteq \Omega'_t$,

and, identifying relations and functions with sets of ordered pairs,

(3) $\rightarrowtail\ =\ \rightarrowtail' \cap (T \times T)$,

(4) for all $t \in T$, $\mathcal{B}_t = \mathcal{B}'_t \cap (\Omega_t \times \Omega_t)$,

(5) for all $t \in T$, $\mathcal{I}_t = \mathcal{I}'_t \cap (\Omega_t \times \Omega_t)$ and

(6) $\mu \subseteq \mu'$.

DEFINITION 14. Fix a chronicle and a pair $(\alpha, t) \in \Omega \times T$ with $\alpha \in \Omega_t$. We say that at (α, t) there is:

- an *A-defect* if there is a formula φ such that $\neg A\neg\varphi \in \mu(\alpha, t)$ and there is no $\omega \in \Omega_t$ such that $\varphi \in \mu(\omega, t)$,

- a *B-defect* if there is a formula φ such that $\neg B\neg\varphi \in \mu(\alpha, t)$ and there is no $\omega \in \mathcal{B}_t(\alpha)$ such that $\varphi \in \mu(\omega, t)$,

- a $\bigcirc^{-1}$-*defect* if there is a formula φ such that $\Diamond^{-1}\varphi \in \mu(\alpha, t)$ and there is no $t' \in T$ such that $(\alpha, t') \hookrightarrow (\alpha, t)$ and $\varphi \in \mu(\alpha, t')$,

- a $\bigcirc$-*defect* if there is a formula φ such that $\Diamond\varphi \in \mu(\alpha, t)$ and there is no $t' \in T$ such that $(\alpha, t) \hookrightarrow (\alpha, t')$ and $\varphi \in \mu(\alpha, t')$.

Note that there is no need to consider the possibility of an *I-defect*, since $(\alpha, t) \models \neg I\neg\varphi$ does *not* mean that there is an $\omega \in \mathcal{I}_t(\alpha)$ such that $(\omega, t) \models \varphi$ but rather that $\mathcal{I}_t(\alpha) \neq \lceil\neg\varphi\rceil_t$. For example, it could be that $\mathcal{I}_t(\alpha)$ is a proper subset of $\lceil\neg\varphi\rceil_t$.

LEMMA 15. *(Repair Lemma). Fix a coherent chronicle $\langle \mathcal{R}, \mu\rangle$ where T and Ω are finite sets. Suppose that there is a defect at (α, t). Then there exists a finite coherent extension $\langle \mathcal{R}', \mu'\rangle$ of $\langle \mathcal{R}, \mu\rangle$ where that defect at (α, t) is no longer present.*

Proof. Let D be a countably infinite set containing T and W a countably infinite set containing Ω.

Suppose first that there is an A-**defect** at (α, t_1), that is, there is a formula φ such that $\neg A\neg\varphi \in \mu(\alpha, t_1)$ and there is no $\omega \in \Omega_{t_1}$ such that $\varphi \in \mu(\omega, t_1)$. By Lemma 9 there is an $\hat{m} \in \mathbb{M}_{\mathbb{L}_0}$ such that $\mu(\alpha, t_1)\mathcal{A}^c\hat{m}$ and $\varphi \in \hat{m}$. Construct the following extension of $\langle \mathcal{R}, \mu\rangle$: $T' = T$; $\hookrightarrow' = \hookrightarrow$; let $\hat{\omega} \in W\backslash\Omega$ and define $\Omega' = \Omega \cup \{\hat{\omega}\}$; for every $t \in T\backslash\{t_1\}$, let $\Omega'_t = \Omega_t$, $\mathcal{B}'_t = \mathcal{B}_t$ and $\mathcal{I}'_t = \mathcal{I}_t$; let $\Omega'_{t_1} = \Omega_{t_1} \cup \{\hat{\omega}\}$ and $\mu'(\hat{\omega}, t_1) = \hat{m}$; for $\omega \in \Omega_{t_1}$, $\mathcal{B}'_{t_1}(\omega) = \mathcal{B}_{t_1}(\omega)$ and $\mathcal{B}'_{t_1}(\hat{\omega}) = \varnothing$; let $\mathcal{I}'_{t_1}$ be defined as follows:

(i) for every $\omega \in \Omega_{t_1}$, $\mathcal{I}'_{t_1}(\omega) = \begin{cases} \mathcal{I}_{t_1}(\omega) & \text{if there is no } \psi \text{ such that} \\ & I\psi \in \mu(\omega, t_1) \text{ and } \psi \in \hat{m} \\ \mathcal{I}_{t_1}(\omega) \cup \{\hat{\omega}\} & \text{if there is a } \psi \text{ such that} \\ & I\psi \in \mu(\omega, t_1) \text{ and } \psi \in \hat{m} \end{cases}$

and

(ii) for every $\omega \in \Omega_{t_1} \cup \{\hat{\omega}\}$, $\omega \in \mathcal{I}'_{t_1}(\hat{\omega})$ if and only if, for some ψ, $I\psi \in \hat{m}$ and $\psi \in \mu'(\omega, t_1)$.

Since $\hat{\omega} \in \Omega'_{t_1}$, $\mu'(\hat{\omega}, t_1) = \hat{m}$ and $\varphi \in \hat{m}$, the A-defect at (α, t_1) is no longer present in $\langle \mathcal{R}', \mu'\rangle$ so defined. We need to show that $\langle \mathcal{R}', \mu'\rangle$ is coherent. Since, by hypothesis, $\langle \mathcal{R}, \mu\rangle$ is coherent and $T' = T$, condition (1) of Definition 11 is satisfied (note that, since $\hat{\omega} \notin \Omega$, $\hat{\omega} \notin \Omega'_t = \Omega_t$ for every $t \neq t_1$). For condition (2) we need to show that if $t \in T' = T$ and $\omega, \omega' \in \Omega'_t$, then $\mu'(\omega, t)\mathcal{A}^c\mu'(\omega', t)$. If $\omega, \omega' \in \Omega$ then it follows from the hypothesis that $\langle \mathcal{R}, \mu\rangle$ is coherent. If $t = t_1$ and $\omega = \omega' = \hat{\omega}$ then it follows from the fact that $\mathcal{A}^c$ is reflexive. If $t = t_1$, $\omega \in \Omega_{t_1}$ and $\omega' = \hat{\omega}$ then it follows from the fact that (i) $\mu(\omega, t_1)\mathcal{A}^c\mu(\alpha, t_1)$, by the hypothesis that $\langle \mathcal{R}, \mu \rangle$ is coherent, (ii) $\mu(\alpha, t_1)\mathcal{A}^c \hat{m}$, by construction, and (iii) transitivity of $\mathcal{A}^c$. Finally, if $t = t_1$, $\omega = \hat{\omega}$ and $\omega' \in \Omega_{t_1}$ then

1. $\mu(\alpha, t_1)\mathcal{A}^c \hat{m}$ by construction
2. $\hat{m}\mathcal{A}^c \mu(\alpha, t_1)$ by 1 and symmetry of $\mathcal{A}^c$
3. $\mu(\alpha, t_1)\mathcal{A}^c \mu(\omega', t_1)$ by coherence of $\langle \mathcal{R}, \mu \rangle$
4. $\hat{m}\mathcal{A}^c \mu(\omega', t_1)$ by 2, 3 and transitivity of $\mathcal{A}^c$.

Condition (3) of Definition 11 is satisfied, since (i) for $t \in T$ and $\omega \in \Omega_t$, $\mathcal{B}'_t(\omega) = \mathcal{B}_t(\omega)$ and by hypothesis $\langle \mathcal{R}, \mu \rangle$ is coherent and (ii) $\mathcal{B}'_{t_1}(\hat{\omega}) = \varnothing$ and thus the condition holds trivially. Now we turn to condition (4a). Suppose that $I\psi \in \mu'(\omega, t)$ and $\omega' \in I'_t(\omega)$. If $t \in T\backslash\{t_1\}$ or if $t = t_1$ and $\omega, \omega' \in \Omega_{t_1}$ then $\psi \in \mu'(\omega', t) = \mu(\omega', t)$ by coherence of $\langle \mathcal{R}, \mu \rangle$. If $t = t_1$, $\omega \in \Omega_{t_1}$ and $\omega' = \hat{\omega}$, then, by construction (since, by hypothesis, $\hat{\omega} \in I'_{t_1}(\omega)$) there is a χ such that $I\chi \in \mu(\omega, t_1)$ and $\chi \in \hat{m}$; since, by hypothesis, $I\psi \in \mu'(\omega, t_1) = \mu(\omega, t_1)$ and, as shown above, $\mu(\omega, t_1)\mathcal{A}^c \hat{m}$, it follows from Lemma 10 that $\psi \in \hat{m}$. If $t = t_1$, $\omega = \hat{\omega}$ and $\omega' \in \Omega_{t_1}$ then, by construction, there is a χ such that $I\chi \in \hat{m}$ and $\chi \in \mu(\omega', t_1)$; since $\hat{m}\mathcal{A}^c \mu(\omega', t_1)$ (shown above) it follows from Lemma 10 that $\psi \in \mu(\omega', t_1)$. Finally, if $\omega = \omega' = \hat{\omega}$ then, by construction, there is a χ such that $I\chi \in \hat{m}$ and $\chi \in \hat{m}$. Since $\hat{m}\mathcal{A}^c \hat{m}$, it follows from Lemma 10 that $\psi \in \hat{m}$. Next we turn to condition (4b). Fix arbitrary $t \in T' = T$ and $\omega, \omega' \in \Omega'_t$ and suppose that $I\psi \in \mu'(\omega, t)$ and $\psi \in \mu'(\omega', t)$. We need to show that $\omega' \in I'_t(\omega)$. If $t \in T\backslash\{t_1\}$ it follows from coherence $\langle \mathcal{R}, \mu \rangle$, since $\Omega'_t = \Omega_t$ and $I'_t = I_t$. Similarly if $t = t_1$ and $\omega, \omega' \in \Omega_{t_1}$. If $t = t_1$ and $\omega = \omega' = \hat{\omega}$ then, since, by hypothesis, $I\psi \in \hat{m}$ and $\psi \in \hat{m}$, it follows, by construction, that $\hat{\omega} \in I'_{t_1}(\hat{\omega})$. Similarly for $t = t_1$, $\omega' \in \Omega_{t_1}$ and $\omega = \hat{\omega}$. Finally, if $t = t_1$, $\omega \in \Omega_{t_1}$ and $\omega' = \hat{\omega}$, then $I\psi \in \mu(\omega, t_1)$ and $\psi \in \hat{m}$ and, by construction, $\hat{\omega} \in I'_{t_1}(\omega)$.

Suppose now that there is a B-**defect** at (α, t_1), that is, there is a formula φ such that $\neg B \neg \varphi \in \mu(\alpha, t_1)$ and there is no $\omega \in \mathcal{B}_{t_1}(\alpha)$ such that $\varphi \in \mu(\omega, t_1)$. By Lemma 9 there is an $\hat{m} \in \mathbb{M}_{\mathbb{L}_0}$ such that $\mu(\alpha, t_1)\mathcal{B}^c \hat{m}$ and $\varphi \in \hat{m}$. Construct the following extension of $\langle \mathcal{R}, \mu \rangle$: $T' = T$; $\rightarrowtail' = \rightarrowtail$; let $\hat{\omega} \in W\backslash\Omega$ and define $\Omega' = \Omega \cup \{\hat{\omega}\}$; for every $t \in T\backslash\{t_1\}$, let $\Omega'_t = \Omega_t$, $\mathcal{B}'_t = \mathcal{B}_t$ and $I'_t = I_t$; let $\Omega'_{t_1} = \Omega_{t_1} \cup \{\hat{\omega}\}$ and $\mu'(\hat{\omega}, t_1) = \hat{m}$; for $\omega \in \Omega_{t_1}\backslash\{\alpha\}$, let $\mathcal{B}'_{t_1}(\omega) = \mathcal{B}_{t_1}(\omega)$; let $\mathcal{B}'_{t_1}(\alpha) = \mathcal{B}_{t_1}(\alpha) \cup \{\hat{\omega}\}$ and $\mathcal{B}'_{t_1}(\hat{\omega}) = \varnothing$; let I'_{t_1} be defined as follows:

(i) for every $\omega \in \Omega_{t_1}$, $I'_{t_1}(\omega) = \begin{cases} I_{t_1}(\omega) & \text{if there is no } \psi \text{ such that} \\ & I\psi \in \mu(\omega, t_1) \text{ and } \psi \in \hat{m} \\ I_{t_1}(\omega) \cup \{\hat{\omega}\} & \text{if there is a } \psi \text{ such that} \\ & I\psi \in \mu(\omega, t_1) \text{ and } \psi \in \hat{m} \end{cases}$

and

(ii) for every $\omega \in \Omega_{t_1} \cup \{\hat{\omega}\}$, $\omega \in I'_{t_1}(\hat{\omega})$ if and only if, for some ψ, $I\psi \in \hat{m}$ and $\psi \in \mu'(\omega, t_1)$.

Since $\hat{\omega} \in \mathcal{B}'_{t_1}(\alpha)$, $\mu'(\hat{\omega}, t_1) = \hat{m}$ and $\varphi \in \hat{m}$, the B-defect at (α, t_1) is no longer present in $\langle \mathcal{R}', \mu' \rangle$ so defined. Since $\mathcal{B}^c$ is a subrelation of $\mathcal{A}^c$ (see Remark 8), the proof that $\langle \mathcal{R}', \mu' \rangle$ is coherent is identical to the previous proof (condition (3) of Definition 11 is satisfied by construction).

Next we consider the case of a $\bigcirc^{-1}$-**defect** at (α, t_1), that is, there is a formula φ such that $\Diamond^{-1}\varphi \in \mu(\alpha, t_1)$ and there is no $t \in T$ such that $(\alpha, t) \hookrightarrow (\alpha, t_1)$ and $\varphi \in \mu(\alpha, t)$. Let T_1 be the set of predecessors of t_1 in T, that is, $T_1 = \{t \in T : t \prec t_1\}$. Suppose first that $T_1 = \varnothing$. By Lemma 9 there is an $\hat{m} \in \mathbb{M}_{\mathbb{L}_0}$ such that $\hat{m} \hookrightarrow^c \mu(\alpha, t_1)$ and $\varphi \in \hat{m}$. Construct the following extension of $\langle \mathcal{R}, \mu \rangle$: let $\hat{t} \in D \backslash T$ and define $T' = T \cup \{\hat{t}\}$; $\hookrightarrow' = \hookrightarrow \cup \{(\hat{t}, t_1)\}$; $\Omega' = \Omega$; for every $t \in T$, $\Omega'_t = \Omega_t$, $\mathcal{B}'_t = \mathcal{B}_t$ and $\mathcal{I}'_t = \mathcal{I}_t$; let $\Omega'_{\hat{t}} = \{\alpha\}$, $\mathcal{B}'_{\hat{t}}(\alpha) = \varnothing$ and

$$\mathcal{I}'_{\hat{t}}(\alpha) = \begin{cases} \varnothing & \text{if, for every } \psi, \psi \notin \hat{m} \text{ whenever } I\psi \in \hat{m} \\ \{\alpha\} & \text{if, for some } \psi, I\psi \in \hat{m} \text{ and } \psi \in \hat{m}. \end{cases}$$

Finally, let $\mu'(\alpha, \hat{t}) = \hat{m}$. Clearly the $\bigcirc^{-1}$-defect at (α, t_1) is no longer present in $\langle \mathcal{R}', \mu' \rangle$ so defined. For $t \in T$, coherence of $\langle \mathcal{R}', \mu' \rangle$ follows from coherence of $\langle \mathcal{R}, \mu \rangle$. Thus we only need to consider $t = \hat{t}$. Condition (1) of Definition 11 is satisfied trivially, since $\hat{t}$ has no predecessors in T'. Condition (2) follows from the fact that, by construction, $\Omega'_{\hat{t}} = \{\alpha\}$ and, by reflexivity of $\mathcal{A}^c$, $\hat{m}\mathcal{A}^c\hat{m}$. Condition (3) is satisfied trivially, since, by construction, $\mathcal{B}'_{\hat{t}}(\alpha) = \varnothing$. Now we turn to conditions (4a) and (4b). If, for every ψ, $I\psi \in \hat{m}$ implies $\psi \notin \hat{m}$, then (4a) is satisfied trivially because, by construction, $\mathcal{I}_t(\alpha) = \varnothing$ and (4b) is satisfied trivially because $\psi \notin \hat{m}$. Suppose therefore that, for some ψ, $I\psi \in \hat{m}$ and $\psi \in \hat{m}$. Fix an arbitrary formula χ and suppose that $I\chi \in \hat{m}$. It follows from Lemma 10, using the fact that $\hat{m}\mathcal{A}^c\hat{m}$, that $\chi \in \hat{m}$. Thus (4a) and (4b) are satisfied since, by construction, $\mathcal{I}_{\hat{t}}(\alpha) = \{\alpha\}$ and $\mu'(\alpha, \hat{t}) = \hat{m}$.

Consider now the case where $T_1 \neq \varnothing$. By hypothesis, for every $t \in T_1$, it is not the case that $(\alpha, t) \hookrightarrow (\alpha, t_1)$.[5] Thus, by Definition 4, it must be that,

$$(3) \qquad\qquad \text{for all } t \in T_1, \ \alpha \notin \Omega_t.$$

Let t_0 be the farthest predecessor of t_1 in T, that is, $t_0 \in T_1$ and, for every $t \in T_1$, either $t = t_0$ or $t_0 \prec t$ (such a t_0 exists because of backward linearity of $\prec$ and finiteness of T). By Lemma 9 there is an $\hat{m} \in \mathbb{M}_{\mathbb{L}_0}$ such that $\hat{m} \hookrightarrow^c \mu(\alpha, t_1)$ and $\varphi \in \hat{m}$. Construct the following extension of $\langle \mathcal{R}, \mu \rangle$: let $\hat{t} \in D \backslash T$ and define $T' = T \cup \{\hat{t}\}$; $\hookrightarrow' = \hookrightarrow \cup \{(\hat{t}, t_0)\}$; $\Omega' = \Omega$; for every $t \in T$, $\Omega'_t = \Omega_t$, $\mathcal{B}'_t = \mathcal{B}_t$ and $\mathcal{I}'_t = \mathcal{I}_t$; let $\Omega'_{\hat{t}} = \{\alpha\}$, $\mathcal{B}'_{\hat{t}}(\alpha) = \varnothing$ and

$$\mathcal{I}'_{\hat{t}}(\alpha) = \begin{cases} \varnothing & \text{if, for every } \psi, \psi \notin \hat{m} \text{ whenever } I\psi \in \hat{m} \\ \{\alpha\} & \text{if, for some } \psi, I\psi \in \hat{m} \text{ and } \psi \in \hat{m}. \end{cases}$$

Let $\mu'(\alpha, \hat{t}) = \hat{m}$. By (3) and Definition 4, $(\alpha, \hat{t}) \hookrightarrow' (\alpha, t_1)$ and therefore the $\bigcirc^{-1}$-defect at (α, t_1) is no longer present in $\langle \mathcal{R}', \mu' \rangle$ so defined. The proof that $\langle \mathcal{R}', \mu' \rangle$

---

[5]If there were a $t \in T_1$ such that $(\alpha, t) \hookrightarrow (\alpha, t_1)$ then, by coherence of $\langle \mathcal{R}, \mu \rangle$, we would have that $\varphi \in \mu(\alpha, t)$, since $\Diamond^{-1}\varphi \in \mu(\alpha, t_1)$, contradicting our hypothesis.

is coherent is the same as in the previous case.

Finally we consider the case of a $\bigcirc$-**defect** at (α, t_1), that is, there is a formula φ such that $\Diamond\varphi \in \mu(\alpha, t_1)$ and there is no $t \in T$ such that $(\alpha, t_1) \hookrightarrow (\alpha, t)$ and $\varphi \in \mu(\alpha, t)$. Let T_2 be the set of successors of t_1 in T, that is, $T_2 = \{t \in T : t_1 \prec t\}$. Suppose first that $T_2 = \varnothing$. By Lemma 9 there is an $\hat{m} \in \mathbb{M}_{\mathbb{L}_0}$ such that $\mu(\alpha, t_1) \hookrightarrow^c \hat{m}$ and $\varphi \in \hat{m}$. Construct the following extension of $\langle \mathcal{R}, \mu \rangle$: let $\hat{t} \in D \backslash T$ and define $T' = T \cup \{\hat{t}\}$; $\hookrightarrow' = \hookrightarrow \cup \{(t_1, \hat{t})\}$; $\Omega' = \Omega$; for every $t \in T$, $\Omega'_t = \Omega_t$, $\mathcal{B}'_t = \mathcal{B}_t$ and $\mathcal{I}'_t = \mathcal{I}_t$; let $\Omega'_{\hat{t}} = \{\alpha\}$, $\mathcal{B}'_{\hat{t}}(\alpha) = \varnothing$ and

$$\mathcal{I}'_{\hat{t}}(\alpha) = \begin{cases} \varnothing & \text{if, for every } \psi, \psi \notin \hat{m} \text{ whenever } I\psi \in \hat{m} \\ \{\alpha\} & \text{if, for some } \psi, I\psi \in \hat{m} \text{ and } \psi \in \hat{m}. \end{cases}$$

Finally, let $\mu'(\alpha, \hat{t}) = \hat{m}$. Clearly the $\bigcirc$-defect at (α, t_1) is no longer present in $\langle \mathcal{R}', \mu' \rangle$ so defined. For $t \in T$, coherence of $\langle \mathcal{R}', \mu' \rangle$ follows from coherence of $\langle \mathcal{R}, \mu \rangle$. Thus we only need to consider $t = \hat{t}$. Condition (1) of Definition 11 is satisfied trivially, since $\hat{t}$ has no successors in T'. Condition (2) follows from the fact that, by construction, $\Omega_{\hat{t}} = \{\alpha\}$ and, by reflexivity of $\mathcal{A}^c$, $\hat{m}\mathcal{A}^c\hat{m}$. Condition (3) is satisfied trivially, since, by construction, $\mathcal{B}'_{\hat{t}}(\alpha) = \varnothing$. Now we turn to conditions (4a) and (4b). If, for every ψ, $I\psi \in \hat{m}$ implies $\psi \notin \hat{m}$, then (4a) is satisfied trivially because, by construction, $\mathcal{I}_{\hat{t}}(\alpha) = \varnothing$ and (4b) is satisfied trivially because $\psi \notin \hat{m}$. Suppose therefore that, for some ψ, $I\psi \in \hat{m}$ and $\psi \in \hat{m}$. Fix an arbitrary formula χ and suppose that $I\chi \in \hat{m}$. It follows from Lemma 10, using the fact that $\hat{m}\mathcal{A}^c\hat{m}$, that $\chi \in \hat{m}$. Thus (4a) and (4b) are satisfied since, by construction, $\mathcal{I}_{\hat{t}}(\alpha) = \{\alpha\}$.

Consider now the case where $T_2 \neq \varnothing$. By hypothesis, for every $t \in T_2$, it is not the case that $(\alpha, t_1) \hookrightarrow (\alpha, t)$.[6] Thus, by Definition 4, it must be that,

$$(4) \qquad\qquad\qquad \text{for all } t \in T_2, \ \alpha \notin \Omega_t.$$

Let t_2 be any successor of t_1 in T with no successors of its own, that is, $t_2 \in T_2$ and, for every $t \in T$, $t_2 \nprec t$ (such a t_2 exists because of finiteness of T). By Lemma 9 there is an $\hat{m} \in \mathbb{M}_{\mathbb{L}_0}$ such that $\mu(\alpha, t_1) \hookrightarrow^c \hat{m}$ and $\varphi \in \hat{m}$. Construct the following extension of $\langle \mathcal{R}, \mu \rangle$: let $\hat{t} \in D \backslash T$ and define $T' = T \cup \{\hat{t}\}$; $\hookrightarrow' = \hookrightarrow \cup \{(t_2, \hat{t})\}$; $\Omega' = \Omega$; for every $t \in T$, $\Omega'_t = \Omega_t$, $\mathcal{B}'_t = \mathcal{B}_t$ and $\mathcal{I}'_t = \mathcal{I}_t$; let $\Omega'_{\hat{t}} = \{\alpha\}$, $\mathcal{B}'_{\hat{t}}(\alpha) = \varnothing$ and

$$\mathcal{I}'_{\hat{t}}(\alpha) = \begin{cases} \varnothing & \text{if, for every } \psi, \psi \notin \hat{m} \text{ whenever } I\psi \in \hat{m} \\ \{\alpha\} & \text{if, for some } \psi, I\psi \in \hat{m} \text{ and } \psi \in \hat{m}. \end{cases}$$

Let $\mu'(\alpha, \hat{t}) = \hat{m}$. By (4) and Definition 4, $(\alpha, t_1) \hookrightarrow' (\alpha, \hat{t})$ and therefore the $\bigcirc$-defect at (α, t_1) is no longer present in $\langle \mathcal{R}', \mu' \rangle$ so defined. The proof that $\langle \mathcal{R}', \mu' \rangle$ is coherent is the same as in the previous case. ∎

---

[6] If there were a $t \in T_2$ such that $(\alpha, t_1) \hookrightarrow (\alpha, t)$ then by coherence of $\langle \mathcal{R}, \mu \rangle$ we would have that $\varphi \in \mu(\alpha, t)$, since $\Diamond\varphi \in \mu(\alpha, t_1)$, contradicting our hypothesis.

The final step in the completeness proof (construction of a perfect chronicle by a countable application of Lemma 15) is entirely standard (see Burgess [5], p. 101) and is omitted.

5 Conclusion

Bonanno [3] considers an extension of logic $\mathbb{L}_0$ obtained by adding several axioms for belief revision[7] and shows that it provides an axiomatic characterization of the theory of belief revision due to Alchourrón et al. [1], known as the AGM theory. It is shown there that the proposed logic is sound with respect to the sub-class of general temporal belief revision frames that satisfy the restriction that $\Omega_t = \Omega$, for all $t \in T$ (see Footnote 1). An open question is whether the completeness result of the previous section can be proved with respect to this class of frames and whether it can be extended to the several logics (extensions of $\mathbb{L}_0$) proposed in [3] and [4].

BIBLIOGRAPHY

[1] Alchourrón, C., P. Gärdenfors and D. Makinson, On the logic of theory change: partial meet con-traction and revision functions, *The Journal of Symbolic Logic*, 1985, 50: 510-530.
[2] Blackburn, P., M. de Rijke and Y. Venema, *Modal logic,* Cambridge University Press, 2001.
[3] Bonanno, G., Axiomatic characterization of the AGM theory of belief revision in a temporal logic, *Artificial Intelligence*, 2007, 171: 144-160.
[4] Bonanno, G., Belief revision in a temporal framework, in: Krzysztof R. Apt and Robert van Rooij (eds.), *New Perspectives on Games and Interaction*, Texts in Logic and Games 4, Amsterdam University Press, 2008, pp.45ï¿½79.
[5] Burgess, J., Basic tense logic, in: D. Gabbay and F. Guenthner (eds.), *Handbook of philosophical logic*, Vol. II, D. Reidel Publishing Company, 1984, pp. 89-133.
[6] Chellas, B., *Modal logic: an introduction*, Cambridge University Press, 1984.
[7] Goranko, V. and S. Passy, Using the universal modality: gains and questions, *Journal of Logic and Computation*, 1992, 2: 5-30.
[8] Katsuno, H. and Mendelzon, A. O., Propositional knowledge base revision and minimal change, *Artificial Intelligence*, 1991, 52: 263–294.
[9] Levesque, H. J., All I know: a study in autoepistemic logic, *Artificial Intelligence*, 1990, 5: 263-309.

---

[7]For example, the axiom $I\varphi \to B\varphi$ which says that if the agent is informed that φ then she believes that φ.

Are the Foundations of Computer Science Logic–Dependent?

WALTER A. CARNIELLI AND FRANCISCO A. DORIA

ABSTRACT. We formalize computer science (Turing machine theory) within a standard axiomatic framework like Peano Arithmetic or Zermelo–Fraenkel set theory and argue that the manifold examples of unprovable sentences that appear in those axiomatizations show that such an axiomatics, or any similar kind of axiomatics, is unable to prove some naïvely simple facts in the theory. We then change the logical background to a nonclassical, paraconsistent one, and obtain a theory which is close to quantum computing. We conclude with an evaluation of possible "natural" frameworks for the axiomatics of computer science.

1 Prologue

Gödel's incompleteness phenomenon is Janus–faced. One of its aspects is a negative one: there are many nontrivial mathematical sentences that turn out to be undecidable with respect to strong axiomatic systems, like *ZFC* (Zermelo–Fraenkel theory, plus the Axiom of Choice) or even *ZFC* plus some large cardinal axiom. On the other side, the Gödel phenomenon opens up the unsuspected wealth and weirdness of the world of nonstandard models or of forcing–extended models — just as a brief aside, which should be the meaning of a non–Cantorian, exotic ("fake") 4–space? (A recent reference on fake 4–manifolds is Scorpan [33].)

How does the Gödel phenomenon affect computer science? In order to examine that question, we will focus on two approaches.

We will first deal here with examples of the incompleteness phenomenon for versions of computer science which are axiomatized within classical theories, that is to say, first–order theories which include arithmetic and are based on the classical first–order predicate calculus. We then add some extra spice to our considerations, and consider a non–classical, paraconsistent, axiomatic framework for computer science.

What do we learn? By far, the most damaging difficulty for axiomatics of computer science which are based on first–order theories like Peano Arithmetic (*PA*) or *ZFC* is the fact that all such theories do not recognize naïvely total recursive functions beyond a certain growth rate. Moreover, as we show in Section 6 this inability is tied to fundamental, inner workings of the theory itself, since if such a theory were to prove that a certain (partial recursive) function we have noted F is total (see Definition 2.6) then it would prove its own consistency. On the other hand, when we change the logical framework where we are doing our axiomatization (Section 7), we may get a theory which is very close to quantum computation, as in the example that we discuss.

We may summarize our results in this paper in a slogan:

Axiomatized computer science is logic–dependent.

2   A "natural" axiomatics for computer science

We begin with some comments that motivate our main concepts. Our goal here is to discuss possible axiomatic treatments for computer science, as embodied in the theory of Turing machines. (For a background on Turing machines and computability see [22].)

Remark 2.1. Our main tool will be axiomatic systems like the one described below: these will be consistent axiomatic theories (noted S or T) based on a first–order classical predicate calculus that moreover

- have a recursively enumerable set of theorems

- include Peano Arithmetic

- have a model where the arithmetic segment is standard. □

We will deviate as needed from this blueprint; exceptions will be made fully explicit whenever we change the above conditions.

Example 2.2. Suppose that there is a Turing machine that enumerates all theorems in a theory S. Since S includes Peano Arithmetic, its set of theorems will be recursively enumerable but not recursive, due to Gödel's incompleteness. Start our enumerating machine and gradually build up a list of theorems of S. Separate those that formally assert "recursive function f_e, which has Gödel number e, is total". Place them in a second list. These are the S–provably total recursive functions.

Remark 2.3. It is possible to have a well–defined function which however isn't recursive. The best example is the Busy Beaver function; another example is the counterexample function to the $P = NP$ hypothesis [15]. □

Now construct by the usual diagonalization procedure a function F over that second list. That function is immediately seen to be recursive (the diagonal procedure over that second listing gives a simple algorithm for it) and total. Yet it cannot be S–provably total.

That it to say, no theorem of S has the form "F is total". Also, as S is sound — a fact that is a consequence of its arithmetic portion having an interpretation in the standard model — the negation of that sentence, "F isn't total" cannot appear among the theorems of S. Therefore such a sentence is independent of the axioms of S. $\square$

(On that function F see Definition 2.6 and Section 6.) Let us clarify a few points in our example:

- A Turing machine M_S that enumerates all theorems of S can operate as follows. First, start a listing of all well–formed formulae in the language of S. Then consider each integer, $1, 2, 3, \ldots, k, \ldots$ and see if each such integer is the Gödel number of the proof of some well–formed formula in the language of S.

- If so, pick it up from the previous listing and place it in a new list, which will be the enumeration of theorems of S.

"Reasonable" axiomatics

Quotation marks are required here, as this is one of the many possibilities, albeit a very reasonable one. We can take as a starting point the following: Turing machine theory can be seen as arithmetic under disguise. Specifically, it can be seen as a privileged domain within arithmetic: the theory of Diophantine equations. So, we "translate" Turing machine theory into the theory of Diophantine equations [18] in order to axiomatize it.

Of course we always keep in mind the usual picture: Turing machines [32] are introduced via their tables which specify elementary operations like moving the head to the left or to the right, erasing or printing a symbol on the tape's square under the head, and so on. However it isn't clear how to reduce that naïve, even if fertile, picture, to an axiomatic treatment that encompasses all of its clear–cut features, and so we will try at first to give an alternative, strictly formal, definition for Turing machines via Diophantine equations, and we will use that picture to stress that we are dealing with a formal construction that moreover has a clearcut, naïve interpretation.

Let $p_U(k, x_0, x_1, x_2, \ldots, x_k)$ be a universal Diophantine polynomial [18, 36] which we suppose to be fixed. We define the partial recursive function $\{e\}$ of Gödel number e that acts on natural number m as its input and has natural number n as its output as:

Definition 2.4.

$$[\{e\}(m) = n] \leftrightarrow_{\text{Def}} [\exists x_0, \ldots, x_k \in \omega \, p_U(\langle e, m, n \rangle, x_0, x_1, \ldots, x_k) = 0]. \square$$

(ω is the set of natural numbers, $\langle x, y, z \rangle = \langle x, \langle y, z \rangle \rangle$ and $\langle \ldots \rangle$ is the usual pairing function; for the computation of e and the construction of p_U see [18, 32, 36].)

Partial recursive function (or the corresponding Turing machine) $\{e\}$ is given by the preceding definition. Of course there is a relation between (the abstract objects) Turing machines and concrete objects of our real world such as computers (which can be best seen as realizations of finite automata) but we restrict our attention to the mathematical object characterized above.

Axiomatic treatment

We use a Suppes predicate to axiomatize the theory of Diophantine equations within some formal system that includes Peano Arithmetic, such as Peano Arithmetic (*PA*) itself, or Zermelo–Fraenkel set theory with or without the Axiom of Choice (*ZFC*, *ZF*). For the explicit construction see [12]; other examples of the application of Suppes predicates in the axiomatic treatment of several theories are to be found in [11].

A first caveat

The next discussion and examples [14, 17] ponder well–known phenomena from the theory of fast–growing recursive functions in the light of our main theme, that is, axiomatization of *CS* (computer science) within *ZFC* or any other standard axiomatics for set theory, based on the classical predicate calculus, with a recursively enumerable set of theorems, and which extend enough of arithmetic to adequately handle Definition 2.4.

The first example we now give starts from a well–known property of intuitively total recursive functions when framed within usual axiomatic systems: *not every intuitively total recursive function can be proved so within a given axiomatic framework.* That is to say, the axiomatic system isn't able to clearly "see" as total functions some recursive functions that are intuitively total.

The point is [13, 21]:

Remark 2.5. Suppose that we are given a prescription so that, for any integer n, we can compute a finite set of numbers S_n. Then put: $F^*(n) = \max S_n + 1$. Most professional mathematicians would immediately agree that F^* is both computable and total. But is that really so? $\square$

Can we construct that function within *ZFC*? We surely can give a Turing program, or a recursive–function definition for it relative to the recursivity of S_n — but we won't then be able to prove that similar functions are total, in the general case, as it is well–known.

Let's take a closer look at a function F as the ones naïvely described in Remark 2.5:

Definition 2.6. For each n, $F(n)$ is the sup of those $\{e\}(k)$ such that:

1. $k \leq n$.

2. $[\mathrm{Pr}_{\mathrm{ZFC}}([\forall x \, \exists z \, T(e, x, z)])] \leq n$.

    That is, there is a proof of $[\{e\}$ is total$]$ in ZFC whose Gödel number is $\leq n$. (For sentence φ, $[\varphi]$ is its Gödel number; T is Kleene's predicate.) $\square$

F^* in Remark 2.5 is always intuitively total, and so is the particular version F in Definition 2.6. The fact that F is intuitively total follows from the formal Löwenheim–Skolem theorem (or "reflection theorem" in the sense of set theory — [28], p. 133ff — not to be confused with Feferman's Reflection Principles; see [7, 23, 24]), which ensures that, for some particular value of n we will be able to construct a model for a collection of cardinality $\leq n$ of sentences of ZFC, and so to compute $\mathsf{F}(n)$. The analogy is: we can *individually* check that, for each natural number n, $\mathsf{F}(n)$ exists and can be computed. However we cannot "join" all those results together to show that F is total.

Granted those remarks:

Proposition 2.7. We can explicitly compute a Gödel number e_{F} so that $\{e_{\mathsf{F}}\} = \mathsf{F}$. $\square$

F is such that:

Proposition 2.8. The formal sentence $\forall x \, \exists z \, T(e_{\mathsf{F}}, x, z)$ cannot be proved or disproved from the axioms of ZFC, supposed consistent, that is,

$$\forall x \, \exists z \, T(e_{\mathsf{F}}, x, z)$$

is independent of the ZFC axioms. $\square$

(We can naïvely suggest that Proposition 2.8 results from the same kind of obstacle that hinders the extension of the Löwenheim–Skolem theorem, from finite subsets of axioms of ZFC to the whole theory.) The preceding propositions for ZFC can be extended to any theory like our S.

So F cannot be proved to be total in ZFC. In other words: this means that even if F is intuitively total (that is to say, it holds of the standard model for arithmetic), there must exist a model for ZFC with a nonstandard part where $[\mathsf{F}$ isn't total$]$ is verified. (This result originates in Kleene's 1936 paper [26]; see also [27], p. 257.)

We will now require Feferman's Σ_1–soundness reflection principle. Naïvely, a system S is Σ_1–sound if, given a proof of $\exists x \, P(x)$, where P is a predicate with values in the natural numbers, then there is some natural number n so that $P(n)$ is true. That concept was formalized by Feferman as one of his reflection principles [23]. Then there is an important result that connects F to Feferman's Σ_1–reflection principles [23, 24]:

Recall that a reflection principle in Feferman's sense [7, 23, 24] has (roughly) two aspects:

- If we prove in S that there is the Gödel number of a proof of φ, then φ.

- If we obtain for each n, $\mathrm{Pr}_S[\varphi(n)] \to \varphi(n)$, then we can collect all those under a universal quantifier, that is, there is a form of the ω–rule at work here.

Σ_1–soundness is such a principle restricted to Σ_1-sentences φ. The chief result is:

Proposition 2.9. $S \vdash$ [F is total] $\leftrightarrow$ [S is Σ_1–sound]. $\square$

For the proof see Section 6.

3 Can we handle arbitrary infinite sets of poly machines in ZFC?

The next two examples have been first discussed in [14, 17]. We may argue that the results about F deal with fast–growing functions, which are objects quite far from the everyday realm of programs and concrete computers. So we move on to another example, that bears on concepts related to the *P vs. NP* question.

The main point here is: the concept of an *infinite set of poly machines* may be quite difficult to handle within an axiomatic system, even a strong one like *ZFC*.

A *poly machine*, or a polynomially time–bounded Turing machine is a total Turing machine M_m that on a binary input x of length $|x|$ outputs a binary word y after less than $p_m(|x|)$ operation steps, p_m a polynomial kept fixed for M_m. Consider the following example, which is an application of the previous results (the trick used is due to [6]):

Example 3.1. Let $\langle m, a, b \rangle$ denote a Turing machine M_m coupled to a clock $C_{(a,b)}$ — another Turing machine — that stops M_m after it executes $|x|^a + b$ cycles over input x of length $|x|$. We note that pair $\langle M_m, C_{(a,b)} \rangle$. We agree that if clock $C_{(a,b)}$ interrupts the operation of M_m then M_m outputs 0 and stops.

Consider the set $A = \{(m, F(a), F(b)) : m, a, b \in \omega\}$, F as in Definition 2.6. Each individual machine $\langle m, F(a), F(b) \rangle$, m, a, b integers like $0, 1, 2, \ldots$, in it is certainly a poly machine, but we cannot prove in *ZFC* that the whole set only contains poly machines. It is in fact undecidable: "A is a set of poly machines" holds of the standard integers, but doesn't hold in some models for *ZFC* with a nonstandard arithmetic part. $\square$

We have shown:

Proposition 3.2. The sentence "A is a set of poly machines" is independent of *ZFC*, supposed consistent. $\square$

Now we may ask: does it help if we add some strong axiom — here noted X — to *ZFC*, so that the resulting theory *ZFC* + X, supposed consistent, proves the consistency of *ZFC* itself? (Think of *X*, say, as some large cardinal axiom, for instance.) No. For there is a set *A′* so that:

Proposition 3.3. The sentence "*A′* is a set of poly machines" is independent of *ZFC* + *X*, supposed consistent.

Proof: It suffices to obtain the function F′ for *ZFC* + *X*, according to the blueprint in Definition 2.6. □

This example is crucial, because among other consequences, any (tentative) proof of the hypothesis *P* < *NP* must include a step that says, "for *every* poly machine, it will give a wrong answer at least once", and a sentence like "*A* is a set of poly machines" is undecidable within consistent *ZFC*, even if we can pick up infinitely many of its elements and individually show each of them to be a poly machine. This raises the following question: will any system like the *S* we have been considering here (see Remark 2.1), be able to prove *P* < *NP* ? The preceding discussion is quite sobering.

4 More examples of incompleteness for computer science in *S*

We now quote two recent versions of some results by Hartmanis and Hopcroft [17, 25]. They start from a formal theory that:

- includes set theory (more precisely, they ask that the theory be of "sufficient power to prove the basic theorems of set theory").

- Also the theory must allow for predicate symbols *P, Q, . . .*.

- It has a recursively enumerable set of theorems.

- Its theorems are "intuitively true". This is too strong and also vague for the whole of set theory with the axiom of choice — for instance, is the Banach–Tarski theorem intuitively true? So, we take this third requirement to be the *arithmetically soundness* condition, that is, *S* must have a model with standard arithmetic.

We can agree that *S* (see Definition 2.1) adequately fits the picture. We now endow this version of *S* with the *ι*–symbol.

Remark 4.1. We notice here the importance of theories like *S*: most results like those described here extend to any such theory, or, in other words, are sort of insensitive to the axiomatic framework where they are constructed, but for the fact that one requires arithmetic to formulate them. □

According to Hartmanis and Hopcroft [25]:

- The first undecidability result in [25] has to do with the BGS relativization result [6]. The BGS result says that there are recursive oracles $A, B, A \neq B$, so that one has (for the relativized versions) $P^A = NP^A$ and $P^B < NP^B$. Hartmanis and Hopcroft show that there is an oracle C so that the assertion $P^C = NP^C$ is undecidable with respect to the axioms of S. Their proof is by a diagonal argument; we have used here an alternative, quite general argument.

- Then they show that there is an algorithm A (a Turing machine) of which it is true that for input x it runs in time x^2, but so that the formal version of the sentence "$\mathsf{A}(x)$ runs in time $t_\mathsf{A} < 2^x$" is undecidable in S.

We present here a general argument for these results which is based on a version of Rice's theorem in fragments of set theory with the ι-symbol (which we suppose to be available) and which stems from an idea first used in da Costa and Doria [11, 16]. More precisely:

Remark 4.2. For consistent S, let Consis S be the usual formal sentence that asserts the consistency of S; $S \nvdash$ Consis S and $S \nvdash \neg$Consis S. Let ξ, ζ be terms in the language of S, so that for some predicate P in the language of S, $S \vdash P(\xi)$ while $S \vdash \neg P(\zeta)$. Then:

$$\lambda = \iota_x\{[\text{Consis } S \wedge x = \xi] \vee [\neg\text{Consis } S \wedge x = \zeta]\}.$$

$S \nvdash \lambda = \xi$ and $S \nvdash \lambda = \zeta$, but if $\mathbf{N} \models S$ and $\mathbf{N}$ has a standard arithmetic part, then $\mathbf{N} \models \lambda = \xi$. Moreover, $S \nvdash P(\lambda)$ and $S \nvdash \neg P(\lambda)$, while $\mathbf{N} \models P(\lambda)$. $\square$

- For the first result, put oracles A, B as $\xi = A$ and $\zeta = B$. Then C:

$$C = \iota_x\{[\text{Consis } S \wedge x = A] \vee [\neg\text{Consis } S \wedge x = B]\}$$

is proved to be a recursive oracle in S, but $S \nvdash C = A$ and $S \nvdash C = B$. So, $S \nvdash P^C = NP^C$ and $S \nvdash P^C < NP^C$.

- For the second result, if P is a polynomial Turing machine, and E is an exponential Turing machine, then:

$$\mathsf{M} = \iota_x\{[\text{Consis } S \wedge x = \mathsf{P}] \vee [\neg\text{Consis } S \wedge x = \mathsf{E}]\}$$

is such that S proves M to be a total Turing machine which has an exponential time bound which cannot be improved in S, but such that it is true of $\mathbf{N}$ that it is time–polynomial.

We may also use the term:

$$\lambda = \iota_x\{[x = \xi \wedge \beta = 0] \vee [x = \zeta \wedge \beta = 1]\}.$$

β [16] is an algebraic expression which can be explicitly constructed such that $S \not\vdash \beta = 0$ and $S \not\vdash \beta = 1$, while $\beta = 0$ holds of the standard model for arithmetic; see the references for details.

Remark 4.3. The second result can be easily extended to a result by Loo [29]: that there is a Turing machine of arbitrarily large complexity when "seen" from within the standard model, but which is only polynomial in an adequate nonstandard model. An analogous example was discussed in 2002 by one of the authors [20]. An alternative argument that clarifies the above discussion stems from the following observation: consider

$$p(x) = 1 + x + \frac{x^2}{2!} + \frac{x^3}{3!} + \ldots + \frac{x^K}{K!},$$

where K is a nonstandard positive integer.

If $\mathbf{N} \models S$, where $\mathbf{N}$ has a standard arithmetic part, while also $\mathbf{M} \models S$, and $\mathbf{M}$ has a nonstandard arithmetic part, with $K \in \mathbf{M}$, K nonstandard, then we notice that the restriction $p(x)|\mathbf{N}$ is an exponential, while $p(x) \in \mathbf{M}$ is a polynomial. $\square$

5 Function F and function G

Now recall that a consistent system S is ω–consistent if S doesn't simultaneously prove $\exists x\, P(x)$ and $\neg P(0), \neg P(1), \neg P(2),\ldots$. ω–consistency implies consistency, while the converse isn't true. Consider PA which is supposed consistent, and add a new symbol ζ to its alphabet, with the (new) predicate $N(x)$ which should intuitively mean, "x is a natural number".

Then, by the usual compacity argument the system $\text{PA} + \zeta \neq 0 + \zeta \neq 1 + \ldots + N(\zeta)$ is consistent. Also, from $N(\zeta) \rightarrow \exists x\, N(x)$ one proves $\exists x\, N(x)$. We then get the ω–inconsistent system $\text{PA} + \zeta \neq 0 + \zeta \neq 1 + \ldots + \exists x\, N(x)$, which however is consistent.

Our F gives rise to such a situation:

Remark 5.1. Notice that given function F as above in Definition 2.6, we can see that [F is total] is independent of S, but the structure of that function for the nonstandard models where $\neg$[F is total] holds isn't clear at all. We immediately get an ω–inconsistency result, for

$$S + \neg[\text{F is total}] + \exists y\, (\text{F}(0) = y) + \exists y\, (\text{F}(1) = y) + \ldots$$

is a consistent theory. Thus Σ_1–unsoundness (or equivalently $\neg$[F is total]) implies ω–inconsistency. $\square$

We can say that for very large values of its argument, F ceases to be defined in such ω–inconsistent systems.

Construction of function G

But nevertheless we can ask: is there a partial recursive function G with a clearly infinite domain and yet with a behavior similar to that of F? We answer this question in the affirmative through the following:

Proposition 5.2. There is a partial recursive function G so that:

1. If $\mathbf{N} \models S$ and has standard arithmetic, then $\mathbf{N} \models F = G$.

2. $S + [F \text{ is total}] \vdash [\text{Domain G is infinite and G is increasing}]$.

3. $S + [F \text{ is total}] \nvdash [G \text{ is total}]$ and $S + [F \text{ is total}] \nvdash \neg[G \text{ is total}]$.

4. If $S_{(\alpha)}$, $\alpha > 1$ an ordinal, is in the Turing–Feferman hierarchy over S extended by Σ_1–soundness reflection principles, then we can choose α as high as we wish in that hierarchy $\leq \omega$, so that $S_{(\alpha)} \vdash [G \text{ is total}]$, but such that $S_{(\beta)} \nvdash [G \text{ is total}]$, $\beta < \alpha$.

Proof: As we have supposed that S has a model $\mathbf{N}$ with standard arithmetic, then so does theory $S' = S + [F \text{ is total}]$:

- We can explicitly obtain a Diophantine polynomial $p(x_1, \ldots, x_k)$ so that:

  1. $S' \nvdash [\forall x_1, \ldots p(x_1, \ldots) > 0]$ while $\forall x_1, \ldots p(x_1, \ldots) > 0$ holds of $\mathbf{N}$.

  2. $S' \nvdash [\exists x_1, \ldots p(x_1, \ldots) = 0]$ and for some model $\mathbf{M}$ with nonstandard arithmetic, $\mathbf{M} \models \exists x_1, \ldots p(x_1, \ldots) = 0$.

- Define: $\zeta = a_1$, given that for some (necessarily nonstandard) model $\mathbf{M}$,

$$\mathbf{M} \models [p(a_1, a_2, \ldots, a_k) = 0].$$

  (We must impose some uniqueness condition on $\langle a_1, \ldots, a_k \rangle$.)

By construction, for all models of S', [F is total].

Remark 5.3. We now informally describe an algorithm for the function we are looking for. Let h be such that S proves h to be total and strictly increasing. Put, for G:

- If $m < \zeta$, $G(m) = F(m)$.

- If $m > \zeta$ and $n = h(m)$, then $G(n)$ is undefined.

- If $m > \zeta$ and $n \neq h(m)$, then $G(n) = F(n)$.

More precisely: If $m \in \{x | x \in \mathfrak{I}(h)$ and $x \geq \zeta\}$, then $G(n)$ is undefined. Or, If $m \in \{x | x \notin \mathfrak{I}(h)$ and $x \geq \zeta\}$, then $G(n) = F(n)$. $\mathfrak{I}$ is of course the image of ....

So G will always have an infinite domain whenever [F is total] holds. And due to the dependence of G on ζ, it cannot be proved total even in a strong theory such as $S +$ [F is total], that is, $S +$ [S is Σ_1–sound]. For the last assertion, given the hierarchy over S plus Σ_1–sound reflection principles, it suffices to choose an adequate p so that $\forall x, \ldots p(x, \ldots) > 0$ is only proved by $S_{(\alpha)}$, but not by any $S_{(\beta)}$, $\beta < \alpha \leq \omega$. (For the Turing–Feferman theorem see [7, 23, 24].) $\square$

We can give a formal expression for G with the help of the ι-symbol, which we then add to our formal background — S and the required extensions.

- We can write down the algorithm for a Turing machine that never stops over any input n. Let $\{e_0\}$ be that machine.

- Consider h as in Remark 5.3. h is a S–provably total recursive function. Then write:

$$H(n) = \iota_x\{[(x = F(n)) \wedge (n \notin \text{Image}(h))] \vee$$

$$\vee [(x = \{e_0\}(n)) \wedge (n \in \text{Image}(h))]\}.$$

Therefore S proves that such a function equals F (which is total for all models of S), but for the values of h, where it is undefined. As the image of h doesn't exhaust all of ω, its complement is infinite, and H will have an infinite domain.

- Now consider the first Diophantine polynomial p above together with $\zeta = a_1$. If we write $\forall\!\!\!\!/x\, p > 0$ for $[\forall x_1, \ldots p(x_1, \ldots) > 0]$, and similarly $\exists x\, p = 0$, then consider the next expression that can be seen to be in the language of S (possibly extended for definitions):

$$G(n) = \iota_x\{[(\forall\!\!\!\!/x\, p > 0) \wedge (x = F(n))] \vee$$

$$\vee [(\exists x\, p = 0) \wedge [(n < \zeta) \wedge (x = F(n))] \vee$$

$$\vee [(n \geq \zeta) \wedge (x = H(n))]]\}.$$

ζ is as above. This construction originates in the (symbolic) form $(1 - \beta)X + \beta X'$, which was used in [11]; for the expression above see the construction in [16], p. 34. It is also akin to the construction in the result known as Kreisel's Lemma [17]. $\square$

6 Σ_1–soundness and F

The next results show that F sort of codes, or represents, deep facts about the structure of the axiomatic system where it is defined. Namely, if it is total, then the axiomatic system T w.r.t. which F is defined, is consistent. We also show that F is total if and only if T is Σ_1-sound.

Remark 6.1. The *local reflection principle* Rfn(T) for theory T is:

$$\text{Pr}_T([\varphi]) \rightarrow \varphi,$$

(that is, if there is a proof for φ, we actually find it in T); and the (first) *uniform reflection principle*, RFN(T) is:

$$[\forall x \, \text{Pr}_T([\varphi(\dot{x})])] \rightarrow [\forall x \, \varphi(x)],$$

all φ with only x free, and $\dot{x}$ standing for the x that can be represented by actual constants in T. This means that once one can list instances $\varphi(0)$, $\varphi(1)$, ... (which are derivable due to the first supposition in the Reflection Principle) for all nameable n, then a restricted application of the ω–rule leads to the principle.

For Σ_1–soundness one restricts φ to all $\exists x \, \varphi(x)$, φ primitive recursive; the corresponding restricted reflection principle is noted $\text{RFN}_{\Sigma_1}(T)$. $\square$

For T as PA, ZFC, and first–order extensions as considered here:

Corollary 6.2. $T \vdash \text{RFN}_{\Sigma_1}(T) \rightarrow \text{Consis}(T)$.

Proof:

- Suppose $T \vdash \neg(0 = 0)$.

- Therefore, $T \vdash \text{Pr}_T[\neg(0 = 0)]$.

- Given Σ_1–soundness, as this is a trivial Σ_1-sentence,

$$\text{Pr}_T[\neg(0 = 0)] \rightarrow [\neg(0 = 0)].$$

- By contraposition,

$$\{\neg[\neg(0 = 0)]\} \rightarrow \neg\text{Pr}_T[\neg(0 = 0)].$$

- Or, $(0 = 0) \rightarrow \neg\text{Pr}_T[\neg(0 = 0)]$.

- However, $T \vdash (0 = 0)$. Then follows:

- $\neg\text{Pr}_T[\neg(0 = 0)]$. A contradiction.

(This proof is due to N. C. A. da Costa.) $\square$

Σ_1–soundness is equivalent to [F is total]

We argue here for *ZFC*, since we want to have as much "elbow room" as possible. However we could have argued for any theory like the one we have denoted by *S*. If we allow for the abuse of language that subsumes the infinitely many sentences of the Reflection Principle in our formulation as a single one in the statement of our result:

Lemma 6.3. $ZFC \vdash [\text{F is total}] \leftrightarrow [\,ZFC \text{ is } \Sigma_1\text{–sound}\,]$.

Proof: Recall that $[\text{F is total}] \leftrightarrow \forall x \exists z\, T(e_\text{F}, x, z)$, where — here — T is Kleene's T predicate and e_F is a Gödel number for F.

($\Leftarrow$). We first prove: assuming $RFN_{\Sigma_1}(ZFC)$, then $\forall x \exists z\, T(e_\text{F}, x, z)$. Given the (recursively enumerable) infinite set of conditions $RFN_{\Sigma_1}(ZFC)$, for T we get:

$$[\forall x\, \text{Pr}_{ZFC}([\exists z\, T(e_\text{F}, \dot{x}, z)])] \to [\forall x \exists z\, T(e_\text{F}, x, z)].$$

Now, for each *ZFC* constant $\dot{n}$ we have that:

$$ZFC \vdash [\exists z\, T(e_\text{F}, \dot{n}, z)].$$

Then there are proofs of each of these sentences, for each $\dot{n}$. Therefore we conclude: $[\forall x\, \text{Pr}_{ZFC}([\exists z\, T(e_\text{F}, \dot{x}, z)])]$, as $\dot{x}$ only ranges over the constants. From the corresponding restriction of $RFN_{\Sigma_1}(ZFC)$ we conclude that:

$$[\forall x \exists z\, T(e_\text{F}, x, z)]$$

holds, by modus ponens.

($\Rightarrow$). For the converse: given $\forall x \exists z\, T(e_\text{F}, x, z)$, we deduce $RFN_{\Sigma_1}(ZFC)$. (We will have to deduce each instance of the Reflection Principle, for each 1–variable $\exists \mathbf{z}\, \psi(\mathbf{z}, x)$, ψ primitive recursive — for the meaning of $\mathbf{z}$ see below.)

Recall that given each e we can explicitly construct a Diophantine polynomial

$$p_e(\langle x, y \rangle, z_1, z_2, \ldots, z_m)$$

so that:

$$[\forall x \exists z\, T(e, x, z)] \leftrightarrow [\forall x \exists y, z_1, \ldots, z_m\, p_e(\langle x, y \rangle z_1, \ldots, z_m) = 0].$$

(Of course $[\exists z_1, \ldots z_m\, p_e(\langle x, y \rangle z_1, \ldots, z_m) = 0] \leftrightarrow [y = \{e\}(x)]$. Since we aren't using a universal equation, m may depend on e.) We will abbreviate the $z_1, \ldots, z_m$ by $\mathbf{z}$.

1. Now, if [F is total], then, for each $n \in \omega$ we have that $F(n)$ is the sup of all $\{e\}(k)$ so that $k \leq n$ and:

$$[\mathrm{Pr}_{\mathrm{ZFC}}([\forall x \, \exists z \, T(e, x, z)])] \leq n.$$

2. Since n is explicitly given, it is a bound on the Gödel number of the proof. Therefore we can also obtain a $n' > n$ so that:

$$[\mathrm{Pr}_{\mathrm{ZFC}}([\exists z \, T(e, x, z)])] \leq n'.$$

3. Or, for another n'', $[\mathrm{Pr}_{\mathrm{ZFC}}([\exists y, \mathbf{z} \, p_e(\langle x, y \rangle, \mathbf{z}) = 0])] \leq n''$.

   This follows from:

   - Every recursive function $\{e\}(n) = m$ can be represented by a predicate $F_e(n, m)$.

     (The algorithm to produce F_e given e is in Machtey and Young [30], p. 126ff.)

   - Given $F_e(n, m)$ we can use the procedure described in Davis' paper on Hilbert's 10th Problem [18] to get a polynomial p_e out of F_e.

4. Since n'' is explicitly given, we can then recover proofs in *ZFC* of:

$$[\exists y, \mathbf{z} \, p_e(\langle x, y \rangle, \mathbf{z}) = 0],$$

   all e under the specified conditions.

5. We then establish that *ZFC* proves, for all such e,

$$\mathrm{Pr}_{\mathrm{ZFC}}([\exists y, \mathbf{z} \, p_e(\langle x, y \rangle, \mathbf{z}) = 0]) \rightarrow [\exists y, \mathbf{z} \, p_e(\langle x, y \rangle \mathbf{z}) = 0].$$

6. We now add the universal quantifier for x, and as it distributes over $\rightarrow$,

$$[\forall x \, \mathrm{Pr}_{\mathrm{ZFC}}([\exists y, \mathbf{z} \, p_e(\langle x, y \rangle, \mathbf{z}) = 0])] \rightarrow [\forall x \, \exists y, \mathbf{z} \, p_e(\langle x, y \rangle \mathbf{z}) = 0].$$

7. This will of course also hold for (due to the implication's properties):

$$[\forall x \, \mathrm{Pr}_{\mathrm{ZFC}}([\exists y, \mathbf{z} \, p_e(\langle \dot{x}, y \rangle, \mathbf{z}) = 0])] \rightarrow [\forall x \, \exists y, \mathbf{z} \, p_e(\langle x, y \rangle \mathbf{z}) = 0].$$

8. Finally the $\exists \mathbf{w} p_e$ provide an enumeration of all Σ_1-relations in *ZFC* of interest for Σ_1–soundness. Notice that in *ZFC*, f and F recursive,

$$[\text{f is total}] \leftrightarrow \{[\text{F is total}] \rightarrow [\text{F dominates f}]\}.$$

9. To show that the enumeration is exhaustive: suppose that for some p.r. ψ one has:

$$\forall x \, \mathrm{Pr}_{ZFC}(\exists y \, \psi(\dot{x}, y)),$$

and that moreover the following sentence is proved:

$$[\forall x \, \mathrm{Pr}_{ZFC}(\exists y \, \psi(\dot{x}, y))] \rightarrow \forall x \, \exists y \, \psi(x, y).$$

10. Put $f_\psi(x) = \min_y \psi(x, y)$. Then:

$$[\forall x \, \exists y \, (f_\psi(x) = y)] \leftrightarrow [\forall x \, \exists y \, \psi(x, y)].$$

11. That is, f_ψ is *ZFC*–provably total recursive, and therefore falls into the preceding case. $\square$

Remark 6.4. The preceding result shows the essential inner connections between the inability of an axiomatic theory like S to "see" the totality of functions like F and beyond, and the usual formal sentences that assert the consistency of S itself. That is to say, the obstructions we have to face when trying to prove the totality of F in S have to do with the impossibility of proving the consistency of S itself with its own tools.

The proof of the preceding result for *PA* can be found in the original Paris–Harrington paper [31]. $\square$

7 Paraconsistent Turing machines and quantum computation

We will just sketch the main concepts and ideas here, and leave details to be checked in the references. The main contention of this paper is that the way we conceive computation theory, or perhaps more specifically, the way we conceive computation procedures, turns out to be logic–relative and logic–dependent. The preceding examples were intended to stress that fact. We now turn to a different, potentially fruitful, approach.

A new model of *paraconsistent Turing machines* (**PTMs**) and its relation to quantum computation was studied in [3], where one shows how the classical Turing machines model can be enhanced by the change of the underlying logic from classical logic to a paraconsistent one. By conveniently interpreting **PTMs** as standard models of quantum computation (quantum Turing machines and the equivalent quantum circuits), some algorithms surprisingly close in spirit to the classical Turing machine formulation, but possessing quantic efficiency, can be designed to solve the so-called *Deutsch's problem* and *Deutsch–Jozsa problem*.

In a general form, such problems (firstly introduced in [19]) consist in determining, for a function $f : \{0,1\}^n \rightarrow \{0,1\}$ whether f is *constant* $f(x) = f(y)$ for all vectors $x, y \in \{0,1\}^n$) or *balanced* (the number of vectors $x \in \{0,1\}^n$ such that $f(x) = 0$ is equal to the number of vectors $x \in \{0,1\}^n$ such that $f(x) = 1$).

In case $n = 2$ there are obviously two constant functions and two balanced functions; classically we need at least two steps to evaluate f two times (in the entry 0 and in the entry 1) to determine if f is constant or balanced. Quantically, however, we can simultaneously evaluate $f(0)$ and $f(1)$ by means of a single application of a quantum operator that evaluates the f), and determine if f is constant or balanced in a single step. (See e.g. [10], or [3] and [4] for detailed explanations.)

In the general case, a quantum circuit is able to solve deterministically the Deutsch-Jozsa problem by performing a single application of a quantum operator, while the classical view requires (in the worst case) $2^{n-1} + 1$ steps. The classical algorithm has exponential complexity, while the quantum algorithm to solve the same problem has polynomial complexity.

But what is crucial for our arguments here is that standard models of quantum computation (quantum Turing machines and quantum circuits) can be simulated by paraconsistent Turing machines, as we shall see in the next section.

From classical Turing machines to quantum Turing machines

The achievement of this insight into relating quantum computation with logic-dependent notions of computation as the **PTM**s inaugurated the main idea that computation may be advantageously seen as *logic-relative*.

Although the first definition of **PTM**s is due to [1] and [5] and has been better developed in logical terms in [2], the potentialities of paraconsistent models of computation as simulating quantum computing were only fully perceived in [3].

Already in the original definition of what became known as a Turing machine (see [37]), the idea of the distinction between *deterministic Turing machines* (**DTM**s) and *non-deterministic Turing machines* (**NDTM**s) was present.

However, outputs of **NDTM**s cannot subsist simultaneously without leading to deductive trivialization of the underlying theories, since classical deduction lumps together contradictions (spawned by the non-determinism) and deductive triviality.

This is unfortunate from the point of view of a serious theory of quantum computation, if it intends to be a theory of computation based upon basilary principles of quantum mechanics such as superposition of states, entangled states and interference.

In 1994, Peter Shor proposed a quantum algorithm (using the model of quantum circuits) devoted to factoring numbers in polynomial time (see [34] and [35]), a problem of crucial importance for cryptography for which no classical algorithm with polynomial complexity is known. Since Shor's algorithm the research in quantum computing grew drastically, and there is a widespread need for alternative models of quantum computation.

Paraconsistent machines

The idea of one such quantum device is thus simply to allow for ambiguous situations (perplexing as they can be from the classical standpoint) wherein the machine simultaneously executes several instructions, engendering a multiplicity of states, a multiplicity of positions and a multiplicity of symbols in certain cells of the tape.

The strategy to avoid deductive collapsing consists in changing the underlying logic behind such theories from classical to paraconsistent logic (leaving the axioms intact), avoiding trivialization and leading to new Turing machine models able to interpret and to take profit of the consequences of such theories.

In principle many solutions can be chosen, selecting underlying logics from the wide family of propositional paraconsistent logics known as 'logics of formal inconsistency' (**LFI**s) (cf. [8]). **LFI**s are paraconsistent logics that internalize the metatheoretical notions of consistency and inconsistency at the object language level, and are also characterized by preserving the positive fragment of propositional classical logic.

In [3] the first-order paraconsistent logic $LFI1^*$ was chosen (treated in detail in [9], see also [8]). Baked up by such a careful logic which, it is opportune to clarify here, provides robust reasoning over uncertain and contradictory theories but *does not* beget any logical contradiction, we may now define **PTM**s:

Definition 7.1. The **paraconsistent Turing machines** are Turing machines defined by means of finite collections of instructions such that:

- We allow for contradictory instructions (viz. instructions the same initial symbols $q_i s_j$ are allowed);

- The machine executes simultaneously all possible instructions in the face of ambiguity. That procedure engenders multiple states as well as a multiplicity of positions and as multiplicity of symbols in some cells of the tape;

- Outputs are to be kept stored for further consideration;

- If the computation halts, each tape cell may contain multiple symbols and any choice of these symbols represents an output of the computation.

Control of unicity versus multiplicity conditions are the essential ingredients in the simulation of the quantum algorithms that solve Deutsch's and Deutsch–Jozsa's problems via **PTM**s, which are shown to simulate quantum algorithms preserving efficiency.

The symbols $\circ$ and $\bullet$ in the language of the logic $LFI1^*$ are used to indicate respectively unicity (consistency) and multiplicity (inconsistency) conditions. These symbols will be written after the first symbol of the instruction, if the condition is on the state, or must be written after the second symbol of the instruction, if the condition is on the reading symbol. For example, instructions of the form $i_k = q_1^\circ s_1^\bullet xy$ will be executed in situations where the machine is in precisely the state q_1 (unicity required), and where one of the reading symbols is s_1 (multiplicity permitted).

In this way **PTMs** are expressed within first-order paraconsistent theories, and the most relevant cases of quantum parallelism can be simulated by **PTMs** by interpreting current states, positions and symbols on the tape as observables in a quantum system. From this perspective, **PTMs** emulate quantum Turing machines (**QTMs**). Constructive proofs of the following theorem are given in [3] and [4]:

Proposition 7.2. There exist paraconsistent Turing machines that solve deterministically any instance of the Deutsch-Jozsa problem in polynomial time.

Nevertheless such **PTM** models do not simulate quantum Turing machines and quantum circuits in full, as they do not allow an utterly adequate simulation of entangled states. Indeed, the construction of different models of **PTMs** which are able to simulate entangled states is a challenging task. This can be done, but the task now requires a non-adjunctive first-order paraconsistent logic (i.e., a logic where the inference from α and β to $\alpha \wedge \beta$ is not guaranteed) rather than $LFI1^*$, as proposed in [3] and [4]. This logic feature would ensure that the execution of distinct simultaneous instructions will be kept apart, permitting a logic control on the entangled configurations of a **QTM**. This fact — together with our examples above which show that highly complex machines in the standard model become even polynomial in non-standard extensions — confirms the view that computational models may be profitably approached as logic-relative devices.

8   Comments

Which is the natural axiomatic setting for Computer Science? That setting is the non recursive theory of all true arithmetical sentences, noted **T**. We can build **T** in several reasonable ways:

- **T** = PA + Shoenfield's recursive ω–rule [23, 24].

- **T** = PA + all naïvely total recursive functions.

- **T** = PA+ a succession of naïvely total recursive functions so that for any such function g there is an f in the succession so that f dominates g.

- **T** = $\cup_\alpha$PA$_\alpha$, where PA$_\alpha$ is an iterate in the Turing–Feferman [7, 23, 24] progression of theories.

- **T** = $\cup_\alpha$PA$_\alpha$, where now each PA$_\alpha$ is as in Turing's original progression of theories [23], plus some axiom that reproduces the jump [32] within each component theory.

  (Roughly what happens here is: the Turing progression proves all true Π_1-sentences. These include all non halting instances of the Halting Problem. If we add those as an oracle to PA, plus the jump, we get all true sentences in the arithmetic hierarchy.)

All theories in those examples are nonrecursive, but they can be seen from the perspective of their "component" or iterate theories, which are, each one, recursive theories. So, the fact that **T** is the natural setting for *CS* is not as bad as it seems — the whole domain is non–recursive, but we can split it up into an infinite staircase of well–behaved theories.

Nonstandard stuff

And how about nonstandard stuff, like those that appear in our examples? How about logic–dependent models of computation devices? Being nonstandard, like being imaginary or being non-Euclidean, is a relative, historical, context–dependent concept. Take imaginary and complex numbers: electrical engineering is unconceivable without them. Non–Euclidean geometry led to today's gravitation theory. We are still waiting for interpretations of non–Cantorian set theories, and of "fake" [33] spacetimes, but there is no reason not to suppose that reasonable ones will eventually be given.

The same applies to all our exotic, non–orthodox theories of Turing machines, including those sketched in the present paper. For mathematics deals with concrete objects, even if under the disguise of high, lofty abstraction. Someday someone will find a sensible, reasonable, concrete interpretation for all the unorthodox examples we have offered here.

9 Acknowledgments

WAC acknowledges support from FAPESP Thematic Research Project ConsRel grant 2004/14107-2; WAC and FAD have also been supported by a CNPq research grant. Grants from PEP-COPPE-UFRJ, through its Group for Advanced Studies are also acknowledged by FAD.

WAC wishes to acknowledge discussions with J. C. Agudelo. FAD thanks support from COPPE-UFRJ and from his colleagues R. Bartholo, C. Cosenza and S. Fuchs; he also thanks the Academia Brasileira de Filosoï¿½a and its chairman J. R. Moderno. Finally the second author wishes to acknowledge discussions and exchanges on the matters dealt with N. C. A. da Costa and M. Guillaume.

BIBLIOGRAPHY

[1] J. C. Agudelo, *Máquinas de Turing paraconsistentes: algunas posibles definiciones y consecuencias*, monografía de especialización en Lógica y Filosofía. Universidad EAFIT, Colombia (2003), available at http://sigma.eafit.edu.co:90/asicard/archivos/mtps.ps.gz.

[2] J. C. Agudelo, *Da Computação Paraconsistente à Computação Quântica*, M. Sc. dissertation, Unicamp (2006).

[3] J. C. Agudelo and W. A. Carnielli, "Quantum algorithms, paraconsistent computation and Deutsch's Problem", *Proceedings of the 2nd Indian International Conference on Artificial Intelligence*, ed. B. Prasad, 1609–1628 (2005).

[4] J. C. Agudelo and W. A. Carnielli,"Quantum computation via paraconsistent computation", arXiv:quant-ph/0607100 v1 (2006).

[5] J. C. Agudelo and A. Sicard, "Máquinas de Turing paraconsistentes: una posible definición", *Matemáticas: Enseñanza Universitaria* **XII**, 2, 37–51 (2004). Available at http://revistaerm.univalle.edu.co/Enlaces/volXII2.html.

[6] T. Baker, J. Gill and R. Solovay, "Relativizations of the P =?NP question", *SIAM J. Comp.* **4**, 431–442 (1975).

[7] L. Beklemishev, "Provability and reflection", Lecture Notes for ESSLLI'97 (1997).

[8] W. A. Carnielli, M. E. Coniglio and J. Marcos, "Logics of formal inconsistency", in D. Gabbay and F. Günthner, *Handbook of Philosophical Logic*, Kluwer (2005).

[9] W. A. Carnielli, J. Marcos and S. de Amo, "Formal inconsistency and evolutionary databases", *Logic and Logical Philosophy* 115–152 (2000).

[10] I. L. Chuang and M. A. Nielsen, *Quantum Computation and Quantum Information*, Cambridge University Press 2000).

[11] N. C. A. da Costa and F. A. Doria, "Undecidability and incompleteness in classical mechanics", *Int. J. Theor. Phys.* **30**, 1041–1073 (1991).

[12] N. C. A. da Costa and F. A. Doria, "Suppes predicates and the construction of unsolvable problems in the axiomatized sciences", in P. Humphreys, ed., *Patrick Suppes, Scientific Philosopher*, II, 151–191 Kluwer (1994).

[13] N. C. A. da Costa and F. A. Doria, "Consequences of an exotic formulation for $P = NP$", *Applied Mathematics and Computation* **145**, 655–665 (2003); also "Addendum", *Applied Mathematics and Computation* **172**, 1364–1367 (2006).

[14] N. C. A. da Costa and F. A. Doria, "On set theory as a foundation for computer science", *Bulletin of the Section of Logic*, (University of Łodz) **33**, 33–40 (2004).

[15] N. C. A. da Costa, F. A. Doria and E. Bir, "On the metamathematics of P vs. NP," *Applied Mathematics and Computation* **189**, 1223–1240 (2007).

[16] N. C. A. da Costa and F. A. Doria, "Computing the future", in K. Vela Velupillai, ed., *Computability, Complexity and Constructivity in Economic Analysis*, Blackwell (2005).

[17] N. C. A. da Costa, F. A. Doria and E. Bir, "On the metamathematics of the P vs. NP question", to appear in *Applied Mathematics and Computation* (2007). (Available online from ScienceDirect.)

[18] M. Davis, "Hilbert's Tenth Problem is unsolvable", in *Computability and Unsolvability*, Dover (1982).

[19] D. Deutsch , "Quantum theory, the Church-Turing principle and the universal quantum computer", *Proc. R. Soc.Lond-A* **400**, 97–117 (1985).

[20] F. A. Doria, "Metamathematics of P vs. NP", talk at the Suppes Fest, Fed. University at Santa Catarina, Florianópolis (Brazil), April (2002).

[21] F. A. Doria, "Informal vs. formal mathematics", *Synthèse*, to appear (2007). (Available online from SpringerLink.)

[22] R. L. Epstein and W. A. Carnielli, *Computability: Computable Functions, Logic and the Foundations of Mathematics, with the timeline Computability and Undecidability*, 2nd ed., Wadsworth/Thomson Learning, Belmont CA (2000).

[23] S. Feferman, "Transfinite recursive progressions of axiomatic theories", *J. Symbolic Logic* **27**, 259–316 (1962).

[24] T. Franzen, "Transfinite progressions: a second look at completeness", *Bull. Symbolic Logic* **10**, 367–389 (2004).

[25] J. Hartmanis and J. Hopcroft, "Independence results in computer science", *SIGACT News*, 13, Oct. Dec. (1976).

[26] S. C. Kleene, "General recursive functions of natural numbers", *Math. Annalen* **112**, 727–742 (1936).

[27] S. C. Kleene, *Mathematical Logic*, Wiley (1967).

[28] K. Kunen, *Set Theory*, North Holland (1983).

[29] K. Loo, "Internal Turing machines", arXiv:math-ph/0407056 v2 (2004).

[30] M. Machtey and P. Young, *An Introduction to the General Theory of Algorithms*, Elsevier North–Holland (1978).

[31] J. Paris and L. Harrington, "A mathematical incompleteness in Peano arithmetic", in J. Barwise, ed., *Handbook of Mathematical Logic*, North–Holland (1977).

[32] H. Rogers Jr., *Theory of Recursive Functions and of Effective Computability*, reprint, MIT Press (1992).

[33] A. Scorpan, *The Wild World of 4–Manifolds*, AMS (2005).

[34] P. W. Shor, "Algorithms for Quantum Computation: Discrete Log and Factoring", *Proc. 35th Symposium on Foundations of Computer Science*, 124–134, IEEE Press (1994).

[35] P. W. Shor, "Polynomial–time algorithms for prime factorization and discrete logarithms on a quantum computer", *SIAM J. Comput.* **26**, 1484–1509 (1997).

[36] C. Smorýnski, *Logical Number Theory, I*, Springer (1991).

[37] A. M. Turing, "On computable numbers, with an application to the Entscheidungsproblem", *Proc. London Math. Society* **50**, 230 (1937).

Physics and Non-Classical Logics

NEWTON C. A. DA COSTA AND DÉCIO KRAUSE

ABSTRACT. In this paper we present a general discussion on some topics involving the use of non-classical logics in the foundational analysis of physics (but our ideas can be extended to other empirical disciplines as well).[1]

Introduction and overview

The relevance of the foundational analysis of physical theories was pointed out explicitly by Hilbert in the sixth of his celebrated 23 Mathematical Problems ([22]). During the XXth century, much was done in this direction, and the mathematical counterparts of the relevant physical theories were made explicit in great detail. Nowadays, the way of realizing that is by axiomatizing the physical theories, a process which makes explicit their underlying mathematical structures, and this is done by the presentation of a set-theoretical predicate what was later termed the "Suppes predicate" of the theory, which corresponds to a species of structures in the sense of Bourbaki ([4]), to honor Suppes' claim that to axiomatize a scientific theory is to present a set-theoretical predicate ([37], p. 10). Suppes is, among others, recognized as having started the approach that succeeded the so called Received View, linked to logical positivism, in no more seeing a physical theory as a linguistic device, but as revealing a class of mathematical structures, the models (in the mathematical sense) of the corresponding "Suppes" predicate (hence, the models of the theory itself).

Thus, taking a physical theory, T, the Suppes predicate of T is a transportable formula (in the sense of Bourbaki) of the language of (informal) set theory which refers to certain set-theoretical structures; those that satisfy the predicate are the *models* of T. In Suppes' approach, informal (or naïve) set theory is used, so he could surpass the difficult problem of making explicit all the "subsidiary" theories the theory T makes use of, which would demand much preparatory work before to reach to the theory itself. For instance, in order to axiomatize non-relativistic quantum mechanics by means of the Hilbert space formalism, one would make explicit the underlying apparatus of part of standard functional analysis, which presupposes, for instance, the axiomatization of the real numbers system, Hilbert spaces,

---

[1]To Shahid, with kind regards for his 52nd birthday and for his tolerance to heterodoxy.

tensor algebra, partial differential equations, set theory proper and, of course, its underlying logic. Much to be done before reaching what is of real interest for the scientist, namely, the theory T or, more in accordance with the "semantic view", its models.

This semantic approach, making emphasizing the models of the theories, was further pushed to other directions, depending on the class of models (for instance, van Fraassen concentrates in the "empirically adequate" models – [38], while Adams spoke of the "intensional models" – see below). Presupposing set theory and classical logic, Suppes was able to sustain that all these step theories used by a specific physical theory T could be subsumed within set theory itself, as far as all these mathematical theories are "reduced" to set theory. If pushed to make explicit this set theory, Suppes could say that it can be assumed to be an adequate version of Zermelo-Fraenkel set theory.[2] Thus, in this approach we are *presupposing* not only standard set theory but also classical logic. In parallel with the Duhem-Quine thesis, according to which we cannot test a scientific hypothesis in isolation, we cannot consider the axiomatic formulation of a theory out of its logical and mathematical postulates and all theorems they imply.[3] Thus, we may ask: is there any case of a physical theory whose axiomatization demands a different logical framework? (We remark that there are also other options even if we restrict ourselves to "classical" environments, as category theory and classical higher-order logics, but we shall continue to speak in terms of first order logic and set theory as the theory's basic framework.) If we say "no" to the above claim, then we are assuming that any physical theory can be axiomatized within classical logic and mathematics. Which option is the most appropriate one? In this paper we discuss this topic in a general way, with the aim of posing some particular remarks and opinions.

1 Physics and non-classical logics

There have been interesting insights, advanced by several important thinkers, concerning the use of non-classical logics in physics, but according to us, not conclusively – or shown necessary – up today. We shall not provide a wide revision here, although this would be interesting from the historical point of view, but we will just mention some few of the most well known cases. For instance, Bressan suggested the use of certain modal logics to lay the ground for an axiomatic calculus for physics, so providing a logical basis which allows the definition of formal counterparts for basic notions of the language of contemporary physics; however, his ideas were not fully developed up today (see [40]). Reichenbach's three valued logic, although interesting from the point of view of the insights and clarification of some basic assumptions underlying quantum mechanics, has been criticized for not providing a full logical basis for such a discipline, mainly due to the lack of

---

[2]By the way, this was his answer, when one of us asked him some time ago.

[3]The role of this metamathematical framework is being investigated in another work.

a detailed discussion of quantification – the propositional level of his logic does not suffice for physical purposes ([29]). The same can be said about Février's "logic of complementarity" ([17]), and also concerning other tentatives of using non-classical logics in this realm (for some further indications, see [8]; [23], chap. 8). Still concerning quantum physics, in the sixties Suppes introduced a "probabilistic argument for a non-classical logic" in this field ([35]), yet the "right" concept of probability as applied to quantum physics seems to be an open problem. Notwithstanding, the tradition suggests that when we hear something about the relationships between non-classical logics and physics, we usually associate the subject with so called "quantum logics", a field that had its birth in Birkhoff and von Neumann's well known paper from 1936 ([23]; [14]). Originally proposed to cope with some problems which originate from quantum mechanics (like the idea related to a possible violation of the distributive law $\alpha \wedge (\beta \vee \gamma) \leftrightarrow (\alpha \wedge \beta) \vee (\alpha \wedge \gamma)$), their fundamental work, which we could say came from an "applied" stance (since their system was motivated by an empirical science), caused the development of a wide field of research in logic. Today there are various "quantum logical systems", which have been usually studied as pure mathematical systems, practically far from applications to the microphysical world and from much of the insights of the forerunners of quantum mechanics.

In our opinion, a "conclusive" application of a non-classical logics to the foundations of physics would be achieved only after we have (really) *proved* that these logics are in fact *necessary* in this field. But perhaps this cannot be done; as Tarski could have said, the language of (standard) set theory is a kind of universal language, where practically every physical concept can be expressed (at least until today). In short: within classical logic and set theory, (apparently) *we always find a way* of expressing what we need, but for sure only the future will decide if these guesses are right. Anyway, as philosophers of science, we should take into account all possibilities, mainly because the professional physicist in general is not occupied with such questions. The use of non-classical logics and mathematics, yet not being proved to be "essential" for the empirical sciences, can be useful from various points of view, mainly in making explicit philosophical and mathematical questions underlying physical theories, which within the standard frameworks could pass without being noticed. We shall enlighten some cases below.

2   Inconsistencies and physics

According to the Cambridge English Dictionary On Line, the adjective "inconsistent" means "not staying the same in behaviour or quality", while "consistent" means to be "in agreement with other facts or with typical or previous behaviour, or having the same principles as something else". A philosophical dictionary, like Runes', is of course more precise ([31]). There, Alonzo Church wrote that *inconsistency*, as applied to formal (logistic, in his words) systems (there is no mentioning of other contexts), means the opposite of consistency; more precisely, says Church, "[a] set of propositional functions is *inconsistent* if there is some propositional function such that their conjunction formally implies both it and its negation ." So, although we can say informally, for instance, that someone is inconsistent if she constantly changes her beliefs, or then that children and politicians are inconsistent, the concept acquires precise meaning within the scope of logically well developed mathematical systems. In present day language of logic, we say that a theory (a set of formulas closed by deduction) is inconsistent if it has two contradictory theses (theorems), α and $\neg\alpha$, and it is consistent otherwise, where α is a formula of the (language of the) theory and $\neg$ is the theory's symbol of negation . In most cases, having such contradictory theses, the system has also their conjunction as a thesis, namely, the expression $\alpha \wedge \neg\alpha$, which is called a contradiction.[4] More is involved, for we need to have a clear idea, for instance, of the meanings of "negation " and "deduction". In physical theories, while keeping informal, that is, as developed as they are in standard textbooks, we really neither have a general concept of negation nor of deduction. We generally use informal logic and mathematics, employing these concepts as if they were clear. Let's see an example.[5]

Bohr's concept of *complementarity* is recognized as both quite important and difficult to understand. The concept was introduced in quantum mechanics by Niels Bohr in his famous "Como Lecture" in 1927.[6] The consequences of his conceptions were fundamental for the development of the Copenhagen interpretation of quantum mechanics and constitute, as is largely recognized in the literature, one of the most fundamental contributions for the development of quantum theory. Notwithstanding their importance, Bohr's ideas on complementarity are controversial. Truly, it seems that there is no general agreement on the precise meaning of his *Principle of Complementarity*. Although some authors like C. von Weizsäcker and M. Strauss tried to elucidate Bohr's principle from a logical point of view, their ideas never received universal agreement from the scientific community. For instance, it is well known that Bohr himself rejected von Weizsäcker's attempt of logically describing his principle (cf. [23], p. 90). Strauss' proposal of a logic in which two propositions, say α and β (which stand for complementary propositions) may be both accepted, but not their conjunction $\alpha \wedge \beta$, was considered as 'inadvisable' by Carnap, although it seems to deserve attention from a present day point of view.[7] Another work in this direction was presented by P. Destouches-Février; in a series of papers, which were later presented also in her book *La Structure des Théories Physiques* ([17]), she sketched a three-valued propositional logic to deal

---

[4]This is posed this way because there are logical systems where from two propositions α and β we cannot infer their conjunction $\alpha \wedge \beta$. But most of the usual logical systems are *adjunctive* in this sense.

[5]The subject of this section has been considered with more details in [5], chap. 5. Here we deal with examples not given there.

[6]All the references of this section, not explicitly referred to here, may be found in [9].

[7]As anticipated above, today we have several "non-adjunctive" logics.

with complementary propositions, termed $L_{c,3}$.[8] But even here the developments were not incorporated into the mainstream of physics, partially because the lack of a corresponding theory of quantification.

Perhaps the reason these systems did not convince physicists and philosophers of science is due to the kind of approach they encompass. Despite their differences, all these systems assume complementary propositions as describing two distinct "realities", say one of them treating particles as such and the other one as waves, while some philosophers prefer to talk in terms of entities of a different kind. Let us sketch the details. Situations of this sort, that is, involving incompatible "realities", can be accommodated within logical frameworks without much hard work, for instance by using a logic termed paraclassical (which is a non-adjunctive logic), whose main ideas can be summed up as follows (for details, see [8]). Let us call $\mathbb{P}$ the propositional logic (it would be easy to extend this system to a quantificational logic) whose language is that of a standard classical propositional calculus $\mathbb{C}$ (whose deduction symbol is $\vdash$; the definition of formulas is that of $\mathbb{C}$). The concept of *paraclassical deduction*, termed $\vdash_{\mathbb{P}}$, is introduced as follows: if Γ is a set of formulas of $\mathbb{P}$, let α be a formula (of the language of $\mathbb{P}$). Then we say that α is a (syntactical) $\mathbb{P}$-consequence of Γ, and write $\Gamma \vdash_{\mathbb{P}} \alpha$, if and only if (P1) $\alpha \in \Gamma$, or (P2) there exists a consistent (according to $\mathbb{C}$) subset $\Delta \subseteq \Gamma$ such that $\Delta \vdash \alpha$ (in $\mathbb{C}$). It is immediate to see that if α is for instance an atomic formula, then $\Gamma = \{\alpha, \neg\alpha\}$ does trivialize any theory based on $\mathbb{P}$; in other words, $\mathbb{P}$ is a paraconsistent logic (according to us, a logic is paraconsistent if it can be the underlying logic of inconsistent – that is, having contradictory theses like α and $\neg\alpha$ – but not trivial theories – that is, there are formulas which are not theorems). Thus, paraclassical logics can be useful in systematizing whatever theory T is obtained by joining two (perhaps incompatible) theories T_1 and T_2, regarding that their language and concept of deduction are treated adequately.[9] Within this schema, regarding some necessary qualifications (given in [9]), two propositions like "to behave like a wave" and "to behave like a particle", which are of course both possible according to quantum physics and entail one the negation of another, can be both accepted, although their conjunction is not derivable as a theorem. Well done, you can say, but some present day researchers think that we should not fight to accommodate ancient ideas (like those of particle and wave), but that the ontology of quantum physics should be changed for the consideration of *another kind of object*, something different from the particles of classical physics (which are more or less those presupposed also in Bohm's version of quantum mechanics) so as also being different from standard waves, something which some of them call

---

[8] In [8], we have presented her system and discussed it, including the presentation of a paraconsistent version of her logic.

[9] Concerning deduction, in some cases we could use another kind of non-classical logics termed *multideductive*, where various concepts of deduction $\vdash_1, \ldots, \vdash_n$ are considered.

quantons (as suggested by Levy-Leblond and Balibar [24], following an indication of Mario Bunge). We shall come back to this point in a moment. Before that, let us make some further remarks on the needs of paraconsistent logics.

According to philosophers called *dialetheists*, there are contradictions that are true ([28]). Thus, since we should avoid trivialization, once we tend to accept that not every statement is true (for instance, to say that Newton was born before Aristotle is not true), the "logic of science" would be a paraconsistent logic. We will not make here a detailed criticism of dialetheism, but only advance some results that show why we don't agree with such a view.[10] Firstly, let us say that in our opinion the concept of truth associated to empirical sciences is better explained by means of the concept of *partial truth* (or *quasi-truth*), and not by the Tarskian correspondence theory defended by dialetheists, which apparently works better within mathematics. The reason for such a choice is not arbitrary, and was explained in full in [5], so that we shall not discuss it here. Secondly, the contradictions dialetheists (the examples are taken from [28]) present as examples do not convince us. Let us explain in brief why.

Some of the examples given by Priest involve visual perception, for instance, involving what is called the "Penrose's figure". This is something quite similar to Escher's "Ascending and Descending" staircase. According to Priest, "[t]his is the case where we can see a contradictory situation" ([28], p. 59). Even without being specialists on these matters, we agree with Erwin Schrödinger in that the reality we experiment is a mental construct of ours ([32]), so it can be "conceived" from different perspectives, including consistent ones (see the last section). But let us base our claims in quoting Brian Davies, who explains us that "[t]hese illusions are possible because our visual system has to make guesses based on incomplete information. It is a fact that if an object exists then a drawing of it will follow the laws of perspective. However, our visual system follows an incorrect rule: that if drawing follows the laws of perspective then a corresponding object exists, or could exist." ([15], p. 10). Thus, Escher's and Penrose's figures would not represent anything *real*, for there could not be a real stair which goes up and goes down, leading to the same point. The *perception* we have is of a contradictory object, but, if these remarks are right, it would be only a mental construct of ours. Apparently, no "real" physical object can be constructed out of obeying the rules of perspective, which are violated by these figures. Priest's second example concerns motion. As he says, "[i]f the visual field is conditioned by viewing continuous motion of a certain kind, say a rotating spiral, when the viewer then looks at a stationary scene, it appears to be moving in the opposite direction. But a point at the top of visual field, say, does not appear to change place." ([28], p. 60) As Brian Davies explains

---

[10]By the way, let us remark that the first of us is one of the founders of paraconsistency and of paraconsistent logics, but he has never sustained that classical logic is wrong and that it should be changed to a paraconsistent one. We emphasize this point here.

(although he uses another example), this is just a property of our peripheral vision ([15], p. 7). Thus, it seems to us that there is no a real contradiction here too. The next case is concerning colours. More specifically, red and green colours. So explains Priest: if a field is presented to someone, half of which is red and half of which is green, so that the two halves are separated by a black line, when the line is removed, some people report that the vacated space is filled by the brain by something which is red and green. Thus, based on these reports, can we say that there are "real" contradictions, just because some people "see" red *and* green colours? Anyway, these reports are not sufficiently strong to entail that a contradiction does exist in something being red and not red; it is, again, a problem of a particular mental construction. And here is Brian Davies again: "[o]bjects are not red, green, or blue in themselves: our impressions are created by neural processing of the very limited information provided by our retinas." ([15], p. 5)

In our opinion, until now contradictions may exist only within certain formal logical systems, for instance in the trivial ones. Priest's examples involving illusions do not convince us (at page 124 of his mentioned book he reveals that the contradictions in the empirical world are all illusions; so, we could ask: in what sense some of them are *real?*). Within formal systems, the situation is of course different, and we really *can* have contradictions, but it is difficult to say that they are *real* without a detailed discussion of this concept. Thus, although paraconsistent logics have today important applications even in medicine (see [11]), they are still just useful mathematical devices, serving for the formulation of alternative perspectives about a certain domain of knowledge, but their real necessity requires that someone proves that contradictions do exist (*pace*, Hegelians). The only way of supposing that paraconsistent logics can be said to be "the logic of science" is to understand "science" in a very broad sense. Scientists really do not work, in most cases, within strict formal systems, by respecting the formal rules of deduction of a certain underlying logic and so on. They use various alternative forms of reasoning in their intellectual activity, like inductive reasoning, non-monotonic forms of reasoning, and so on, as when they formulate hypotheses and theories. In a certain sense, we agree with Feyerabend when he says that in the process of creation, "anything goes" ([16]). Saying in brief, in dealing with science *informally*, scientists work inconsistently, or within a framework so that, if we could do the super-human effort of putting it within a formal system, probably it would be an inconsistent system, encompassing various theories and conceptions which are incompatible with one another. This does not entail that contradictions do exist, but just that incompatible conceptions were put side by side. If someone argues that such a procedure is not coherent, leading to inconsistencies, we can shelter in paraclassical logic delineated above. Nonetheless, paraconsistent logics contribute to show that the consistency of reality cannot be properly proved by logical arguments only.

In the next section we shall turn to another case (apparently) involving non-classical logics in the context of quantum physics, dealing with quantum objects.

3   Quantum objects, their logic, and more

It is well known that there is not just one "quantum mechanics". Even if we accept the standard formalism of Hilbert spaces, there are different formulations of the two basic questions, which M. Redhead termed "the quantization algorithm" and the "the statistical algorithm" ([30]).[11] Furthermore, there are the first quantization and the second quantization approaches (quantum field theories, in the plural) and, today, even more (string theories and so on). Which one is *the* quantum mechanics? Difficult to say. All of them, in a sense. None of them, in another. Anyway, we should recognize that whatever interpretation we choose, in most cases we shall be dealing, even indirectly, with certain *entities* (for the lack of a better term), or *physical objects*, which are those things theories make reference to (even if indirectly); as an example, just take a look at a book on particle physics, like the introductory [3]. There we can find a wide discussion involving electrons, protons, neutrinos, and so on. These are the "elementary particles", although the term *particle* is already of a misuse here (so as "elementary", in a sense). To avoid misunderstandings, we could use the word "quanton(s)" as mentioned above, but we prefer "quantum objects" ("q-objects" for short). Notwithstanding, the philosophy of physics is not obliged to strictly follow what physicists believe to be valid, but has to criticize the naïve language of physics, trying to base physics on a reasonable foundation.

But this talk of q-objects may dissatisfy someone, who can provide a counter-argument by saying that according to the Copenhagen school (mainly in Bohr's account), there are no *real* q-objects, for the aim of quantum physics would just be to provide ways of relating results of measurements in order to calculate probabilities. Even so, we could reply, the basic concept of this conception, namely, the uncertainty principle, says in one of its formulations that position and momentum *of a q-object* cannot be *both* measured with sufficient accuracy, out of a factor depending on Planck's constant. OK, she could say, but according to Schrödinger's picture, there are no "particles" but just waves – an *ontology of waves*. Well, we respond: although in the final years of his life Schrödinger himself has questioned his own defense of *the reality of the wavefunction* ([2]), we can say that, even within his picture, there are q-objects being considered, as we shall argue below. Our concept of q-object is relative to the interpretation/theory we are dealing with. We don't ignore that they exist, such that independently of the particular formu-

---

[11] Informally speaking, they are (respectively): "What values are possible measurement results for any given observable?", and "For any given state and any given observable, what is the probability that one of the possible measurement results will actually turn up when a measurement is performed?" [30], p. 5).

lation of the theory, they have some peculiar traits which are maintained in all of them (more on this below). Q-objects, according to us, are the basic stuff of the considered quantum theory, and their nature may change in some detail, depending on the theory being considered, although whatever conception of them may have some common features, like to enter in superposition states and to obey quantum statistics. We can assume that q-objects do exist, and further that we have no "complete access" to them; all we have are our theories about them, which describe them in a certain way, and these descriptions vary according to the theory being considered. The fundamental point in our view is that we do not deny their existence. Thus, according to their very nature, there are two basic positions to be sustained: either to accept that their characteristics *remain* veiled *for us*, which introduces a kind of hidden variable theory, or – as we prefer – we can accept them as a kind of *vague* objects, to which no sharp account can be provided even in principle (see [19]). The first option is of an epistemological nature, while the last one is ontological (modulus a theory). Let us emphasize that we do not deny the existence of q-objects, but we also do not immediately and naïvely accept them. Our aim is to investigate such objects according to a theory, or modulus a theory. In doing so, we are not obligate to assume that quantum theories are incomplete, being unable to describe q-objects as they are, but just that they are vague entities that defeat any tentative of sharp description: using another terminology, due to Toraldo di Francia, they are *nomological objects*, given by physical law, ascribed by the particular theory we are handling ([39]; see [20] for a discussion).

Well done, the skeptic can say, before finally questioning: what about Ludwig's approach, where no microscopic reality is assumed? In his approach, Ludwig takes as basic the physical apparatuses, which are the directly experienced tools of the physicist, so that the particles (q-objects) appear only as carriers ("action carries", in his terms) of interactions between two types of macrosystems, *preparation* and *measurement* apparatuses ([25], p. 148ff). The physical theory results from statistics of experiments with such directly observable macroscopic tools, described by classical physics, and quantum mechanics properly keeps grounded entirely on a macroscopic-phenomenological reality. Can we say that in this approach there are no q-objects? Even here, where microscopic objects are not *assumed directly*, we think not. Ludwig has *expressed* quantum physics in a way that depends of macroscopic apparatuses, grounded in classical physics, but as a quantum theory, the resulting quantum mechanics needs to refer to q-objects having the same typical features of the q-objects of the other theories. In fact, although his physical objects are action carriers, he uses the word "quanta" to refer to them, "for the sake of physical intuition" ([25], p. 157). Furthermore, Ludwig develops his view grounded in the belief that "[d]eep physical laws are never read off from experience" ([25], p. 158), something which for sure Einstein would disagree with. Thus, we can understand Ludwig's (and other) approaches as just approaches to quantum physics, and not as the only possible one – see the last section. As said above, q-objects encompassing characteristics enabling superpositions, the validity of quantum statistics, and so on, *do appear* indirectly even in approaches like Ludwig's, and if we deny their existence, we need to distinguish between the mathematical formalism of a physical theory from its (some) interpretation, and then we will be doing pure mathematics, not physics.

This last point is of course quite important, and we remark that it marks a point where we disagree in part with Suppes (and perhaps with Ludwig), for he doesn't see any theoretical way "of drawing a sharp distinction between a piece of pure mathematics and a piece of theoretical science" ([37], p. 33). This is based on his claim that there are not differences between a set-theoretical definition embodying concepts of pure mathematics and one "involving concepts" of some particular science. We agree with the part referring to "concepts", for physical concepts are expressed in mathematical terms, hence within mathematics properly. According to Díez y Moulines, the lack of a clear explanation of how such a piece of theoretical science is linked to physical experience, which he tried to explain in his well known paper "Models of Data" ([34]), has lead his disciple E. Adams to introduce critical reflections on the problem of characterizing the empirical models of scientific theories, resulting in the idea that we should distinguish among the models which are *intensional applications* of the theory from the mere mathematical models of it. Adams' view goes in the direction of assuming that the intensional models (which are part of the mathematical models of the theory) of a physical theory are fixed by the scientific community – for a discussion, see [18], chap. 10; [36]. By the way, Díez y Moulines also claim (as we are doing here) that what lacks in this "semantic" approach is "something which expresses 'the things'" the theories claim to be so and so ([18]). Then their response goes to the *structuralist approach*, initiated by Sneed, but we will not pursue this point here. We sum up by saying that physical theories encompass interpretation even implicitly, or meaning if you wish,[12] and *every* particular theory in the wide field of quantum physics refer to q-objects of a type, so presupposing them in a certain sense, yet sometimes not incorporating them in the theories themselves. Let us be more specific on this last point.

Take for instance the discussion in the philosophy of quantum mechanics about the interpretation of the state vector (out of its probabilistic interpretation – Born's rule). As it is well known, in the standard presentations, to every physical system it is attributed a suitable (complex!) Hilbert space H, and the states of the system are given by the unitary vectors of that space, while the observables are represented by auto-adjunct operators on the same space. This makes physics works well, but

---

[12]This shows that we ought to reconsider the claim of the logical positivists in speaking of "rules of correspondence", but perhaps grounded on other considerations than those of them. This is a point to be developed further, but also not here.

such a pure mathematical formalism needs interpretation, for instance, when we say that a certain vector (certain symmetric vector) of the Hilbert space $H \otimes H$ (the space for two indistinguishable systems) represents the state of *two q-objects* called bosons. What are *bosons*? Usually, we interpret the term "bosons" in connection with certain characteristics (some of them of an experimental nature) the theory (even if implicitly) postulates. Q-objects, like bosons, are not properly represented in the standard formalisms, since we go *directly* to Hilbert spaces and no semantics (in the formal logical sense of the word) is provided any more – the interpretation remains informal – (this is the link to Díez and Moulines phrase mentioned above, asking for the needs of expressing the things the theory makes reference to). Really, we simply take a certain vector which is symmetric with respect to the exchange of the coordinates of the q-objects and say, by ostension: this is a vector representing a pair of bosons. Really, it seems that quantum theories have no formal semantics.

Going back to our point, once we have accepted that quantum theories suggest that there are (even implicitly) q-objects of a kind, that is, things which make marks in the apparatuses and tracks in a Wilson Chamber, enter in superposition, and that have properties that can be measured, like spin and momentum, among others, we can ask for the kind of logic such entities would obey, and which kind of models are those which represent such q-objects more accurately, encompassing their characteristics mentioned above. Important to say that it is not necessary, at least in a first approach, to ask for the very nature of these entities: their relevant characteristics are prescribed by the considered theory. As put by Maudlin, metaphysics, "the most generic account of what there is" and, as such, "must be informed by empirical science" ([27], p. 461). So, in order to answer the question about their nature and main characteristics, we ought to turn to the physical theories proper.

In Bohm's approach, as it is well known, the q-objects are entities like the particles described by classical physics, but accompanied by "pilot waves", which are responsible for the interference patterns, typical of quantum entities.[13] There is also the possibility of considering q-objects as *individuals* of a kind, similarly with classical physics, but at the expenses of imposing restrictions on the states they may be in and in the observable we regard as possible to measure. But this hypothesis entails that their individuality must be ascribed by something which is not a property of them, and so we enter the realm of the so-called theories of *substratum*, which philosophers of physics tend to avoid ([20]). The widespread conception, notwithstanding, is that which regards q-objects as *non-individuals*, that is, as entities devoid of individuality. This idea came from the beginnings of quantum theory. Planck, when deriving his radiation law, admitted the distribution

---

[13]The dynamics of the formalism has, besides the Schrödinger equation, a certain "guiding equation" depending on the positions of the particles ([21]).

of indistinguishable *non-individual* quanta over oscillators. In 1926, Born, while defending the corpuscular as opposed to wave-like conception, acknowledged that these corpuscles could not be identified as individuals. In the same year, Heisenberg noted that Einstein's theory of the ideal quantum gas implies that the "individuality of the corpuscle is lost" (see [20], p. 85ff; [8]). Weyl and Schrödinger are other important physicists who defended the non-individuality of elementary 'particles".

Thus, in a possible way of providing a semantic for quantum physics, the q-objects would have no identity criterion, in the sense that despite they can be aggregated into amounts, they cannot be identified by names or labels, for if indistinguishable, they are not invariant by permutations. Whatever permutation of q-objects of the same species conduce to the same observable results. *Individuals*, on the contrary, can be supposed to be always associated to a baptism name, or to a description, which individualize them in all possible worlds as *being that individual*.[14] It seems clear that such non-individuals cannot obey the rules of classical logic. In short, the main reasons are the following: firstly, collections of them may have a cardinal, which expresses the "quantity" of q-objects there are in the aggregate, as when we say that there are two protons and two neutrons in the nucleus of an Helium atom, although they cannot be differentiated from one another (those of the same kind, of course). But these collections do not have an associated ordinal. In some situations, for instance when we consider relativistic quantum physics, even the cardinal of the collection cannot be defined, due to the creation and annihilation processes. Collections such that the permutation of one element by "another" of similar species do not conduce to a distinct aggregate, as bosons, according to quantum statistics, do not obey the axiom of extensionality of standard set theories (for us here, logic encompasses set theory). Other restrictions can be mentioned, as for instance the definition of a function, which no more distinguishes between arguments or values, if they have no individuality. All these details were incorporated in the *theory of quasi-sets* (see [20] for details). Another alternative is to pursue directly Schrödinger's idea that identity does not make sense for q-objects in the direction of higher-order logics (this is also supposed in quasi-set theory); our Schrödinger logics ([6]; [7]) were elaborated in this direction (see also [20]). Thus, there are "other" non-classical logics distinct from the usual systems of "quantum logics" presented to the literature (for an updated overview in the field, see [14]) which seem to be useful for expressing the characteristics of q-objects, according to a certain view of them.

But even here we cannot say that the logic of q-objects *is* a non-classical logic if our aim is just to formulate a theory which gives physics a way to work. Let us explain what we mean. As we have seen, the conception of q-objects (the basic

---

[14]The question on what confers individuality to an individual is subtle and important in these contexts, but we shall not touch on this problem here, accepting an intuitive understanding of the concept.

entities considered by the particular theories) depends on the theory under consideration. But almost every quantum theory, Bohm's inclusive, accommodates some basic ideas, typical of quantum physics. For instance, uncertainty, nonlocality, and the superposition phenomena (the two slits experiment). Furthermore, the basic entities assumed by the theory (its q-objects) obey a process of counting known as "quantum statistics". In considering these "postulates" (the underlying ideas of these claims are actually incorporated in the postulates of most quantum theories). Despite we have no grounds to make assertions about the *nature* of q-objects, at least one lesson we learn: if we are not disposal to accept restrictions on states and observables and accept that Bohm's theory (or other theories admitting individuals) does not provide the final word about the subject, we may accept that q-objects are not *individuals*. Even in the "Schödinger's picture", where the wavefunctions are the basic entities of the theory, they refer to non-individual q-objects. In fact, take the wavefunction of a system consisting of two q-objects (aha!). Then, each point of the configuration space specifies the location of both objects. As put by Maudlin, "[o]ne cannot, in this case, sensibly ask what the value of the wavefunction is *here* (indicating a point in physical space); one must rather ask what the value of the wavefunction is for the configuration in which 'one' particle is here and 'the other' at some other particular location." ([27], p. 463)

So, we are in a position of asking whether q-objects can be viewed as non-individuals of a kind, that is, saying in brief, entities (this word should be understood in wide sense) which can form collections, produce marks and so on, as explained above, but which do not maintain their identity if its actual position is blurred.[15] But, do we really need non-classical logics to work with them? Let us address this point a little bit. In the first section we have suggested that the language of set theory is a kind of universal language, where almost all mathematical and scientific concept can be expressed.[16] Such a language is so powerful that even a view of q-objects as non-individuals can be accommodated within this framework, although, as we have said, the resulting "objects" cause some gaps in the physical and philosophical intuition about quantum entities. Thus, for our discussion we can suppose Zermelo-Fraenkel set theory (ZF, supposed consistent) with the Axiom of Foundation, so that we may stay inside of the constructive hierarchy of sets (with our without *Urelemente*; this is not important here). Let us suppose now that we have axiomatized a particular quantum mechanics, say that presented by Mackey in his well known book [26], by means of a set theoretical

---

[15]Although it should be remarked that discernibility and individuality are distinct concepts – see [20]. A slight but fundamental difference between a q-object in this sense and an ant we "lost" in a swarm of ants in an ants' nest, is that the ant can be marked (say by painting it) in order to re-identify it later in the swarm, but this cannot be made with q-objects.

[16]Note that we are speaking of the *language* of set theory, and not of a particular set theory. Even category theory can be expressed in this language, once we introduce, say to the postulates of Zermelo-Fraenkel, suitable postulates admitting the existence of *universes* of a kind.

predicate. A "model" of such a theory is a set theoretical structure $\mathfrak{A}$,[17] hence something existing *inside* the well founded universe, called $\mathcal{V}$. The indiscernibles of $\mathfrak{A}$ are the objects of its domain which are invariant by the automorphisms of $\mathcal{A}$ (which form the Galois group of $\mathfrak{A}$) – see [12]. In other words, two objects a and b are indiscernible in $\mathcal{A}$ if there is an automorphism h or $\mathfrak{A}$ such that $h(a) = b$, and they are discernible otherwise. This idea underlies the standard formalisms of the quantum theories, when symmetric and anti-symmetric vectors or wave functions are assumed to represent the only relevant physical systems, or when it is assumed to hold the Indistinguishability Postulate, which says that the expectation value of any observable has the same value before and after a permutation (in the physical sense) of indiscernible q-objects (in symbols, using the standard notation, $\langle\psi|Q|\psi\rangle = \langle P\psi|Q|P\psi\rangle$ for every vector ψ, observable Q and permutation operator P).

This means that *we can represent* non-individual q-objects within classical frameworks, as physicists actually do, in order to make physics work, but this asks for ignoring the role of the labels a and b initially attached to the entities, which make them individuals. Within $\mathfrak{A}$, all happens as if a and b were indistinguishable, but really they are not. The reason is that any structure (built in ZF) can be extended to a rigid structure, that is, to a structure where the only automorphism is the identity function. As a structure, $\mathcal{V}$ is rigid in this sense. Hence, by adding to $\mathfrak{A}$ adequate relations, we can always distinguish between a and b. In short, in the *whole* ZF (that is, in $\mathcal{V}$), every entity is an individual. So, physics works *as if* q-objects where non-individuals. For really expressing them, we need an alternative formalism (this would be quasi-set theory, or something like that).

The lesson is that if we aim at to make physics work, the classical logic-mathematical framework seems to suffice (at least up today), but we need to make concessions, like to add laws of symmetry in the quantum case, or a difficult concept of *substratum*, or something else. Informally, that is, if one is not occupied with logical questions, this procedure apparently works, as the success of present day science shows. What is then the role of non-classical logics in this realm? This leads us to our final section.

4 The final stop: pluralist constructivism

Let us mention in brief our pluralism. A wide field of empirical knowledge can be "perceived" theoretically from different points of view, or "perspectives", each of them capturing part of the field and from a certain point of view (generally, influenced by the previous background of the scientist). We do that as far as science is a conceptual activity, depending on (in these days, in which Kant's apriorism was apparently surpassed) the background of the scientists, as well as on several other

---

[17]A discussion on physical structures is presented in [13].

factors.[18] These distinct perspectives are in general posed informally, that is, out of axiomatization or formalization, as most of the physical theories were presented for the first time. But each informal theory can be axiomatized, or even formalized, and this can be made in different ways (we are assuming here that if we change language and postulates, we change the theory, yet they have the "same" class of theorems). Furthermore, each axiomatized theory (say by presenting its set-theoretical predicate) can have distinct mathematical models (structures that satisfy the postulates of the theory). Of course we can distinguish among subclasses of this class of models, say by selecting those which are empirically adequate (van Fraassen) or intensional (Adams). This does not matter us now. These models, only in part, and indirectly, given by some informal semantics (assumed even implicitly by the physicists, *via* certain rules of connection), reflect the original empirical field. We can say that the chosen theory (informal, axiomatized or a model) stands for the "theory of the field" only indirectly, say by means of experiences. But other people can chose another perspective, chose another informal theory, axiomatized theory and a corresponding model as the intensional one. Well done. Out of a debate, in general (but not always) sustained by empirical evidence, it is difficult to sustain a perspective in detriment of another one. This is what makes science so rich. These different perspectives may require distinct logic and mathematical frameworks, but no one will be able to prove, without agreement with experience, that *her* perspective is the better one. Thus, experience is important in this view, and hence semantics, the way to link theory with "the world".

The interesting thing is that, in some cases, as concerning the microscopic realm (perhaps celestial bodies would be another example), sometimes we have no direct access to the basic entities that are the raw stuff of the considered field. Even so, we have the theories, and perhaps we can arrive at a situation were the field itself is so blurred that all we have are the theories themselves: the entities, that is, the theory's basic ontology, are prescribed by them, as nomological objects referred to above. Thus, we are like an engineer that needs to construct a certain piece of material without having it to copy, but just from the top view, the front view and one of the side views of the object (in oblique drawings). Each view acts as one of the informal theories mentioned above. None is "the theory" of the field, as none view is the exact picture of the object, but all of them, taken jointly, provide to the engineer a quite precise idea of the object. Thus, perhaps in considering all perspectives (different theories and models) of a certain field, we will have a better idea about it, without the needs of defending one perspective over the others, except by local pragmatic issues, like to say that the frontal view is better to see

---

[18]We also will not pursue this topic here, but an interesting view, which we partially endorse, is Schrödinger's, grounded on his "principle of objectivation" – [32]. According to Schrödinger, in a certain sense, the "reality" is a mental construct of ours, but not from fixed categories in Kant's sense – see also [1].

some particularity of the object.

Thus, concerning science, we might suppose that the "non-classical views", grounded on non-classical logics and mathematics, are in a certain sense as licit as the "classical" one, and only pragmatic criteria, or the evolution of science itself will show the right way to look to our intended field of knowledge, although we strongly believe that even in the future there will be no "right way" to theorize about a certain (sufficiently wide) field. So, how to deal with such a plethora of possibilities? This is our final remark: at the end, as a supra meta-rule, there is a kind of *constructive* activity. The different perspectives may act as pieces of a mosaic we use to cover the terrain (the field of knowledge under analysis). We join the parts as a child does with her corner puzzle.

BIBLIOGRAPHY

[1] Ben Menanhen, Y. (1992), "Struggling with realism: Schrödinger's case", in Bitbol, M. and Darrigol, O. (eds.), *Erwin Schrödinger: Philosophy and the Birth of Quantum Mechanics*, Paris: Frontières, 25-40.

[2] Bitbol, M., (1996), *Schrödinger's Philosophy of Quantum Mechanics*, Dordrecht: Kluwer Ac. Press (Boston Studies in the Philosophy of Science, Vol. 188).

[3] Coughlan, G. D. and Dodd, J. E. (1993), *The Ideas of Particle Physics: An Introduction for Scientists*, Cambridge: Cambridge Un. Press, 2nd. ed.

[4] da Costa, N. C. A. and Chuaqui, R. (1988), "On Suppes' set theoretical predicates", *Erkenntnis* 29, 95-112.

[5] da Costa, N. C. A. and French, S. (2003), *Science and Partial Truth: A Unitary Approach to Models and Scientific Reasoning*, Oxford: Oxford Un. Press.

[6] da Costa, N. C. A. and Krause, D. (1994), "Schrödinger logics", *Studia Logica* 53 (4), 533-550.

[7] da Costa, N. C. A. and Krause, D., (1997), "An intensional Schrödinger logic", *Notre Dame Journal of Formal Logic* 38 (2), 179-194.

[8] da Costa, N. C. A. and Krause, D. (2006), "Remarks on the applications of paraconsistent logic to physics", in Pietrocola, M. and Freire Jr., O. (orgs.), *Filosofia, Ciência e História: uma homenagem aos 40 anos de colaboração de Michel Paty com o Brasil*, S.Paulo, Discurso Editorial, 357-359.

[9] da Costa, N. C. A. and Krause, D. (2006a), "The logic of complementarity", in van Benthem, J., Heinzmann, G., Rebuschi, M. and Visser, H. (eds.), *The Age of Alternative Logics: Assessing Philosophy of Logic Today*, Springer, 103-120.

[10] da Costa, N. C. A. and Krause, D. (2007), "Logical and Philosophical Remarks on Quasi-Set Theory", *Logic Journal of the IGPL* 15, 1-20.

[11] da Costa, N. C. A., Krause, D. and Bueno, O. (2006), "Paraconsistent logics and paraconsistency", in in D. Jacquette, D.M. Gabbay, P. Thagard and J. Woods (eds.), *Philosophy of Logic*, Elsevier, 2006, in the series Handbook of the Philosophy of Science, v. 5, 655-781.

[12] da Costa, N. C. A. and and Rodrigues, A. M. N. (2007), "Definability and invariance", *Studia Logica* 86(1), 1-30.

[13] Dalla Chiara, M. L. and Toraldo di Francia, G. (1981), *Le Teorie Fisiche: Un' Analisi Formale*, Torino: Boringhieri.

[14] Dalla Chiara, M. L., Giuntini, R. and Greechie, R. (2004), *Reasoning in Quantum Theory: Sharp and Unsharp Quantum Logics*, Dordrecht: Kluwer Ac. Press (Trends in Logic, Vol. 22).

[15] Davies, Brian E. (2003), *Science in the Looking Glass: What do Scientists Really Know?*, Oxford: Oxford Un. Press.

[16] Feyreabend, P. (1975), *Against Method*, London: NLB.

[17] Février, P. D. (1951), *La structure des Théories Physiques*, Paris, Presses Un. de France.

[18] Díez, J. A. and Moulines, A. U. (1999), *Fundamentos de Filosofí de la Ciencia*, Barcelona: Editorial Ariel, 2a. ed.

[19] French, S. and Krause, D. (2003), "Quantum Vagueness", *Erkenntnis* 59, 97-124.

[20] French, S. and Krause, D. (2006), *Identity in Physics: A Historical, Philosophical, and Formal Analysis*, Oxford: Oxford Un. Press.

[21] Goldstein, S. (2006), "Bohmian Mechanics", *The Stanford Encyclopedia of Philosophy* (Summer 2006 Edition), E. N. Zalta (ed.), URL = http://plato.stanford.edu/archives/sum2006/entries/qm-bohm/.

[22] Hilbert, D. (1901)[1976], "Mathematical problems", reprinted in Browder, F. E. (ed.), *Mathematical Problems Arising from Hilbert Problems*, Proceedings of Symposia in Pure Mathematics, Vol. XXVIII, Providence, American Mathematical Society, 1-34.

[23] Jammer, M. (1974), *Philosophy of Quantum Mechanics*, New York: John Wiley.

[24] Levy-Leblond, J.M. and Balibar, F. (1996), *Quantics? Rudiments of a Quantum Physics*, North-Holland

[25] Ludwig, G. (1985), *An Axiomatic Basis for Quantum Mechanics*, Vol. I, Springer-Verlag.

[26] Mackey, G. W. (1963), *Mathematical Foundations of Quantum Mechanics*, New York, Benjamin.

[27] Maudlin, T. (2003), "Distilling metaphysics from quantum physics", in M. J. Loux and D. W. Zimmermann (eds.), *The Oxford Handbook of Mataphysics*, Oxford: Oxford Un. Press, 461-487.

[28] Priest, G. (2006), *Doubt Truth to be a Liar*, Oxford: Clarendon Press.

[29] Reichenbach, H. (1944) [1998], *Philosophic Foundations of Quantum Mechanics*, Mineola: Dover Pu.

[30] Redhead, M. (1987), *Incompleteness, Nonlocality, and Realism: a Prologomenon to the Philosophy of Quantum Mechanics*, Oxford: Clarendon Press.

[31] Runes, D. D. (org.) (1942), *The Dictionary of Philosophy*, New York: Philosophical Library.

[32] Schrödinger, E. (1992), "The principle of objectivation", in Schrödinger, E., *What is life? & Mind and Matter*, Cambridge: Cambridge Un. Press.

[33] de Souza, E. (2000), "Multideductive logic and the theoretic-formal unification of physical theories", *Synthese* 125, 253-262.

[34] Suppes, P. (1962), "Models of data", in E. Nagel, P. Suppes and A. Tarski (eds.), *Logic, Methodology and Philosophy of Science: Proceedings of the 1960 International Congress*, Stanford: Stanford Un. Press, 252-261.

[35] Suppes, P. (1966), "The probabilistic argument for a non-classical logic of quantum mechanics", *Philosophy of Science*, 33, 14-21. Reprinted in C. A. Hooker (ed.) (1975), *The Logico-algebraic Approach to Quantum Mechanics*, Dordrecht, Reidel, 341-350.

[36] Suppes, P., (1994), "A brief survey of Adams' contributions to philosophy", in E. Eells & B. Skyrms (eds.), *Probability and Conditionals: Belief Revision and Rational Decision*, Cambridge: Cambridge University Press, 201-204.

[37] Suppes, P. (2002), *Representation and Invariance od Scientific Structures*, Stanford: CSLI Pu.

[38] van Fraassen, B. (1980), *The Scientific Image*, Oxford: Clarendon Press.

[39] Toraldo di Francia, G. (1985), "Connotation and denotation in microphysics", in Mittelstaedt, P. and E. W. Stachow (eds.), *Recent developments in quantum logics*, Mannheim, 203-214.

[40] Urchs, M. (1995), "Comments on Aldo Bressan's paper", *Logic and Logical Philosophy* 3, 37-42.

The Programme of Aristotelian Analytics

MICHEL CRUBELLIER

ABSTRACT. In this paper, I submit an overall interpretation of Aristotle's *Analytics* (*Posterior* as well as *Prior*) which I could express, to put it in a nutshell, by saying that the *Analytics* are analytic. That is, they do not lay out progressive or constructive processes, in which, given certain fundamental premises, terms or rules, one would go ahead and draw conclusions or even build a systematic body of knowledge on the basis of these principles. Rather they describe a backward movement, starting from a proposed or provisional conclusion and asking which premises could (or could best) be used in order to deduce, support, prove or explain it.[1]

1   On the title *Analytics*

These views arose from certain perplexities I have long had about the title *Analytics* (which is most certainly genuine, to judge from the numbers of mentions in the Corpus and from the fact that many of them could hardly have been later additions made by an ancient editor). This title raises immediately two problems:

(1) It is common to two quite different works, one of which is supposed to be a treatise of formal logic (the so-called "syllogistic"), while the other one is about the epistemology and methodology of exact sciences. There are some very clear indications that Aristotle, even if he wrote them at distinct (and maybe distant) times of his life, did regard them as two parts of one and the same project. See for instance the beginning of *An. Pr.* I 4: "Having made these determinations, let us now say through what premises, when and how every deduction comes about. We will need to discuss demonstration later.

---

[1] I have expounded the ideas that I develop here before various audiences, in Paris and Nanterre in 2001, and in Créteil in 2005. I want to thank my hosts, Pierre Pellegrin, Francis Wolff and Souad Ayada, for giving me these opportunities to test my views and (I hope) to improve them on some important points. During the same period, I had opportunities to present them to my students and to some colleagues in Lille, and I benefited greatly from long and exciting discussions with Shahid Rahman and his logician friends and students. I also want to thank the two referees who reviewed this paper before its publication, one of which proved to be no other than Robin Smith; I felt much encouraged by their positive comments, which made my own ideas much clearer even to myself. Of course, the remaining faults are mine.
An outline of the same views, much shorter and less systematic, was published in [8], chapter 2.

Deduction should be discussed before demonstration, because deduction is
more universal: a demonstration is a kind of deduction, but not every deduction is a demonstration."[2] This passage is echoed by the first lines of the
last chapter of the *Post. An.*: "Now as for deduction and demonstration, it is
evident both what each is and how it made up."[3]

(2) The word *Analutika* itself might appear surprising. If "deduction" and "de-
    monstration" were to be conceived of – as it is generally admitted – as con-
    structive processes, starting with the assumption of certain propositions and
    then proceeding on to the conclusions that the premises logically entail, how
    strange it would be to call these treatises *Analytics*! The verb *analuein*, on
    the contrary, indicates the splitting-up of a whole into its component parts.
    One use of the word, probably the most frequent in the Aristotelian corpus,
    means the way in which all natural objects may be decomposed into the four
    simple bodies, earth, water, air and fire. Aristotle also knows the "analytic"
    method used in geometry for the resolution of some problems.[4] The third
    important use is the one I want to elucidate in this paper.

In the following pages, I will use the word *analytics* (without a capital) to indicate the peculiar ability (which is not a science) that one may acquire through
reading and working up the *Analytics* (the treatises), just as Brunschwig ([4]) once
proposed to call *topics* the peculiar ability which you may acquire through reading
and practice of Aristotle's *Topics*. The person who practises the analytical ability
I will sometimes call "the analyst".

2   The plan of the *Prior Analytics*

What is the project of Aristotle's analytics? The opening pages of the first treatise
give us very few indications about that question. After having given in the first

---

[2]*Pr. An.* I 4, 25b 26-31. – I quote the *Prior Analytics* in Smith's translation ([13]), and other
Aristotelian texts in the "Revised Oxford Translation"(ROT), into which I made such changes as were
required (1) to match the interpretations that I want to defend and (2) to obtain sufficient terminological homogeneity between different treatises. To avoid making my footnotes too cumbersome, I did
not attempt, except on very few occasions, to indicate and justify these changes. I hope that readers
who would like to compare my citations with Smith and the ROT will easily understand what I have
changed and why. For the same reason, I tried to leave aside, as far as possible, the many linguistic
and philological questions raised by these passages. I only wish to mention here that although I did not
keep Barnes' rendering of *episteme* through "understanding", I do think that there are good reasons to
support it (Burnyeat [5]); but I thought that the traditional translation through "science" would be more
evident to those readers who are not sedulous Aristotelians.

[3]*Post. An.* II 19, 99b 15-16. – We may be tempted to infer from this that the last chapter of *Post. An.*
is entirely distinct from the rest of the two treatises; or even that, indeed, the whole is divided into two
distinct parts (although these would be quite unequal in length), the first one dealing with deduction
and the second one with induction. But on this see section 6 below, pp.140-141.

[4]See for instance *Nicomachean Ethics* III, 1112b 21-22.

chapter, without any comment, a small number of inaugural definitions, Aristotle immediately enters into his subject matter. But later on in Book I, there are two important passages that, while emphasizing a transition, do disclose a larger plan:

Transition between chapters 26 and 27 of Book I

"[a] From what has been said, then, it is clear how every deduction comes about, both through how many terms and premises and what relationship they are in to one another, and furthermore what sort of problem is proved in each figure, and what sort in more, and what in fewer figures. Now it is time to explain [b] how we may ourselves always be supplied with deductions about what is set up and the route by which we may obtain the principles concerning any particular subject. For surely one ought not only study the origin of deductions, but also have the power to produce them."[5]

Transition between chapters 31 and 32 of Book II

"It is evident from the things which have been said, then, [a] what all demonstrations come from, and how, and [b] what things one should look to in the case of each problem. But after these things, we must [c] explain how we can lead deductions back into the figures stated previously; for this part of the inquiry still remains. For if we should [a] study the origin of deductions, and [b] also should have the power of finding them, and if, moreover [c] we could resolve (*analuoimen*) the existing deductions into the figures previously stated, then our initial project would have reached its goal. It will also result at the same time that what we have said previously will be rendered more secure, and it will be clearer that this is how things are, by what we are now about to say; for all that is true must in all ways be in agreement with itself."[6]

These passages suggest that the *Prior Analytics* divide into three main parts, corresponding to three distinct competences characteristic of analytics, two of which ([a] and [c]) are theoretical, while the other one is more practical:

[a] to know "how" or "from what" deductions are "produced" (or "what their inner structure is" – for the verb *gignetai*, which is used in both passages, does not necessarily mean a real process of production, but might refer only to the form of the so-called modes of syllogism);

---

[5] *Pr. An.* I, 43a 16-24.
[6] *Pr. An.* I, 46b 38 - 47a 9.

[b] to be able to find appropriate deductions for any "problem" whatever (that is, every proposed conclusion);

[c] to understand how existing types of inference do conform to the "syllogistic" structures set forth in part **[a]**, thus providing an *a posteriori* corroboration for the results of that part. This part of the treatise, by far the longest one, is based on the conviction that all our reasoning, or the largest part of it, can be explained through the models established in section **[a]**. Witness this declaration towards the end of Book II:

> "Now, it should be explained that not only dialectical and demonstrative deductions come about through the figures previously mentioned, but also rhetorical ones, and absolutely any form of conviction whatever, arising from whatever discipline."[7]

(The text goes on: "For we have conviction about anything either through deduction or from induction", on which see below, section 5.)

3   *Analytics* **and** *Topics*

The programme of the "practical" part **[b]** immediately reminds of the beginning of the *Topics*:

> "Our treatise proposes to find ways of proceeding (*methodos*) whereby we shall be able to reason deductively from reputable opinions about any subject presented to us, and also shall ourselves, when putting forward an argument, avoid saying anything contrary to it."[8]

The notions of "ways of proceeding" (*methodos*)[9] and of "any proposed problem" indicate the same kind of prospect as in the *Analytics*. Even the definition of deduction is exactly the same in both treatises:

> "A deduction is a discourse in which, certain things being supposed, something different from the things supposed results of necessity because these things are so"[10]

– a fact which is generally recorded with some embarassment by the commentators, since the standard model of a syllogism is obviously absent from the *Topics*. But anyway, the aim of the *Topics*, seen as a whole and, so to say, "from outside",

---

[7] *Pr. An.* II 23, 68b 9-13.

[8] *Topics* I 1, 100a 18-21.

[9] The term *methodos* is echoed in the *Analytics* passage by that of a "way" (*hodos*). This same word recurs at the beginning of *Pr. An.* I 30 to summarize the contents of chapters 27-29 – on which see below, section 9.

[10] *Pr. An.* I 1, 24b 18-20; *Topics* I 1, 100a 25-26.

appears to be the same as that of the *Analytics*: to teach ways of finding premises for a given conclusion.

Nevertheless, there are also some significant differences between the two treatises. To begin with, the *Topics* passage makes two important qualifications: first, the aimed-at ability finds its place in a situation of contest or struggle, where one will have not only to find arguments, but also to defend oneself as well as to attack one's opponent; second, the *Topics* says that the relevant deductions should be based on "reputable opinions"[11], while in the *Prior Analytics* no particular requirement is expressed as to the origin or status of the premises. (In fact, the general notion of "premises for a given conclusion" seems to call for such specifications, according to the epistemic character of various types of conclusions and the contexts in which they are to be used.[12] It is all the more noteworthy that the *Prior Analytics* seems to disregard that point.) Moreover, while the search for a method of finding premises seems to be the sole aim of the *Topics*, it represents a comparatively small part of the *Prior Analytics*.

This raises the problem of the relationship existing between *Topics* and *Analytics*. The most ancient view on that subject commands the general plan of the *Organon* as we know it since Andronicus of Rhodes: after two short introductory treatises on simple terms (the *Categories*) and on propositions (*De Interpretatione*), comes a general, mainly or purely formal, theory of deduction (the *Prior Analytics*) followed by two special treatises: the *Posterior Analytics* on "demonstration" (*apodeixis*), i.e. scientific deduction, based on true premises, and another one about deduction from "reputable" premises, i.e. beliefs which are current among educated people or among specialists. One weakness of this model is that it breaks down the unity of the *Analytics*, which seems to me beyond reasonable doubt.

One might also consider the case from a historical point of view. It is generally admitted that the *Topics* was written at an early stage of Aristotle's career. But there are several cross-references between it and the *Analytics*, a fact that suggests not only that the treatise was still read by the time Aristotle was writing the *Analytics*, but also that the subject matter itself was probably still a part of the teaching programmes of the Lyceum.

---

[11] On this notion of "reputable opinions" (*endoxa*), see *Topics* I 1, 100b 21-23.

[12] The *Rhetoric* too is a handbook of argumentative discourse, and could be described (at least partly) by the same formula. Indeed, Aristotle stresses that his conception of rhetoric as an "art" (i.e. a technical study) owes much to his (presumably earlier) discoveries in the domain of dialectic and analytics. There is one important difference: rhetorical argumentation deals with objects that are actions and decisions, and therefore particular and contingent items. The distinctive inferential moves of rhetoric, enthymemes and examples, are closely corresponding to deductions and inductions, but the fact that they deal with particulars, makes them somewhat different from the standard analytical models of induction and deduction. On this see [7].

We shall have to come back again to these issues in the last two sections of this paper (§9 and §10). Although they may seem to be of merely historical interest, I believe that a more exact assessment of the relationships between Aristotle's different projects in the field of the theory of argumentation may greatly improve our understanding of their philosophical meaning.

4   The analytic method

Now, in what sense do the three parts that we have distinguished in the *Prior Analytics* implement one method, which may be aptly described as "analytic"? The verb *analuein* has here a strong and precise meaning: it indicates one definite act, by which the analyst divides the conclusion into two distinct propositions, which will be the premises, by chosing a convenient "middle term". The *problema* or "proposed conclusion" is a proposition, which – in the simplest cases – is made out of two terms. The middle term, freely introduced by the analyst, combines separately with each one of the two terms of the *problema* to give two new propositions. These propositions may – if the middle term has been aptly chosen – become the premises of a regular deduction or "syllogism". Thus, when Aristotle writes in the *Metaphysics* that "the parts are causes of the whole, and the hypotheses are causes of the conclusion, in the sense that they are that out of which these respectively are made",[13] we have to take it literally. It appears (from some characteristic phrases that occur frequently in the *Analytics* and throughout the Corpus) that Aristotle often illustrated his "logical" explanations with diagrams in which a proposition of the form "A belongs to B" was figured by a vertical line, the predicate being at the top of it and the subject at the bottom (fig. 1):

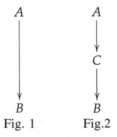

Fig. 1   Fig.2

Then the essential analytical act is the setting of point C (fig. 2), the middle term, producing two new propositions, "A belongs to C" and "C belongs to B". Of course, further important points must be taken into account, which are well known by every person who has at least a basic knowledge of "syllogistic":

- The middle term need not necessarily be placed between A and B, as it is in our example (corresponding to the Aristotelian "first figure"); it may also

---

[13]*Metaphysics* Δ 2, 1013b 20-21.

be said to belong to B and to A as well (and thus be placed "above" A, as is the case with the "second figure"), or to be a subject to which both A and B belong, and then be placed "under" B ("third figure").[14]

- The relation between two given terms may be specified as affirmative or negative, universal or particular, and modalized in different ways.

- Not every combination of three terms (the so-called "moods") gives a valid deduction (i.e. conforms to the above-mentioned definitions of deduction, i.e. such as the affirmation of the conclusion follows necessarily from the assumption of the premises). The role of part [a] of the *Prior Analytics* is precisely to determine which of them are valid and which are not. This is done in chapters 4 to 6 for non-modalized propositions, and in chapters 8-22 for modalized ones. It is worth noticing that this section is not "analytical" in the sense just defined, since it does not proceed backwards, from the proposed conclusion to the premises. But neither is it a constructive process, starting from simpler or more self-evident axioms or rules to infer from them more complex or less evident results. In fact, Aristotle generates all possible deduction models through a strictly combinatory process, then he checks each of them separately. He then rejects those that are not valid, giving an invalidity proof for each one, and gives proofs for the existence of a deduction for all the other cases. Even when he shows, in chapter 7, that all valid moods of the second and third figures can be reduced (*analuesthai*) to valid moods of the first, these are not meant to play the role of "axioms", the truth of which would ensure the validity of the second- and third-figure moods, but the aim of this (analytical) operation is just to corroborate the results that have been first obtained through the "combination plus elimination" procedure. The method is basically the same, although its application is often more complicated, in the exposition of modal deductions.

Thus, if the interpretation that I am advocating here is correct, it implies – surprising as it might seem to some people – that these first chapters are not the core of Aristotle's project, but rather a preliminary step to the use of the analytical method in order to find [b] convenient starting-points for argumentation and to understand [c] our current modes of inference.

Another striking aspect of Aristotle's way of putting these matters is his insistence on figures, while modern accounts of "syllogistic" tend to give more importance to the moods. See for instance this judgment by Łukasiewicz: "The distribution of syllogisms into figures seems to have no other than a practical purpose:

---

[14]That may provide a sufficient explanation for the fact that Aristotle did not consider a "fourth figure".

the point is to make sure that no true syllogistic mood is left out."[15] On the contrary, Aristotle has no technical word for the "moods", while he refers constantly to the three figures in his analyses and explanations of kinds of inferences or errors, and so does he in the *Posterior Analytics* as well. This aspect of analytic method certainly deserves more attention from the scholars than it has received till now. Commentators have spent much ingenuity on the – purely formal – question of the so-called "fourth figure". But what strikes me most is the fact that figures are a really effective tool for the classification and assessment of various kinds of inferences, and for the tactical construction of some proofs. Aristotle seems to think that the differences between the figures do reflect significant differences in reality.[16]

5   The third part of the *Prior Analytics*

Besides the act of fixing the convenient middle term, or resting on this act, there are three main operations to which Aristotle refers by the verb *analuein*, all of which are brought into play in the third and longest part of the *Prior Analytics*:

1. To bring back a complex argument to a chain of elementary deductions matching the models set forth in part [a]. This is only an extension through iteration of the basic analytical move. One can find an interesting illustration of such a process in chapters 19-23 of Book I of the *Posterior Analytics*. The aim of this long and complicated argument is to show that there are complete scientific demonstrations;[17] even more, that every conclusion which is susceptible of demonstration can be brought back to a finite number of premises, i.e. that one can insert a certain number of middle terms (let us say C_1, C_2, C_3, ...) between the terms of that conclusion, but that the process cannot go on indefinitely: the demonstration can be brought to a stage at which God himself could not find anything more to demonstrate in it.

2. To translate a piece of argument expressed in natural language, into syllogistic formulas. The following lines offer an example of a situation in which the analyst might hesitate on how to translate conveniently:

> "Something extra duplicated in the premises should be put with the
> first extreme, not with the middle. I mean, for instance, if there should
> be a deduction that there is knowledge in that it is a good of justice,
> then the expression 'in that it is good' (or '*qua* good') should be put
> with to the first extreme. For let A be 'knowledge in that it is a good',

---

[15][11], §9.

[16]"– What do you mean by "reality"? Our real processes of thinking, or relations between the things themselves, or between concepts thought of as objective realities? – Well, given Aristotle's epistemological realism, it might well be the three of them together."

[17]See below, section 6.

B stand for 'good', C stand for 'justice'. Then it is true to predicate
A of B: for of the good there is knowledge in that it is good. But it
is also true to predicate B of C: for justice is essentially a 'good'. In
this way, then, an *analusis* comes about. However, if the expression
'in that it is a good' is put with B, then no *analusis* will be possible.
For A will be true of B, but B will not be true of C: for to predicate
the term 'good in that it is a good' is incorrect and not intelligible."[18]

In order to extract the elementary terms of a deduction from the more complex
structure of natural discourse, the difficulty is to individuate them correctly. The
point at issue is the allocation of the phrase '*qua* good' (or: 'that it is good'). The
correct *analusis* produces a regular first-figure syllogism, with the predicate A =
'knowledge of x *qua* good'. In the mistaken *analusis*, A becomes A' = 'knowledge', and B becomes B' = 'the good *qua* good'. Then it is indisputably true that
there is a knowledge of the good *qua* good, but the minor premise: 'justice is a
good qua good' is "false", in fact because the phrase 'good *qua* good' does not
make any sense.[19]

**3. To explain or show the effectiveness of various kinds of inference, even of
those which are clearly distinct from deduction, for instance proofs by reduction to impossibility, or inductive inferences.** As to the reduction to impossibility, Aristotle says[20] that, taken as a whole, it cannot be "analyzed" in the sense
explained here. But one can distinguish two parts in it: (1) the general frame of the
proof, i.e. the claim that if a given proposition p is shown to lead necessarily to an
impossibility, then p is false and not-p is true; (2) the derivation of this impossible
consequence. The latter can and must be given the form of a regular deduction, and
thus it is analysable, while the effect of the former (i.e. that the opponent's claim
proves untenable) rests on a dialectical rule: the impossibility to assume contradictory propositions. More generally, inferences that rest on a "hypothesis" with
the effect of eliminating the possibility to admit certain claims are not analysable.

The case of inductive inference (*Prior An.* II 23) is perhaps the most interesting,
to show the scope and distinctive efficiency of the "analytic" procedure. Aristotle
describes it thus:

"Induction, then, or rather a deduction from induction,[21] is deducing
one extreme to belong to the middle through the other extreme, for
example, if B is the middle for A and C, proving A to belong to B

---

[18] *Pr. An.* I 38, 49a 11-22.

[19] Let me draw the attention of readers of the *Metaphysics* on the fact that this remark may concern
the famous opening sentence of Book Γ: "There is a science which considers being *qua* being", so
that one is not allowed to extract from it the phrase "being *qua* being", as if it were the designation of
something.

[20] *Pr. An.*. I 44, 50a 29-38.

[21] More on this apparently strange phrase in the next section.

by means of C (for this is how we produce inductions). For example, let A be long-lived, B stand for not having bile, and C stand for a particular long-lived animal, as man, horse, mule. Now, A belongs to the whole C (for whatever is bileless is long-lived)[22]; but B (not having bile) belongs to every C. If, then, C converts with B and the middle does not reach beyond the extreme, then it is necessary for A to belong to B (...). But one must understand C as composed of all the particulars; for induction is through them all."[23]

Two features are particularly striking in this description:

(1) Here, contrary to the original presentation of *analusis*, the subject of the conclusion is called "the middle term". This departure from Aristotle's own constant usage may seem surprising. Why didn't he use rather the model of third-figure deductions, in which the middle term is set "outside" the [A, B] interval and below B? The reason for this is easily found: third-figure moods give only particular conclusions, while in the case of induction what is supposed to be inferred must be a universal proposition. (Another justification lies in Aristotle's express claim that this kind of inference is appropriate to a "primary and unmiddled premise; for the deduction of those premises of which there is a middle term is by means of the middle term; but the deduction of those of which there is not a middle term is by means of induction."[24] But with this last remark he leads us beyond a mere theory of inference in general, since the notion of "primary and immediate propositions" is crucial to his theory of science, and maybe does not mean anything outside that context – on which see the next section.)

(2) This apparent oddity is made good by the fact that the proposition: "B belongs to C" (e.g. "man, horse, mule are bileless") is assumed to be convertible. Thus we obtain the following regular first-figure deduction:

(The) bileless animals are {man, horse, mule}

{Man, horse, mule} are long-lived
_____

Bileless animals are long-lived

Here we may recognize the standard model *Barbara*, except that the middle term is the list {man, horse, mule}. Thus one could describe inductive inference as a kind of "deduction" in which the role of the middle term is

---

[22]This parenthesis, which has shocked many scholars, must be read as a sort of commentary from an "external" point of view, so to say, and not as a proper part of the inference itself (since that would make it circular).

[23]II 23, 68b 15-29.

[24]II 23, 68b 30-32.

played not by a single term, but by a list. This characteristic implies two weaknesses, which give rise to the "induction problem": (a) the unity of this middle term, and (b) its completeness, are not guaranteed. This is not the place to discuss these issues. But it is important to have in mind, while reading this chapter, that Aristotle is not claiming to introduce another type of deduction matching the standards of necessity and accuracy of the so-called "syllogistic". He is just trying to account for the fact that we do make such inferences as that of the bileless animals, and for the (limited and risky, but real) efficiency of that spontaneous activity of human mind. (Besides, he also says[25] that it is more particularly appropriate to rhetoric than to science or technical reasoning – but we need not take him at his word, since, as we shall see, induction plays an important role in his theory of science). Seen that way, it is no misfortune, but rather a good point for Aristotle's *analusis*, if it raises interesting questions about this kind of inference.

In the same spirit, in one section of Book II of the *Prior Analytics*[26] he "analyzed" various kinds of faults in reasoning by means of the three figures of deduction, with the intention to make his reader more able to avoid such faults, or to detect them in his opponent's discourse. In other places, he even went further in that direction, since he made use of the "middle term" in order to explain and discuss some mental processes which have little or nothing to do with inference, such as the structure of definitions,[27] the doubtful guesses of physiognomony,[28] the formation of voluntary decisions,[29] and even the practical tricks of mnemonics.[30] Clearly, he considered his discovery of the middle term to be a great achievement (and a great achievement it was, as we shall see). But fond of it as he may have been, his fondness did not lead him to uncritical enthusiasm, as we can see from the fact that he honestly admits that such – indisputably conclusive – modes of inference as proof by reduction to impossibility cannot be analyzed in that manner.

6 Analytic method and the foundations of science

I hope that what I have said so far may account in a satisfactory way for the contents and order of the *Prior Analytics* (please refer to the general plan given in the *Appendix*). But what about the *Posterior Analytics*? The textual data about the aim and plan of this treatise are not so clear as those that we have found in the *Prior Analytics*. The most obvious fact is the division into two main parts correspond-

---

[25]Compare *Pr. An.* II 23, 68b 9-13, quoted above.

[26]II, 16-21.

[27]*Post. An.* II, 3-10; on which see section 8.

[28]*Pr. An.* II, 27.

[29]*On the Soul* III, 10-11; *Movement of Animals*, chap. 7; *Nicomachean Ethics* VI, 1142a 22-23; VII, 1147a.

[30]*On Memory and Recollection*, chap. 2.

ing to the two books, the first one on demonstration and the second on definition – although its last six chapters (at least) are dealing with rather different topics, namely the method for setting problems (14-15), causal explanations (16-18) and the knowledge of general terms (19). Each of the two books has a finely worked-out introduction, which sets out its subject in a methodical and philosophically illuminating manner, but both seem to end rather confusely, in a loose series of notes more or less closely related to their main topic (a feature which is not un-common in Aristotle's treatises). Even the inner order of each book is far from being clear. The most clearly programmatic text in the *Posterior Analytics* is the beginning of Book II, which classifies the objects of knowledge under four head-ings: "the fact, the reason why, if it is, what it is" – which are straightafter reduced to two, since Aristotle tells that knowing the fact is just a preliminary step to the knowledge of the "why", and knowing "that it is" is a step towards the knowledge of "what it is". Thus we are left with two paths of scientific knowledge, the first one dealing with propositions, and the second one with simple terms. This in turn seems to correspond well enough to the subject matter of the two books, demon-strations on the one hand and definitions on the other. It suggests the picture of an independent treatise on science and its methods, with no or little connection with the problems of formal logic which form the whole substance of the *Prior Ana-lytics*, despite the two passages that I mentioned at the beginning of this paper,[31] which seemed to presuppose a strong unity between the two treatises. In this case the title *Analytics* would have been given to the *Posterior* owing to the fact that demonstration is a kind of deduction, and nothing more (as in the commentators' vulgate).

Yet one can find in the *Posterior Analytics* clear indications that the problematic of finding a middle term is quite relevant to Aristotle's presentation of science. (1) At the beginning of Book II, immediately after having set out his typology of the objects of scientific inquiry, he adds: "in all our searches we seek either if there is a middle term or what the middle term is."[32] This implies (2) that definition is susceptible of a kind of analysis: "we have said (...) how the 'what it is' is ex-plained into the terms",[33] and that in some sense definition might be considered as a demonstration of a kind, or at least is closely related to a possible demonstration: "and ⟨we have said⟩ in what way there is or is not demonstration or definition of it [= the 'what it is']."[34] Besides, I have already mentioned (3) the long demon-stration of Book I, chap. 19-23, to the effect that a scientific demonstration can be brought to a state in which it is complete, i.e. that there is a finite number of middle terms between its predicate and its subject.

---

[31] *Pr. An.* I 4, 25b 26-31; *Post. An.* II 19, 99b 15-16.
[32] II 2, 90a 5-6.
[33] II 13, 96a 20.
[34] II 13, 96a 21.

There are still other places in which there are limited, but often quite explicit resorts to analytic method.[35] But in order to complete that discussion, we should now concentrate on the first chapters of Book I, in which the relevance of analytics is not so immediately obvious. These chapters set out a comprehensive character-ization of scientific knowledge, which seems to be drawn up on the basis of the notions that educated people of Aristotle's time, informed of the new advances of sciences (especially of geometry) and of the philosophical developments[36] on the subject, had of a science. The starting-point of this account is the following description of scientific knowledge:

"We think we have scientific knowledge of something (...) whenever we are aware both of the explanation because of which the object is (i.e., that this is the explanation of that object), and that it is not possible for it to be otherwise"[37]

– which in turn is spelled out in the following way:

"If, then, scientific knowledge is as we posited, it is necessary for demonstrative knowledge in particular to depend on things which are true and primitive and immediate and more familiar than and prior to and explanatory of the conclusion." [38]

I have kept Barnes' "things" to translate the neutral plural adjectives of the Greek original; but the reference to the "conclusion" makes it clear that these "things" are premises. The special character of its premises distinguishes scientific knowledge, and thus it turns out that in a significant way the *Posterior Analytics* fit into the programme inaugurated by the *Prior*. One could even be tempted to go farther and consider them to be nothing but a part (although expanded at great length) of the original plan, in keeping with 99b 15-17. There is indeed some continuity between the end of the *Prior Analytics* and the *Posterior*: after having dealt with the application of *analusis* to dialectical discussions (*Pr. An.* II, 16-21), next to rhetorical proofs (23-27), Aristotle would then come to apply it to scientific demonstrations and definitions. In other words, instead of opening in my *Appendix* a separate section 1.2, dedicated to Aristotle's theory of science, maybe I should rather have prefixed all the contents of the *Posterior Analytics* (with the exception of the last chapter) with 1.136, parallel to my numbers 1.134 (dialectic) and 1.135 (rhetoric).

But matters are slightly more complicated, since the overall view of scientific knowledge given at the beginning of Book I includes the statement of an impor-

---

[35]E.g. I 14, I 15-16, or II 16-18.

[36]Mainly Plato, whose conceptions and vocabulary are clearly perceptible in these chapters. Compare *Phaedo* 99-100, or *Republic*, Books VI and VII.

[37]I 2, 71b 9-12.

[38]I 2, 71b 19-22.

tant objection to the very possibility of such a knowledge, a "skeptical" objection already faced by Plato,[39] which is best expressed in the form of a dilemma: either the scientific demonstration of any given proposition is an infinite task, or one will have to assume some premises without demonstration, which would amount to make scientific knowledge rest on arbitrary assumptions:

> "Now some think that because one must have scientific knowledge of the primitives, there is no science at all; others that there is, but that there are demonstrations of everything."

Aristotle adds:

> "Neither of these views is true, nor is it necessary that one of them should be the case."[40]

The whole theory of science expounded in the *Posterior Analytics* develops under the pressure of this objection, with the result that it has to deal not only with premises *tout court*, but with principles (*archai*), i.e. absolutely first premises (and first terms). Since the scientific character of demonstration depends on the truth, explanatory value and epistemic status of its premises, a theory of demonstration cannot consider a single deduction separately; in fact, it has to introduce a more architectonic point of view, in order to provide an account of the ultimate foundations of science. Aristotle has to dismiss the "skeptical" objection, which he does partly in Book I, as we have seen, by showing that there are complete demonstrations, and partly in the last chapter of Book II – which is explicitly dedicated to that question and relates the formation of universal notions through perception, memory and language. The latter should itself be completed by *Prior Analytics* II 23, with its mention of "primary and immediate propositions". Although Aristotle speaks of "induction" in both contexts, there is a crucial difference between them: in *Pr. An.* what emerges from induction is a conclusion, i.e. a proposition (that is why he allows himself the phrase "inductive deduction"), while in *Post. An.* it is a simple term.

Other consequences of this concern with the foundations of science are (1) the presence in the *Posterior Analytics* of some reflections upon the systematic order of sciences[41] and the possibility of a unified science,[42] and (2) a celebrated classification (in chapter I 10) of the various types of "first principles", which is strongly

---

[39] The adjective "skeptical" is most certainly anachronistic, since Pyrrho, the founder of the Skeptic school, was much younger than Aristotle; but the above-mentioned dilemma became a standard of the skeptics' arsenal. Its occurrence in the *Meno* suggests a Sophistic origin.

[40] I 3, 72b 5-7. In another place, Aristotle alludes to the *Meno* paradox: "you will learn either nothing or what you know" (I 1, 71a 30, compare *Meno* 80d).

[41] See for instance I 11, 77a 26-35, I 27-28, II 15.

[42] A perspective that Aristotle rejects; see *Post. An.* I 32.

reminiscent of Euclid's *Elements*.[43]

Because of these distinctive features, the *Posterior Analytics* cannot be called a mere part of the *Prior Analytics* project. The last chapter on the origin of universal notions will show how intricate the situation is: while this chapter lies clearly outside the scope of the analytical programme, since the psychical processes that it describes do not involve anything like a middle term, it is undoubtedly an essential part of Aristotle's answer to the skeptical objection, and thus of the *Posterior Analytics*. So I am inclined to think that these were written independently of the *Prior Analytics*, most probably after them and perhaps many years after, but that Aristotle was conscious that he was deepening and increasing the kind of research he had initiated in his former treatise. That may be the reason why he referred to both of them by the same general title.

7   An overview of the *Posterior Analytics*

We are now in a position to get a clearer picture of the contents of the *Posterior Analytics*.

In the first 14 chapters of Book I, as we have seen, Aristotle characterizes scientific demonstration by means of a series of constraints bearing on the premises. I leave it to the reader to see how these constraints are related to the general description of scientific knowledge, and to follow their systematic exposition in chapters 2 to 15, since this first section is comparatively well ordered, and has been thoroughly examined by Aristotelian scholars.[44] Notice that it ends in chapter 14 with the typically analytic remark that the first figure is more fit for scientific knowledge than the other ones.

Next come three chapters (16-18) on "ignorance" (*agnoia*), which is certainly introduced here as a counterpart of knowledge. The interesting fact is that it is defined as "error coming about through deduction",[45] and analyzed in order to locate precisely the original source of the mistake, following the schemes of the three figures.

In my plan, I have treated chapters 19 to 26 as one section, under the general heading of "the ideal demonstration" (1.213). Maybe it would have been more accurate to distinguish and separate its two constituents, since the first part (19-23)

---

[43]Euclid's books were probably written some decades after Aristotle's. But since he seems to have summed up the results of a tradition of geometrical treatises called *Elements*, it is hard to decide if there has been an influence in one or the other direction, or even a reciprocal influence. Euclid himself does not comment on his method, neither does he express epistemological views of his own; the oldest commentary that we have on the *Elements* is the work of Proclus who, as any good Neoplatonist, had had a thorough training in the *Organon*. At any rate, the *Posterior Analytics* and the *Elements*, like two symmetrical monumental pillars, were to form the portico of the temple of scientific method for generations of Western philosophers.

[44]Non-Aristotelian and Greekless readers will find Jonathan Barnes' commentary [2] particularly helpful.

[45]*Post. An.* I 16, 79b 24.

upholds a particularly strong thesis (the existence of finite complete demonstrations) which is crucial to the possibility of science, while the following chapters (24-26) intend only to compare the demonstrative values of different kinds of inferences.

The rest of Book I may be considered as a rather discontinuous stretch of additional notes, related in various ways to the main contents of the book: some of them concern the kinds of facts which are object of scientific knowledge (chapters 30-31, 33), others discuss larger issues about the systematic order of sciences (27-28, 32), and the last chapter is about "exactness of mind" or "*acumen*".

The main part of Book II is about definitions. As we have seen, this large division of *Posterior Analytics* into two parts is explicitly based on a systematic classification of the objects of scientific inquiry.[46] There is no problem about that, but it is worth noticing that in Aristotle's view, a definition is a piece of knowledge and may be true or false. Which is more, a real definition is not an immediate piece of knowledge: if you define thunder as "a noise in the clouds" or an eclipse as "privation of light from the moon", you will have captured only the fact "that it is"; to express what it is in a precise and intelligible way, you have to tell that thunder is a noise caused by the quenching of fire in the clouds,[47] and eclipse a privation of light from the moon by the earth's screening.[48] Thus a scientific definition can be analyzed. Nevertheless, it has not the same status as a demonstrable proposition. Here Aristotle's discussion may fairly seem somewhat embarrassed: sections 3-7 and 8-10 are near to contradict each other, and the final result is that definitions (at least some of them, and the most interesting ones) are demonstrations of a kind, "oblique" demonstrations, so to say, "differing in aspect from demonstration".[49] Maybe one should say that such a definition "contains" a demonstration, either because it recapitulates a demonstration or because a demonstration is potentially present in it.

Another problem with this section on definitions is to mark exactly where it ends. Since chapter 13 gives directions on how to find definitions, it may seem that the section extends as far as that (and this is the division I adopted in the *Appendix*). But chapters 11 and 12 deal with the middle term as an expression of the cause, and this discussion is not necessarily limited to the middle terms of definitions (especially in the case of chapter 12, which raises an issue about the retrodictive vs. predictive character of causal explanations). So it might be more

---

[46]Barnes [2] explains the general division of the *Posterior Analytics* by saying that Book I puts forward the necessity to base scientific knowledge on first principles, and that definitions are among these first principles (p. xiii). This is true, but why put the stress on definitions rather than on some other kind of principles? One explanation might be that of all the principles, definitions are not as "first" as the other ones, since they feature a middle term.

[47]*Post. An.* II 10, 94a 5.

[48]*Post. An.* II 2, 90a 15-16.

[49]*Post. An.* II 10, 94 12-13.

appropriate to consider chapters 11 to 18 of Book II as a larger heuristic section, parallel in a way to the heuristic section of *Prior Analytics* (1.12 in my plan) and bearing on all the matters examined in the *Posterior*, i.e. precepts for (1) finding causes (chapters 10-11), (b) searching definitions (13) and (c) setting a problem clearly (14-18).[50]

8   Why so few formal syllogisms in the Aristotelian treatises?

Incidentally, grasping the distinct purpose of analytic method may cast a new light on a notorious puzzle about the *Posterior Analytics* and help, if not to solve it, at least to show there is not so much harm in it as is commonly believed. It has been noticed long ago that, while Aristotle expressly claims that scientific knowledge must take the form of demonstrations, which are themselves a kind of syllogisms, there are astonishingly few recognizable syllogisms in the rest of the Corpus. While some ancient and medieval commentators painstakingly tried to rephrase Aristotle's bright, creative and intuitive analyses into regular formal syllogisms, some of the moderns tried other ways out. Weil [14] suggested that the treatises pertained to the province of dialectic – but that would seem to pass over what Aristotle explicitly says of the epistemological hierarchy of science and dialectic, and over the many places in the corpus where he calls his reader's attention on the fact that he will proceed dialectically (*logikôs*) as opposed to *phusikôs*, which must mean: "according to the standards of natural science."[51] Barnes added[52] the suggestion that the description of complete and logically faultless demonstrations based on true premises was meant to hold only as the ideal picture of an achieved science. But it seems that we are not much better off with that, since Aristotle himself glosses the phrase "scientific deduction" with these words: "by 'scientific' I mean one *in conformity with which, by having it*, we have scientific knowledge of something."[53] There are some facts of which there is a demonstration, and this is an objective property that they have: i.e., their notion can, and indeed must, be analyzed into more fundamental terms, which manifest their cause. Thus, to have a scientific knowledge of them is to know them under the form and in the order that the demonstration exhibits. Aristotle specifies, "by having it" to exclude the case in which someone would come by chance to think of those things in that order, but without being aware that this is a demonstration. If this is so, to say that we are not yet in possession of such demonstrations means that we have, at least for the time being, no science at all. Aristotle would not have shared such pessimism or

---

[50]A third alternative, of course, would be to admit that here too we have just a disordered series of endnotes.

[51]There is even one place where *logikôs* is opposed to *analutikôs* (although analytics is not properly a science): *Post. An.* I 22, 84a 7-8.

[52]Barnes [2], pp. xii-xiii (these pages date from the first edition, 1975. Barnes qualified his position in an addition in the second edition, pp. xviii-xix).

[53]*Post. An.* I 2, 71b 18-19 (italics mine).

modesty.

Now, once it is clear that the distinct epistemological contribution of the *Analytics* does not consist in prescriptive rules concerning the logical form of scientific discourse, but rather in an explanation of the specific nature of scientific knowledge, there is no reason to consider that an argument is unscientific because it has not the canonical form of a syllogism. It is scientific as soon as it is conclusive, rests on necessary true premises and has some explanatory value. It may be interesting and useful (in order to show that it is conclusive, among other things) to translate it later on into a canonical deduction, but this is not necessary, neither does it add anything to its scientific value.

I would like to illustrate my point in reference to a case which is commonly considered to be particularly disadvantageous to Aristotle, namely mathematical reasoning. He is often blamed for having given a distorted picture of mathematics by claiming that geometrical proofs were demonstrations in his sense of the word, i.e. syllogisms. In fact, he even goes so far as to claim that the proofs of "arithmetic, geometry and optics" are first-figure syllogisms.[54] Contrast for instance Kant's clear-cut distinction between "philosophical" (i.e. syllogistic) deductions and geometrical reasoning "through construction of concepts".[55] The interpretation that I have sketched here could help to face this difficulty. For he may have meant that geometrical proofs are demonstrations because they bridge the interval between a given object (e.g. the triangle) and a given property (e.g. to "have two right angles") by means of a middle term: the three angles "disposed around one point" so as to make appear that they are equal to two right angles.[56] This is in keeping with what he writes about *analusis* as a standard procedure for the resolution of geometrical problems. For all that, I do not wish to claim that Aristotle had a clear understanding of the nature of geometrical proofs and their often complex structure, nor that his analytics, with its linear topology, could have given the means of an adequate description of mathematical reasoning. But in any case he must not be charged with blind dogmatism with regard to these questions; at the most, he may seem to be guilty of some vagueness or carelessness.

9   Is there an *ars inveniendi* in the *Analytics*?

Alongside this role of accounting (a posteriori) for our effective processes of inference, the *Analytics* claim to pursue another methodological goal, which is to make the reader, or the practising student, able to find premises by himself so that he "will never fall short" (*euporêsei*) of deductive arguments. This part of the analytic programme is unquestionably prescriptive; but does that mean that there is a

---

[54] See *Post. An.* I 14 – a claim which he justifies by saying that the first figure is the only one that can provide universal affirmative conclusions.

[55] *Critique of Pure Reason*, "Transcendental Methodology", part I, section 1.

[56] *Metaphysics* Θ 9, 1051a 24-25.

systematic method, i.e. a set of rules establishing a specific way of proceeding and possibly ensuring, under optimal or even under normal conditions, a successful outcome?

Something like that[57] appears in the central section of the *Prior Analytics* (I, chapters 27-31, labelled **[b]** above). At the juncture of chapters 29 and 30, for instance, Aristotle writes:

> "It is evident from what has been said, then, not only that it is possible for all deductions to come about through this route (*hodos*), but also that this is impossible through any other (...). The route is the same with respect to all things, then, whether concerning philosophy or concerning any kind of art or study whatever."[58]

Notice Aristotle's insisting on the fact that this method is the same for all kinds of disciplines, that is, probably, that it is independent of the distinction between science and dialectic, and thus of any restrictive condition concerning the epistemic status of the premises. The general scheme is as follows.[59] Let the proposed conclusion be of the type "A applies to E". Then you have to make six different lists of terms: three lists of terms related to A (the predicate):

(B) terms which follow from A

(Γ) terms of which A follows

(Δ) terms which are incompatible with A

three lists of terms related to E (the subject):

(Z) terms which follow from E

(H) terms of which E follows

(Θ) terms which are incompatible with E.

Now, a term which is common to one of the predicate-related lists and one of the subject-related lists may (with certain restrictions that I will not consider here) be used as a middle term in order to demonstrate that A applies (or: does not apply) to (every, or a certain) E. For instance, if you have to demonstrate that vine is deciduous, you have to find a term such as broad-leaved, which is implied by

---

[57]There is also a very similar passage about the search for definitions in *Posterior Analytics* II, chapter 13.

[58]*Prior Analytics* I 29-30, 45b 36 - 46a 2.

[59]In the following lines, I am freely paraphrasing chapter 28 from 44a 11 on. I kept Aristotle's Greek letters to name the different terms and lists (except M, the middle term, which is my own addition), but I limited myself to one example, while Aristotle, of course, examines at length all possible cases.

the notion of vine (so that broad-leaved is a member of the list Z) and implies the character deciduous (and thus is also a member of the list Γ). In chapter 28, 44a 17-19, Aristotle establishes that if a term M belongs to both Γ and Z, then it follows that A applies to every E, and that this can be demonstrated by means of M in the first figure, according to the standard *Barbara* formula:

> A applies to every M
> M applies to every E
> ───────────────────
> A applies to every E.

A similar determination of the middle term may be given for every other valid syllogistical mode.

To what extent can this be called a method for finding demonstrations? Aristotle's own assessment is strikingly balanced, as we may see from the last lines of chapter 30:

> "Consequently, if the facts concerning every subject have been grasped, from then on we are prepared to bring the demonstrations readily to light. For if nothing that truly belongs to the subject and to the predicate has been left out of our collection of facts, then concerning every fact, if a demonstration for it exists, we will be able to find that demonstration, and demonstrate it, while, if by its nature it does not admit of a demonstration, we will be able to make that evident."[60]

That may sound proud and self-confident. Indeed, Aristotle has brought out a very important result, namely, that (1) every (direct) demonstration must have the form of one of his syllogistic models, and that (2) the middle term always satisfies the condition of being at the intersection of two different lists, one of predicate-related terms, and the other of subject-related terms. Now, in order to establish this theorem, he had to suppose that we dispose of six complete lists B, Γ, Δ, Z, H and Θ. The difficulty lies less in the fact that some of these lists may be very long[61] than in the lack of any general or formal mark to ascertain that a given term belongs to a given list and, still the more, to ascertain that the list has been completed. That is probably the reason why, in the same chapter, Aristotle reminds his reader that all knowledge rests ultimately on experience:

---

[60] *Prior Analytics* I 30, 46a 22-27.

[61] This must be the case with Δ and Θ, i.e. the predicates that cannot apply to either the subject or the predicate of the proposed conclusion. The other four lists (terms that necessarily imply, or are necessarily implied by, the subject or the predicate) should be relatively short, since in Aristotle's view the essence of any real being is finite and can be known very precisely. But even the "negative" lists Δ and Θ might well be finite, (though probably very long in most cases) since they should contain only properties that are always incompatible with A (or E), which implies some necessary relation to A's (or E's) essence.

"The majority of principles for each science are peculiar to it. Consequently, it is for our experiences concerning each subject to provide the principles. I mean, for instance, that it is for astronomical experience to provide the principles of the science of astronomy (for when the appearances had been sufficiently grasped, in this way astronomical demonstrations were discovered); and it is also similar concerning any other art or science whatsoever."[62]

Thus, in the space of a few lines, Aristotle utters two strong and seemingly conflicting theses: (1) he claims that there is a universal "method" to determine with certainty the appropriate premises for any given conclusion; (2) he maintains the essentiality of experience. Here we meet, in fact, with the main difficulty for every project of an *ars inveniendi*: it has to stipulate rules and procedures that must be at least partly independent from the material import of that conclusion. Otherwise the very idea of a special, "instrumental" method would be meaningless, and the only necessary and sufficient requirement for finding convenient premises would be a thorough and accurate knowledge of the subject matter. But on the other hand one cannot discover anything without some substantial knowledge of the investigated objects, especially (in Aristotle's view) of their essence. It is interesting to notice that both theses occur together in an apparently anti-Platonic context.[63] He attacks Plato, so to say, from two opposite sides: as regards formal analysis, he claims the superiority of his own method for finding the premises, saying that Plato's method of division is "only a small part" of it; at the same time, he stresses that you cannot discover anything without experience, which implies that a purely formal method would be fruitless. In doing so, he was not necessarily inconsistent or unfair. He may have had in mind the idea that Plato's dialectic was unduly mingling formal features (i.e., based on logical relations between the terms) and elements of contents based on empirical knowledge. The interesting move he made in chapter 30 consists in distinguishing the formal from the empirical element, although he appears to have thought that both are always necessary in order to acquire any piece of real knowledge whatever.

Thus the "method" explained in chapters 27-29 of *Prior Analytics* I has its limts. It is not entirely devoid of heuristic efficiency, since it specifies which kind of relationship the middle term must have with one each of the terms of the conclusion. But it cannot go any futher, because every research must incorporate some amount of experience, i.e. some acquaintance with the things themselves, plus a special aptitude to discern the essence of each one from the collection of all its properties. To describe this particular combination, Aristotle often uses the metaphor of hunt-

---

[62]*Prior Analytics* I 30, 46a 17-22.

[63][13, 157]. The anti-Platonic scope is confirmed by the next chapter (I 31), which criticizes the Platonic "method of division".

ing.[64] Although hunting is not a random activity (it has certain rules and can be practised in a more or less rational way), it can never be made entirely independant of chance. Admittedly, the metaphor may refer to quite different situations. The method described in *Prior Analytics* I 27-30 looks more like beating over a large area in order to rouse the game; but this may take a long time, while there are also more direct ways of doing, such as following a track or standing on the watch in a place where one knows that the game is likely to show up. Not by chance does the first book of the *Posterior Analytics* end with a note on "exact mind", by which some people prove able to "hit upon the middle term in an imperceptible time."[65]

So I have to refine (and perhaps to complicate a little) the overall picture of the *Prior Analytics* I gave in section 2 above. Section **[b]**, although it seemed to be introduced as the practical part of the programme, has also a theoretical aim, i.e. to sum up the results of part **[a]** by means of a new and shorter proof. Symmetrically, practical concerns often come near to the surface of part **[c]**, since it contains a lot of instructions and dwells often on the strong or weak points of the various kinds of inference it examines, as well as on the mistakes to avoid and the causes of error.

10   Back to the relationship between *Analytics* and *Topics*

As we have seen, both treatises follow a backward path. Both aim at discovering models of inference with some degree of generality, but in two very different ways.

The *Topics* is mainly a long catalogue of "places" (*topoi*),[66] a term which belongs to the special idiom of dialectic and rhetoric. A "place" is, in Brunschwig's illuminating phrase, "a premise-making device",[67] which tells us how to construct, on the basis of the proposed conclusion, another proposition which entails it, so that once our opponent has granted this proposition, he cannot escape the conclusion.[68] The making of such devices presupposes the identification of certain

---

[64] See for instance : "...in what way one ought to hunt for the principles of deductions" (*Pr. Anal.* I 30, 46a 11-12), and in a very similar context, about definitions : "...how one ought to hunt for the predicates that are contained in the what-it-is" (*Post. An.* II 13, 96a 22-23).

[65] *Post. An.* I 34, 89b 10-11.

[66] This literal translation is to be preferred to the less surprising "commonplace", since in some contexts (e.g. in the *Rhetoric*) Aristotle makes a distinction between "common" and "special" topoi. The metaphor seems to refer to the "places" where a tidy person is sure to find the objects that he has stored there ([10], §373); Aristotle himself suggests a connection with mnemonics (*Topics* VIII 14, 163b 28-32).

[67] [4], p. xxxix.

[68] This account must be oversimplified, because it is not possible to set out conveniently here the problems raised by this kind of argumentative structure. It is to be specified at least that a "place" is not always a positive argument as in the example given above; it may also be used in view of a refutation (in cases in which T_2 implies T_1, so that the negation of T_1 entails the negation of T_2). Other interesting but difficult questions are (1) how the rule determining the content and form of T_1 on the basis of T_2 can be discovered and validated, and (2) the motives and meaning of the general classification of "places" offered by Aristotle in the *Topics*, on the basis of the "four predicables", namely accident,

standard types of proposition, with some characteristic features, which allow stating, as a general rule, that a proposition of type T_1 entails a proposition of type T_2. This will appear more clearly through a few examples:

(1) "That which is in itself the cause of good is more desirable than what is so per accidens, e.g. virtue than luck, for the former is in itself, and the latter per accidens, the cause of good things."[69]

(2) "If one predicate is asserted of two subjects, then if it does not belong to the subject to which it is the more likely to belong, neither does it belong where it is less likely to belong; while if it does belong where it is less likely to belong, then it belongs as well where it is more likely."[70]

(3) "If justice is knowledge, then injustice is ignorance: and if 'justly' means 'knowingly' and 'skilfully', then 'unjustly' means 'ignorantly' and 'unskilfully'; whereas if the latter is not true, neither is the former, as in the instance given just now – for 'unjustly' is more likely to seem equivalent to 'skilfully' than to 'unskilfully'.".[71]

These are "places". Some of them are not unfamiliar to us: perhaps the reader recognized in (2) our mode of reasoning *a fortiori*, and in (3) one version of our *a contrario*. Each of these examples shows a particular argument, which is supposed to be an instance of a more general rule. But, as one can easily see in (3), this rule itself is not necessarily incontrovertible. It may be only "likely", and its relevance to the case in point may also be questioned. This is one important difference with the deductive moods of the *Prior Analytics*, which are necessarily "true". Thus dialectical inferences are not only based on simply plausible premises, they also need not be conclusive.

But this is in fact the consequence of another, more fundamental difference: "topic" types of argument, general as they may be, still retain some elements of content, while the deductive moods of the *Prior Analytics* are purely formal. (Or, more precisely, they are syntactically formal. For, in a sense, it might be said that the universal "places" of the *Topics* are formal, as opposed to the more special arguments which are proper to some particular science; but the relevant form is "semantical" form.) So the topic way of looking for premises must end in a limited kind of universality, while the analytical formulas are not only "more universal" than the places; in fact they are universal in an entirely different way. Between the *Topics* and the *Prior Analytics* Aristotle made a grand discovery: he discovered logical form.

---

genus, proper, and definition.

[69] *Topics* III 1, 116a 1-2.
[70] *Topics* II 10, 115a 6-8.
[71] *Topics* II 9, 114b 8-13.

I would like to go a little farther and try a plausible guess about how Aristotle, starting from the *Topics*, eventually came to his notion of formal models of deduction.[72] Places such as those I have described above, are in fact anagogic arguments (or arguments "by reduction") which in principle involve two distinct moves: (1) a "shift" from proposition T_2 (the *problema*) to proposition T_1, and (2) an argument to make T_1 hold. The *Topics* concentrates on the first move, and seems to assume that the dialectician will somehow manage to have his opponent grant proposition T_1. But Aristotle may have wondered how this could be done. The most immediate solution is that, in the course of a dialectical discussion, the opponent be asked: "Do you admit that T_1, or not?" and give an affirmative answer. But one cannot be sure that this will work. So Aristotle may have been led later on to contemplate the possibility of a special proof for T_1, and then to find the device of the two premises with one common (middle) term.

A variant of this guess would be as follows. At a most abstract level, the general formula of a *topos* is:

> "[...] if the proposition that you want to establish (or refute) has the form C (in which convenient variables are substituted for the terms referring to the special subject-matter), then it will be necessary and/or sufficient to establish another proposition of the form P, where the same variables occur and may be replaced by the relevant material terms."

Now this description, with one important but limited modification (i.e.: "... it will be sufficient to establish *a pair of propositions* P_1 and P_2 ..."), could fit the syllogistic figures; so that the figures could have been discovered as *topoi* of a certain sort, particularly effective but requiring the determination of an appropriate middle term.

Be that as it may, the discovery of purely formal models must have marked a turn in Aristotle's philosophy of inference. From then on, the *Topics* was outclassed by the *Prior Analytics*; that is the reason why it was not properly integrated into the new analytical project. But neither was it altogether discarded, as we have seen. Why did Aristotle keep it alive? The most plausible answer is that he recognized its specific value as a means of finding new arguments. As we have seen, the "heuristic" parts of the *Prior* and *Posterior Analytics* are far from being as rich and fruitful as the *Topics*, because the *Topics* benefits from the resources of semantical analysis, while in the *Analytics* the "hunting" for premises or definitions draws mainly from past experience and memories of acquired knowledge,[73] or depends on the good luck of an "exact mind".

---

[72] I am freely drawing this hypothesis from a suggestion made (though on the basis of quite different presuppositions and concerns) by Hintikka [9]

[73] See *Pr. An.* I 27; *Post. An.*, I 34 and II 13.

Appendix: a general plan of the Analytics

1. *On Deduction*

1.1 **A general theory of deduction** (*Prior Analytics*)

1.11 *Systematic exposition of the elements* (or: "How deductions are constituted", Bk I, 1-26):

    1.111 Basic propositions and their conversions (I, 2-3)

    1.112 Construction of elementary deductions, non-modal (I, 4-7) and modal (I, 8-22) (classified according to the three figures)

    1.113 Explanation of the structure of elements (I, 23-25)

    1.114 Another classification of elementary deductions: "which problems are easy/difficult to solve" (I, 26)

1.12 *Heuristic* (or: "How to find appropriate deductions for any proposed conclusion", Bk I, 27-31):

    1.121 Precepts for the choice of premises (I, 27-30)

    1.122 Criticism of Plato's "method of division" (I, 31)

1.13 *Analysis of existing processes of inference* (Bk I, 32-3, and Bk II):

    1.131 Some precepts for the translation of natural-language sentences into "syllogistic" formulas (mainly negative precepts, i.e. aimed at avoiding errors) (I, 32-43)

    1.132 Cases of deductive inference that cannot be (or cannot be entirely) analysed (I, 44-45)

        – (A remark about negative conclusions – on the difference between "not being A" and "being not-A", I, 46)

        – (A remark about universal and particular syllogisms, II, 1)

    1.133 Analysis of some remarkable cases of deduction (deduction of true from false, circular reasoning, reduction to impossibility) (II, 2-15)

    1.134 Analysis of some faults in arguing (in dialectical situations) (II, 16-21)

    (Two remarks (II, 22):

        – about cases in which these extremes are coextensive

        – about axiological reasoning)

    1.135 Analysis of rhetorical models of inference (induction, example, *apagôgê*, objection, enthymeme) (II, 23-27)

1.2 **Analysis of scientific deduction (or demonstration)** (*Posterior Analytics*)

1.21 *Theory of science* (Bk I)

  1.211 Definition of science (I, 1-15)

    1.2111 Definition of science; statement of the "skeptical" objection (I, 1-3)

    1.2112 Further development of the definition: specific constraints for the premises of scientific demonstrations (I, 4-13):

      1.21121 They must be necessarily true and universal (I, 4-9)

      1.21122 They must be 'first' (undemonstrable) and 'proper' (I, 9-12)

      1.21123 They must indicate the cause (I, 13-14)
          (Remark: there are immediate negative propositions, I, 15)

  1.212 Analysis of ignorance (I, 16-18)

  1.213 The ideal demonstration (I, 19-26):

    1.2131 Demonstration of the possibility of a complete demonstration (= solution of the skeptical objection) (I, 19-23)

    1.2132 Comparison between the different types of demonstration: universal demonstrations are better than particular ones, affirmative better than negative and direct better than indirect (I, 24-26)

  1.214 Some consequences for the theory of science:

    1.2141 There are ordered series of sciences (I, 27-28)

    1.2142 There may be several demonstrations of the same proposition (I, 29)

    1.2143 Chance events and sensible facts are not objects of science (I, 30-31)

    1.2144 It is impossible that all demonstrations could have the same principles (?) (I, 32)

    Two further remarks:

    – about science and opinion (I, 33)

    – about "exact mind" (I, 34)

1.22 *Theory of definition* (II, 1-13):

  1.221 The four objects of knowledge: the fact and the 'why', existence and 'what it is' (II, 1-2)

  1.222 A definition cannot be reduced to a demonstration (II, 3-7)... nevertheless it can be analysed as the setting of an appropriate middle term (II, 8-10)

  1.223 The middle term indicates a cause (II, 11-12)

2. *The discovery of the first principles of science, and how we know them* (II, 19)

BIBLIOGRAPHY

[1] Aristotle: *The Complete Works of Aristotle*, 2 vols., Princeton University Press: Princeton 1984.

[2] Jonathan Barnes: *Aristotle's Posterior Analytics*, Clarendon Press: Oxford 1994 (2nd edition).

[3] Jacques Brunschwig: "L'objet et la structure des Seconds Analytiques d'après Aristote" in Berti, Enrico (ed.): *Aristotle on Science. The Posterior Analytics*, Antenore: Padova 1981, pp. 61-96.

[4] Jacques Brunschwig: *Aristote, Topiques I-IV*, ed. and translated by J. B., Les Belles Lettres: Paris 1967.

[5] Myles Burnyeat: "Aristotle on Understanding Knowledge", in Berti, Enrico (ed.): *Aristotle on Science. The Posterior Analytics*, Antenore: Padova 1981 pp. 97-139.

[6] John Corcoran: "Aristotle's natural deduction system", in Corcoran, John: *Ancient Logic and its modern Interpretations*, Reidel: Dordrecht-Boston 1974, pp. 85-131.

[7] Michel Crubellier: "Aristotle on the ways and means of rhetoric", forthcoming (2008).

[8] Michel Crubellier and Pierre Pellegrin: *Aristote. Le philosophe et les savoirs*, Seuil: Paris 2002.

[9] Jaakko Hintikka: "Socratic Questioning, Logic and Rhetoric", *Revue Internationale de philosophie* 184 (1993-1), pp. 5-29.

[10] Heinrich Lausberg: *Handbuch der literarischen Rhetorik*, 2 vols., Max Hueber: München 1974.

[11] Jan Łukasiewicz: *Aristotle's Syllogitic from the Standpoint of Modern Logic*, Clarendon Press: Oxford 1957 (2nd edition).

[12] Ian Mueller: "Greek Mathematics and Greek Logic", in Corcoran, John: *Ancient Logic and its modern Interpretations*, Reidel: Dordrecht-Boston 1974, pp. 35-70

[13] Robin Smith: *Aristotle: Prior Analytics*, translation and commentary by R.S., Hackett: Indianapolis 1989.

[14] Eric Weil: "La place de la logique dans la pensée aristotélicienne", *Revue de Métaphysique et de Morale* 56 (1951), pp. 283-315; English translation: "The place of Logic in Aristotle's Thought" in Barnes, J., Schofield, M. and Sorabji, R.: *Articles on Aristotle*, vol. I, Duckworth: London 1975, pp. 88-112.

Probability in the Law:
A Plea for not Making Do

DOV M. GABBAY AND JOHN WOODS

'If one's concept of probability is limited to the estimation of chances in games played with cards and dice, and similar situations, the scope for using probability theory in the courts will be considered very limited indeed.'

Richard Eggleston

1 Logic and the law

This is one of a series of papers on probability and the law, in progress or planned. Our purpose here is to chart the general direction of our thinking about legal probability, and to do so while keeping technical considerations to a minimum. We cannot put all that we have in mind to say about probabilistic reasoning in the law into a single paper. But it is our intention, and our hope, that what we say in the present piece will launch our project in ways that will hold some interest for the relevant research communities.

Logic and the law share a number of fundamental concepts, not the least of which are evidence, probability, relevance, reasonableness, precedent (analogy), inference, presumption, plausibility and proof. Some historians of logic remark that most of these logical notions, if not all, originated in the legal usages of antiquity. Given this rather substantial conceptual commonality, it is striking that legal theorists and logical theorists have virtually nothing to say to one another, although the topic of probability is something of an exception. A number of explanations press for consideration. One is that commonalities are merely lexical, and that in their respective disciplinary orientations, these are simply different concepts. Another is that logicians and legal scholars operate with conflicting methodological presuppositions. A third is that the lack of theoretical interaction is inadvertent, a casualty of the over-specialization that passes for expertise in the contemporary university.[1] There is reason to think that all three possibilities are implicated in

---

[1] A further explanation is that, for well-over a century, mainstream logic has been a branch of pure mathematics, having little or no interest in reasoning as it occurs in real-life conditions. It is true that mathematical logic has been the dominant face of logic since the latter part of the 19th century. But for most of that period, especially the last forty years, there has occurred, both within logic itself as well as computer science, considerable pressure to re-attach logic to its historic role as a science of reasoning.

the explanation of this alienation. But it would be premature to suppose that the explanation justifies the mutual indifference. It is not only possible but highly likely that the legal items on our present list possess conceptual textures that no extant logical theory takes due notice of. But it should not be inferred from this short-fall that logic is incapable of enriching its theories with these peculiarities and nuances firmly in mind. Concerning most of these concepts, it is also true that the law favours an implied epistemology of tacitness. Although a lawyer's argument or a judge's finding can be lengthy, detailed, highly technical and wrought with scholarly care, nevertheless with regard to foundational concepts such as relevance, reasonable doubt, balance of probabilities, and *ratio decidendi*, there is an inclination among legal practitioners and theorists alike to think that definition and analysis equal conceptual distortion.[2] Among logicians, however, analysis is precision, an attitude which logicians prize highly.

Is this an unbridgeable methodological divide? Possibly it is, but saying so should await the results of attempts to overcome it. A case in point is the criminal proof standard, concerning whose meaning we find the law at its most tacit, and which has recently been shown to admit of a welcome elucidation in the logic of abduction [18, chapter 8]. Perhaps, then, the better course is to keep an open mind; the principle of best evidence would seem to require it. As for the alienations that flow from the territorial fixities of the contemporary organization of the academic professions, it would do no harm to any logician having an interest in this conceptual commonality (real or apparent) to take some courses at his home instituition's law school.

2   Probability

Probability is a fundamental concept of reasoning in every system of jurisprudence since the *Talmud*.[3] From antiquity to the Renaissance, it has been the object of theoretical elaborations of considerable interest [14]. Probability is no less a part of common sense reasoning and scientific enquiry. As science entered its modern period in the 17th century, probability was caught up in the general drift toward the mathematical. By the end of that century Blais Pascal (1633 - 1662), Pierre Fermat (1601 - 1665), and Christiaan Huygens (1629 - 1695) had succeeded in mathematizing a conception of probability applicable to games of chance – or *aleatory* probability [4, 8].[4] This, the probability calculus, has been considerably refined in the ensuing centuries, and in its present form is without question the most suc-

---

[2] 'Reasonable doubt is a term in common use as familiar to jurors as to lawyers. As one judge has said, it needs a skillful definer to make it plainer by multiplication of words.' [48, p. 517]. '[Given well-known disputes over the legislative status of case-law], the scene is set for over a century and a half of more or less fruitless pursuit of the elusive notion of ratio decidendi – the notion of the clear, valid rule of law, discoverable in any binding precedent.' [37, p. 60]

[3] See, for example, [44].

[4] Aleatory' derives from the Greek word for game.

cessful and complete formal articulation of probability yet attained.[5] A further attraction of the probability calculus is the promise it offers of informing the logic of ampliative (i.e., other than strictly deductive) reasoning in a quite general way. 'Bayesianism' is a term that names the most dominant of these contemporary approaches to probabilistic reasoning [42, 1].[6] In recent years a dispute has arisen between those who hold that Bayesianism is the canonical theory of probability, including probability in the law, and those who see its range as limited to the peculiarities of games of chance. Anti-Bayesians tend to look with a certain wistfulness at the four thousand year history of learned commentary on probability. They tend to regard the probability calculus as a 'Johnny-come-lately', too self-enamoured for its own good [14, 6]. In what follows we will assume that readers have a general familiarity with the probability calculus. However, when there is occasion to cite an axiom, we will provide an informal formulation of it in a footnote.

One of the places where concepts of logic and the law overlap most conspicuously is the standard of proof in civil actions. In such cases, the plaintiff must prove his claim 'on the balance of probabilities'. Probability also has a place in criminal proceedings, albeit not in the official wording of its proof standard (guilt beyond a reasonable doubt). For example, in Canadian criminal law, if an accused wishes to exclude evidence on grounds that it was acquired unconstitutionally, the burden is his to prove on the balance of probabilities.[7] In another example, in what the law calls the 'logical' definition of relevance, evidence is relevant when it alters the probability of some matter at hand.

It can scarcely be doubted that the probability calculus is the correct theory of the concept of chance embedded in the rules and practices of gaming. It is also true that, if other conceptions of probability exist, no attempt to give them a theoretical articulation comes close to the power and sophistication with which the probability calculus elucidates the aleatory conception. This fact alone creates a substantial temptation to assume the applicability of the probability calculus in what appear to be non-aleatory contexts. On the principle that an imperfect theory is better than

---

[5] Key to this development is the axiomatic treatment of [35].

[6] More technically, a Bayesian is someone who holds that an agent's belief function changes by way of 'Bayesian conditionalization'. Bayesianism is named after Thomas Bayes (1702 - 1761), discoverer of the famous probability theorem. For readers who may not be familiar to it, we state Bayes' theorem as follows.

Let E be a given state of affairs and F be some possible state of affairs. Suppose we want to determine the probability of F given the state of affairs E, symbolized as $\Pr(F/E)$. According to Bayes' theorem

$$\Pr(F/E) = \frac{\Pr(E/F)\Pr(F)}{\Pr(E)}$$

That is to say: the probability of F given E (or F on E) is the probability of E on F multiplied by the probability of F alone, which product is then divided by the probability of E alone. The conditional probability axiom asserts that the probability of p on q $\Pr(p \mid q)$ is the probability of p and q divided by the probability of q (provided that the probability of q is not zero). For reservations about unconstrained application of the theorem to probabilistic contexts in legal proceedings, see [61].

[7] Such burdens are 'assigned by the rule of law' [9, p. 27].

no theory at all, it is easy to see the *a priori* attractiveness of the aleatory model.

The impulse to invoke interpretations of phenomena that currently enjoy the support of our best theories is sanctioned by what might be called the *Can Do Principle* (discussed further in [38]). *Can Do* is a principle of economic conservatism. It bids the enquirer to make use of tools he has already mastered, and of knowledge he has already acquired. It counsels against re-invention of the wheel. *Can Do* is also a significant motivator of scientific reductionism and the unity of knowledge. It is a force for interpreting new phenomena in the light of what we already know. It is almost impossible to overstate the power of the grip of *Can Do* and of the benefits that flow from it. Even so, like all virtues, *Can Do* possesses a degenerate case, which we call *Make Do*. *Make Do* is *Can Do* put to inappropriate uses – to engagements that exceed its reach – simply on grounds that the resources it calls into play are well understood, ready to hand and, in *some* contexts, successful. As strong a virtue as *Can Do* is, *Make Do* is equally a vice. It drives the engines of dogmatism and it brokers deep misconceptions in the research programmes of human enquiry. It is an especially odious kind of 'procrusteanism' [51].[8] There is a particularly entrenched version of it that should be guarded against. It is expressed by the proposition noted just above, that it is better to have a defective theory of something rather than no theory at all. This is not the place to dwell on what might explain the appeal of this sentiment. But perhaps it might quickly be observed that for anyone drawn to epistemological holism, there is on occasion *some* case to be made for saving strong theories at the cost of discounting or reconceptualizing dissident data. This alone makes it extremely difficult to mark with reliable precision the point at which the virtue of *Can Do* collapses into the vice of *Make Do*. Still as best we can we should be on guard against their confusion.[9]

This raises an important question for legal theory. *Is the legal conception of probability aleatory?* Although the subject of an aggressive dissensus among legal theorists, our own view is that, on balance, the answer to this question must be in the negative. If this is right, we shall have brought to light a large open problem in legal theory, any good answer to which will require the formulation of a suitably rigorous theory of *legal* probability. Let it be clear, however, that if the civil standard and the law's definition of logical relevance do embed the aleatory notion of

---

[8]See [1] on Bayesian modeling: 'Indeed it is easy to distort the problem under consideration by making implausible assumptions that make the modeling easier. And it is often all too easy to get carried away by the formulae. Not every partial derivative is meaningful - no matter how pretty the result is.' (p. 130)

[9]A further complication is the progressiveness of theory. The history of science discloses lots of cases in which an imperfect theory is retained not only because its virtues outweigh its defects, but because efforts to correct it are inducements to the eventual construction of superior successor-theories. Here, too, we meet with the necessary but vexed distinction between theories whose imperfections do and theories whose imperfections don't necessitate their abandonment or at least the curtailment of their would-be reach. For all its lack of transparency and sharpness, it is a distinction for whose application we have some, if not perfect, aptitude.

probability, the logic of the legal concept of probability will already have received precise articulation in a mature theory. And, if that were in fact so, it would be well worth knowing by all concerned, by the legal theorist and the logical theorist alike.

This would be a good place to enter a disclaimer. It is not our position that consideration of aleatory probabilities is never a factor in trials at common law. Our question here is whether in its duty to determine whether something has been proved on the balance of probabilities or has been proved beyond mere probability, triers of fact have a kindred obligation to reach their findings by manipulating the concept of probability in the manner set down by the aleatory axioms and rules. It is not to our purpose to deny – nor is it in fact true – that both witnesses and counsel may engage material issues in aleatory ways. A pathologist may compute the probability of the time of death on the physical evidence in a thorough-going aleatory way, and counsel for the defendant may emphasize this conditional probability in his closing argument. But these are not the issues in question here. Our concern rather is, in the first instance, to determine whether, in rendering a verdict on the balance of probabilities, a trier of fact makes his way by aleatory computation, and, in the second, whether, in rendering a verdict tied to a standard greater than probability, it is also the aleatory concept that is in play.

3   Getting oriented

In suggesting that the legal concept of probability does not satisfy the aleatory theorems, we invoke a distinction between the *mathematical analysis* of probability and a *conceptual analysis* of it. There is nothing new in this distinction, and nothing new, either, in the idea that a mathematical theory of something need not require a unique conceptual interpretation of it. So, then, although there is a widely-accepted mathematical theory of probability, the nature of its subject matter remains a matter of philosophical dispute. This sets up a pair of interesting alternatives:

1. Notwithstanding their differences, different conceptions of probability all satisfy the theorems of the probability calculus.

2. Some conceptions of probability do not satisfy the theorems of the probability calculus.

Among probability theorists, (1) is the favoured option by a considerable margin. We will briefly review (1) in the light of the four different interpretations of probability that are thought to favour it. These are the degree-of-belief, degree-of-support, frequency and objective interpretations of the probability concept.[10]

---

[10]A good survey is [29].

Consider first construing probabilities as *degrees of belief* [45, 11]. If we interpret a degree of belief as betting behaviour, it can be shown that persons who violate the axioms of the probability calculus lie exposed to a Dutch Book. A Dutch Book has been made against a bettor if he accepts odds and makes bets in such a way that a winning the bet is impossible no matter what. If it is accepted that bets subject to this difficulty are irrational, then it may be claimed that degrees of belief conform to the calculus if they are rational.

One can also think of probability as the measure of a *degree of support* lent by a body of evidence to a proposition[11] [28, 31, 34]. Some theorists are of the view that this interpretation can be connected to the degree-of-belief interpretation in a straightforward way. We could say that the degree of support evidence E gives to hypothesis H just is the degree of belief that it is rational to give to H if E is all they know. If this is right, we could then re-apply the Dutch Book Argument to show that the degree-of-support conception must also obey the theorems of the calculus. According to the *frequency* conception of probability, the probability of H given E equals the limit that relative frequency of Hs among Es would approach were there indefinitely many Es [55, 43, 56]. One problem with this account (which leads some theorists to say that the frequency conception cannot be reconciled to the degree-of-support conception) lies in giving a satisfactory characterization of 'the long run'.[12]

A further conception is the *objective* interpretation (also called the *chance* interpretation) [38, 36]. On this view, probability is a postulated property. It is the property that explains relative frequencies; but it is also a property for which the relative frequencies are themselves evidence. However, the reasonableness of the postulation is a matter of contemporary controversy.

Some theorists interpret the conceptual gap between the frequency and degree-of-belief interpretations as casting doubt on whether the frequency conception can satisfy the theorems of the calculus. A like scepticism would also apply to the objective, or chance, interpretation.

Sketchy though these remarks have been, perhaps enough has been said to make the following point:

> Not even among the dominant mainstream approaches to probability is there unanimity that option (1) is true, i.e., that all *bona fide* concepts of probability honour the theorems of the probability calculus.

Even so,

---

[11] Degrees-of-support theories usually define probabilities over sentences, whereas the calculus treats of events. It turns out to be rather tricky to get sentential probabilities to dance to the aleatory axioms. But it can be done. See [57, pp. 401-404].

[12] Concerning which, Keynes famously quipped that in the long run we are all dead.

• It is widely held by those who take a mainstream approach to probability, that the failure of a concept of probability to comport with the theorems of the calculus is only a matter of degree and is, in any event, at least something to regret. Accordingly, even if (2) is true, i.e., that some probability interpretations are not quite captured by the probability calculus, the mainstream view is that this, to some extent, holds the interpretation open to question.[13]

By these lights, the probability calculus is seen as the theoretical core of any tenable concept of probability. Probability interpretations that wholly comport with the theorems of the calculus are advantaged by that fact. Deviations are possible, but when they occur, the benefits had better be good.

Some inductive logicians, notably Jonathan Cohen [4, 6, 7, 8], hold that option (2) is both true and nothing to regret. In particular, Cohen finds that degree-of-support probability[14] does not yield satisfactorily to a betting-behaviour analysis, and thus loses that direct tie to the Dutch Book constraint that would link it to the theorems of the calculus. According to Cohen, degree-of-support neither has, nor should it be expected to have, the conceptual structure of aleatory probability.

Another dissenting example is [52]. Although Toulmin does not invoke it explicitly, we see here the presence of another important distinction. It is the distinction between probabilistic expressions and probabilistic concepts. In a rough and ready way, whenever someone is using a probabilistic *vocabulary* in what he utters or writes, it would seem reasonable to suppose that he has put in play some or other *concept* of probability. Not only is this a reasonable hypothesis to entertain, it is tacitly endorsed by most supporters of the mainstream probability conceptions (nor is there anything in Cohen's dissenting view that requires him to deny it). But Toulmin demurs from it, observing that there are lots of quite ordinary instances of the use of a probabilistic vocabulary in which *no* concept of probability is at play. For example, the idioms of probability sometimes work as assertion-weakeners (as in, 'I'll probably increase my donation to the University'). In these cases it isn't necessary to postulate the speaker's manipulation of a probability concept as a condition of correctly interpreting his remarks. Another way of saying the same thing is that for such 'pragmatic' issues of probability idioms there is no feasible presumption that their correct interpretation requires us to attribute to their utterer real-valued calculations, however tacit.[15]

---

[13]See [54].

[14]Which he also calls *inductive* or *Baconian* probability. Roughly speaking, inductive probability is the concept of inductive support captured by Mill's methods [39] minus the 'method of residues' [4, p. 163].

[15]This is not to overlook that what logicians see as the 'traditional' notion of probability admits of a distinction between the qualitative, the comparative and the quantitative. Both the qualitative and comparative conceptions – especially the comparative – have yielded to axiomatizations as a kind of pretheory of the quantitative or mathematical theory of probability. Certainly a good many commonsense invocations of the probability idiom can be reconciled to the qualitative or comparative conceptions.

We are now in a position to set our course with greater precision. In what follows, we shall argue for the following two theses.

Thesis one. *If probabilistic reasoning in legal contexts does indeed involve the manipulation of a probability concept, then there are situations in which the legal probability concept neither satisfies, nor should be expected to satisfy, the aleatory axioms. In other words, pressing the aleatory theorems on the legal concept of probability is at risk for* Make Do.

Thesis two. *Nonetheless, there is reason to suppose, in the manner of Toulmin, that probabilistic reasoning in certain legal contexts does not involve the manipulation of an embedded probability concept. In other words, there is reason to believe that probabilistic discourse in those settings is not conceptually probabilistic. If so, then pressing the aleatory theorems into play would be an especially extreme form of* Make Do.

4   The civil standard

As in criminal prosecutions, the defendant in a civil trial is presumed not to be liable until a contrary finding is rendered. Since the plaintiff has the burden to prove his case on the balance of probabilities, we shall simply assume, first, that the lexical character of the probability standard embeds a concept of probability and, second, that this concept is that of aleatory probability. If, as we proceed, these assumptions lead to difficulties, we may wish to contemplate their abandonment.

Imagine a case in which a customer is suing a pharmaceutical company for damages. What the plaintiff must prove is

D = The company is liable for damages.

As in criminal trials, the company need not offer a defence, but usually a defence is offered for at least tactical reasons, and it takes the form of attempting to show that

$\sim D$ = The company is not liable for damages.

What does the civil standard require? It seems to demand that the winning side meet the following two conditions:

But Toulmin's point (and we agree with him) that there are uses of 'probably' whose sole function is to serve as *commitment-weakeners*. Consider again 'I'll probably increase my contribution', and compare with the following exchange. 'Hello, Mr. Smith. I trust that you'll be increasing your contribution this year.' 'Oh, I suppose I will.' Although Mr. Smith is using the language of supposition, no one would think that he is entertaining a supposition. He is hedging a commitment. We think that there are contexts in which 'I'll *probably* increase it' and 'I *suppose* I'll increase it' are equivalent. If this is right, then it should be the case that Mr. Smith's hedged commitment can equally well be conveyed in the language of comparative probability without its being a judgement of comparative probability ('Oh, it's more probable than not that I'll increase the contribution').

- *It must carry a probability greater than chance.*

- *It must carry a probability greater than that of the opposition's case.*

This would be a good place to take note of an apparent difficulty. If we find it reasonable to hold that D's probability is n if and only if ~ D's probability is 1 - *n*, then we will find our second condition to be redundant. This intuition would be entirely correct for cases in which opposing counsel accept the same evidence. This, of course, hardly ever happens. Evidence E is the sum total of any testimony the judge allows the trier of fact to hear. Typically, E is an inconsistent set of propositions, and even in those rare cases in which all the propositions attested to by all the witnesses are logically consistent with one another, it is frequently the case that some of them (if not all) will be given different *weight* by opposing counsel. As these matters tend actually to play out in real-life, it is perfectly routine for counsel to ground their interpretations of the evidence in different and conflicting proper subsets E' and E" of E rather than E in its entirety, and to give to the subsets to which they happen to agree different weightings.

Consider any case in which this so. Then there is an E' relative to which D's probability is reckoned and a different E" relative to which ~ D's probability is reckoned. It now is easy to see that our two conditions might well have application here. Not only might Pr (D | E') and Pr (~ D | E") both exceed 0.5, but the one might well be higher than the other. We may accordingly set out the proof standard schematically as follows.

- A civil suit succeeds if Pr (D | E') is > 0.5 and Pr (D | E') > Pr (~ D | E"). It fails if Pr (D | E') < 0.5 or if Pr (D | E') > 0.5 and Pr (~ D | E") > Pr (D | E'), where E' and E" are distinct proper subsets of E and further, if E' ∩ E" ≠ ∅, the propositions in common may be subject to differential weightings by the two sides.[16]

We note in passing that where E' and E" are incompatible it cannot be 'deemed' that the competing conditional probabilities are reckoned on the common superset E. The reason for this is technical. If E is an inconsistent set, then the conjunction of its propositions is a contradiction. In that case, the probability of any proposition conditional upon E is non-computable. What this means is that in the standard case opposing counsel argue from irreconcilably different bodies of evidence. Alternatively, it means that the probability of the balance-of-probabilities standard is not the aleatory conception. For the time being, we shall assume the former interpretation. Recall that our plan is to leave the aleatory presumption in place until we come upon reasons to give up on it.

---

[16] "E' ∩ E" ≠ ∅" expresses the proposition that some of the sentences in body of evidence E' are also in body of evidence E".

As the schema makes plain, the competing probabilities are conditional upon what each side takes as evident. In the probability calculus, conditional probabilities are not computable unless the relevant *prior* probabilities are known or can safely be presumed. In particular, Pr (D | E') and Pr (~ D | E") are compatible only when the probabilities of D, ~ D, E' and E" are at hand. The presumption of non-liability is such that D has no positive probability prior to the trier of fact's assessment of it in the light of what he (or they) take as evident. In light of these considerations, it might strike one as intuitive to put the prior probability of D at 0 and of ~ D at 1. Notwithstanding, we can see at once that if we do so we will have ruined the case for the aleatory interpretation of the embedded probability concept. While it may well be the case that prior to the trial D has no positive probability, to assign it 0 requires that we treat it as a contradiction, hence as a claim which cannot in any sense be proved on a balance of probabilities. Similarly, if the prior probability of ~ D is 1, it is a tautology that the company is not liable, in which case the suit against it cannot possibly succeed.[17]

Some will see this as a merely technical difficulty, and will propose accordingly that we set the prior probability of D as some arbitrarily small positive value and of ~ D as some arbitrarily large number less than 1. In the spirit of this proposal, let us agree to set the prior probability of D at 0.1 and of ~ D at 0.9.

With D as the presumption of non-liability with a stipulated posterior probability of 0.1 let E' be construed by the following propositions attested to by witnesses. For ease of exposition, we set their values at 0.8 each.

A = The company did not withdraw the product.

B = The product was red-flagged by the FDA as unsafe.

C = All competitors withdrew the product.

F = There was a 93% adverse reaction to the product by those who used it.

Intuitively, we see that E' considerably increases the conditional probability of D. In calculating the probability of this conjunction, we must free ourselves of the assumption that D and E are probabilistically independent. This requires in turn that the relevant conjunctive probability be calculated by the general (or un-

---

[17]Against this it might be argued that in the standard literature on the coherence of testimony, e.g. [1], evidence sets consist of reports in the form "Witness W attested that p" and "Witness W^* asserted that $\sim p$". Suppose that for a number of q, W swears that q and W^* swears that $\sim q$. Let E be the sum total of sentences in the form "W attests that q" and E^* be the sum total of sentences of the form "W^* attests that $\sim q$". Clearly E and E^* are not inconsistent with one another. So the problem of how to model testimonial inconsistency doesn't arise in the form in which we are addressing it here.

We demur from this suggestion. It is true that in legal proceedings the evidence is, without exception, given by testimony. When W testifies that p, it is p that is introduced into evidence, not W's testifying to it. What counts here is not whether a juror accepts p, nor whether he accepts that W attested to it. It is quite true that if a juror has (or thinks he has) reason to doubt W, he may well find that he does not accept p. But if the present suggestion concerning the reconstruction of evidence were sound, the juror would be obliged to disbelieve that W asserted that p, which is surely not what happens (or should).

restricted) conjunction axiom,[18] which as applied to our present parameters would give

$$Pr (D \wedge A \wedge B \wedge C \wedge F) = Pr (D) \times Pr (A \wedge B \wedge C \wedge F \mid D) \qquad (1)$$

Reapplying the general conjunction axiom to deal with the three remaining embedded conjunctions gives a calculation of daunting complexity, one that certainly is beyond the unaided combinatorial powers of human reasoners. Bearing on this is a well-known difficulty. [27] points out that, as new evidence is heard, the requisite conditional probabilities need to be updated. Consider a case in which, on a given day, thirty new pieces of evidence are introduced. Integrating this new information would require 2^{30} computations – slightly over a billion. This is far from demonstrating that our probabilistic practice cannot produce intuitively arrived-at results that broadly conform to the results that such calculations would also produce. But if it were indeed the case that human reasoner's instinctively arrived at credible approximations of probability estimates they could not themselves compute, this would leave us with the interesting and pressing task of trying to explain this remarkable coincidence.

Here, too, we meet with a problem that requires a choice. Either we abandon the assumption that the balance-of-probabilities standard embodies the aleatory conception of probability, or we cease interpreting the non-liability presumption probabilistically. This latter we could do by aping criminal practice by presumptively assigning it the truth value *false* (and of assigning to $\sim$ D the truth value *true*). By taking these measures, we would solve the problem of having given to D and $\sim$ D arbitrary priors, since now they would in law be presumed to have no probabilities whatever.

It won't work. If D and $\sim$ D lack probabilities, then the conditional probabilities D | E' and D | E" cannot be computed. So it would seem that the only course is to reason as follows:

- In criminal trials, it is taken as given that evidence can transform a presumed falsehood (namely, that the accused is guilty as charged) into a proposition proved to be true beyond a reasonable doubt (namely, that self-same proposition that the accused is guilty as charged). Whatever the details of the complex relation between the presumption of innocence and the evidence heard, in virtue of which a verdict of guilty is sustained beyond a reasonable doubt, some such evidential link exists and is describable in principle.

---

[18]The basic idea is that probabilistic conjunction is multiplication. When p and q are mutually exclusive (i.e., cannot both be true), then the restricted conjunction axiom says that the probability of p and q is the probability of p multiplied by the probability of q. There is another form of the axiom which allows us to drop the qualification of mutual exclusivity. By the generalized conjunction axiom, the probability of p and q is the probability of p multiplied by the conditional probability of q on p.

Why should it be any different in civil trials? Prior to a verdict, D is presumed to be false. At the point of decision by the trier of fact, D may be judged to be more probable than not and more probable than ~ D. Whatever the details of the complex relations between D and E' and ~ D and E", some such complex of relations exists and is describable in principle. But, as we now have reason to see, it is far from obvious that any such description must attribute to the trier of fact manipulations of the aleatory conception of probability.

If these observations are sound, they will have lent some encouragement to our first thesis; that is, they will have given us reason to think that if the civil proof standard embeds a probability concept, the concept it embeds is not aleatory probability. What we now wish to reconsider is whether a case might not be made for thesis two, that is to say, for the thesis that the probability *idioms* in which the civil standard and the notion of relevance are formulated do not embed, or presuppose the manipulation of, a concept of *probability*.[19]

Jonathan Cohen discusses similar difficulties. Let X, Y, Z be facts proof of which is individually necessary and jointly sufficient for a plaintiff to succeed. If we put the degree of proof of each of the three at 0.75, their conjunctive probability is 0,42, which fails the civil standard. Cohen's reservations are summarized in [4, pp. 35-36]. A similar problem is that of convergent evidence. Suppose that for facts L, M, N, each individually lends to the plaintiff's claim D a probability less than 0.5, but that together they raise the support of D to well beyond 0.5. But the conjunction of low probabilities is lower than the lowest of them.

As will become clear before long, the principal difference between Cohen and ourselves lies in the diagnoses of these difficulties. According to Cohen, what they show is that a non-aleatory concept of probability is at work in these contexts. Against this, we will suggest that in ranges of such cases, there is reason to think that *no* concept of probability is in play. Let us turn to that suggestion now.

5   Linda

In a famous experiment, Amos Tversky and Daniel Kahneman ([32, 53]) presented their subjects with the following set of facts E. A certain Linda is 31 years of age, single, outspoken, and extremely intelligent. As a student she majored in philosophy and was deeply involved in movements for anti-discrimination and social justice. She was also active in anti-war campaigns. Subjects were then asked to select the more probable claim from a list containing the following:

---

[19] A similar point can be made about the Allais Paradox concerning preference selection under risk. As [30] points out that it is perfectly possible that the apparent mistake involved here is not actually a mistake, and that 'naive expected utility is inappropriate for modeling many cases of rational decision making – including this one – since it is insensitive to factors of marginal utility and risk aversion.' (p. 143)

1. Linda is a bank-teller.

2. Linda is a bank-teller active in the feminist movement.

A large percentage of the responses favoured proposition (2) over (1). The experimenters took this as evidence that their subjects had committed a probabilistic conjunction fallacy. Although E is irrelevant to the task set for the subjects it is possible that, since it was introduced as a preamble to the experiment, the subjects interpreted their instructions as requiring the choice of the most probable of a list containing

1.' Pr (Linda is a bank-teller | E)

2.' Pr (Linda is a bank-teller $\wedge$ Linda is active in the feminist movement | E.)

Here, too the selection of (2') as more probable than (1') would be an aleatory error. Be that as it may, these are not the choices imposed by the experiment. The experiment requires a choice between (1) and (2), not (1') and (2'). Subjects familiar with the rules of the probability calculus would know that for *any p* and *q* having non-integral probability values, the probability of *p* always exceeds the probability of *p* $\wedge$ *q*. Further, they would know that this ratio held no matter what E says. In other words, they would know that the ratio holds for arbitrary E. So the cause of the subjects' difficulty is E. Its very introduction presupposes its relevance, when in fact it is of no relevance at all to the *stated* problem set by the experimenters.[20] A related factor is E's undoubted relevance to a problem that the subjects were not (expressly) asked to solve. We might call this the problem of selecting the more plausible hypothesis. Putting 'Pl (*p*)' as 'the plausibility of *p*', if E is indeed a factor in the exercise, it is easy to see that, for some *m* and *n*,

- *Pl* (Linda is a bank-teller | E) = *n*

whereas

- *Pl* (Linda is a bank-teller $\wedge$ Linda is a feminist activist | E) = *n* + *m*.

Not much is yet known in a formal way about the structure of plausibilistic inference (see [46] for an important pioneering effort; cf. [18, chapter 7] and [21].) Even so, two things stand out as fundamental.

1. If propositions are internally consistent but jointly incompatible, they can have the same plausibility values but not the same probability values.

---

[20]Similar reservations may be found, for example, in [22], especially pages 19, 169, 197, 242, 248-250.

2. Unlike aleatory probability, conjunctive plausibility is not multiplicative. In particular, it could easily be the case that *Pl* (Linda is a bank-teller and a feminist activist) is greater than the plausibility of its leftmost conjunct and less than the plausibility of its rightmost conjunct.

The admissibility of E as a factor is key to the plausibilistic solution of the Linda-problem. If E is relevant to its solution, so too might various inferences from E or from E together with the subjects' background knowledge. Given that background, we may take it that E tells us that Linda was something of an ideologue. She disliked on principle institutions that discriminate against women, a matter concerning which she judges that banks have not had an especially good track record. That Linda now works as a teller – one of the bank's more low-level positions – suggests that she doesn't now dislike banks or that she simply has been unable to find other employment. That Linda is a bank-teller and also active in the feminist movement is also explicable. The bank may have improved its anti-discrimination practices, and the bank may now consider Linda's feminism to be an institutional asset. Both explanations are conjectural, needless to say. But of the two the second accords to E the greater explanatory force, which is one of the reasons that the conjunctive hypothesis stands out. Another way of saying this is that E lends this second explanation greater *explanatory coherence* than it does the first.[21]

It is well to note that neither historically nor currently is the Linda experiment an investigation of its subjects' understanding of probabilistic conjunction. Rather the experiment investigates the distortive impact of set-up information. If all that we wanted to know about these subjects is whether they understood probabilistic conjunction, the experiment would have been quite different – something along the following lines:

- Subjects are given two hypotheses. One is that Bill is a banker (B). The other is that Bill is a banker with an interest in football (B ∧ F). Which has the greater probability, B or B ∧ F?

In fact, of course, the set-up information about Linda is crucial. Clearly its very presence in the experiment presupposes its relevance. One possibility is that it is not in fact relevant, hence that the presupposition is false; in which case, the failure of the subjects was not the failure to appreciate probabilistic conjunction, but was rather an absence of sensitivity to the falsity of the presupposition. Another possibility is that the presupposition is true. If so, it cannot be the case that the problem posed by the experiment is one of comparative probability estimates in conjunctive

---

[21]For the notion of explanatory coherence see [49]. See also [18, section 6.4]. We may note here that, unlike conditional probability, explanatory coherence is symmetric. I.e., whereas $Pr(H \mid E) \neq Pr(E \mid H)$, E coheres with H if and only if H coheres with E.

contexts. One version of this possibility, as we have seen, is that since the set-up information would be clearly relevant to the subjects if they were attempting to solve a plausibility problem, subjects interpreted the question 'Which is the more probable?' as 'Which is the more plausible?'. If so, the answer that favours the conjunction over the conjunct is the right answer.

A different possibility has long been recognized in the psychological literature. Here is one version of it. Suppose that the set-up information and the attendant instruction to subjects is taken to have been 'rendered conversationally'. A stretch of discourse from one party to another occurs conversationally when it is subject to the precepts, presumptions and presuppositions of 'a logic of conversation', in the manner of, say, [25]. If so, then we may presume that in the absence of indications to the contrary, when the experimenters declare the two possibilities that Linda is a teller and Linda is a teller who is active in the feminist movement, these utterances are subject to the Quantity Maxim. In the present case, the Quantity Maxim provides that in each of the two hypotheses there is a close concurrence of utterance-meaning and speaker-meaning. In the particular case of 'Linda is a teller', the Quantity Maxim excludes what would be conveyed by 'Linda is a teller and is an active feminist'. Accordingly, there is a conversational implicature to the effect that the possibility that Linda is a teller is the possibility that Linda is a teller and not an activist. Thus if what is 'really' being said by the utterance of 'Linda is a teller' is that Linda is a teller and not an activist, then the comparisons the subjects are required to make are between

- (1) Linda is a teller who is not an activist

and

- (2) Linda is a teller who is an activist.

But clearly it is quite possible for (2) to have a higher probability than (1).

Although we ourselves have shown a certain fondness for the plausibility solution of the conjunction paradox, we don't mean to rule out the 'Gricean' solution. It suffices for our purposes that the plausibility solution itself has a certain plausibility. Our general caution is that the idiom of probability sometimes encapsulates a concept of plausibility. That the Linda problem at least admits of a plausibility solution doesn't decisively demonstrate that thesis, but it does lend it some degree of support.

Before quitting this section, it would be appropriate for us to admit to a tactical liability. As we have already noted, the mathematical theory of probability is a robust, deeply dug-in and highly successful theory. Its range is equally impressive. There is virtually no aspect of ampliative reasoning that it has not influenced in some principled way. By what, in section 2, we called the Can Do Principle,

there is ample reason to attempt to extend and deepen the influence of the probability theory. After all, we are inclined to prefer successful mature theories over
alternatives that are fledgling and callow. Nothing succeeds like success.

What we are hoping to do in this paper is to plead a plausible case for plausibility. In so doing, we encumber ourselves with the liability that next to aleatory
probability hardly anything is known of plausibility in a theoretical way. This
matters. We are bidden by the Can Do Principle to try to cover the behavioural
and linguistic data which, we say, calls for plausibilistic treatment probabilistically. For if it's a theory we are after, probability offers us a theory and plausibility
doesn't. So how can it be preferable to play the plausibilistic card?

Against this we have two cards to play. One is that it is that ordinary probabilistic reasoners do not in the general case achieve their ends by making the
calculations required of them by the aleatory theory of probability. Of course, as
already noted, this alone doesn't show that probability theory has nothing to tell
us about successful probabilistic reasoning of the everyday sort. For it is possible
that whenever an ordinary probabilistic reasoner reasons successfully, he produces
– somehow – results that would independently be sanctioned by a calculus that he
himself has no chance of being able to implement. We say again that, if true, this
would be serendipity of the highest order, and perhaps a coincidence too good to
be true.

A second point in our defence is the quite general fact that any theory no matter
its track record can be extended beyond its natural reach. In section 1, we said the
same thing, pointing out that Can Do is always at risk for the degenerate case which
we call Make Do. It is not our position in this paper that the probability calculus
has collapsed into Make Do, but more circumspectly that it might be at risk for
it, that this is a possibility we should be on guard against. Hence Linda. We are
not interested in joining the throngs of theorists who seek the scalps of Kahneman
and Tversky. If that were our purpose, we would have to examine the already
huge literature on Linda alone. Ours is a less dramatic purpose. Suppose, we say
that the Kahneman and Tversky subjects had been responding to the experimental
inputs as if they were to solve a plausibility problem. Then the good news is that
they did solve it. That, for us, is reason to take plausibility seriously as a candidate
for eventual theoretical articulation. The day is long off before plausibility claims
the level of theoretical articulation presently claimed by probability. But we have
to start somewhere.

6   The relevance of Linda for legal probability

Probability is one of a trio of expressions (hereafter, *P*-concepts) which in conditions of actual linguistic usage admit of a certain degree of interchangeability. For
example, something can be said to be a *strong possibility*, a turn of phrase which,
in some uses, has no distinguishable difference in meaning from a *high probabil-*

ity. Hypotheses that are described as *not very plausible* might also without relevant loss be described as *not very probable*. *Bare possibilities* have *minimum probability*. The *most probable* are among the *most highly possible*. And so on.

If in everyday speech these *P*-interchangeabilities are a commonplace, among logicians they are not. If in everyday speech the three idioms of the possible, the probable and the plausible, tend to leech into one another, among logicians they are strictly disjoint.[22] We have already made the point that most logicians identify probability with aleatory probability. In like fashion, logicians associate possibility with sentence operators of the kind studied by standard systems of modal logic. Although logicians have yet to bring the idea of plausibility to the levels of theoretical articulation that one finds in the probability calculus or in modal logic, it is widely held that plausibility is disjoint from the other two.

There are cases galore in which competent speakers interact with one another in ways that give lexical indication that *one or other* of the associated *P*-idioms is in play. Lexical clues include expressions such as 'in all probability', 'not very plausible', 'the merest possibility', and so on. In the presence of such indicators, it is reasonable to attribute to discussants the use of a *P*-concept. In making such attributions, it is important to recognize that the attributor's default position is that the discussants are engaging in everyday speech, since that is what most speech is. In the absence of indications to the contrary, the attributor must assume the discussant's *P*-idiom to be one in which the associated *P*-ideas are not necessarily disjoint. For example, if someone is heard to say that there is the strongest possibility that Albania will not win the gold medal in Olympic ice-hockey or that it is a practical impossibility that it will, it would be madly inapposite to take him as manipulating, say, the S5-modality.[23] The surrounding context of the remark might embody remarks about betting odds, making it likelier that although his *idiom* is that of the possible, his *concept* is that of the probable.

It is the same way with the idiom of the probable. If our comments on the Tversky-Kahneman experiment are sound, then, although the subjects formulated their choices in the language of probability, there is reason to think that the concept at play in their decisions was the concept of plausibility or – mindful of our own point about the interchangeability of the *P*-idioms in everyday speech – that the

---

[22]This is something that the 19th century algebraic logician Augustus De Morgan insisted upon ([12]).

[23]S5 is one of the standard propositional logics of logical possibility and logical necessity, but they are not concepts that come in degrees. What is more, when we say that it is hardly possible that Albania will win, it is more likely that we are attributing physical or causal impossibility, rather than logical impossibility. Here, too, we might usefully recur to an earlier point about *Can Do* and *Make Do*. It is widely agreed among logicians that no extant theory of physical possibility enjoys the power and sophistication of standard theories of logical possibility. Suppose, again, that someone were to reason as follows: 'Everyone knows that powerful and sophisticated theories are better than weak and naive ones. So let us force the theorems of S5 on the concept of physical possibility.' Doing so would be procrustean; it would be caving in to *Make Do*.

concept in play was one or other of the *P*-concepts for which the probability calculus *is not the right theory*. The reason for thinking this is quite straightforward.
The experimenters instructed their subjects to be guided by their intuitions about
the probable. That is to say, they were instructed to make their decisions in light of
how the idiom of the probable operates in ordinary speech. But in ordinary speech
the idiom of the probable leeches into idioms of the possible and the plausible.

Let us briefly consider a second, non-probabilistic, example. In actual linguistic
practice, speakers frequently use terms that suggest that a concept of implication is
in play. These lexical indicators include expressions such as 'hence', 'so', 'therefore', 'since', 'if', 'only if', and so on. In the presence of such indicators it can
reasonably be supposed that the speaker is manipulating a concept of implication
or consequence. Some consequence relations are well-understood at the level of
theory. Classical and intuitionistic consequence have been deeply plumbed, as
have relevant and other forms of paraconsistent consequence. Counterfactual consequence lacks the same kind of solid consensus perhaps, but its general outlines
are theoretically well-understood. This is getting to be a rather long list, and we
have not yet considered even the standard varieties of inductive, causal and abductive consequence.[24]

When someone is reasoning consequentially, it is safe to say that *some* kind of
consequence relation is in play.[25] Often it cannot be said with the same security
which consequence relation that is. Accordingly, a similar diffidence is required
for any assessment of that reasoning. The problem is compounded by the fact that
the question, 'What concept of consequence is in play?' is not one to which a
definitive answer reposes in a competent speaker's linguistic intuitions. So, for the
most part, it won't much help directing this query to the speaker. The concept in
play will have to be inferred from his behaviour, with due regard for contextual
clues. It may also help to ask the speaker for the reasons that support his thinking,
or to challenge it outright. Somehow we manage to be hermeneutical adepts.[26]
Even untutored speakers experience no general difficulty in framing responses to
consequential and conditional utterance. Lying at the heart of this capacity is a
question, one of the answers to which is called (misleadingly) the Principle of

---

[24]First-order (or classical) logic is the standard logic of deductive consequence. Intuitionistic logic
rejects certain classical laws such as the claim that p or not-p is always the case. Relevant logic holds
that when q is a consequence of p, there must be a relevant connection between p and q. Paraconsistent
logic permits logically correct inferences from inconsistent databases. Counterfactual logic examines
sentences in the form, '*If p were* the case, then q *would be* the case'. Inductive logic investigates
non-deductive generalizations from samples. Causal logics interpret the causal relation as a kind of
modality. Abductive logic examines, among other things, inference to the best explanation.

[25]Depending on content, it might be that several consequence relations could be combined ('fibred')
together. We might also consider taking fibred consequence as our general consequence relation. For
fibring, see [15].

[26]Interpretational aptitudes are discussed in greater detail in [18, chapter 8] as forms of abductive
reasoning.

Charity (see, e.g., [26]). The question is this: in attributing a given consequence relation to a speaker, should one eschew those candidates which, were they in play, would expose the speaker's performance as defective?[27] By the Principle of Charity, the answer to this question is that, except where indications to the contrary present themselves, one should indeed avoid such an attribution. It also bears repeating that such hermeneutical decisions should not be over-influenced by what we happen to have a good theoretical knowledge of. Otherwise, one would radically over-attribute first-order (or classical) consequence .

As with the idioms of possibility, so too with the idioms of consequence; sometimes the idioms are not attended by the concept. Accordingly, 'There's juice in the refrigerator if you'd like some', although decked out in the consequentialist idiom 'if', actually embeds no concept of consequence. It is just a way of inviting someone to help himself to some juice.

We learn from these examples a valuable lesson about the assessment of human linguistic behaviour.

- *If the behaviour in question violates rules that regulate some concept K, the assessor must have independent reason to interpret the behaviour as K-behaviour, hence as an (implicit) undertaking to comport with the K-rules.*

We now see the relevance of the Tversky-Kahneman experiments to the analysis of probability in the law. When a juror tries to determine where the balance of probabilities lies, he is performing his task not as an expert, but as an ordinary reasoner reasoning in the way of ordinary reasoners. Judges don't include in their instructions to juries (or to themselves) a short course on the probability calculus. Triers of fact are required to rely on their common sense, on their intuitions about what is probable and what is not, guided by how the probability idioms fare in everyday speech. This being so, there are two mistakes that one can make in interpreting the behaviour induced by these instructions.

- *One might assume (incorrectly) that the concept that these instructions put in play is a concept of probability disjoint from the other P-concepts.*

- *One might also assume (incorrectly) that the disjoint concept is that of aleatory possibility.*

The moral of the past few paragraphs is that, without proper prior attribution, judgements of success and failure are groundless. When we note that the trier of

---

[27]For example, if in the general case, we put it that consequentialist utterance betokened the manipulation of a relation of *deductive* consequence, then by and large one could only find that the utterance in question is defective. For, in the general case, the consequences that are actually drawn by a speaker are not deductive [41]. Still, deductive consequence is a Circean enchantment for many logicians and computer scientists. The sheer power of deductive theories drives logicians to make utterly suspect attributions of deductive intent to human reasoners. More *Make Do* still.

fact has a duty to find where the balance of probabilities lies or that a judge must exclude evidence if it would not affect the probability of something material, these observations are bound by antecedent requirements. One is correctly to identify the *P*-concept that is in play in such contexts. The other is to determine, as best as one can, the conditions under which the *P*-concept is properly manipulated. Another way of saying this is that, in ascribing to a trier of fact or to a judge some probabilistic behaviour, the attribution is *descriptively underdetermined* in the absence of specification of the relevant *P*-concept. Similarly, an assessment of the adequacy of a person's manipulation of that *P*-concept is *normatively underdetermined* in the absence of the requisite norms for its proper employment. In light of the obvious difficulty of these tasks, it is easy to appreciate the seductive appeal of aleatory assumptions. We actually know what the aleatory norms actually are. Specifying an agent's *P*-concept as probability in that sense is a huge step in completing the second of these tasks. Still, the temptation to aleatorize probabilistic behaviour should not be yielded too casually. As we have tried to make clear, there is little to be said for aleatorizing the balance-of-probabilities proof standard in civil law.

7   A brief rejoinder

Against the present suggestion, one might argue as follows. Tversky and Kahneman re-tested their original subjects. The experimental contest was now quite different. Subjects had been made aware of their original (alleged) errors, and appeared to have accepted that they were errors. It was explained to them that the majority choice in the first experiment violated the conjunction rules of the probability calculus. Subjects were then given new information and a new list of possible choices, in which was embedded the design of the original problem. Again they were asked to choose the most probable of a list of propositions. Although the group average was better than the first time out, there was a significant degree of recidivization. What makes the subsequent experiment important is that there was reason to believe that the subjects believed that giving $(p \land q)$ a higher probability than p alone is virtually always an error; yet numbers of them did so anyhow.

Tversky and Kahneman took this as even more compelling evidence of the probabilistic irrationality of even quite intelligent human subjects. Subjects knew they were being asked to play the aleatory game. They knew that, with few (integral) exceptions, $\Pr(p \land q)$ is always lower that $\Pr(p)$. But when it came to the decision, numbers of subjects were unable to stay on track. The diagnosis of Tversky and Kahneman cannot be rejected out of hand. It is possible that human subjects are probability-misfits. But another possibility is that human subjects find it unnatural to apply the aleatory concept in contexts in which information such as E is given a set-up role. It is possible that, when primed by E, the reasonable course is to operate with a different concept of probability. It is also possible that, in such circumstances, it is more reasonable to operate plausibilistically. A key factor is the precise wording of the experimenters' instructions in the subsequent experiments. If subjects were told to *compute* or to *determine* the higher probability, perhaps they can be found guilty of not paying attention. If, on the other hand, they were told to choose what *felt* like the more probable, it is very much harder to pin on them the rap of aleatory misconduct. Doing so would seem to breech the Principle of Charity.

8 The criminal standard

Earlier we noted that if the presumption of non-liability $\sim$ D is expressed by a probability value, problems arise in computing the conditional probability of D on the evidence sworn at trial. Having noted that there is no intuitively satisfactory expression of the non-liability presumption in probability terms, we considered an alethic interpretation on which $\sim$ D is presumed to be true and D false. ('Alethic' derives from the Greek word for truth.) However, what such an interpretation gains in intuitiveness, it loses in its capacity to engage the machinery of probability. For if D and $\sim$ D lack probability values, their conditional probabilities on the evidence cannot be computed (which is a technical way of saying that they don't exist). Mainstream theorists of probability also tend to take a mainstream approach to logic, in which the alethic values are *true* and *false*. This makes for ungainly admixtures of probability values and alethic values. Probability is a matter of degree; truth and falsity are not (in mainstream logic). Except for the probabilities of 1 and 0, probability assignments are independent of alethic assignments. Although 1 is identified with logical truth and 0 with logical falsehood, no other probability value is identifiable with an alethic value. On the face of it, then, a falsehood might have a very high (non-integral) probability and a truth a very low one. Since no one would say that the presumption of non-liability is reasonably captured by making D a contradiction (and $\sim$ D a tautology), we seem to have landed in an intractable difficulty. In its intuitive meaning, the civil proof standard requires that the truth of the proposition that expresses the presumption of non-liability be overturned by the probability of its opposite. In common sense terms, the presupposition that $\sim$ D is abandoned when the probability that D is high enough. But when we try to cash these intuitions in the probability calculus we find that no probability assigned to D *does* overturn the truth of $\sim$ D. So we have a problem.[28]

It would be instructive at this juncture to consider the criminal proof standard. For the most part, judges shy away from defining the standard when giving juries their instructions. An exception (of ludicrous unhelpfulness) comes by way of a

---

[28] Against this it might be argued that the initial default assumption is not that the defendant is not liable ($\sim$ D), but rather that it is highly improbable that he is. If this is right, then the problem of overturning the truth of $\sim$ D doesn't arise. But it isn't true. The presumption of liability is categorical. The presumption is that it is *true* that the defendant is not liable.

mock instruction formulated by the Canadian Supreme Court in *R. v. Lifchus*.[29]
On the one hand it provides that jurors need not have absolute certainty of the accused's guilt but, on the other, that his probable guilt is not enough. Even believing
that he is guilty is not enough. In a subsequent case it was averred that it would
'be of great assistance for a jury if the trial judges situates the reasonable doubt
standard appropriately between the two standards.' (*R. v. Starr*, (2000), 2 S. C. R.
144 at para. 242.) Certainty here we might reasonably equate certainty with the
impossibility of the opposite. Thus it is certain that the accused is guilty (*G*) if on
the evidence, it is not possible that he is innocent. Let us express this as $\Pr (F \mid
E''') = 1$. This is one on the standards that *Lifchus* rules out. The other is that $\Pr
(F \mid E''') = n$, where n is less than 1 and greater than 0. This too is ruled out. *Starr*
proposes that judges instruct jurors to find the appropriate standard somewhere in
the gap between $\Pr (G \mid E) = 1$ and $\Pr (G \mid E) = n$ (for non-integral n). Probabilities
are real numbers. 1 is a real number. 0 is a real number. The real numbers have
lots of interesting properties. One is that they are *everywhere dense*. Between any
two real numbers, there is always a third. The criminal standard says in effect that
no probability other than certainty suffices for *G* and that certainty is too strong
as a general requirement. Jurors are obliged to find something less than certainty
but stronger than any probability less than certainty. The trouble is that if certainty
is 1, there is nothing less that it and greater than any thing less than it. At least
there is nothing in between that is *comparable* with the two standards that *Lifchus*
excludes. Clearly, however, comparability is precisely what the *Lifchus* charge demands. For what is wanted is something *weaker* than certainty and *stronger* than
probability. But if probability is interpreted aleatorily, the only comparable interpretation of certainty is 1. And if certainty and probability are aleatory, there is no
comparable standard in between. We may conclude, therefore, that the probability
of the standard that *Lifchus* excludes is not aleatory probability.

It is axiomatic that the civil standard is weaker than the criminal standard. That
weakerliness is constituted by the fact that probability suffices for the civil standard but not for the criminal standard. Since the two standards are comparable
(after all, one is weaker than the other), since they part company on the issue of
probability, since the probability of the criminal standard is not aleatory probability, then neither is it likely that the probability of the civil standard is aleatory. The
model charge of *Lifchus* also contains the following clause.

- In short, if based upon the evidence . . . you are sure that the accused
  committed the offence you should convict since this [i.e., the conviction]
  demonstrates that you are satisfied of his guilt beyond a reasonable doubt.
  (13-14)

This is extraordinarily interesting. It is, in effect, an *operational* definition of

---

[29](1997), 9 C. R (5th) 1 (S. C. C.)

reasonable doubt: If upon considering the evidence by thinking and reasoning in the way of ordinary (i.e. untutored) thinkers you convict, then you have met the required proof standard. In other words, *reflection followed by conviction = ascertainment of guilt beyond a reasonable doubt.*

Also of interest is the word 'sure'. The *Lifchus* charge is that when one is sure of the accused's guilt then one must convict. 'Sure' is Lifchus' own stab at identifying something midway between certainty and probability. Since the criminal standard is weaker than certainty and stronger than possibility, it is comparable with the more extreme pair. Will this work? Presumably it will. With 'sure' as the basic term, 'certain' could be interpreted as 'absolutely sure'. Since all three of the standards are comparable, there should be a comparable interpretation of 'probable'. There is: it is satisfied'.

It does not stand to reason that criminal and civil law would have incomparable proof standards. It does not stand to reason that the probabilities required by the civil standard be incomparable with the probabilities excluded by the criminal standard. As we have attempted to demonstrate, the criminal standard will not brook the aleatory interpretation of probability. By present reasoning, the same must be said of the civil standard. If this is right, then the civil standard should be representable as a *variation* of the criminal standard. It can.

- In short, if based upon the evidence . . . you are [satisfied] that the accused [is liable] you should [find against him] since [doing so] demonstrates that you are satisfied of his [liability on the balance of probabilities].

This clearly is a variation of the criminal standard, a variation in which every change (indicated by square brackets) is driven by the core variation of 'satisfied' on 'sure'. Further evidence of comparability is that in its present interpretation we also have an operational definition of the balance of probabilities. If upon considering the evidence by thinking and reasoning in the way of ordinary [i.e. untutored] reasoners, you find for the plaintiff, then you have met the required proof standard. In other words, *reflection followed by finding for the plaintiff = ascertainment of liability on the balance of probabilities.*

We shall bring this section to end with the observation that it is in no sense required that in the application of these proof standards that triers of fact be manipulators of any concept of probability, aleatory or otherwise.

9 Probabilistic relevance

In a widely-used textbook on the law of evidence, we find the following remark:

- One fact (conveniently called an evidentiary fact) is relevant to another when it renders the existence of the other fact probable or improbable. Relevancy therefore is a matter of common sense and experience rather than law [9, p. 148].

This is the definition of what lawyers call 'logical' relevance. The notion of 'legal' relevance is different. The quotation from Cross and Wilkins adumbrate this distinction. They say expressly that relevancy (i.e. logical relevance) is not a matter of law, correctly leaving the quite correct inference that legal relevance is a matter of law. Our task here is to discern the concept of probability embedded in the definition of logical relevance. To that end, it is not necessary to expound further the difference between logical and legal relevance. It suffices to take to heart Cross and Wilkins' assertion that logical relevance is 'a matter of common sense and experience.' If this is so, the embedded notion of probability is also a matter of common sense and experience, which suggests in turn that it is not the aleatory interpretation that is here in play. Accordingly, it is prudent to repeat the admonition that the ordinary idioms of probability underdetermine their associated P-concepts. When a judge excludes evidence on grounds of its (logical) irrelevance, he is invoking *some* kind of P-concept. Given that the axioms are not part of a competent speaker's linguistic intuitions as they affect idioms of probability, it may be doubted that judges are applying the regulae of the probability calculus.

Even so, we should not lose sight of the fact that some legal scholars are of the view that, to the extent that someone is a sound probabilistic reasoner, his or her reasoning (whatever the details) produces the same result as would be achieved by express application of the full-bore calculus of probabilities. If this were so, a considerable advantage would flow to the legal theorist, as opposed to the legal practictioner. He would be able, at least in principle, to calculate the degree of relevance a piece of evidence would have if admitted, never mind that the rules by which he made his calculation may not have been employed by the judge when functioning as an intuitive probabilistic reasoner.

How might this suggestion be answered? Of course, this is a possibility that cannot be rejected out of hand. But, on reflection, we are drawn back to a point about coincidence. We have already made the point that concurrences of such striking serendipity call out for explanations, none of which has yet to come forth in a suitably detailed way.

This is not a point that we make with dismissive intent. The idea that in the transaction of their sundry cognitive agendas, beings like us satisfy the following condition:

- *When individual agents perform their cognitive tasks, there is a substantial likelihood that they will produce results that at least greatly resemble results that would have been produced had those individuals been executing the rules or fulfilling the conditions mandated by a normatively ideal theory (e.g., first order logic, rational decision theory, the probability calculus, and so on).*

What, then, would explain this amazing concurrence?

- *The best explanation of the concurrence is that, when they are operating properly, individual agents perform their cognitive tasks by executing the requirements of the requisite normative theories* tacitly.

We have two reservations, both of them serious. The first is that, with regard to the ideal models cited by it, the first proposition is certainly false. For first order logic and probability theory, see again [27, chapter 1]; and independently [58]. Thus the force of the claim is this: that when you and I operate as we should, there is *some or other* normative model whose provisions 'cover' our own results. But now the claim is vacuous.

A second reservation pertains to the second claim. It is a claim that embodies an interesting idea. The idea is that behaviour that is well-beyond our conscious reach – for example, behaviour that exceeds our computational capacities – falls well within our unconscious reach. If, for example, I succeed in balancing the probabilities properly, I have executed the requirements of the probability calculus, without knowing it, so to speak. In previous writings, we have emphasized the importance of subconscious inference and have urged the development of what might be called the logic of 'down below' ([16, 18, 2, 23]). We have no doubt that in its general form, the present idea has much to recommend it. But it strains credulity that any theory that gave as correct the same results as vouchsafed be the untutored behaviour of an individual when operating as he should, must be the theory that, when so behaving, he tacitly executes. If the tacitness thesis hides the operational theory from the operating agent, how could it not likewise hide the operational theory from the cognitive theorist?

Before quitting this section, we might touch on a point of related importance. A survey of the standard theories of rational belief-formation and decision-making (notably, again, first order logic, probability theory and rational decision theory), we see a readiness on the part of the theorist to characterize the rules of the theory as normative idealizations which real-life reasoners and decision-makers only approximate to. This certainly is the approach of legal theorists such as Richard Eggleston to the analysis of probabilistic reasoning in legal contexts. As he sees it, 'we must observe that no matter how we analyse the logical effect of [for example] the combined testimony of witnesses, tribunals are not strictly rational in their assessments' ([13, p. 208]). This is a judgement that has no traction whatever independently of an independent verification of the normative soundness of the rules, in whose failure to comport with, an ideal-life reasoner is judged to be irrational. In our view, this requirement to verify the normative legitimacy of a theory's rules of reasoning is not in the least met by pleading the *ideal* legitimacy of the proferred rules. There is not space to develop this point in any detail here, but we will note in passing our recent discussion of this problem in [17]. For now it suffices to say that while deviations from the requirements of some or other ideal model is sometimes evidence of a subpar performance, on other occasions it is evidence of the untenability of the norms in question. In our view, this is an extremely tricky issue to get to the bottom of. Certainly that a jury's natural behaviour sometimes discomports with the rules of the probability is not *just so* sufficient reason to judge him negatively.

10   Evidence

Earlier we made the point that in criminal and civil trials it is common for opposing counsel to lodge their respective arguments in different and incompatible subsets of the evidence heard, and that, where those conflicting subsets intersect, to give the intersecting propositions different weights. On the face of it, this is quite wrong. Justice requires that the evidence driving the opposing objectives of conviction and acquittal be commonly shared by all the parties concerned. To preserve this requirement of justice, the law is drawn to a kind of fiction. Since all parties must *hear* the same evidence, arguments by counsel and verdicts by the triers of fact must be *based on* the same evidence. Speaking this way masks an equivocation between two quite different states of affairs. Let E stand for the totality of evidence heard. Then the two states of affairs are these:

- A: An argument or verdict is grounded in E if and only if all propositions of the ground are members of E.

- B: An argument or verdict is grounded in E if and only if every proposition of E is a proposition of the ground.

As we have seen, given that E is routinely inconsistent, (B) must be rejected if the probability concept of the civil proof standard is aleatory probability. But that aside, it is not a very plausible interpretation of the requirement that all parties work from the same evidence. It blurs the distinction between *hearing* the same evidence (which is a reasonable requirement) and *accepting* the same evidence (which is not a reasonable requirement).

The state of affairs represented by (A) is, hands-down, the more descriptively accurate in relation to what actually happens in trials. It serves as a useful reminder that the proof standard is never just a matter of the evidential relationship between the evidence selected and the party's argument or finding. Of at least equal importance are the standards in play for the forming of the respective subsets E' and E''. Exposing these standards is a complex and thorny matter. We will mention just two features of it here.

First point. It is necessary to emphasize that in determining their respective E' and E'' it is the function of opposing counsel to operate *tendentiously*. Unlike the inquisitorial traditions of many European countries, the Anglo-American legal system has a deeply adversarial character. This imposes upon counsel the duty to advance one-sided arguments and, more generally, to eschew the ideal of impartiality. To the extent that they can do so with a degree of credibility, counsel will pick their evidentiary subsets E' and E" with a view to the support they lend to their respective pre-selected targets D and ~ D.

If counsel had their way, finders of fact would confine their roles to the adjudication of just two arguments, or theories of the evidence, viz., D | E' and ~ D | E". In fact, however, any judge or juror worth his salt will at least consider a third subset E"', on the basis of which he himself will evaluate D and ~ D. (If the trier of fact is a jury, the construction of still further evidentiary subsets will surely be produced.)

This turns out to be rather consequential. It wholly destroys the formal definition of the civil proof standard set out in section 4. We said there that a plaintiff succeeds when $Pr(D \mid E') > Pr(\sim D \mid E')$ and $Pr(D \mid E') > Pr(\sim D \mid E")$. As we now see,[30] a plaintiff succeeds when $Pr(D \mid E"') > Pr(\sim D \mid E"')$, a requirement that can be met even if

- $Pr (D \mid E"') < Pr (D \mid E') (1)$

- $Pr (\sim D \mid E") > Pr (D \mid E"') (2)$

Second point. In drawing his subset E"' from E, it is neither necessary nor sufficient that a trier of fact restrict himself to what of E he *believes*. On the one hand, in as much as a proposition might be included in E"' for its strong plausibility, and a finding of plausibility does not necessitate belief, belief is too strong a condition. On the other hand, since belief often outruns the force of the evidence, belief is too weak a condition. What, then, is the criterion against which triers of fact are expected to winnow out E"' from E? If belief is not the standard, what is?

We can take it as given that, unlike each side's counsel, the trier of fact is not at liberty to pick his E"' tendentiously. He cannot pick his E"' for the comfort it gives an already hoped-for result. The trier of fact has a duty to proceed impartially and to let the evidence determine his finding. This in turn requires that he accepts testimony which he finds credible and discounts (or ignores) testimony to the degree that he finds that it lacks credibility. In making these judgements of credibility, it is not necessary that the trier actually believe the propositions he finds to be credible. It suffices that given his understanding of the burden and standard of proof, supplemented by any further instructions from the judge as to inferences it is necessary or permissible to draw or weight that might be given to various parts of evidence

---

[30]If the trier of fact is a n-person jury, then the plaintiff succeeds if and only if, for each of up to *n* subsets of E each possibly distinct from E' and E" and from each other, D is favoured over ~ D to the requisite degree.

heard, the trier of fact places into his E''' propositions he believes will enable him to render a fair verdict. Accordingly, the trier of fact has two duties to perform. He must determine how D fares in relation to the evidence. He must also determine what the evidence *is*. The civil proof standard gives instruction about the first of these duties, but has nothing to say about the second. Given the language of the proof standard, it is reasonable to suppose that, in some sense or other, the finder of fact is to determine which of D, ~ D the evidence makes more probable. But there is nothing in the nature of the second task that comes close to sustaining an aleatory interpretation of what it requires. How, then, does a dutiful and impartial trier of fact select the propositions for his E'''?

This, too, is an issue of considerable complexity. Since it is the business of the present essay to elucidate the notion of probability in the law, we shall confine ourselves to some negative remarks. Intuitively, one selects for E''' from the propositions in E by appeal to what one finds convincing. We say again that it is not a condition of one's finding a witness's testimony that P convincing that one believe that P. Strictly speaking, it is not *propositions* that are convincing, but rather it is *people* who are. Convincingness reposes in the manner of a response as much as in its content. It is influenced by eye-contact, inflection, skin-tone and the myriad other aspects of body language. It is not at all easy to construe the notion of convincingness in probabilistic terms (except where the idiom of probability camouflages a concept of plausibility; see below). Consider the suggestion that in selecting the propositions to place into E''' one should confine oneself to the most probable members of E.

Do we then pick propositions with the highest conditional probability? But then, if so, the probability *of* what *on* what? In the standard approaches, the answer to the first question is D and to the second is E'''. But here it is the probability of E''' that is (allegedly) in question, and that on which that probability is reckoned conditionally cannot be E''' itself or anything else in E (since, by definition anything in E but not in E''' is rejected). This, too, is rather consequential. It shows that judgements of what is convincing are grounded not just in the testimony of witnesses but rather on whatever background considerations may enable one quite generally to assess the convincingness of what others tell us. At the heart of such judgements is our capacity for determining whom to trust, albeit that witnesses are usually strangers to finders of fact. It hardly needs saying that a full analysis of trust, supposing such were possible, lies well beyond what we have space for here.[31] Even so, there are grounds on which to suggest that the estimates of trust that underlie judgements of convincingness are dominantly plausibilistic in character rather than probabilistic. One of the reasons that support such a suggestion is the peculiar ambiguity of plausibility. One may characterize someone as a plausible *witness* even though the content of her testimony is not a plausible *proposition*.

---

[31] See [3], [24] and [16, chapter 2]. See also [8].

Equally, the proposition embedded in a witness's response might be highly plausible and yet the person whose proposition it is might not be a plausible witness. Actual practice indicates that plausibility of manner tends to trump plausibility of content, although clearly enough the optimum is plausible propositions attested to by plausible witnesses.

In [18], we emphasized the connection between the plausible and the characteristic. This, too, is more than we have time for here. It may be, however, that there is a manner of attestation that is characteristic of its truthfulness[32] just as there appear to be behaviours that are characteristic of lying. Accordingly, a person may be said to be a plausible witness to the extent that his testimonial behaviour has the characteristics of truthful attestation.

Characteristicness also plays a role in judgements of propositional plausibility. We may say that P is a plausible proposition in a context K to the extent that, were it the case that K obtained, it would not be out of character that P. It is true that the idiom of the characteristic has nothing like the clarity and hard edges that one finds in the probability calculus. Some people will be drawn to the idea that, even so, the concept of characteristicness is an essentially probabilistic one. We ourselves are contrary-minded. Suffice it to say that for any sense of probability and plausibility for which the probable and the plausible are disjoint concepts, the probable cannot be a defining condition on any defining condition of the plausible. So we may with reason conclude that in constructing any E''' the notion of conditional probability is not essentially in play.

A similar caution applies to the idea that one constructs E''' by selecting from E propositions with sufficiently high *prior* probabilities. Even if the propositions of E came clearly coded with prior probabilities, improbable statements might be redeemed by plausible witnesses. Beyond that, the very idea of prior probabilities is a bit of a tangle. In the probability calculus a prior probability is a wholly unconditional probability. What this means is that P's stand-alone probability is determined without reference to anything that might be considered evidence for or against it. But consider a paradigm case. A true coin has an equal probability of showing a head or showing a tail on any fair toss. The prior probability of each is 0.5. However, that this is so depends on a number of facts in whose absence the probabilities would falter. If the coin weren't designed to have just two sides, or if it were not a true coin, then the probabilities of heads (and of tails) would be disturbed. These are clearly evidential considerations. Given that the coin is both two-faced and true, given that any toss is fair, then Pr (H) = Pr (T) = 0.5. But this is conditional probability.

---

[32]Note: 'truthfulness' rather than 'truth'. Someone is truthful when he thinks he is telling the truth. What is at issue here is the extent to which there are behavioural and other markers of someone's thinking that he is telling the truth.

Probability theorists are disposed to remove this problem by expedient definition or by manipulating a contrast between propositions that express identity conditions and propositions that express evidential considerations. By these lights,
what in some contexts would be evidential considerations are built in as individu-
ating traits of the objects in the theory's event space. Accordingly, the probability
calculus is not about coins and tosses. It is about *true* coins and *fair* tosses. And,
since real coins are rarely perfectly true and real tosses rarely perfectly fair, the
probability calculus pays for the expedient suppression of evidential considerations at the level of prior probability by idealizing its objects.

As far as coin tosses go, this is not something to get over-excited about. The
coin in my pocket is an approximation of a coin defined to be true, and any toss I
have in mind for it is a toss that would approximate to a toss defined to be fair. If
I think that such approximations preserve close approximations of what the probability calculus asserts of ideal coins ideally tossed, then I might be able to satisfy
myself that prior to the next actual toss, Pr (H) is pretty close to 0.5. Good science
is replete with idealizations for which approximation relations are mathematically
definable. The law, too, has its idealizations, or anyhow its artificialities. But there
is nothing in the law of evidence that sanctions triers of fact to conceptualize the
events led in evidence on the model of ideal coins ideally tossed. Natural objects
and natural events strain the distinction between propositions expressing identity
conditions and propositions expressing evidential considerations. It is a consequence of this strain that the very idea of prior probability is a strained idea for
natural happenings.[33]

11 Corroboration

One of Jonathan Cohen's dissatisfactions with the aleatory interpretation of probabilistic reasoning in legal contexts centres on the notion of corroboration. Intuitively speaking, the probability that a given proposition is true increases with the
number of requisitely placed witnesses who independently attest to it, and the lack

---

[33]In some approaches, an event's prior probability is relative only to given bodies of evidence. So
conceived, a proposition p will have a prior probability relative to q iff it has a probability reckoned
independently of q, but which might well be conditional on other propositions r. Accordingly, the prior
probability of the propositions in E could be the probabilities they chance to have whose calculation is
independent of E itself but is conditional on some r. Whatever we make of this suggestion generally,
it gives little encouragement to the idea that the propositions of E have prior probabilities relative to E
but conditional probabilities relative to other bodies of information I, I', I'', and so on. For one thing,
given that I, I', and I'' are not subsets of E, they are not in evidence. Triers of fact are sorely limited
in what they can accept beyond what is led in evidence. To some extent, triers of fact may allow
themselves to be guided by what they (and we) take as common knowledge. One requires no evidence
to permit the inference that water is wet. The wetness of water is common knowledge. But equally
it is common knowledge that people are hardly ever guilty of serious crimes, yet common knowledge
as well that people are usually guilty of the serious crimes with which they've been charged. So the
common-knowledge test of the admissibility of extra-evidential information is hardly reliable.

of those, similarly placed, who speak against it. Consider the case of eye-witness testimony. It is important to keep in mind that, while eye-witnesses purport to testify about what they saw or otherwise perceived of some given event, their testimony is based on what they now remember of what they then saw or otherwise perceived. It is easy to see, therefore, that a person's eye-witness testimony is subject to three kinds of difficulty. One is that the witness may be lying. The other is that, while the witness's observations were completely veridical, he has now misremembered them in some material way. The third is that, while his present memory of what he then thought he observed is impeccable, his observation at the time was defective – what he thought he observed, he did not in fact observe in all material aspects.

It falls to the trier of fact to determine whether such testimony is to be accepted. Aside from the afore-mentioned kinds of defectiveness, further considerations bear on this determination. One is, as we say, the plausibility of what is attested to; another is the plausibility of the attestor. Also salient are the factors of tendentiousness with which actions at common law are imbued. A witness who strikes the trier of fact as 'on the side' of one party as against the other is more readily dismissed than one who, as it would appear, has no axe to grind. It can hardly be said that witnesses never achieve the desired level of impartiality, but that they have been called in the first place – that they will be heard at all – is the result of a highly partial strategic decision on the part of counsel. A further set of considerations involves the defaults that normally attend the transmission of information from a second party. The principal default is that in the absence of reasons to the contrary, what someone tells you is something that you should accept. More importantly still, there is a *causal counterpart* of this rule. It provides that in the general case, being told by someone that P is causally sufficient for the production of your belief that P. In other words, if you want someone to believe that P, simply tell him that P is the case. Doing so will make him believe it. A final consideration in the assessment of eyewitness testimony is the impossibility of independent direct confirmation. Although lawyers and judges talk loosely about facts, in strictness, verdicts are reached in the total absence of facts. They are reached rather on the basis of *attestations* of facts.

The basic idea seems to be that corroboration is a likelihood-enhancer. Something attested to by one witness gains in credibility when supported by a second witness, even more so by a third, and so on. It seems a rather simple and straightforward idea, but in fact it is one that embeds a good deal of complexity. Central to the efficacy of one's assurance that P is that one is in a position to give it, i.e., that P is something that one already knows. In run-of-the-mill situations of information-passing, the very fact that you tell me that P is reason for me to think that you are in a position to tell me that P. Acceptance of sayso is also a defeasible matter. If you tell me that P I might come to believe P without qualification; but I might instead come to hold to P somewhat more circumspectly. However, in either case, my subscription to P is open to revision in the face of new information. Corroboration seems tailor-made for cases in which there is room for the belief that P to be strengthened. If sayso is assurance, corroboration is re-assurance. If this is how corroboration works as a belief-strengthener, it should also be noted that it works differently in those cases in which my belief that P on your sayso is full and unqualified. In that case its function is to cancel possible counter-evidence. My acceptance of P on your sayso presumes that anyone else, in your place would attest to P as well. In the ideal case, every such person is known and each corroborates the original sayso. In actual practice, such exhaustive testimony is out of the question. In these cases the role of a solitary witness is that of a randomly selected member of the total set of witnesses. The efficacy of that particular person's attribution that P pivots on the presumption that any suitably situated person would attest to the same thing. Corroboration by a second witness is two things at once. It is very slight *confirmation* of that presumption. And it is a much less slight *strengthener* of it. There is a considerable gap between confirmation and strengthening. The presumption that any suitably positioned of the attested-to situation would say that P only if P were true is given miniscule confirmation by the solo fact of your attestation, and very little additive confirmation by the like-testimony of a second party. Even so, leaving its confirmation status aside, the presumption is strong enough in the first instance to ground my acceptance of P on your sayso - which is to say that it is strong enough to override a negatory confirmation status. Corroboration strengthens an already strong presumption, and by a factor that further stretches the gap between confirmation and strengthening.

The fact remains that in the absence of indications to the contrary two heads are usually better than one in the game of sayso. Probability theorists are minded to think that this betterness is adequately captured by the aleatory constraints on probability. This can't be so, as Cohen observes, if corroboration is construed as the conjunction of the probabilities that attach to the sayso of the two (or more) witnesses individually. So if the probability that what witness W1 says is true is 0.3 and that what W2 says is true is 0.4, then their conjunctive probability is a lower, not higher, number, namely 0.22 ([4, p. 95]).

Eggleston argues (rightly, in our view) that Cohen has misconceived the probabilistic structure of the corroboration-situation ([13, pp. 204-209]). But in estimating the efficacy of a second witness's sayso, Eggleston's own view is that the following considerations need to be considered and given due weight in the ensuing calculations. Let P be the proposition attested to in the first instance by W1 and in the second instance by W2. Then

1. The increased likelihood of P on the testimony of W2, is construed as the probability that W2 confers on P *given* the probability conferred on it by W1.

2. In calculating the requisite probabilities, one must consider both the likeli-hood that P is true, as well as the likelihood that P is false, i.e., that $\sim P$ is true.

3. The likelihoods in question require that the ways in which P would be false be factored into the probabilistic reckoning.

Accordingly, when we apply the odds form of Bayes' theorem for the following values, assuming that there are five ways to lie about P:
$\Pr(W1^{P}/P) = 0.3$; $\Pr(W2^{P}/P) = 0.4$; $\Pr(W1^{\sim P}/\sim P) = 0.7 \times 0.2$; $\Pr(W2^{\sim P}/\sim P) = 0.6 \times 0.2$

Then, substituting into the equation of the theorem, we avert the charge that since corroboration is conjunctive, it is a probability depressor. Eggleston con-cludes that 'the phenomenon of corroboration is compatible with, rather than in-consistent with, the dictates of mathematical probability theory' (p. 207).

Eggleston is right. At least, he is right if he has also got the conceptual struc-ture of corroboration right. But, in fact, there is reason to think that in his effort to structure the concept of corroboration in such a way as to engage the odds version of Bayes' theorem, he has actually *distorted* the concept of corroboration. If this is so, he has fallen prey to Make Do. The imprint of Make Do shows in some of the details of the Eggleston analysis, and also in a matter of more general impor-tance. In the category of detail, the analysis represents the corroboration-assessor as calculating the probability that witnesses would attest to P, on the assumption that P is in fact true, and, on the further assumption that P is in fact false. Each way, the corroboration-assessor strives to probabilify the *production* of the testi-mony at hand under these complementary assumptions. But there isn't the slightest empirical reason to suppose that this is a general feature of such assessments.

A further detail involves the ways in which an attested-to P might be false. It is doubtless true that when an eyewitness reports his observation that P, an assessor of the testimony will sometimes consider certain contextually-cued rivals of P. For example, if a witness reports hearing the accused threaten the victim with the words, 'I'll get you if it's the last thing I do', he may consider the possibility that, while those words were in fact uttered, they were uttered jokingly or ironically.

Of course, in trial settings, there are additional and quite significant constraints that attend the general presumption of the reliability of sayso. If we suppose that someone, W1, observed the incident in question and on that basis attests that P is the case, his evidence is *not* that P, but rather that he was in a position to know whether P is the case and that indeed he does know that P is the case. Conversa-tionally formulated, W1's evidence is not that P but rather that he *observed* that P. It is his observation that satisfies the requirement that he provide reason to believe that he was in a position to know that P. Schematically, this may be put as follows:

- *W1 gives eye-witness testimony that* P *iff*

1. *W1 attests that he was in a position to know whether* P.

2. *W1 attests that, being in that position, he did (and does) know that* P.

It is worth emphasizing that if we add as a third condition that

3. *W1 attests that* P

we are adding a *redundancy*, since that P follows from W1's knowledge that P. Thus, in all strictness, W1's evidence that P does not require that W1 say that P and would be wholly unconvincing were 'P' all that W1 did say.

For expository convenience we might say that, in giving his evidence in connection with P, W1 *declares his bona fides*. Again, he does not assert that P. He asserts that he observed that P. He tells how he came to know that P. It is noteworthy that in the general case, an informant does not, and is not expected to, declare his *bona fides*. It is taken for granted that they exist and that they are presentable upon demand. So while W1's evidence that P is subject to the presentation of *bona fides* in the manner of clauses 1) and 2) of our schema, the disclosures contained in those clauses are in their own right forwarded on the sayso of W1 where W1's testimony in these respects is *not* subject to the declaration of *bona fides*. What clauses 1) and 2) convey is rooted in W1's present memory of the observation of the P-situation. If his memory is reliable, his evidence is telling. His memory might not in fact be reliable in this case, but that it is so is not something that W1 can efficaciously attest to.

All of this raises the question of what it is that W2 corroborates when W2 also gives eyewitness evidence that P. Like W1, W2 implements the basic schema:

1. *W2 attests that he was in a position to know whether* P.

2. *W2 attests that, being in that position, he did (and does) know that* P.

Like W1, W2's evidence that P involves the declaration of his own *bona fides*. Like W1, W2's declaration of his *bona fides* is achieved by his assertion of sentences in the manner of 1) and 2), but in testifying that 1) and 2) he does not, and could not, declare his *bona fides*. And, like W1, W2's evidence that P does not require that he assert that P and would be unconvincing if all that he did were to assert that P. It is also the case that sometimes W2 is in a position to attest to the truth of the information embedded in W1's clauses 1) and 2). For sometimes, in being positioned himself to know that P, W2 is positioned to know that W2 is positioned to know that W1 is positioned to know that P. Sometimes indeed W2 is positioned to propose that, being thus positioned, W1 could not have failed to know that P. But clearly this does not happen in the general case. In corroborating W1's evidence that P, W2 is not required to corroborate anything that W1 is required to say as constitutive of his testimony that P.

How, then, does it come to be the case that the testimony of W2 corroborates the testimony of W1? What is the *structure* of this corroboration?

The facts (which is to say, the *purported* facts) attested to by eyewitnesses come in varying degrees of particularity. If W1 tells us that it was snowing in Regina at that hour of the day in January, that is not much of a surprise. But it is still a quite particular fact. It doesn't snow everyday in January in Regina; in fact, it doesn't snow on most days there and then. Even on days when it does snow, for much of the day it doesn't snow at all. Other facts are more particular still. That the accused hit the deceased on the head with a snow shovel is a very particular fact. Although there is some variation here, the facts introduced by eye-witness testimony are facts whose prior probability ranging from not much to hardly any. This is a general feature of eyewitness testimony, not a peculiarity of it in contexts of legal proceedings. The principal contribution of an informant's sayso is its production of high credibility in the face of low prior probability. Take the commonplace fact of Zenon Tyrlowski eating a ham sandwich at 1:45 p.m. in Fred's Diner on January 8th, 2007, on Lexington and 12th Avenue. For the arbitrarily selected individual, there is next to no reason to believe this to be true. Construed as subjective probability, what such a person knows of the world lends vanishingly little evidence to this claim. But if I told you that this is what I saw at that time in that place, there is in default of indications to the contrary a massive spike of credence. Now you know it too. If we assigned some imaginary probabilities, your sayso about Zenon converts a probability of 0.1 to a probability of 0.9.

Part of what makes an eyewitness's evidence credible is how what he says implicates his position to know: 'I turned the corner and nearly ran into the accused, who was battering the deceased with a snow shovel. He was as close to me as you are now.' What does this tell us? It tells us that any randomly selected individual actually or counterfactually in that position, or a relevantly similar position, would have seen the same thing. It is easy to see that if he satisfies the requisite conditions of testimonial independence, W2 fulfills the role of that arbitrarily selected individual.

This turns out to be an important general feature of corroboration. One element is the veracity of the witness. Another is the structure of his situation. In attesting that *P*, W1 is telling us something critical about his situation. He is telling us that it is a situation in which anyone would observe that *P*. It is important that W1 tells us that he observed that *P*. It is even more important that he tells us that anyone thus placed would have observed that *P*. This embodies a generalization, albeit in generic form:

- *G*: Situation *S* is such that anyone x in it, or a relevantly similar position, would know that *P*.

What W2's testimony gets us to see that W2 himself is a value of x in this generalization. Likewise part of W1's own testimony is to this same effect. Both are committed to *G* and each attests that he himself is a value of x. From the point of view of confirmation theory, this is pretty small beer. It strongly suggests that the structure of corroboration is significantly underdetermined by the probability calculus.

What this suggests is that the structure of corroboration is essentially abductive and that, this being so, it is more a phenomenon of plausibility than of aleatory possibility. In giving his evidence, W1 implicates that *G*. This is a kind of prediction. In giving his evidence, W2 implicates that he is a value of x in *G* and thus that his situation is a positive instance of it. The gross structure of this can be set out as follows.

1. *In testifying that* P, *W1 implicates that* G.

2. *In testifying that* P, *W2 instantiates* G.

3. *In the absence of indications to the contrary, a trier of fact will take W2's positive instantiation of* G *as reason to believe that* G, *which in turn, is reason, albeit provisionally, to accept W1's assurance that* P.

What (3) clearly suggests is that the trier of fact has made an abduction in the form:

i) *W1 implicates that if an arbitrarily selected individual is placed in the manner of* G, *he will attest to* P.

ii) *By his own testimony, W2 was placed in the manner of* G *and did attest to* P.

iii) *The best explanation of this is that W1's implicature is true.*

We are now in position to state our more basic reservation about the Eggleston effort to retain a Bayesian approach to corroboration. It is that corroboration is not an inductive but rather abductive matter; that corroboration is not about the bunching of probabilities, but rather the fibring of plausibilities; that the conceptual contours of corroboration are significantly more complex than anything envisaged by a Bayesian revisionism. Eggleston's attempt to cleave to the mathematics of probability in corroborative contexts is evidence of Make Do.[34]

---

[34] We note in passing that expert testimony has some of the same features found in eye-witness testimony. In particular, expert testimony is never allowed without a declaration of the witness's *bona fides*. These include the witness's formal qualifications as an expert, as well as the exposure of the witness's particular position from which his preferred opinion was generated (e.g., 'I examined the wound under a fluoroscope and determined that...'). Expert testimony also carries a presumption that resembles *G*, roughly to the effect that 'any suitably qualified expert positioned as I was positioned would have found as I found.'

12   Conclusion

Old habits die hard. The entrenchment of the probability calculus in the natural and social sciences is deep to the point of dogmatism. This is tantamount to an accusation that should not be shirked, notwithstanding its apparent harshness. It is that the sheer scope of the entrenchedness of the probability calculus betokens the presence of *Make Do*. When one considers the vastness of the sweep of the phenomena explored by the sciences, both natural and social, the presumed scope of aleatory probability is simply breath-taking. From statistical mechanics and parts of quantum mechanics to physical chemistry, from neo-classical and welfare economics to decision theory, from statistical thermodynamics to judgements in the form 'Yes, the company probably broke the contract', aleatory probability's claim to universality is striking enough to attract at least a modicum of scepticism, as would any vigorously cross-type putative universality.[35] Aleatory probability has long knocked at the doors of legal theory, as witness [50]. It is a request for admittance that should not be lightly acceded to. It is quite true that it is better to have a good theoretical grasp of a centrally important concept than not. It is also true that to date only aleatory probability can boast a mature and deep theory. But having a superb theory that doesn't fit is no improvement on having no theory at all. It has been the burden of this paper to point out the lack of fit between the calculus of probability and the conceptual contours of legal probability, such as they may be. Aleatory probability does indeed knock at the doors of legal theory. Admittance should be refused. Admittance would be capitulation to *Make Do*.

We said at the beginning that should the case against aleatorizing the concept of legal probability hold up, it will have brought to our attention an open problem in the research programme of legal studies. The problem is that, at present, the concept of legal probability has no theory. We may say that a problem X is an *open* problem in a research programme only to the extent that a theoretical articulation of X is achievable in principle. If the judgement that X does indeed give rise to an open problem, there is inducement to seek out its theoretical articulation. In so saying, we come full circle to a tension we took note of in section 1, where we remarked upon the law's latent distrust of theoretical analyses of fundamental concepts, especially analyses that pivot on explicit definitions, on definitions in the form 'Something is X *if and only if* conditions Y, Z, W hold'. This serves as a useful reminder. It gets us to attend to the possibility that legal probability is unamenable to any form of theoretical treatment typically favoured by logicians, and certainly to any form of theoretical treatment of the sort typified by the probability calculus. This might well be true, but before we acquiesce to it, there are two moderating considerations to take note of. One is that if a theory of legal probability exists, it need not have the character of a mathematical system. The other is that it

---

[35] Some critics question the presumed provenance of aleatory probability even in physics ([51]) and economics ([40, 41]).

is intellectually sounder practice to ground one's abandonments on tries that have manifestly failed, rather than on not trying at all. This being so, there is reason to take on sufferance the issue of probability as a genuinely open problem for legal studies. In future work we hope to reduce the size of this openness. In particular, we are drawn to the view that once there is a suitably stable logic of plausibility up and running, we will have the wherewithall to press hard on the suggestion that even where the lexicon suggests a probabilistic enterprise, finders of fact in civil and criminal trials are charged with the task of performing plausibilistic, rather than probabilistic, tasks. See here [21].[36]

BIBLIOGRAPHY

[1] Luc Bovens and Stephan Hartmann, *Bayesian Epistemology*, Oxford: Oxford University Press, 2003.
[2] P.D. Bruza, D. Widdows and John Woods, 'A quantum logic of down below'. In D. Gabbay, K. Engesser, and D. Lehmann, editors, the *Handbook of Quantum Logic and Quantum Structures*, volume 2. Amsterdam: Elsevier to appear.
[3] C.A.J. Coady, *Testimony: A Philosophical Study*, Oxford: Oxford University Press, 1992.
[4] L. Jonathan Cohen, *The Probable and the Provable*, Oxford: Clarendon Press, 1977.
[5] L. Jonathan Cohen, 'Psychology of prediction. Whose is the fallacy?', *Cognition*, 7, p. 385-407, 1979.
[6] L. Jonathan Cohen, 'Some historical remarks on the Baconian conception of probability', Journal of the History of Ideas, volume 41, pp. 219-231, 1980a.
[7] L. Jonathan Cohen, 'Whose is the fallacy: A rejoinder to Daniel Kahneman and Amos Tversky', Cognition, volume 8, pp. 89-92, 1980b.
[8] L. Jonathan Cohen, *An Introduction to the Philosophy of Induction and Probability*, Oxford: Clarendon Press, 1989.
[9] Rupert Cross and Nancy Wilkins, *An Outline of the Law of Evidence*, London: Butterworths, 1964.
[10] Rupert Cross and Nancy Wilkins, *An Outline of the Law of Evidence*, sixth edition, London: Butterworths, 1986.
[11] Bruno de Finetti, 'Foresight. Its logical laws, its subjective sources'. In *Studies in Subjective Probability*, edited by H.E. Kyburg and H.E. Smokler, pp. 93-158, New York: John Wiley, 1964.
[12] Augustus De Morgan, *Formal Logics, or the Calculus of Inference, Necessary and Probable*, London: Taylor and Walton, 1847.
[13] Richard Eggleston, *Evidence, Proof and Probability*. Second edition. London: Weidenfeld and Nicolson, 1983. First published in 1978.
[14] James Franklin, *The Science of Conjecture: Evidence and Probability before Pascal*, Baltimore MD: The Johns Hopkins University Press, 2001.
[15] Dov M. Gabbay, *Fibring Logics, Oxford Logic Guides 38*, New York: Oxford University Press, 1998.
[16] Dov M. Gabbay and John Woods, *Agenda Relevance: A Study in Formal Pragmatics*, Volume 1 of *A Practical Logic of Cognitive Systems*, Amsterdam: Elsevier, 2003a.

[36]For advice and criticism we would like to express our gratitude to S.B. Armstrong, Q.C., the late Jonathan Cohen, Andrew Irvine, Jan Willem Romeyn, Ori Simchen, Bas van Fraassen, John Williamson, C.L. Woods, Q.C., K.A. Woods, and, especially, the editors' astute and helpful anonymous referee. Funds supporting this research derive in part from The Abductive Systems Group, University of British Columbia and the Engineering and Physical Sciences Research Council of the UK, for which our thanks. For able technical support we also thank Carol Woods in Vancouver and Jane Spurr in London.

[17] Dov M. Gabbay and John Woods, 'Normative models of rationality: The disutility of some approaches', *Logic Journal of IGPL*, volume ii, pp. 597-613, 2003b.

[18] Dov M. Gabbay and John Woods, *The Reach of Abduction: Insight and Trial*, Volume 2 of *A Practical Logic of Cognitive Systems*, Amsterdam: Elsevier, 2005a.

[19] Dov M. Gabbay and John Woods, 'The practical turn in logic'. In *Handbook of Philosophical Logic* edited by Dov M. Gabbay and F. Guenther, volume 13, pp. 15-122. 2nd revised edition, Dordrecht and Boston: Kluwer, 2005b.

[20] Dov M. Gabbay and John Woods, *Seductions and Shortcuts: Fallacies in the Cognitive Economy*. Volume 3 of *A Practical Logic of Cognitive Systems*. London: College Publications, to appear, 2009.

[21] Dov M. Gabbay and John Woods, *The Ring of Truth: Towards a Logic of Plausibility*, Volume 4 of *A Practical Logic of Cognitive Systems*, London: College Publications, 2010.

[22] Gerd Gigerenzer, *Adaptive Thinking: Rationality in a Real World*, Oxford: Oxford University Press, 2000.

[23] Gerd Gigerenzer, *Gut Feeling*. New York: Viking USA, 2007.

[24] Trudy Govier, *Dilemmas of Trust*, Montreal and Kingston: McGill-Queen's University Press, 1997.

[25] Paul Grice, *Studies in the Way of Words*, Cambridge MA: Harvard University Press, 1989.

[26] Nicholas Griffin, 'Through the Woods to Meinong's Jungle'. In *Mistakes of Reason: Essays in Honour of John Woods*, edited by Kent A. Peacock and Andrew D. Irvine, pp. 15-32, University of Toronto Press, 2005.

[27] Gilbert Harman, *Change in View: Principles of Reasoning*, Cambridge, MA: MIT Press, 1986.

[28] Carl G. Hempel, 'Studies in the logic of confirmation'. In Carl G. Hempel, *Aspects of Scientific Explanation and Other Essays in the Philosophy of Science*, pp. 3-51, New York: The Free Press, 1970.

[29] Colin Howson, 'Theories of probability', *British Journal for the Philosophy of Science*, volume 46, pp. 1-32, 1995.

[30] A.D. Irvine, 'How Braess' paradox solves Newcomb's problem', *International Studies in the Philosophy of Science*, volume 7, pp. 141-160, 1993.

[31] E.T. Jaynes, *Probability theory: the logic of science*, http://bayes.wustl.edu/etj/prob.html. 1998.

[32] D. Kahneman and A. Tversky, Oxford 'On the interpretation of intuitive probability: A reply to Jonathan Cohen', Cognition, 7, p. 409, 1979.

[33] Daniel Kahneman, 'New challenges to rationality assumptions'. In *Choices, Values and Frames*, edited by D. Kahneman and A. Tversky pp. 758-774, New York: Cambridge University Press, 2000.

[34] John Maynard Keynes, *A Treatise in Probability*, New York and London: Harper and Row and Macmillan, 1921.

[35] A.N. Kolmogorov, *The Foundations of the Theory of Probability*, New York: Chelsea Publishing Company, 1950. Oxford

[36] David K. Lewis, 'A subjectivist's guide to objective chance'. In David Lewis, *Philosophical Papers Volume II*, pp. 83-132, 1986.

[37] Neil MacCormick, *Legal Reasoning and Legal Theory*, Oxford: Clarendon Press, 2003.

[38] David H. Mellor, *The Matter of Chance*, Cambridge: Cambridge University Press, 1971. Oxford

[39] J.S. Mill, *A System of Logic*, London: Longman's Green, 1843.

[40] Philip Mirowski, *Against Mechanism*, Towata, NJ: Rowman and Littlefield, 1988.Oxford

[41] Philip Mirowski, *More Heat than Light*, Cambridge: Cambridge University Press, 1989.Oxford

[42] Judea Pearl, *Probabilistic Reasoning in Intelligent Systems: Networks of Plausible Inference*, San Mateo CA: Morgan Kaufmann, 1988.

[43] K.R. Popper, *Conjectures and Refutations. The Growth of Scientific Knowledge*, London: Routledge and Kegan Paul, 1974.

[44] Nachum L. Rabinovich, 'Studies in the history of probability and statistics. XXII Probability in the Talmud', *Biometrika*, 56 (1969), 437-441.

[45] Frank Ramsey, 'Truth and probability', in *Studies in Subjective Probability*, edited by H.E. Kyburg and H.E. Smokler, pp. 61-92, New York: John Wiley, 1964.

[46] Nicholas Rescher, *Plausible Reasoning: An Introduction to the Theory and Practice of Plausible Inference*, Assen and Amsterdam: Van Gorcum, 1976.

[47] Michael Scriven, *Reasoning*, New York: McGraw-Hill, 1976.

[48] John W. Strong, *MacCormick on Evidence*, 5th edition, St. Paul, MN: West Group, 1999.

[49] Paul Thagard, *Conceptual Revolutions*, Princeton NJ: Princeton University Press, 1992.

[50] Peter Tillers and Eric D. Green, *Probability and Inference in the Law of Evidence: The Use and Limits of Bayesianism*, Dordrecht and Boston: Kluwer, 1988.

[51] Stephen Toulmin, *The Philosophy of Science: An Introduction*, London: The Hutchinson University Library, 1953.

[52] Stephen Toulmin, *The Uses of Argument*, Cambridge: Cambridge University Press, 1958.

[53] Amos Tversky and Daniel Kahneman, 'Extensional versus intuitive reasoning: The conjunction fallacy'. In *Heuristics and Biases - The Psychology of Intuitive Judgment*, edited by T. Gilovich, D. Griffin and D. Kahneman, pp. 19-48, Cambridge: Cambridge University Press, 2002.

[54] Bas van Fraassen, 'The day of the dolphins'. In Kent Peacock and Andrew Irvine editors, *Mistakes of Reason: Essays in Honour of John Woods*, pp. 111-133, Toronto: University of Toronto press, 2005.

[55] Richard von Mises, *Probability, Statistics and Truth*, London, Allen and Unwin, 2nd edition, 1957.

[56] Jon Williamson, 'The actual frequency interpretation of probability', philosophy.ai report pai_jw_99_b,http://www.kcl.ac.uk/philosophy.ai.1999.

[57] Jon Williamson, 'Probability logic'. In *Handbook of the Logic of Argument and Inference: The Turn Towards the Practical*, edited by Dov M. Gabbay, Ralph H. Johnson, Hans J. Ohlbach and John Woods, pp. 397-424, Amsterdam: North Holland, 2002.

[58] John Woods and Douglas Walton, 'On fallacies', *Journal of Critical Analysis*, 5, pp. 103-111, 1972.

[59] John Woods, *Paradox and Paraconsistency: Conflict Resolution in the Abstract Sciences.* Cambridge and New York: Cambridge University Press, 2003.

[60] John Woods, 'Eight theses reflecting on Stephen Toulmin'. In David Hitchcock and Bart Verheij, editors, *Arguing on the Toulmin Model: New Essays in Argument Analysis and Evaluation*, pp. 379-397. Amsterdam: Springer Netherlands, 2006.

[61] John Woods, Should we legalize Bayes theorem?'. In Hans V. Hansen and Robert C. Pinto, editors, *Reason Reclaimed: Essays in Honor of J. Anthony Blair and Ralph H. Johnson*, pp. 257-267. Newport News, VA: Vale Press, 2007.

A Non Common Sense View of Common Sense in Science

GERHARD HEINZMANN

1

In the Blue Book, Wittgenstein rejected two well-known positions which both consider the philosopher's task to be the determination of common sense, the expression "common sense" understood according to the Reidian tradition as those beliefs that are accepted by most speakers in all societies (thus probably excluding all fables from common sense). The first position that Wittgenstein rejected is ordinary language philosophy, whose aim it is to determine the common use of language by conceptual analysis (one obtains the common use of language by eliminating the elaborate code (*Bildungssprache*) from natural language [5, 60]; the second position that Wittgenstein rejected was Russell's aim to determine, by logical analysis of language, the description of a non-linguistic reality, which was thought to be accessible for any rational mind. In other terms, Wittgenstein rejected the linguistic as well as the ontological common sense positions. I quote:

'There is no common sense answer to a philosophical problem. One can defend common sense against the attacks of philosophers only by solving their puzzles, i.e., by curing them of the temptation to attack common sense; not by restating the views of common sense [against ordinary language philosophy]. A philosopher is not a man out of his senses, a man who doesn't see what everybody sees; nor on the other hand is his disagreement with common sense that of the scientist disagreeing with the coarse views of the man in the street. That is, his disagreement is not founded on a more subtle knowledge of fact [against Russell].' [14, 58-59]; quoted from [5, 128-129]

Wittgenstein's emphasis on the impossibility of justifying common sense sentences [13, §124] does neither concern the reduction to an ultimate common sense language nor the reduction to ultimately evident facts. Wittgenstein's language games do introduce linguistic expressions of common sense language, but they do not necessarily capture the ordinary *de facto* use of language:

'Our ordinary language [...] holds our mind rigidly in one position, as it were, and in this position sometimes it feels cramped, having a desire for other positions as well' [14, 59].

In this context, Wittgenstein's most important observation was that common sense language games provide the framework for the justification of our beliefs. But these games cannot themselves be in need of justification or in doubt. Therefore common sense is not only a postulate concerning beliefs by the observer, but a pre-condition for beliefs [2, 156-157]. Nevertheless, according to Wittgenstein, common sense can lead to disagreements with the scientist's practice as this practice is based on a more subtle knowledge of facts than the coarse views of the man in the street. Today, the admission of the different roles of common sense in philosophy on the one hand and in science on the other hand seems to be the standard view. One of its unfortunate consequences is the idea that a conference in philosophy shall be comprehensible for any reasonable person, a demand which is certainly never made for a conference on, say, quantum physics.

2

At first sight, the Wittgensteinian position concerning common sense and its different roles in philosophy and in science can be attacked along two lines of argumentation: according to the first, one should overrule common sense in science *and* in philosophy; according to the second, science rests on hypotheses that are justified on a common sense basis. Most scientists either agree with Wittgenstein or endorse the first objection: In science and in philosophy concepts and propositions of common sense can yield absurdities when taken literally. For example, the fact that the sun rises can lead to such a paradoxical situation. But what are the tests of absurdity? [10, 81] asks, and rightly so. Carnap [1, p. 67] opts for the other objection. He considers knowledge by acquaintance (*Elementarerlebnisse*) to be the basis for scientific activities, which themselves involve knowledge by description. According to his reconstruction, science rests on a common sense basis, too. Although both opinions are rather extreme, they help to understand how Wittgenstein's conception about the role of common sense in philosophy can be extended to science.

3

At first sight, in science it seems to be hopeless to maintain the conception of common sense as those beliefs that are commonly held. As perceptions are not resemblances of physical things, we are forced to abandon the common sense level with regards to knowledge of objects. Then, it seems to be a conceptual fact that the perception of 'looking square or round' can not resemble the abstract property of 'being square or round'. As Poincaré [8, 56] remarks in his reconstruction of geometry, the resemblance exists with phenomenal elements, constructed by ourselves, but not with pictures of the physical [12, 107]. A theory about things looking round and a theory about things being round are two completely different things [9] and [12, 108]. The theory about round things corrects some classifications of things as round because they look round. This corresponds to the traditional picture according to which a common sense *theory* that rests on common sense realism is corrected by a scientific theory resting on abstract realism or nominalism. Indeed, I don't know of any arguments that yield a substantial insight in the relations between common sense theory and scientific theory. This difficulty seems to be insurmountable. Both, nominalists and realists often support the thesis that it is hopeless to look for a conception of the relation between science and common sense once these two levels have been separated from each other.

4

However, Strawson [12] presents a strong argument against the view that there is a common sense theory, that is in disagreement with a scientific theory. It is related to Russell's distinction between knowledge of acquaintance and knowledge by description but there are differences: according to Russell [9], we do not genuinely know common-sense objects like tables and books, because in a certain sense they are only "inferred" from shapes and colors with which we are acquainted. Strawson disputes the doctrine that the common sense view of the world "has, for any man, the status of a theory in relation to sensible experience, a theory in the light of which he interprets that experience in making his perceptual judgements" ([12, 97]). For it is impossible to describe the data that support the theory independently of this theory. "The 'data' are laden with the 'theory'" [12, 96]. In this sense, the scientific view, i.e. the 'theoretical' view, cannot go 'beyond' the common sense view because on the pre-theoretical level peculiar to common sense we do not perceive objects at all. The reason is that on this level from the subject's perspective there is no distinction between perception and perceived objects. (I exclude the reflection about common sense from the proper common sense level.)

If this right we cannot speak of a common sense theory in the sense of a theorization of data. On the level of common sense expressions we cannot even speak of trueness and falsehood because there is not yet a clearly defined semantic structure that could serve as a bearer of the relevant aspects of meaning, those aspects that contribute to the expressions being true or false. Furthermore, what is asserted is not known. The reason is that knowledge is at least a triadic relation. Its most simple form is: P identifies x as y, reducing the unknown x to y [11]. If the common sense view has only a pre-theoretical status, such an identification relation cannot be had on the level of common sense and then common sense cannot constitute knowledge, of course. This coincides with Wittgenstein's claim that the content of common sense sentences is something that cannot be said to be known: "I should like to say", he writes in *On Certainty*, §151, "Moore does

not know what he asserts he knows, but it stands fast for him, as also for me [...]."
Even more importantly, a fortiori the contrast between common sense and science
cannot be captured via non-scientific vs. scientific knowledge. Once again, the
point is not that common sense propositions may be false from a scientific point of
view or always true from a practical point of view, but rather that at the common
sense level we don't even dispose of trueness or falsehood. Common sense is a
sort of evidence concerning a life situation, a language-game in the broadest sense
(Wittgenstein), a *doctrine préalable* [3, 36].

5

By means of the Wittgenstein-Strawson argument we have seen that the issue is
not about the distinction between the common sense theory (for example the theory
of phenomenal pictures) and the scientific theory, but rather concerns the distinction between common sense and scientific theory. In other words, there is not an
elementary or evident first level theory on what there is independent from all scientific activity, which only gets further substantiated and developed afterwards by
scientific explanations. Rather, the scientific explanations already concern the very
articulation and expansion of common sense. More exactly, the scientific activity
not only aims at knowledge by description (knowledge that) but already involves
"knowledge" by acquaintance (knowledge of), i.e. in science we are also trying to
get acquainted with further objects. Indeed, this competence now includes semiotic aspects and can no longer be classified as purely phenomenal, pertaining only
to our sense organs [6, 3]: Without doubt there are common sense expressions that
have the form of propositions although their meaning cannot determine trueness
or falsehood. That's why the term "knowledge of" should be replaced by Lorenz's
expression "object-competence": this new term underlines that there are not two
different kinds of knowledge but only two sides of one and the same coin: object-
competence and semiotic competence. The fact that, strictly speaking, pure data
can never be had in the framework of an empirical approach, was an important
insight of scientists and philosophers at the end of the 19th century. Nevertheless,
there can be common sense sentences, that are either true or false from a higher
level or scientific point of view. Indeed, their truth or falsity can now be seen as
expressing their adequacy with respect to a specific chosen system. Thus, trueness
became hypothetical in principle. Such a position has been denounced as scientis-
tic [4, 12]. But, I think one can avoid total relativism by the adoption of a dialogical framework whose criteria of rationality are interaction and belief revision (and
which supplies the celebrated reflexive equilibrium). Then, rationality also applies
to mere practical uses of signs and in a fundamental dialogical game one can single
out the skills of object-competence and meta-competence which are a source and
pre-condition for knowledge without yet being knowledge: I call this stage "the
signs-action-level of common sense". At the common sense level, language is just

used. This is the starting-point for a dialogical abstraction procedure. The common sense level as a whole determines the ontological commitments of the higher-level activities. As the scientific abstraction process advances and the theoretical framework is further developed, new common sense levels emerge, with further domains one is ontologically committed to. This way, one should reconstruct the development of the concept of space or the geometrisation of numbers. Looking back at the common sense use of language in retrospect has nothing to do with ultimate foundations for trusting the reliability of our scientific activities: it only means that raising questions of validity presupposes a more basic practical level that does not yet involve questions concerning validity itself. Such a view is in full agreement with the Wittgensteinian position concerning the relation between philosophy and common sense. Or put differently, the role of common sense that has been worked out here cannot serve to draw a demarcation line between scientific knowledge and philosophy. From this point of view, the classical question concerning realism – "Can we coherently identify immediately perceptible things which the common sense supposes (common sense theory), with the configurations of unobservable ultimate abstract particulars (scientific theory)?" [12, 108] – simply makes no sense. One reason is that the absolute distinction between concrete and abstract objects now has become an unjustified myth; the distinction is at best a relative one. Every domain of objects – and not only the domain of microscopic objects – has to be introduced as a domain of abstract objects as a common access to such a domain requires a theoretical level of reflection. The myth of the abstract-concrete distinction and the myth of the existence of brute facts are related to each other: there is no pure information, and therefore all objects and the relations between them always are schematic constructions. Doubtlessly, not only the scientific content evolves, but with it also the epistemic means to attain knowledge, and thus the references to common sense can change, too. Of course, I do not deny that in this respect there is a difference between the natural sciences and the humanities (*Geisteswissenschaften*) with their poorly developed experimental standards. In the natural sciences the means of description are more stable, whereas in the humanities the historical perspective always has been emphasized. But also in the natural sciences there is no reason to neglect the "principles in terms of which the attribution of 'hardness' to specific data is to be evaluated" [7, 33].

Acquaintance with a life situation and its involved beliefs can take place on a highly scientific level, too. There are various common sense levels that no longer correspond to the Reidian interpretation of common sense as the sum of those beliefs that are accepted by most speakers in all societies. What really had been broken was not the relation between science and common sense but the relation between highly developed science and first tangible theories, for example the Euclidean one.[1]

---

[1] I am grateful for comments I received from Helge Rückert.

BIBLIOGRAPHY

[1] Carnap, Rudolf, Der logische Aufbau der Welt (1928), Frankfurt/Berlin/Wien: Ullstein.

[2] Descombes, Vincent, L'idée d'un sens commun, *Philosophia Scientiae* 6 (2), 2002, 147-161.

[3] Gonseth, Ferdinand, Les conceptions mathématiques et le réel, in: *Le science et le réel*, ASI 1061, Paris: Hermann, 31-60.

[4] Haller, Rudolf, Die Vernünftigkeit des Common Sense, *5. Int. Kolloquium Biel*, Manuscript, 1981.

[5] Lorenz, Kuno, *Elemente der Sprachkritik. Eine Alternative zum Dogmatismus und Skeptizismus in der analytischen Philosophie*, Frankfurt: Suhrkamp, 1971.

[6] Lorenz, Kuno, *Rationality in Science*, talk given at the van Leer-Foundation, Jerusalem (Dec 1984), manuscript.

[7] Nagel, Ernest, Russell's Philosophy of Science, in: *The Philosophy of Bertrand Russell* (ed. P.A. Schilpp), New York: Happer Torchbooks, 1963, 319-349.

[8] Poincaré, Henri, *La science et l'hypothèse*, Paris. Flammarion 1968.

[9] Russell, Bertrand, *The Problems of Philosophy* London: Oxford University Press.

[10] Ryle, Gilbert, Categories, in A.G.N. Flew (ed.), *Logic and Language. Second Series*, Oxford: Blackwell, 1966, 65-81.

[11] Schlick, Moritz, 'Is there Intuitive knowledge?, in: *Philosophical Papers I (1909-1922)*, Dordrecht/London: Reidel, 141-152.

[12] Strawson, Peter, Perception and its Objets, in: *Perceptual Knowledge*, Oxford: University Press, 1988, 92-112.

[13] Wittgenstein, Ludwig, *Philosophical Investigations*, Oxford: Blackwell, 1953.

[14] Wittgenstein, Ludwig, *The Blue and Brown Books*, Oxford: Blackwell, 1964.

[15] Wittgenstein, Ludwig, *On Certainty*, Oxford: Blackwell, 1969.

IF Logic in a Wider Setting: Probability and Mutual Dependence

JAAKKO HINTIKKA

ABSTRACT. IF logic supplemented by a sentence-initial contradictory negation is the correct basic first-order (**FO**) logic. Algebraically it is a Boolean
algebra with a closure operator. Geometrically, dual negation corresponds to
orthogonality. A probability measure for a sentence S of this logic can be defined in terms of an equilibrium point in the corresponding semantical game
$G(S)$. In the resulting logic an ordinary function becomes a function from
probability distributions to probability distributions. In the ensuing logic-
cum-probability theory and mathematics, one can interpret c^*algebra and also
interpret irreducibly mutual dependence of variables of each other (of the kind
found in quantum theory).

1 Extended IF logic as the natural basic logic

Independence friendly (IF) logic has opened interesting perspectives into several
different problem areas in pure and applied logic. (See e.g. [10] and [8].) In this
paper, it will be seen how it can contribute to our understanding of the idea of
probability and its behavior, and conversely how the introduction of probability
enriches IF logic. IF logic starts from the insight that the only way of expressing
on the first-order level the dependence or independence of a variable on another
one is by the formal dependence or independence of the quantifiers to which they
are bound. In the received first-order languages, this formal dependence between
quantifiers is expressed by the nesting of their scopes. This nesting relation is
antisymmetric and transitive. Since the possible patterns of dependence and independence between variables do not all involve only such relations as satisfy these
conditions, they cannot all be expressed by means of scope relations and hence
cannot be all expressed in the conventional first-order logic. In order to make
such expressibility possible, IF logic was created by introducing a slash notation
in which (Q_2y/Q_1x) expresses the independence of the quantifier (Q_2y) of the
quantifier (Q_1x) in whose syntactical scope it occurs. Its semantics can be formulated especially naturally in the framework of game-theoretical semantics, in
that *de facto* dependence and independence are expressed simply by informational
dependence and independence in the usual game-theoretical sense. In the game
rules, negation can be taken to indicate a switch in the roles of the two players.

In IF logic, this treatment of the notions of dependence and independence extended also to propositional connectives in their interplay with quantifiers and with each other. The idea of independence applies to them, too, and is expressed in the same way.

The resulting logic differs from the received first-order logic in many ways. It can express several logical and mathematical notions not expressible in the received first-order logic. Also, the law of excluded middle does not hold in it. Sentences of an IF language can thus be indefinite, that is, neither true nor false. This is because the semantical game interpreting an IF sentence may not be determinate. The simplest example of such a sentence is $(\forall x)(\exists y / \forall x)(y = x)$.

A richer logic (called extended IF logic) can be constructed by introducing a sentence-initial contradictory negation $\neg$ over and above the strong (dual) negation $\sim$. The latter is defined by the usual semantical (game) rules for negation. The contradictory negation $\neg S$ is true iff S is false or indefinite, otherwise false. The result is algebraically speaking a Boolean algebra with the operator $\neg\sim$.

The extended IF logic is the natural basic logic. It is the symmetrical between its two halves. IF logic in the original narrow sense is equivalent to the Σ_1^1 fragment of second-order logic. For this fragment, we have a complete disproof procedure but not a complete proof procedure. The other half consists of the contradictory negations of IF formulas. It is equivalent with the Π_1^1 fragment of second-order logic. There exists a complete proof procedure but not a complete disproof procedure for this fragment.

In the extended IF logic, most of the usual concepts of classical mathematics can be expressed. Already in the narrower IF logic, we can express equicardinality, infinity, and topological continuity. If one uses IF logic as the logic of a system of elementary arithmetic, one can prove the consistency of the system by (arguably) elementary means. (See [12].)

2   On the geometry of the logical space of the extended IF logic

One of the important features of extended IF logic is the presence of two different negations. The semantics of the strong (dual) negation $\sim$ is obtained in a game-theoretical framework simply by keeping all the traditional "classical" rules of semantical games. The semantical game $G(\sim S)$ associated with $\sim S$ is the same as $G(S)$ with the roles of the two players exchanged. The dual negation $\sim$ is thus in a sense the true "classical" negation. However, the result is a negation different from the contradictory negation $\neg$ which is often (but misleadingly) considered the "classical" one. Since the classical rules yield $\sim$ rather than $\neg$, there cannot be any natural game rules for the contradictory negation $\neg$. As long as we are using game-theoretical semantics, it can therefore occur only sentence-initially.

But what is the interpretation of the different ingredients of extended IF logic? (Cf. here [6].) It is easily seen that, algebraically speaking, extended IF logic is a Boolean algebra with an operator that is expressed by $\neg \sim$. Hence Tarski's and Jónsson's results on such algebras apply to it. (See [14] and [15].) One of them is that there is always a set-theoretical ("geometrical") interpretation for such an algebra. In the resulting interpretation, $\neg$, & $\vee$ have their usual alter egos in the form of complementation, intersection and set-theoretical union. This result can be developed further. The proof of this interpretational pudding is naturally in the behavior of $\sim$. Not only is the algebra of extended IF logic a Boolean algebra with an operator. This operator can be interpreted as a closure operator. Hence this algebra is a closure algebra. Hence it has the same structure as the S4 system of modal logic. Now it is known that intuitionistic propositional logic is interpretable in S4. Hence the extended IF logic can be considered as a formulation of intuitionistic logic, as far as its propositional structure is concerned. Hence IF logic turns out to be closely related to the ideas of intuitionists, as might indeed be expected on the basis of the failure of *tertium non datur* in it.

But what is the interpretation of the strong negation? It has been claimed that it does not have one. This is true only if one arbitrarily restricts one's interpretational options to received ones. There is in fact an eminently natural geometrical interpretation of $\sim$. It can be taken to express orthogonality. More explicitly, two propositions A and B can be taken to be "orthogonal" to each other iff it is the case that $(\sim A \vee \sim B)$. This illustrates the fact that the notion of orthogonality does not necessarily presuppose the notion of angle. (Cf. [4].)

From the geometrical viewpoint, the interpretability of $\sim$ as expressing orthogonality means that we can define the notion of dimension by its means for our logical space. Here we restrict our attention to the finite-dimensional case. Logical space has d dimensions iff there are d mutually orthogonal propositions $B_1, B_2, \ldots B_d$ such that for any $i \neq j$

$$(2.1) \qquad\qquad (\sim B_i \vee \sim B_j)$$

and that for any proposition A we have

$$(2.2) \qquad \neg((\sim A \vee \sim B_1)\&(\sim A \vee \sim B_2)\& \ldots \&(\sim A \vee \sim B_d))$$

It can be shown that this notion of dimensionality is independent of the choice of the base $B_1, B_2, \ldots B_d$.

We can also define coordinate representation in a logical space. For the purpose, note that from (2.2) it follows (by putting $(A \vee \sim A)$ for A) that

$$(2.3) \qquad\qquad (\neg\sim B_1 \vee \neg\sim B_2 \vee \ldots \neg\sim B_d)$$

Hence any proposition A is equivalent with

$$(2.4) \qquad (\neg A \& \neg\sim B_1) \vee (A \& \neg\neg\sim B_2) \vee \ldots \vee (A \& \neg\neg\sim B_d)$$

which is equivalent with

(2.5) $\neg(\neg A \vee \sim B_1) \vee \neg(\neg A \vee \sim B_2) \vee \ldots \vee \neg(\neg A \vee \sim B_d)$

This is a kind of componential (coordinate) representation of A. It suggests that we can consider $(A \& \neg \sim B)$ as the projection of A on B. (Cf. (2.4).) In a propositional language with two atomic propositions A, B the constituents $(A \& B)$, $(A \& \sim B)$, $(\sim A \& B)$, $(\sim A \& \sim B)$ are orthogonal to each other and hence constitute a coordinate system. The coordinate representation of an arbitrary pure IF proposition (i.e. a proposition without $\neg$) is obviously its distributive normal form. This can be extended to the case in which $\neg$ is also present. There is thus a close connection between the geometry of logical space and the theory of constituents and distributive normal forms.

The presence of two different negations gives extended IF logic some extra flexibility. One interesting thing is the behavior of the two negation combinations $\neg\sim$ and $\sim\neg$. The main facts are the following: The explanations given above enable us to interpret any sequence of the two negations, proceeding from outside in. They show that the meaning of all the sequences of $\neg$ and $\sim$ can be specified in terms of truth-values. Thus $\neg S$ is true if S is false, false if S is true, and indefinite when S is. The double negation $\sim\sim S$ equals S, whereas $\neg\neg S$ expresses strong truth: it is true iff S is true, otherwise false. The same goes for $\sim\neg S$. In contradistinction to these, $\neg\sim S$ expresses nonfalsity, in that it is true iff S is not false (true or indefinite) false only wherever S is false. Hence what it expresses is in effect what is called no-counterexample interpretation. This interpretation can thus be expressed explicitly in extended IF logic. Furthermore, all the further $\neg$ and $\sim$ combinations are equivalent with shorter ones in that we have equivalences like the following.

(2.6) $\neg\sim\neg\sim S \leftrightarrow \neg\sim S$

(2.7) $\sim\neg\sim\neg S \leftrightarrow \sim\neg S$

(2.8) $\neg\sim\sim\neg S \leftrightarrow \neg\neg S$

(2.9) $\sim\neg\neg\sim S \leftrightarrow \neg\neg S$

3   Functions in IF logic

The extended IF logic is thus an interesting structure and a powerful tool in logical analysis. Developed as far as has been done here, it nevertheless involves problems that prompt us to develop it further.

One of the most characteristic features of IF logic is that the law of excluded middle fails in it for the strong negation $\sim$. This is what forces us to allow sentences to have an indefinite truth-value besides "true" and "false". Otherwise expressed, predicates (complex and simple) must be allowed to have truth-value gaps.

This causes no problems as far as predicates are concerned. But what can be said about functions? Their logic does not reduce to that of predicates. The main reason is that it is a part of the meaning of a function that it specifies a definite value for a given argument. But if truth-value gaps are allowed indiscriminately, a function apparently cannot any longer be required to do so.

The obvious solution is to require that a function $f(x)$ specifies a value $f(a)$ for a given argument a, but only probabilistically. Let this probability be $P(f(a) = b)$. Then it will have to be required that the object determined by the function is still unique. This is what distinguishes functions from arbitrary functions. What this means is obviously that we have for a given x

$$(3.1) \qquad \sum_y P(f(x) = 1)$$

Here the sum is taken over all admissible values y. In a continuous case where P expresses probability density instead of probability, we must likewise have

$$(3.2) \qquad \int P(f(x) = y)dy$$

and the integral taken over all admissible values of y.

What this means is that in order to capture adequately the logic (semantics) of functions in IF logic, we must have a notion of probability which assigns a probability (or probability density) to sentences that are neither true nor false. This will be done in sec. 5 below.

4   The problem of mutual dependence

There is another problem about the extended IF logic so far developed that turns out to be solvable by means of a notion of probability.

Our independence (slash) notation enables us to express all sorts of patterns of dependence and independence between variables that were not expressible in the received first-order languages. A case in point are branching quantifier structures. It turns out that the patterns of dependence and independence they can express have frequently been used in traditional mathematics, however unwittingly. (Cf. here e.g. [13].)

An important problem nevertheless remains. It concerns situations in which two variables depend on each other mutually. Needless to say, such dependence has to be irreducible in order to be *sui generis*. For instance, two variables x and y are not irreducibly interdependent if they are both separately functions of a third variable.

How can the mutual dependence of two variables x and y be expressed? Let us assume that they both depend on t and that their overall behavior is governed by the law $F[t, x, y]$. Then it might be possible to express the fact that x and y depend on each other by

$$(4.1) \qquad (\exists f)(\exists g)(\forall t)(\exists x)(\exists y/\exists x)(x = f(t, y) \,\&\, y = g(t, x) \,\&\, F[t, x, y])$$

But then for some f and g

$$(4.2) \qquad\qquad\qquad\qquad x = f(t, g(t, x))$$

Hence the different possible values of x for a given t are obtainable as the solutions of this equation. In this sense x is a function of t alone. By symmetry, the same holds of y. Accordingly, x and y are not irreducibly interdependent.

More fully expressed, the different possible values of x are the *eigenvalues* of the operator that takes us from $g(t, x)$ to $f(t, g(t, x))$, and likewise for y. The fact that x has one of these possible values does not determine which of its possible values y has.

Independence-friendly notation does not seem to help us here, either. In such a notation, (4.1) could be written as

$$(4.3) \qquad (\forall t)(\forall x)(\forall y)(\exists z/\forall x)(\exists u/\forall y)((x = z) \,\&\, (y = u) \,\&\, F[t, x, y])$$

But this is beset by the same problems as (4.1).

We have thus reached here a surprising general result. An irreducible mutual dependence between two variables cannot be mediated by any functions in the ordinary sense of the word. Yet such mutual dependence seems to be a perfectly possible state of affairs. We are thus facing an interesting general conceptual problem: Is a genuine mutual dependence (of two variables on each other) conceptually possible? If so, in what way?

The problem of mutual dependence can be solved by means of a notion of probability extended to IF logic.

5   Probability in extended IF logic

In a language using extended IF logic there are propositions that are neither true nor false. Can we define probabilities for them in a natural sense?

This account can be generalized by allowing an atomic sentence A to be true only with a certain probability p. Thus the payoffs of a play of a semantical game that ends with A will be $(2p - 1, 1 - 2p)$.

This semantics associates with each proposition A of an interpreted language a two-person game $G(A)$ (with well-known rules). The players will be called the verifier and the falsifier. Each play of the game ends up after a finite number of moves with an atomic sentence. If this sentence is true, the verifier wins and the falsifier loses. If it is false, vice versa. The payoffs (utilities) for the two players in these two cases can be taken to be $(+1, -1)$ and $(-1, +1)$, respectively. Hence the game $G(A)$ can be assumed to be a zero-sum game.

A is true iff there exists a winning (dominating) strategy for the verifier. A is false iff there exists a winning strategy for the falsifier. The law of excluded middle asserts that the one or the other player has always a winning strategy, i.e. that semantical games are all determinate. In traditional first-order languages, this is always the case. In an IF language, it can happen that a semantical game $G(\Lambda)$ is not determinate. This is the case when the law of excluded middle fails for Λ.

By strategies, we have so far meant pure strategies in the sense of game theory. One important fact about these pure strategies is that even when the domain of individuals is infinite, in the game $G(S)$ associated with a given sentence S there will be only a finite number of different (nonequivalent) strategies. Indeed, the distributive normal form of S (at its own quantificational depth) can be taken to represent the extensive form of the game $G(S)$. By the same token, the number of nonequivalent strategies is finite also whenever the model on which semantical games are played is finite. In the following, the probability assignment is first defined on the basis of an assumption of the finiteness of the strategy sets and then extended to the case of infinite models.

In any case, it makes perfect sense to speak also of mixed strategies in semantical games. Then we know from game theory that all semantical games, being finite two-person zero-sum games, have a solution in terms of mixed (or in limiting cases pure) strategies. That is, in a game $G(A)$ there is a pair of pure and mixed strategies for the verifier and for the falsifier such that, given the other player's strategy, one of the players cannot improve her outcome by choosing a different strategy.

We will call the verifier's strategies α_i and the falsifier's strategies β_j. The payoff to the verifier u_{ij} of a play of the game with these strategies is $+1$ if the verifier wins, otherwise -1. For the falsifier the corresponding payoffs are -1 and $+1$. In addition to these payoffs, we will consider another kind of value of the outcome of a semantical game for the verifier. It will be called the p-value. It is defined as $\frac{1}{2}(u + 1)$, where u is the utility of the outcome. Thus when the verifier wins, the p-value of the outcome is 1. When the falsifier wins, the verifier's p-value is 0. The falsifier's p-values are defined dually, that is, as $\frac{1}{2}(u - 1)$. Thus when the verifier wins, the falsifier's p-value is 0. When the falsifier wins, the falsifier's p-value is -1.

Because the p-values are monotonic functions of the utilities, dominance relations and solutions are the same for semantical games independently of whether the payoffs are original utilities or p-values. Hence there exists a solution in the sense of game theory also when the payoffs are p-values.

The expected p-value of the game for the verifier using the strategy α_i is thus $\sum_j q_j u_{ij} = \sum_k q_k$ where k ranges over those values j for which $u_{ij} = 1$, i.e. for which the verifier wins. The expected p-value of the entire game for the verifier is

(5.1) $$\sum_{ij} p_i q_j u_{ij}$$

In other words, $\sum q_j u_{ij}$ is the sum of the probabilities q_j of those strategies of the falsifier against which the verifier's strategy α_i wins. We can now define the probability $P(A)$ of A as the expected p-value of the game for the verifier, that is, as (4.1). This presupposes that $P(A)$ so defined satisfied the usual laws of the probability calculus. In order to show this, we have to show the following:

(P.1) For any proposition A, $0 \le P(A) \le 1$

(P.2) If A is logically true, $P(A) = 1$

(P.3) If A logically implies B, $P(A) \le P(B)$

(P.4) If A and B are logically incompatible, $P(A \lor B) = P(A) + P(B)$

Here (P.1) and (P.2) are obvious. In order to prove (P.3), it can first be shown easily that two logically equivalent propositions are verified by the same strategies. Now if B is a logical consequence of A, A is logically equivalent to $(A\&B)$. Hence all the strategies verifying A also verify B, which suffices to show that $P(A) \le P(B)$.

In order to prove (P.4), we first assume that A and B can be represented as disjunctions of constituents in the usual sense familiar from the theory of distributive normal forms. (For this theory, see [11].) Since A and B are logically incompatible, at some sufficiently great depth they are representable as disjunctions of different constituents. Then it suffices to prove (P.4) for the case in which A and B are incompatible and hence different constituents of the same depth.

This can be done by induction on the quantificational depth of these constituents. The basis of the induction is the case of depth zero. Then the normal form of any proposition is a disjunction of constituents C each of which is of the form

(5.2) $(C_1 \& C_2 \& \ldots)$

where each C_i is a different atomic formula or the negation of one. In dealing with (5.1) in a semantical game, the falsifier chooses the conjunct C_j with the lowest probability. Then the payoff of the play of the game is $2p_2 - 1$. The value of the disjunction of the conjunctions (5.1) in the normal form of A is the sum of these, and likewise for B and for $(A \lor B)$. But since A and B are logically incompatible, these two sums have different addenda, and the value of $(A \lor B)$ is here the sum of the values of A and B.

In the inductive step, it likewise suffices to consider two incompatible constituents, say

(5.3) $$\wedge_i(\exists x)C_{1i}[x] \quad \& \quad (\forall x) \vee_i C_{1i}[x]$$

and

(5.4) $$\wedge_i(\exists x)C_{2i}[x] \quad \& \quad (\forall x) \vee_i C_{2i}[x]$$

The fact that these two constituents are different means that the two universally quantified disjunctions have partially different disjuncts. A strategy of the falsifier's in the overall game includes choosing a $(\exists x)C_{1i}[x]$ or $(\forall x) \vee_i C_{1i}[x]$ and also includes choosing a $(\exists x)C_{2j}[x]$ or $(\forall x) \vee_j C_{2j}[x]$. These choices must be followed among other things by the verifier's choices of an individual b such that $C_{1i}[b]$ and a choice $C_{2j}[x]$ such that $C_{2j}[b]$.

The only way in which the falsifier can hope to win is to choose $(\exists x)C_{1i}[x]$ in such a way that $C_{1i}[x]$ does not occur in $\vee_j C_{2j}[x]$, or vice versa. Hence the falsifier's strategy in the equilibrium (solution) case must be of this kind. It suffices here to consider the first of these two possibilities.

In this case, the game is continued with respect to $C_{1i}[b]$ and $C_{2j}[b]$. These are two different constituents of depth lower by one than those of (5.3) and (5.4). By the inductive hypothesis, the value of the continued game with the disjunction of $C_{1i}[b]$ and $C_{2j}[b]$ is therefore the sum of the p-values for the verifier of the games and these two disjuncts. But these two disjuncts come from the two different constituents. Hence the probability (p-value) of the original game with the disjunction of (5.3) and (5.4) is the sum of the probabilities (p-values) of the games with (5.3) and with (5.4). This proves the inductive step and hence (P.4).

This proves (P.4) for propositions A, B without slashes, that is, for propositions for an ordinary first-order language extended by allowing atomic sentences to be merely probable. The notion of probability so defined can be extended to the propositions of IF first-order logic. Let A be such a proposition (not equivalent with any proposition of ordinary first-order logic). Then because of the separation theorem (see e.g. [10], p. 61) there is a proposition B of the ordinary first-order such that

(5.5) $$\vdash (A \supset B) \qquad \vdash (\sim A \supset \neg B)$$

The probability $P(A)$ is then defined as the minimum (g.l.b) of all the probabilities $P(B)$. It follows from this definition that the probability of $\sim A$ is the minimum (g.l.b) of $P(\neg B)$.

The probability so defined satisfies the characteristic requirements (P.1)-(P.4) of a probability measure. Most of those are obvious. That (P.4) is satisfied can be seen as follows: Assume that two logically incompatible formulas A, B of IF first-order logic are given. Then there are by the separation theorem formulas C, D of the ordinary first-order such that

(5.6) $\vdash (A \supset C)$ $\vdash (B \supset \neg C)$

(5.7) $\vdash (B \supset D)$ $\vdash (A \subset \neg D)$

Clearly it can also be required that C and D are logically incompatible. Then $P(A \lor B) = minP(C \lor D) = minP(C) + minP(D) = P(A) + P(B)$

A new feature of the probability measure so defined is its asymmetry. In an IF language in the narrow sense there are not necessarily any propositions A such that $P(A) = 1$, in that for any B, $(B \lor \sim B)$ may not be true. This means that for any strategy of the verifier there is some strategy of the falsifier which defeats it. Then according to the definition given above $P(B \lor \sim B)$ is smaller than 1. In contrast, for any B, $P(B \,\&\, \sim B) = 0$

Thus it might look as if the probability logic we have reached is inherently fallibilist. This impression may be based on an interesting truth, in that a game-theoretical viewpoint naturally leads to a fallibilist epistemology. (See [7].) However, there is no necessary connection here. For there exists a dual probability measure $P^o(A)$ that can be defined as $P(\neg \sim A)$. For the dual measure we may not have any propositions B of an unextended IF language such that $P^o(B) = O$.

The definition of probability measures presented in this section pertains to finite languages. In order to build a perfect bridge to the usual measure-theoretic foundations, they should be extended to languages with countably long formulas. Indications as how this could be done can be gathered from Scott and Krauss [19] and from other work along the same lines, beginning with Jónsson and Tarski ([14], [15]).

6   A new probability calculus

Thus we can define a probability measure for the propositions of a language using extended IF logic. The upshot is in effect a new probability logic, practically a new probability calculus which is richer than the conventional one.

One interesting thing about the probability measures just defined is that they are relative to a given model of the language in question. They do not involve any considerations of the other models ("worlds"). A true attribution of such a probability is an objective fact about the actual world. Hence the kind of probability defined here is relevant to the idea of probability as propensity. It shows that there is an objectivistic interpretation of probability that is not frequentistic.

The notion of probability that has been defined opens new possibilities of theorizing in several different directions. Only a few of them can be mentioned here.

Some of these new openings are due to the fact that extended IF logic has a richer geometrical content than conventional logic. For instance, from the possibility of a coordinate representation of propositions in a dimensional logical space it follows that a similar componential analysis of probabilities is possible in them. For one prominent example, from the equivalence of A and (2.4) it follows that

$$(6.1) \qquad P(A) = \sum_i P(A \&\neg\sim B_i)$$

for any two disjuncts in (2.4) are easily seen to be logically incompatible.

Actually, this result does not depend on the definition of $P(X)$, only on the fact that it is a probability in the sense of satisfying the laws of the probability calculus. Because of this independence, (6.1) is related to Gleason's theorem that in a Hilbert space any probability measure is determined by the probabilities of the unit vectors in all different dimensions. (See [3].)

A related connection is to the so-called quantum logics. It was shown in [9] that the Birkhoff-von Neumann quantum logic is interpretable in the extended IF logic. (Cf. [1].) Moreover, the interpretation is almost predictable. Birkhoff's and von Neumann's logic included in effect the connectives $\sim$ and &. However, if we add to them the usual disjunction, we do not obtain a lattice as Birkhoff and von Neumann obviously wanted. Hence they used as their ersatz disjunction what for us is the defined connective $(\neg\sim A \lor \neg\sim B)$. This has the inevitable consequence that some of the usual distribution laws fail. Those that remain are the laws characteristic of orthomodular lattices.

For another instance of the new possibilities, we can now define what it means for two propositions A and D to be parallel. It can be defined to mean that for any C and E

$$(6.2) \qquad \frac{P(A \&\neg\sim C)}{P(D \&\neg\sim C)} = \frac{P(A \&\neg\sim E)}{P(D \&\neg\sim E)}$$

Intuitively, this means that the projections of A and D on any two propositions always have the same ratio λ. In particular, if $B_1, B_2, \ldots$ is a basis of an orthogonal coordinate system, we have for each B_i

$$(6.3) \qquad \frac{P(A \&\neg\sim B_i)}{P(D \&\neg\sim B_i)} = \lambda$$

Then it is natural to express the relationship between A and D as

$$(6.4) \qquad A = \lambda D$$

If $\lambda = 1$, then A and D coincide. Moreover, it is natural to assume that for each A and a positive real number λ there is a proposition D such that (5.3) holds. Then

$$(6.5) \qquad P(A) = \lambda P(D)$$

For instance if $\frac{P(A \& \neg B_i)}{P(B_i)} = \lambda$, then there is a proposition $\lambda_i B_i$ such that

(6.6) $(A \& \neg \sim B_i) = \lambda_i B_i$

Then the coordinate representation (2.4) can be rewritten in the form

(6.7) $A = \vee_i (\lambda_i B_i)$

7   IF logic and star algebras

The introduction of probability into extended IF logic enables us to study the algebraic structure of this logic in greater depth. The main result that can be established here is that IF logic includes in effect the structure of c^*-algebra. (For this algebra, see e.g. [17], [18].) The only qualification needed is that we are here considering only spaces over real numbers, not complex numbers.

In order to show this, we first have to interpret in extended IF logic the notions that characterize a c^*-algebra. Such an algebra is first of all a Banach algebra. A Banach algebra has an addition which is here interpreted as disjunction and a multiplication $A \cdot B$. Hence we have to interpret this multiplication. It turns out that this first step in relating IF logic and c^*-algebra is the least obvious one. The product $(A \cdot B)$ can in fact be defined as the proposition ("vector") which is parallel with

(7.1) $(\neg \sim A \& \ \sim \neg B)$

and whose length is

(7.2) $(P(\neg \sim A \& \ \sim \neg B))^2$

The norm $\|A\|$ of A is defined as $P^o(A)$. In a Banach space, the product and the norm have to satisfy the condition $\|A \cdot B\| \leq \|A\| \cdot \|B\|$ which now means

(7.3) $(P^o(\neg \sim A \& \ \sim \neg B))^2 \leq P^o(A) \cdot P^o(B)$

This is satisfied because

$$P^o(\neg \sim A \& \ \sim \neg B) \leq P^o(\neg \sim A) = P(\neg \sim \neg \sim A) = P(\neg \sim A) = P^o(A)$$

Likewise:

$$P^o(\neg \sim A \& \ \sim \neg B) \leq P^o(\sim \neg B) = P(\neg \sim \sim \neg B)$$
$$= P(\neg \neg B) \leq P(B) \leq P(\neg \sim B) = P^o(B)$$

Hence, $(P^o(\neg \sim A \& \ \sim \neg B))^2 \leq P^o A \cdot P^o(B)$.

The product must be associative. This can be verified as follows:

(7.4) $A \cdot (B \cdot C) = \neg{\sim}A \,\&\, {\sim}\neg({\neg}{\sim}B \,\&\, {\sim}\neg C) = \neg{\sim}A \,\&\, \neg\neg B \,\&\, {\sim}\neg C$

(7.5) $(A \cdot B) \cdot C = \neg{\sim}({\neg}{\sim}A \,\&\, {\sim}\neg B) \,\&\, {\sim}\neg C = \neg{\sim}A \,\&\, \neg\neg B \,\&\, {\sim}\neg C$

(Cf. here equivalences (2.6)-(2.9) in sec. 2 above.) The other conditions on a Banach space are straightforward. It is easy to verify the following:

(7.6) $(P(A) \le P(A) = 0$ iff A is contradictory
(7.7) $P(\lambda A) = \lambda P(A)$ by definition
(7.8) $P(A \vee B) \le P(A) + P(B)$

Banach star algebras are Banach algebras having an involution operation which takes us from A to A^*. This operation is interpreted as the exchange of ${\sim}$ and $\neg$ everywhere. It can be seen to satisfy the defining conditions of a Banach star space which are the following

(7.9) $(A + B)^* = (A \vee B)^* = (A^* \vee B^*) = A^* + B^*$
(7.10) $(\lambda A)^* = \lambda A^*$
(7.11) $(A \cdot B)^* = (\neg{\sim}A \,\&\, {\sim}\neg B)^* = {\sim}\neg A^* \,\&\, \neg{\sim}B^* = B^* \cdot A^*$
(7.12) $(A^*)^* = A$

A Banach star algebra is a c^*-algebra if $\|A^* \cdot A\| = \|A\|^2$. Now

(7.13) $A^* \cdot A = \neg{\sim}(A^*) \,\&\, {\sim}\neg A = ({\sim}\neg A)^* \,\&\, {\sim}\neg A$

Here $({\sim}\neg A)^*$ can be shown to be implied by ${\sim}\neg A$. This follows essentially from the fact that $\neg$ can occur only sentence-initially, plus the equations (2.6)-(2.9). They show that whatever $({\sim}\neg A)^*$ is, it is equal to one of the propositions A, $\neg\neg A$, $\neg{\sim}A$ or ${\sim}\neg A$. Of these, ${\sim}\neg A$ implies the other three. This representability of c^*-algebras is significant in view of the role of these algebras in mathematical physics. Among other things, they have the structure of the closed subspaces of a Hilbert space, and hence relevant to the Hilbert space formalism of quantum theory. (Cf. [16].)

8   On q-logic and q-mathematics

The introduction of probabilities into extended IF logic solves the problem of accommodating functions in this logic. However, doing so by the systematic use of probabilities in our basic logic forces us to face an interesting major challenge. What was argued in sec. 3 is that in order to do justice to the meaning ("logic") of functions we must allow values of functions be indeterminate objects, in effect probability distributions. This is now seen to be feasible. Bu since we clearly want to be able of handle nested functions and inverse functions, we must allow also the arguments of functions to be indeterminate ("fuzzy") objects. More generally, such indeterminate objects (probability distributions) must be capable of serving as values of our individual variables. Thus we face the task of developing not only a logic but a mathematical theory for the generalized functions with probability distributions as their argument and as their values. This is a bigger task than can be attempted in a single paper. Some of the basic features of the new theory will nevertheless be outlined here.

By admitting probability distributions as arguments and as values of numerical variables, we are in effect generalizing the concept of number. I propose to call the new "numbers" q-numbers. The objects they are numbers of will be called q-objects. Whenever the distinction is relevant, ordinary numbers are called c-numbers. They are a subset of q-numbers, in that the entire probability mass can be concentrated on one argument value. Analogously, we will be speaking of q-functions and c-functions. Each q-function is an extension of a c-function to a longer class of arguments. Hence we are generalizing not only the notion of number but the notion of function.

9   Mutual dependence vindicated

The development of q-mathematics is too large an enterprise to be dealt with adequately here. Even without a detailed discussion of their properties, we can nevertheless put q-numbers and q-variables to work. In this section, it will be indicated how probabilistic concepts can solve the problem of mutual dependence discussed in sec. 4 above. By means of an example, it will be shown that two q-variables can be irreducibly mutually dependent.

In order to outline such an example, consider the q-number Q_1 that assigns to each of a finite number of values $a_1, a_2, \ldots, a_d$, the respective probabilities $\psi_1, \psi_2, \ldots, \psi_d$ (and zero to all other values). We can consider Q_1 as the value of some suppressed variable. These probabilities are the probabilities $P(A_i)$ ($i = 1, 2, \ldots, d$) of the propositions

(9.1) $$A_i = (x = a_i) \qquad (i = 1, 2, \ldots, d)$$

They can be summarized in a vector

(9.2) $$\Psi = \begin{pmatrix} \psi_1 \\ \psi_2 \\ \vdots \\ \psi_d \end{pmatrix}$$

Now the different propositions (9.1) are orthogonal in the sense of sec. 2, in other words, for any $i \neq j$ it is the case that $(\sim A_i \vee \sim A_j)$. Then the propositions (9.1) form a base for a coordinate system for a d-dimensional space. For any other vector B_j (proposition) in the same space we have

(9.3) $$B_j = \vee_k g_{jk} \cdot A_k$$

for suitable c-numbers g_{jk}. (Cf. sec. 6.) From (9.3) it follows that

(9.4) $$P(B_j) = \sum_k g_{jk} P(A_k)$$

Assume now that there is another similar q-number Q_2 which to each of $b_1, b_2, \ldots, b_d$ assigns the probability $\psi_1, \psi_2, \ldots, \psi_d$, respectively. Assume further that the propositions $B_j (j = 1, 2, \ldots, d)$ span the same d-dimensional space as the A_i. Then we have for each B_j (9.4)

(9.5) $$P(B_j) = \varphi_j = \sum_k g_{jk} \psi_k$$

Putting

(9.6) $$\Phi = \begin{pmatrix} \varphi_1 \\ \varphi_2 \\ \vdots \\ \varphi_d \end{pmatrix}$$

(9.7) $$G = \begin{pmatrix} g_{11} & g_{12} & \cdots & g_{1d} \\ g_{21} & g_{22} & \cdots & g_{2d} \\ \vdots & \vdots & \vdots & \vdots \\ g_{d1} & g_{d2} & \cdots & g_{dd} \end{pmatrix}$$

(9.5) becomes in a matrix format

(9.8) $$\Phi = G \cdot \Psi$$

Analogously, we have for some $d \times d$ matrix F

(9.9) $$\Psi = F \cdot \Phi$$

Here (9.8) and (9.9) show that the q-variables Q_1 and Q_2 depend on each other. The matrix G expresses how Q_2 depends on Q_1 (via Φ) and F shows how Q_1 depends on Q_2 (via Ψ).

Moreover, the mutual dependence of Q_1 and Q_2 (i.e. of Φ and Ψ) is not trivial. For one thing, F and G do not simply express inverse dependence. In order to see this, consider the product $F \cdot G$, that is,

$$(9.10) \qquad \left(\sum_k f_{ik} g_{kj} \right)$$

Since F changes Φ to Ψ and G changes Ψ to Φ, we must have

$$(9.11) \qquad \sum_k f_{ik} g_{ki} = 1$$

$$(9.12) \qquad \sum_k f_{ik} g_{kj} = 0 \qquad \text{if } (i \neq j)$$

Hence $F \cdot G$ is the matrix

$$(9.13) \qquad \begin{pmatrix} 100\ldots00 \\ 010\ldots00 \\ 001\ldots00 \\ \vdots \\ 000\ldots01 \end{pmatrix}$$

In the same way, one can see that

$$(9.14) \qquad G \cdot F = \begin{pmatrix} 00\ldots001 \\ 00\ldots010 \\ 00\ldots100 \\ \vdots \\ 10\ldots000 \end{pmatrix}$$

Hence $F \cdot G - G \cdot F$ is not zero but

$$(9.15) \qquad \begin{pmatrix} 100 & \ldots & 00-1 \\ 010 & \ldots & 0-10 \\ 001 & \ldots & -100 \\ & \vdots & \\ 0-1 & \ldots & 010 \\ -10 & \ldots & 001 \end{pmatrix}$$

If d is odd, the element of (9.15) with the indexes $\frac{(d+1)}{2}, \frac{(d+1)}{2}$ is 0. This shows that F and G are not simply inverses of each other.

Thus we can see that even though two c-variables cannot be irreducibly mutually dependent, two q-variables can be. This result puts the entire notion of mutual dependence to an interesting light.

BIBLIOGRAPHY

[1] George Birkhoff and John von Neumann, "The logic of quantum mechanics", *Annals of Mathematics* vol. 37, pp. 823-843 (1936).

[2] Rudolf Carnap, *The Continuum of Inductive Methods*, University of Chicago Press: Chicago 1952.

[3] Andrew M. Gleason, "Measures on closed subspaces of Hilbert spaces", *Journal of Mathematics and Mechanics* vol. 6, pp. 885-893 (1957).

[4] Robert Goldblatt, *Orthogonality and Spacetime Geometry*, Springer Verlag: New York and Heidelberg 1987.

[5] Jaakko Hintikka, "Mathematics for IF logic", forthcoming.

[6] Jaakko Hintikka , "What is the true algebra of logic?", in Vincent Hendricks et al., editors, *First-Order Logic Revisited*, Logos Verlag: Berlin 2004, pp. 117-128.

[7] Jaakko Hintikka, "On the epistemology of game-theoretical semantics", in Jaakko Hintikka et al., editors, *Philosophy and Logic in Search of the Polish Tradition: Essays in Honour of Jan Wolenski*, Kluwer Academic: Dordrecht 2003, pp. 57-66.

[8] Jaakko Hintikka, "Hyperclassical logic (aka independence-friendly logic) and its general significance", *Bulletin of Symbolic Logic* vol. 8, pp. 404-423 (2002).

[9] Jaakko Hintikka, "Quantum logic as a fragment of independence-friendly logic", *Journal of Philosophical Logic* vol. 31, pp. 197-209 (2002).

[10] Jaakko Hintikka, *The Principles of Mathematics Revisited*, Cambridge U.P.: Cambridge 1996.

[11] Jaakko Hintikka, "Distributive normal forms in first-order logic", in J.N. Crossley and Michael Dummett, editors, *Formal Systems and Recursive Functions*, North-Holland: Amsterdam 1965.

[12] Jaakko Hintikka and Besim Karakadilar, "How to prove the consistency of arithmetic", *Acta Philosophica Fennica*, vol. 78, pp 1-15 (2006).

[13] Wilfrid Hodges, "The logic of quantifiers", in R. Auxier, editor, *The Philosophy of Jaakko Hintikka*, Open Court: LaSalle 2006, pp. 521-534.

[14] Bjarni Jónsson and Alfred Tarski, "Boolean algebras with operators I", *American Journal of Mathematics* vol. 73, pp. 891-939 (1951). Also in Tarski 1986, vol. 3, pp. 369-419.

[15] Bjarni Jónsson and Alfred Tarski, "Boolean algebras with operators II", *American Journal of Mathematics* vol. 74, pp. 127-162 (1952). Also in Tarski 1986, vol. 3, pp. 421-358.

[16] Miklós Rédei, *Quantum Logic in Algebraic Approach*, Kluwer Academic: Dordrecht 1998.

[17] C.E. Rickart, *General Theory of Banach Algebras*, Van Nostrand: New York 1960.

[18] Shôichirô Sakai, *C*-Algebras and W-Algebras*, Springer: New York and Heidelberg 1971.

[19] Dana Scott and Peter Krauss, "Assigning probabilities to logical formulas" in Jaakko Hintikka and Patrick Suppes, editors, *Aspects of Inductive Logic* North-Holland: Amsterdam 1966, pp. 219-264.

[20] Alfred Tarski, *Collected Papers* ed. by Steven R. Givent and Ralph N. McKenzie, Birkhäuser: Basel 1986.

Functional Anaphora

JUSTINE JACOT AND GABRIEL SANDU

ABSTRACT. Our aim in this paper is to provide a referential account of functional anaphora within a Skolem functions framework. We will give an interpretation of indefinite NPs as Skolem terms in order to show that the referential link established between an anaphoric pronoun and its antecedent is a descriptive one. Then we will argue that functional anaphora can be understood as a particular kind of E-type pronouns, in the sense that, for a large corpus, the pronoun can be replaced by a descriptive expression of a first-order language. It will next be shown that the functional framework can be fruitfully applied to other kinds of sentences involving pronominal anaphora, especially in modal contexts.

1 Introduction

Consider the following example:

(1.1)   Every graduate student received a job offer. Some accepted them.

The first sentence says that all the graduate students received a job offer and the second that only a certain number of them accepted the job they were offered. These two sentences introduce the idea that to each graduate student corresponds a single job offer, and that one can build a function mapping graduate students to job offers. The problem here is that the entity referred to by *them* changes according to the different arguments of the function. In addition to that, the second sentence introduces an anaphoric link between the two sentences via the pronoun *them*. The symbolization of these sentences should consequently focus on those two points: accounting on the one hand for the functional dependency, and on the other hand for the anaphoric link. If one tries to formalize (1.1) in a first-order language, it would translate as something like this:

(1.2) $\forall x \exists y(Sx \rightarrow (Jy \land R(x,y))) \land \exists x(Sx \land A(x,y))$

Obviously (1.2) does not capture the meaning of (1.1) because the final occurrence of the variable y is not bound: the values assigned to y in the first conjunct do not persist beyond the scope of the initial universal quantifier.

We will see that a functional analysis of the anaphoric pronoun is precisely what allows such a resolution. We will deal in the following sections with such examples of functional anaphora, all involving standard universal quantifiers:

1. Every child received a present. Jim opened it (his) immediately.

2. Every child received a present. Some children opened them (theirs) immediately.

3. For every merchandise there is a price. For meat it is 10 euros.

4. For every merchandise there is a price. For some it is 10 euros.

5. Every man owns a donkey. He beats it. He feeds it rarely (Fine [3]).

6. Every female professor has an office computer. They are financially responsible for them.

7. Every country has a tyrant. The bigger the country, the more powerful its tyrant (Barwise [1]).

8. Every pie has a price. The smaller the pie, the lower the price (Fox [4]).

Let us sketch the general idea: a discourse involving functional anaphora introduces in the first sentence or first part of the sentence a functional relation, and the second sentence or second part says something about this functional relation. More precisely, in examples 1-6, the second sentence gives further information about some arguments of the function, whereas in examples 7-8, the second sentence is more specific about the function itself. The point here is that such a functional relation cannot be expressed in terms of the "classical" analysis of anaphoric pronouns as bound variables. The anaphoric pronoun is a particular case of an E-type pronoun. When we say "a particular case", the emphasis is on "particular": such a pronoun can be replaced, in many cases, by a "description" (understood in the broad sense), but the most natural way in which the descriptive content is specified is in terms of a Skolem function which expresses the relevant functional dependence.

Let us give a short overview of the reasons why the two main theories of pronouns are unable to give a satisfactory account of functional anaphora.

For an anaphoric pronoun to be interpreted as a bound variable, it has to satisfy the following condition: being assigned the same values as its antecedent. In other words, in order to express the meaning of an anaphoric pronoun as the values assigned to a variable bound by a quantifier, one has to choose the values that pick up the same individual as the antecedent. This is what is defined as

co-variation in semantics such as DPL (Dynamic Predicate Logic).[1] In those semantics, the meaning of an anaphoric pronoun is the one of its antecedent within
a process of cross-sentential binding: the existential quantifier can bind variables
outside its scope, especially in further conjuncts which represent the continuation
of a discourse. The pronominal anaphora is consequently resolved at the level of
meaning: it is a synctactically free variable, but semantically bound.

Other difficulties arise when one has to deal with interpreting anaphoric pronouns as definite descriptions, as in the E-type interpretation of pronouns developed by Evans, and we will see that this approach does not function either for our
purpose. In those cases, anaphoric pronouns are disguised definite descriptions (E-
type pronouns), having a reference of their own. Taking the following example:

(1.3)   A girl smiles. She is happy.

we can say that the anaphoric pronoun *she* stands for "the girl who smiles", i.e. *she*
has exactly the same informative content as its antecedent, what would be rendered
as:

(1.4) $\exists x(Gx \wedge Sx) \wedge \exists y(Gy \wedge Sy \wedge Hy)$

which is logically equivalent to:

(1.5) $\exists x(Gx \wedge Sx \wedge Hx)$

In other words, in this case, the bound variable analysis and the descriptive analysis
are equivalent.

But there are cases where the two analyses are not equivalent, and the pronoun
cannot be interpreted as a variable bound by its antecedent, as in the famous example of Evans [2]:

(1.6)   Few congressmen admire Kennedy, and they are very junior.

In this case, the pronoun *they* cannot be bound to *few congressmen* since it would
lead to the wrong reading:

(1.7)   Few $x(x$ are congressmen and
                     x admire Kennedy and x are very junior)

because what (1.7) says is that "there are few congressmen who both admire
Kennedy and are very junior", which is obviously not what says (1.6), which is:
"there are few congressmen who admire Kennedy and *all of them* are very junior".
In other words, the pronoun here takes its reference from a set of individuals, a

---

[1] See Groenendijk & Stokhof [6].

restricted domain over which the quantification operates, as it is the case in functional anaphora.

We will not discuss further the E-type approach, suffice it to say that, in its "standard" form, this analysis cannot account adequately for the cases we are dealing with, because of the great rigidity of the link established between the pronoun and its antecedent, the former being understood as a definite description. Yet the descriptive approach as a whole should not be totally given up, and can be partially vindicated, as will be seen later.

It should be clear by now that in cases of functional anaphora, neither the variables interpretation nor the E-type approach would yield a satisfying answer in their standard form. The values taken by the anaphoric pronoun cannot be the same as those of its antecedent precisely because it takes its values from a set of entities. Going back to our previous example (1.1), we can see that the anaphoric resolution claims that the pronoun *them* would take its values relatively to the set of all graduate students. In other words, *them* does not get its reference from a definite set of offers, but from a certain number of pairs ⟨a student, his/her own offer⟩. Then it would be impossible to resolve the anaphora neither by assigning values to the corresponding variable nor by replacing it with a standard E-type pronoun, such pronouns requiring the unicity of their antecedent. However, the E-type approach provides a valuable interpretation of functional anaphora in terms of a description: the point to be spelled out in those cases is the nature of the description standing for the anaphoric pronoun.

We will in the next sections give an account for the functional approach to anaphora, beginning by sketching some insights from GTS (Game-theoretical semantics [2]) and CCG (Combinatory Categorial Grammar [3]). We shall then handle with the main theme of this paper: the recovering of descriptional anaphora from the use of Skolem functions. We shall eventually suggest a way of broadening functional analysis to intentional identity, in order to show that the Skolemization of sentences where existential quantification occurs in the scope of modal operators can offer a means to rise independent choices of the indefinite terms upon propositional attitudes.

2   GTS, CCG: some foundational issues in functional interpretation of pronominal anaphora

2.1   Extensive form vs. strategies in GTS

As well-known, Game-theoretical semantics provides a definition of truth and falsity in terms of winning strategies. The truth (or falsity) of a formula is given as the existence of a winning strategy for the existential (or the universal) player in a

---

[2]Hintikka & Sandu [10].
[3]Steedman [15, 16].

game G. Roughly speaking, this means that, once the game tree is built up for a game G, each maximal branch of G is a win for the existential player and a loss for the other. It is a win for the existential player if the chosen elements satisfy all the formulae of the maximal branches, otherwise a loss (and a win for the universal player).

Such a notion of strategy has to be taken into account at the level of the game-trees building. Let us go back to example (1.3), repeated here:

(2.1)   A girl smiles. She is happy.

In the semantical game G associated with (2.1) and a background model M, the existential player chooses an individual from the model. The game continues with respect to:

$$G(x) \wedge S(x). \text{ She is happy.}$$

Then the universal player chooses one of the formulae $G(x)$, $S(x)$. The game continues with the formula "She is happy". The first subsentence has thus been converted into a game tree G_1, that is, a set of branches in which each maximal branch has the form:

$$\bullet \longrightarrow a \longrightarrow P(x)$$

where $\bullet$ is the root, a stands for any of the individuals which can be chosen by the existential player, and $P(x)$ stands for the formula chosen by the universal player. The convention is that a is the interpretation of the variable x.

Next the sentential operator indicates a procedure to continue G_1: prolonge every maximal branch of the tree with the choices corresponding to the processing of the second subsentence of (2.1). The rule for the anaphora is: the existential player chooses the unique element a introduced during the play. The play goes on with respect to the formula $H(y)$: following the rule for the anaphora, a is to be the interpretation of the variable y. The end result is the game tree G in which each maximal branch has the form:

$$\bullet \longrightarrow a \longrightarrow P(x) \longrightarrow a \longrightarrow H(y)$$

and represents a complete play of the game. A player's winning strategy is a method which leads him to win against any choice of his opponent.

What can be noticed here is that the resolution of the pronominal anaphora is Evansian in spirit. The element a which interprets the anaphora and satisfies $H(y)$ must be the same individual which satisfies also $G(x)$ and $S(x)$. For this reason, the truth of (2.1) amounts to that of the sentence:

(2.2) $\exists x(Gx \wedge Sx) \wedge \exists x(Gx \wedge Sx \wedge Hx)$

It is obviously noticed here that the first subsentence of (2.2) is recovered as a subformula of the second subsentence, interpreting the anaphora as a definite description, which is logically equivalent with its interpretation as a bound variable:

(2.3) $\exists x(Gx \wedge Sx \wedge Hx)$

   Two points have to be emphazised here. First, a strategy for a player in the game is a set of functions, one for each move of the player, defined on all the earlier possible sequences of choices and giving the player the choice for his or her move. As we said, the strategy is a winning one if whenever used (that is to say against all possible moves of the other player) it results in a win for the player in question. Game-theoretical truth and falsity of a formula φ in a model M with respect to an assignment s are then defined as follows:

   - $M, s \models \varphi$ iff the existential player has a winning strategy in $G(\varphi, M, s)$
   - $M, s \not\models \varphi$ iff the universal player has a winning strategy in $G(\varphi, M, s)$

For instance, for φ being the formula $\forall x \exists y \forall z \exists w S(x, y, z, w)$, the truth and falsity are defined as follows:

   - $M, s \models \forall x \exists y \forall z \exists w S(x, y, z, w) \Leftrightarrow M \models \exists f \exists g \forall x \forall z S(x, f(x), z, g(x, f(x), z))$
   - $M, s \not\models \forall x \exists y \forall z \exists w S(x, y, z, w) \Leftrightarrow M \models \exists f \exists x \forall y \forall w \neg S(x, y, f(x, y), w)$

   Second, the resolution of anaphora is made in the process of building up the game-tree: the anaphora is resolved by choosing an individual that satisfies the formula independently of matters of truth. The fact that the anaphoric link is prior to matters of truth and falsity is precisely what we wanted to work out in this case, because of the existence of the anaphora in the sentence (2.1), independently of what is effectively the case in reality. Indeed, if we represent a semantical game as a zero-sum extensive game of perfect information, we see that the interpretation of sentences of natural language is made in a procedural way, in the spirit of dynamic semantics such as DRT [4] or DPL. The extensive form of a semantical game shows that strategies are generated as the game moves on. Once the game-tree is completely built up, the winning strategies can be determined and thus the truth in a model be defined. Hence the anaphora resolution in the GTS manner amounts to the same distinction between meaning and truth as in dynamic semantics.
   In that sense, technically speaking, extensive form of games gives individual moves priority over strategies. In the case of functional anaphora, a strategy for the existential player will consist of a function applied in the construction of the game-tree, available later on in order to yield the correct interpretation whenever a new individual is introduced as an argument for this function. Encoding strategies in such a way is precisely what Skolem functions stand for.

---

[4]See Kamp [11] and Kamp & Reyle [12].

2.2 Skolem terms as arbitrary objects in CCG

We suggest to extend, through the functional framework, a referential interpretation of terms that are commonly analyzed as quantified variables. Namely, that is to replace the interpretation of some anaphoric pronouns as variables bound by a quantifier, by terms that refer of their own, as E-type pronouns do. In this sense, Skolem terms can be seen as a certain kind of E-type pronouns, not in the sense that they are definite descriptions, but inasmuch as they are descriptions in a broader sense, namely some descriptive expressions of the language, which will allow to give an account of functional anaphora at the first-order level.

This is related to Steedman's work in Combinatory Categorial Grammar. His first aim was to account for scope ambiguities in sentences such like:

(2.4)   Everybody loves somebody

which provides two different readings:

(2.5) $\forall x \exists y (Person(x) \rightarrow (Person(y) \land loves(x,y)))$

(2.6) $\exists x \forall y (Person(x) \land (Person(y) \rightarrow loves(y,x)))$

(2.5) says that for each person, there is somebody that he/she loves, whereas (2.6) says that there is one person who is loved by everybody.

The question is why a single English sentence should have two such different readings, in other words, why the surface-composition should yield two different interpretations at the level of logical form. To solve the problem Steedman uses Skolem terms instead of existentially quantified variables to interpret indefinite NPs. The idea is the following: interpreting indefinite NPs as Skolem terms, themselves understood as denoting arbitrary objects, and therefore giving a purely referential interpretation to indefinite NPs rather than an existential interpretation in terms of quantifiers, as it is the case in the standard Frege-Montague framework.

For instance, Steedman proposes the following readings for (2.5) and (2.6):

(2.7) $\forall x [person' x \rightarrow (person(sk_1' x) \land loves'(sk_1' x) x)]$

(2.8) $\forall x [person' x \rightarrow (person' sk_2' \land loves' sk_2' x)]$

where all instantiations of existentially quantified variables are replaced with Skolem terms, which leads to the following readings: (2.7) says that every person loves the individual that the Skolem function sk_1 maps them onto, while (2.8) says that every person loves the single individual identified by the Skolem constant sk_2. Steedman points out the interesting result that these two formulae are identical, apart from the details of the Skolem terms themselves, which capture the fact that the referent of *somebody* is dependent or not upon the individuals quantified over by *everybody*.

Now turning to *donkey-sentences* such as

(2.9)   Every man who owns a donkey beats it

the problem lies in the following facts: the pronoun *it* looks like it might behave as a variable bound by an existential quantifier interpreting the indefinite NP *a donkey*, but the existential quantifier cannot both remain in the scope of the universal quantifier and bind the pronoun at the same time. Steedman assumes that the interpretation of indefinite NPs like *a donkey* as existential quantified terms should be given up, and replaced with a purely referential interpretation, in order to unify the semantics of indefinite NPs in natural language. Skolem terms seem good candidates to do such a work.

While in the example sketched above, the Skolemization of a first-order formula was operated in a "standard" way, namely with properties separately predicated over Skolem terms, Steedman argues that in the case of *donkey-sentences*, the indefinite NP should be replaced by what he calls a *generalized Skolem term*, where nominal properties must be associated with the Skolem term itself, in order to produce an underspecified translation of *a donkey* as: *skolem'donkey'*, the Skolem term denoting here an arbitrary object,[5] that is to say an "object with which properties can be associated but whose extensional identity in terms of actual objects is unspecified." [6] Roughly speaking, the specification of the Skolem term is achieved when the term is applied to all the variables bound by a universal quantifier or other operator in whose scope the Skolem term falls. [7] This yields a generalized Skolem term of the form: $sk^{(x)}_{donkey'}$, understood as "a term which associates a standard Skolem term made up of a Skolem functor and some bound variables with a nominal property." [8]

The anaphoric pronouns like *it* possibly occurring in sentences will translate as uninterpreted constants written *it'*. The interpretation of those pronouns will be accomplished by a DRT-like mechanism of the form *pro'x* where x is a discourse referent and *pro'* the identity function allowing x to copy its antecedent expression.

We can then account for Steedman's reading of *donkey-anaphora* (2.9):

(2.10) $\forall x[(farmer'x \land own'sk^{(x)}_{donkey'}x) \rightarrow beats'(pro'sk^{(x)}_{donkey'})x]$

Three facts are to be noticed in this interpretation. First, the Skolem term $sk^{(x)}_{donkey'}$

---

[5] For the notion of "arbitrary object", see Fine [3].
[6] Steedman [15], p. 303.
[7] For details, see Steedman [16].
[8] *Ibid.*

associates farmers with distinct dependent donkeys. Secondly, the universal quantification is made over farmers, not farmer-donkey pairs, which frees our hands from the proportion problem in the sense that the reading of (2.10) constrains each farmer that is quantified over to beat each donkey he owns. This leads us to the third point, which is the fact that (2.10) is given a "strong" reading and does not allow a "weak" one.

Although Steedman's picture of Skolemization is valuable in its translation of indefinite NPs as Skolem terms, we will not follow him on assuming that Skolem terms can be represented as arbitrary objects such as Fine's ones. The ontological "cost" seems too harsh to pay, since introducing arbitrary objects in the semantics requires an explanation of the referential link to such entities. Furthermore, it seems to us that mapping strategies in Skolem functions provides us with a quite efficient tool to analyse the phenomena at stake here.

3   Skolem functions

We will develop in this section an analysis of functional anaphora from Steedman's insights, keeping in mind the approach of GTS which can be summed up as follows: "By the Skolem functions of a first-order formula, we mean the choice functions for its several existential quantifiers."[9] In the examples 1-8 given above, the function or relation so introduced is a specific one: in 5 it is an owing relation and in 8 it is the function of having a price.

3.1   Skolemization for formal languages

Given a first-order sentence in prenex normal form, its Skolem form is obtained algorithmically: when an existential quantifier occurs in the scope of a universal quantifier, the existentially quantified variables must be replaced with a Skolem function of the universally quantified variables. In a first-order formula, for instance $\forall x \exists y P x y$, the choice of a value for y is dependent on the choice of a value for x since the formula asserts that for each x there is an appropriate value for y. In this case, the variable y would be replaced with a Skolem function of x and the resulting formula would be: $\forall x P(x, f(x))$, where $f(x)$ is the Skolem term standing for the value of y. The Skolem function can be quantified over, which is rendered as: $\exists f \forall x P(x, f(x))$. In either case the choice of a value for x is independent of the choice of a value for y.

Here is another example:

(3.1) $\forall x \exists y \forall z \exists w\, S(x, y, z, w) \leftrightarrow \exists f \exists g \forall x \forall z\, S(x, f(x), z, g(x, z))$

The reader can notice that (3.1) amounts to the winning strategy of the existential player in GTS, as said in the previous section.

---

[9]Hintikka [9], p. 308.

Now we can apply this method to natural language. Consider the following example:

(3.2)   Every man reads a book

which is rendered in a first-order formula as:

(3.3) $\forall x \exists y (Mx \rightarrow (By \wedge R(x, y)))$

And here is the Skolem form of (3.3):

(3.4) $\exists f \forall x (Mx \rightarrow (B(f(x)) \wedge R(x, f(x))))$

3.2   Recovering the descriptional analysis of pronominal anaphora from Skolem terms

Each of the sentences 1-8 has a natural interpretation in terms of Skolem functions. We will list below, for the first four sentences, their Skolem representation and the first-order sentence which is equivalent to them. Each sentence is first translated into a skolemized formula, where the indefinites are translated by Skolem terms, as required by Steedman's analysis. Then a first-order translation of the Skolemization is given, in order to bring to light the mechanism of anaphoric resolution.

(3.5)   Every child received a present. Jim opened it immediately.

(3.6) $\exists f [\forall x (P(f(x)) \wedge R(x, f(x)) \wedge O(j, f(j)))]$

(3.7) $\forall x \exists y (Py \wedge R(x, y)) \wedge \exists z (Pz \wedge R(j, z) \wedge O(j, z))$

(3.8)   Every child received a present.
                               Some child opened them immediately.

(3.9) $\exists f [\forall x (P(f(x)) \wedge R(x, f(x)) \wedge \exists z\, O(z, f(z)))]$

(3.10) $\forall x \exists y (Py \wedge R(x, y)) \wedge \exists x \exists z (Pz \wedge R(x, z) \wedge O(x, z))$

In both examples, the first-order universal quantifier ranges over children. In the second example, the existential quantifier $\exists x$ ranges over children.

(3.11)   For every merchandise there is a price. For meat it is 10 euros.

(3.12) $\exists f[\forall x(Pr(f(x)) \land Has(x, f(x)) \land f(m) = 10)]$

(3.13) $\forall x \exists y(Pr(y) \land Has(x, y)) \land \exists z(Pr(z) \land Has(m, z) \land z = 10)$

(3.14)   For every merchandise there is a price. For some it is 10 euros.

(3.15) $\exists f[\forall x(Pr(f(x)) \land Has(x, f(x)) \land \exists z(f(z) = 10)]$

(3.16) $\forall x \exists y(Pr(y) \land Has(x, y)) \land \exists x \exists z(Pr(z) \land Has(x, z) \land z = 10)$

In both examples, the first-order universal quantifier ranges over merchandises. In the second example, the existential quantifier $\exists x$ ranges over merchandises.

In each of the sentences (3.5)-(3.14), the indefinites are translated by Skolem terms and the pronominal anaphora by appropriate instantiations of these terms. We can mention the following points.

First, it should be observed that here the Skolemization is a two-step operation: first, one must replace every existentially quantified term that occurs in the scope of a universal quantifier by a Skolem term, and then introduce quantification over functions. This operation shows that sentences involving functional anaphora introduce first a functional relation and then the function is applied to some arguments of the function in order to yield the appropriate value for them. For instance, in (3.6) and (3.9), the function is first applied to the variable x which falls into the universal quantifier's scope, and then it is applied in (3.6) to the constant j and in (3.9) to the variable z, that is to say it is re-used in order to give further information about some entities of the domain. It shows that each term of the two sentences is dependent upon the previous one: the functional relation states that presents are dependent upon children, and the facts stated about the arguments of the functional relation are dependent upon this functional relation.

Second, the Skolem form of these sentences rises the fact that the universal quantification is made over restricted domains. Indeed, in all cases, the universal quantifiers range over definite sets of individuals rather than over the entire domain. This is what is precisely the case in sentences involving E-type pronouns.

Third, in the cases (3.7), (3.10), (3.13) and (3.16), there is a descriptive content that works as a first-order subformulae. In the first-order formula recovered from the Skolem form of the sentence, the first existential quantification is repeated in the second conjunct of the formula. Thus in (3.7), $\exists y(Py \land R(x, y))$ is retrieved in $\exists z(Pz \land R(j, z))$, and in (3.10), $\exists y(Py \land R(x, y))$ is retrieved in $\exists z(Pz \land R(x, z))$. That is exactly what was pointed out by Steedman in his account of the pronoun *it* in the *donkey-sentence* analysed above: the pronoun acts as a copy of its antecedent, namely the appropriate Skolem term which was quantified over by *every farmer* and which is repeated as a translation of this pronoun.[10] In our examples, the pronoun is translated as a copy of a first-order subformula which occurs in the first conjunct, and therefore refers to the same entity as its antecedent.

We can then conclude here that the first-order recovering is of an E-type kind, in the sense that the resolution of pronouns in functional anaphora requires a plain copy of the antecedent in order to be interpreted. However, such copies are not to be understood as definite descriptions like in the standard E-type approach (the requirement of uniqueness does not hold here), but rather as descriptive expressions of the language itself. Neither are they to be interpreted as standard *pronouns of laziness*, since the pronouns thus far interpreted have a reference of their own, being plain Skolem terms. For instance, in the second conjunct of (3.16), $\exists z(Pr(z) \wedge Has(x,z))$ stands for the descriptive account of the pronoun and $z = 10$ for the proper instantiation of the variable z. (3.16) could then be paraphrased as:

(3.17)   For every merchandise there is a price. For some

                merchandise there is a price and this price is 10 euros.

If we now turn back to the Skolem form of (3.14), we see that the phenomenon is nicely rendered in terms of Skolem entities, since all Skolem terms can be thought of as translating referential terms of the language, in the sense that they stand for individuals with which nominal properties are associated within the Skolem term itself. Thus, both the antecedent and the pronoun refer to the same individual, since they can be associated with the same descriptive content. This amounts to the same result as Steedman's, what can be easily verified in the translation of the *donkey-sentence* (2.9):

(3.18) $\exists f[\forall x(Fx \wedge D(f(x)) \wedge O(x, f(x)) \rightarrow B(x, f(x)))]$

which is equivalent to Steedman's interpretation of the sentence, (2.10) repeated here:

(3.19) $\forall x[(farmer'x \wedge own'sk^{(x)}_{donkey'}x) \rightarrow beats'(pro'sk^{(x)}_{donkey'})x]$

In both examples, the term referring to *a donkey* in the antecedent of the implication is translated as a Skolem term, which is re-used in the consequent as a translation for the pronoun. And, as can be seen in the first-order translation of (3.18)

(3.20) $\forall x[(Fx \wedge (\exists y(Dy \wedge O(x, y)))) \rightarrow \forall z((Dz \wedge O(x, z)) \rightarrow B(x, z))]$

---

[10]Steedman [16] even says in a note that "the use of pro-terms is not essential: we could use indices or plain copies of the antecedent, and assume that the binding conditions are consequences of the process of anaphora resolution itself."

we retrieve the interpretation of the anaphoric pronoun as a descriptive expression: in the consequent, $\forall z((Dz \wedge O(x,z))$ stands for the descriptive content of the pronoun *it* with a universal force, as required for the strong reading of the sentence.

What remains here to cope with is the interpretation of Skolemization itself. If Skolem terms are used to translate each occurrence of a referential term in a sentence, it does not appear that these Skolem terms need to denote arbitrary objects as Steedman thinks. Indeed, one can think of what is going on here by turning back to the GTS framework mentioned above. As required by the GTS interpretation, sentences (3.5)-(3.14) are first translated into first-order formulae, where the existentially quantified terms are translated by Skolem terms. Thus, (3.5) would translate as:

(3.21) $\qquad \forall x(P(f(x)) \wedge R(x, f(x)) \wedge O(j, f(j)))$

(3.8) would translate as:

(3.22) $\qquad \forall x(P(f(x)) \wedge R(x, f(x)) \wedge \exists z O(z, f(z)))$

and so on.

Now we apply to (3.21) and (3.22) the GTS interpretation given in the previous section. The definition of truth in a model as the existence of a winning strategy in a semantical game will render the truth of (3.5) and (3.8) equivalent to those of the sentences (3.6) and (3.9) respectively. Thus in semantical games, Skolem functions amount to the use of a strategy. A given function is available later on, so that whenever a new individual is introduced as an argument for this function, the function yields the appropriate interpretation for the anaphora as a Skolem term.

Consequently, a semantical game associated with a sentence involving functional anaphora will show up one strategy of the existential player. The point here is that the notion of strategy must not be thought of anymore in terms of winning strategies, as far as anaphora are at hand. Indeed, the Skolemization of sentences like (3.5)-(3.14) rises the fact that the existential player chooses an individual as the value for the function f (and another individual if it is the case that a new functional relation has to be introduced) and repeats her choice in order to pick up the same individual later on. Thus, in a sentence where an anaphoric resolution has to be made, the individual first chosen as a value through the function f and standing for the antecedent of the anaphoric pronoun, will be picked up again as a value standing for the pronoun, independently of matters of truth.

For example, in (3.8), the first sentence introduces a functional relation and, accordingly, in the corresponding game, a function that captures a strategy of the existential player is introduced. This strategy says that, for each child, the existential player must choose as a value the present which matches with him. Then the game proceeds with another sentence to be interpreted, where the existential player must choose again an individual in order to resolve the anaphora. So she will follow her previous strategy and apply again the same function, picking up the present which corresponds to each child. The game over, the anaphora has been resolved before the truth in a model has been verified.

To conclude here, the interpretation of indefinite NPs and anaphoric pronouns remains referential, and has to be understood as an E-type account, in the sense that the Skolem function allows to pick up the adequate individual for the antecedent and for the pronoun, which does not have to be the same one but must correspond to the correct value of the same function. In other words, the pronoun is viewed as a descriptive expression in the largest sense, that is to say an expression that carries sufficient information to give a means to identify the object it refers to. This remains also independent of matters of truth and is only suborned to matters of meaning, in the sense that the function is one of the strategy of the existential player, a strategy applied as many times as required in the course of the game, and defined afterwards as a winning strategy if it happens that the sentence is true in the model.

3.3   Extension and limits of the functional anaphora framework

The functional analysis of anaphora should be broadened to several types of phenomena where the functional relation is not obvious at first glance. For example, the Skolem function approach seems to extend quite naturally to the analysis of so-called *paycheck-sentences*:

(3.23)   I give my paycheck to my wife, but John gives it to his mistress.

(3.24)   John published a paper in *Nature* in 1986.
                        Peter published one in *Science*.

(Examples taken from Hess [7].)

In (3.23)-(3.24), a functional or relational correlation has been implicitly introduced in the context, and then it is applied over and over again to particular arguments. Game-theoretically, the strategy of the existential player is "remembered" later on, as seen before. In (3.23), for instance, a functional relation between persons and paychecks is introduced: what (3.23) says implicitly is that everybody receives a paycheck, thus the universal quantifier ranges over persons. Then more facts are stated about this particular relation: a person gives his paycheck to his wife, and another one gives it to his mistress. The "paycheck" is interpreted as a Skolem term, and the function expressed in this way can therefore be repeated to different arguments, as in our previous examples.

In order to account for the Skolem form of (3.23), we need to make a little change in the natural language example, stating explicitly the functional relation between persons and paychecks:

(3.25)   Everybody receives a paycheck. I give it to

                          my wife, but John gives it to his mistress.

And here is the Skolem form of (3.25):

(3.26) $\exists f[\forall x((Pers(x) \rightarrow R(x, f(x))) \wedge$

$\exists y(Wy \wedge G(i, f(i), y)) \wedge \exists z(Mz \wedge G(j, f(j), z)))]$

The reader can check that a first-order translation of (3.26) would lead, again, to a reading where the pronoun is understood as a description.

However the method consisting in recovering the descriptive content of the pronoun from the Skolemization in a first-order formula does not always work, as was pointed out by Barwise [1]. Such examples as 7-8 do not have a first-order equivalent. We give below their Skolem form:

(3.27)   Every country has a tyrant. The bigger

                          the country, the more powerful its tyrant.

(3.28) $\exists f[\forall x \forall y(((Cx \wedge Cy) \rightarrow (T(f(x)) \wedge T(f(y)) \wedge Has(x, f(x))$

$\wedge Has(y, f(y)))) \wedge ((x > y) \rightarrow (f(x) > f(y))))]$

(3.29)   Every pie has a price. The smaller the pie, the lower the price.

(3.30) $\exists f[\forall x \forall y(((Px \wedge Py) \rightarrow (Pr(f(x)) \wedge Pr(f(y)) \wedge Has(x, f(x))$

$\wedge Has(y, f(y)))) \wedge ((x < y) \rightarrow (f(x) < f(y))))]$

Contrary to the examples examined so far, in these sentences, the anaphora cannot be resolved by a descriptive approach like the one examined above, that is to say by copying the descriptive content of the antecedent in a first-order formula in order to account for the coreference. In, e.g. (3.27), *its* cannot stand for a *pronoun of laziness*, neither for a definite description, because what the second sentence introduces about the function defined in the first one is not a partition on the restricted domain of countries but a claim on the function itself, namely choosing, as values for the individuals belonging to the set of the countries, tyrants whose power holds in proportion with the size of the country. In other words, the fact stated about the function is not a particular but a general one. Therefore it seems to us that in such cases, the purpose is not to pick up some individuals in the domain, in order to make a specific predication upon them, but to give a general predication over the entire domain, what should be dealt with in a higher-order translation. It is furthermore impossible to translate (3.28) or (3.30) in first-order formulae since those sentences talk about entities that are independent from each other, when a first-order translation would yield a reading where quantifiers and other operators would be embedded.

However, the Skolem form of these sentences shows that a function f is drawn in order to account for the functional relation existing between the sets of individuals in the first sentence. Thus it is possible to interpret the pronoun *its* as a Skolem term. We saw that such terms can stand for every indefinite NPs and anaphoric pronouns in order to give a referential interpretation of them. Thus, if we agree on the fact that Skolem terms translate correctly indefinite NPs and pronouns in a way that makes them purely referential, the phenomenon at stake in Barwise's and Fox's examples seems though to allow for a referential interpretation too. Indeed, according to the interpretation of Skolem terms developed so far, the value $f(x)$ of such a function f can be seen as the unique individual with which is associated the nominal property born by both the antecedent and the pronoun.

4   A solution to intentional identity within the functional interpretation

4.1   The unsolved problem of modal subordination

A puzzle well-known in the literature is that of intentional identity, which was firstly pointed out by Peter Geach. Let us quickly remind the problem at stake: "We have intentional identity when a number of people, or one person on different occasions, have attitudes with a common focus, whether or not there actually is something at that focus." [11] The issue was exemplified in such kind of sentences:

(4.1)   Hob thinks a witch blighted Bob's mare, and Nob

                 thinks she (the same witch) has killed Cob's sow.

First, in what we will henceforth call the Geach sentence, the existence of an anaphoric link between the pronoun and its antecedent does not demand that witches exist or even that Hob or Nob may have in mind a precise individual that one might refer to by the term *a witch*. Thus the first point to deal with is the translation of the indefinite NP *a witch* by an existential quantifier: how can the variable bound by the quantifier be assigned a value while the intended reading does not ask for it, or rather does not ask for this value to be a precise one? That is to say the existential quantifier has a *de dicto* reading. Hence (4.1) hardly translates as

(4.2) $\exists x (Wx \wedge T_H Bx \wedge T_N Kx)$

---

[11] Geach [5], p. 627.

or

(4.3) $\exists x(T_H(Wx \wedge Bx) \wedge T_N Kx)$

because the sentences (4.2)-(4.3) say that there is some particular witch whom is believed to have cast a spell on the animals.

The same difficulty would arise if treating the anaphora as a *pronoun of laziness*, because such pronouns act as copies of their antecedent, that is to say they go proxy for them. But what would be copied there, since no particular reference is at hand?

Secondly the Geach sentence does not presuppose Hob and Nob having peculiar thoughts about each other's beliefs. In other words, no iteration of propositional attitudes is required for the sentence above to have the meaning Geach asserts to be the right one. The two propositional attitudes cannot be represented by embedded modal operators, since a different content occurs in their respective scope and need not be connected in any sense. Likewise, it does not seem plausible to formalize (4.1) like this:

(4.4) $T_H(\exists x(Wx \wedge Bx \wedge T_N Kx))$

because what (4.4) says is:

(4.5)   Hob thinks that a witch blighted Bob's mare and
                   that Nob thinks that she has killed Cob's sow.

which is not, again, the intended reading of the sentence for it presupposes that Hob thinks something about Nob's beliefs.

The remaining solution seems to be splitting the conjunct into two separate blocks:

(4.6) $T_H \exists x(Wx \wedge Bx) \wedge T_N \exists x(Wx \wedge Kx)$

However this does not work either, because the two quantified variables do not have to bear the same individual as a value for the sentence to be satisfied.

Introducing an identity clause would not do either, because in this case, there would be a free variable in the second conjunct:

(4.7) $T_H \exists x(Wx \wedge Bx) \wedge T_N \exists y(x = y \wedge Wy \wedge Ky)$

To bind this free variable, one has to widen the scope of the existential quantifier in the first conjunct:

(4.8) $T_H \exists x(Wx \wedge Bx \wedge T_N \exists y(x = y \wedge Wy \wedge Ky))$

But, again, this formula leads to an improper reading, inasmuch as the existential quantification is resumed, as well as the nesting of propositional attitudes, what was precisely intended to be avoided.[12]

The puzzle lies here in the linear interpretation of the modal operators and quantifiers. Indeed, since the sentence is interpreted from the first modal operator T_H and the first existential quantifier $\exists x$, the following operators and quantifiers will depend on them. A solution may be provided by our functional framework, in the sense that, as we saw it, Skolem functions offer a way to translate correctly indefinite NPs such as *a witch*, and to choose adequate individuals within a functional relation established between several sets of individuals. We can thus think of the Geach sentence as establishing such a functional dependence between the set of beliefs of an agent (understood as a set of possible worlds) and the individuals the function allows him to pick up in this set.

4.2 May propositional attitudes be skolemized?

For matters of simplicity, we will assume the example (4.1) being of the form:

(4.9) $$\Box_1 \exists x W(x) \wedge \Box_2 P(x)$$

where $\Box_1$ and $\Box_2$ stand for the two modal operators. The conjunction could be interpreted here as a conjunction, because the existential quantifier has to reach over the first modal operator, the conjunction and the second modal operator, to bind the variable x in $P(x)$, allowing binding outside its scope.

The Skolem function allows the following: for every choice of a possible world w, yielding an individual which is W at w, and for every choice of a possible world w', which is P at w'. Thus the Skolem form of (4.9) amounts to:

(4.10) $$\exists f \exists g [\forall w \forall w' (W(f(w)) \wedge P(g(w')) \wedge f(w) = g(w'))]$$

argument where $\forall w$ and $\forall w'$ range over the relevant successors of $\Box_1$ and $\Box_2$ respectively. Notice the identity clause about the two Skolem terms: it allows them to pick up the same individual in the two possible worlds without presuming an existential quantification. As in previous functional anaphora, it accounts for the fact that the NP *a witch* is given a referential interpretation, but this time *within* the different possible worlds w and w'. This means that the two Skolem terms $f(w)$ and $g(w')$ have the same descriptive content in the sense that, although standing for the values of two different functions, the identity clause holding on these terms insures that both functions pick up the adequate individual relatively to each world.

Going back to the Geach sentence, it means that the individual of which Hob thinks as being a witch does not have to be a particular one or even exist, to be

---

[12]For further discussion of these questions, see Pietarinen [13], who delivers an entire overview of the problem and a solution within an *IF-logic* framework close to ours.

the same individual of whom Nob believes she has killed Cob's sow. No need
of a quantification ranging existentially over individuals to ensure that the witch
chosen in Hob's possible world is the same individual as the one Nob has in mind.

Furthermore, (4.10) yields an interpretation where modalities are not embedded: indeed each of the two functions chooses a single individual in each of the
possible worlds, respectively Hob's and Nob's, independently of each other. Thus
in this interpretation, Hob does not have thoughts about Nob's beliefs, for the
identity of individuals is insured by the identity of the Skolem terms, not by an
assumption on the existence of a specific individual.

The thing to be emphasized here, again, is that the E-type interpretation is re-
coverable from the functional interpretation. (4.10) is equivalent with:

(4.11) $\forall w \exists x (W_w(x)) \wedge \forall w' \forall w \exists y (P_{w'}(y))$

which can be paraphrased as:

(4.12)   Hob thinks a witch blighted Bob's mare and Nob thinks

             that the witch of whom Hob thinks she blighted

             Bob's mare has killed Cob's sow.

We see here the plain descriptive content of the pronoun *she*: "the witch of
whom Hob thinks she blighted Bob's mare". In other words, the instantiation of
an individual at world w' is independent of whether or not this individual exists
but is dependent on the anaphoric link holding between the antecedent and the
pronoun.

Hintikka introduced in the sixties a distinction between proper and improper
individuals.[13] The former, typically associated with a "rigid designator", is a constant function which picks up the same (instance of an) individual from each member of a class of possible worlds. An improper individual, on the other side, typically associated with *de dicto* propositional attitudes, is a function whose value
varies from one world to another.[14] Now, one way to look at (4.10) is to see the
Skolem functions f and g as providing an improper individual. $f(w)$ and $g(w')$ being improper individuals, the predication relation varies from one world to another,
and so does the anaphoric link.

5   Conclusion

We saw that in many cases of functional anaphora, we can account for the anaphoric
resolution in terms of Skolem functions, provided we are ready to allocate Skolem
terms a particular referring status. The process can be summed up as follows: in

---

[13]Hintikka [8].

[14]For a recent discussion about proper and improper individuals, see Sandu [14].

a discourse, each time an indefinite NP within the scope of a universal quantifier is encountered, replace it by a Skolem term associated with the nominal property expressed by the NP, which Skolem term remains underspecified until the sentence is closed by all instantiations of the variables which fall under the scope of the universal quantifier, as was pointed out by Steedman. When a pronoun occurs, repeat the instantiation of a Skolem term with the adequate property. But instead of interpreting Skolem terms as arbitrary objects, we interpret them as ordinary terms associated with properties, that is to say carrying a certain descriptive content, and referring to the correct individuals the function maps them onto.

In some cases, the Skolem terms can be translated in first-order formulae, which shows explicitly the descriptive account of the pronominal anaphora. But there are cases where a first-order translation is not available, as in Barwise's and Fox's examples. Yet, the account in terms of Skolem functions is clearly the same in all cases. Thus, we can see Skolem terms as providing ordinary individuals rather than arbitrary or abstract ones. The descriptive content of a Skolem term standing for a pronoun is either specified in a first-order subformula retrieved from the Skolem form of the sentence, either left underspecified where such a recovering is impossible in first-order language.

But the main fact to be noticed is that the functional relation is prior to the descriptive content in the interpretation of sentences involving functional anaphora. Indeed, we know from Barwise's result that the first-order recovering is not always available. However, that does not prevent us from accounting for the main point in such cases, which is the functional relation. In that sense, the semantic content of functional anaphora can be left unspecified, since the Skolem function provides a way to pick up the right individual in the anaphora resolution. In this way, cases of functional anaphora can be thought of as particular cases of E-type pronouns, because what had been highlighted is a referential content of the pronoun. The link between the pronoun and its antecedent is accounted for in the functional relation itself, what is shown in a GTS account of functional anaphora, provided games are described in extensive form, where can be seen that the use of a function as a strategy, when remembered and repeated, amounts to a winning strategy.

BIBLIOGRAPHY

[1] BARWISE, J., 1979: On Branching Quantifiers in English, *Journal of Philosophical Logic*, **8**, 47-80.
[2] EVANS, G., 1980: Pronouns, *Linguistic Inquiry*, **11**, 337-362.
[3] FINE, K., 1985: *Reasoning with Arbitrary Objects*, Oxford, Oxford University Press.
[4] FOX, C., 1999: Vernacular Mathematics, Discourse Representation, and Arbitrary Objects, unpublished ms.
[5] GEACH, P., 1967: Intentional Identity, *The Journal of Philosophy*, **64**, 627-632.
[6] GROENENDIJK, J. AND STOKHOF M., 1991: Dynamic Predicate Logic, *Linguistics and Philosophy*, **14**, 39-100.
[7] HESS, M., 1989: *Reference and Quantification in Discourse*, Habilitationsschrift, Zürich, unpublished ms.

[8] HINTIKKA, J., 1969: *Models for Modalities*, Dordrecht, D. Reidel Publishing.

[9] HINTIKKA, J., 1998: Truth Definitions, Skolem functions and Axiomatic Set Theory, *The Bulletin of Symbolic Logic*, **4**, 303-337.

[10] HINTIKKA, J. AND SANDU, G., 1997: Game-theoretical Semantics, in van Benthem, J. and ter Meulen, A. (eds.), *Handbook of Logic and Language*, Amsterdam, Elsevier Science Publishers, 361-410.

[11] KAMP, H., 1981: A Theory of Truth and Semantic Representation, in Groenendjik, J., Janssen, T. M. V. and M. Stokhof (eds.), *Formal Methods in the Study of Language*, part 1, Amsterdam, Mathematisch Centrum Tracts, 277-322.

[12] KAMP, H. AND REYLE, U., 1993: *From Discourse to Logic*, Dordrecht, Kluwer Academic Publishers.

[13] PIETARINEN, A., 2001: Intentional Identity Revisited, *Nordic Journal of Philosophical Logic*, **6**, 147-188.

[14] SANDU, G., 2006: Hintikka and the Fallacies of the New Theory of Reference, in Auxier, R. E. and Hahn, L. E. (eds), *The Philosophy of Jaakko Hintikka*, The Library of Living Philosophers, vol. 30, Chicago & La Salle, Open Court, 541-554.

[15] STEEDMAN, M., 1999: Quantifier Scope Alternation in CCG, in *Proceedings of the 37th Annual Meeting of the Association for Computational Linguistics (ACL'99), College Park, MD*, 301-308.

[16] STEEDMAN, M., 2007: Surface-Compositional Scope-Alternation without Existential Quantifiers, unpublished ms.

Against Possible Worlds

REINHARD KAHLE

Diese Semantiken für die Modallogik benutzen allesamt [...] den Begriff der möglichen Welt als einen grundlegenden, undefinierten Begriff. Aus diesem Grund darf man sie lediglich als algebraische Strukturen ansehen, die die logischen Wechselbeziehungen zwischen verschiedenen modalen Komponenten klären, aber uns keine Kriterien zur Verfügung stellen, was möglich ist und was notwendig ist.

Dagfinn Føllesdal [7]

1 Introduction

In this paper we argue against the use of possible worlds in the analysis of necessity and possibility.[1] Possible worlds have a dignified tradition and we will not question their importance as a philosophical concept. We will focus just on one particular use of this concept: their function as *explanans* in a theory of necessity. In fact, we do not only question the rôle of possible worlds in the analysis of necessity, but also the rôle of *modal logic*, as far as it is understood as their syntactic counterpart.

Modal logic and possible worlds are considered as the standard logical tools to study necessity, on the syntactic and semantic level, respectively. Therefore, a criticism of them needs a justification. The first part of this paper is devoted to such a justification. In fact, most of the problems of modal logic and possible worlds discussed below are well-known as technical troubles. However, here we like to emphasize their *philosophical* significance as criticism.

Often, criticism of modal logic and possible worlds is rejected by the plain argument that there is no alternative approach to necessity. To answer this response we will propose at the end of this paper an alternative account to necessity based on a proof-theoretic view. In fact, this analysis allows to reintroduce possible worlds in the analysis, but now only in a secondary place.

---

[1] The defense of this position is surely outside of the scope of the logical and philosophical mainstream. Thus, we hope it is in the spirit of SHAHID RAHMAN to whom this volume is dedicated. The polemic character in the following is intended to fuel the discussion; by no means would we like to affront the protagonists of possible worlds semantics personally. In contrast, we value them as philosophical antagonists. This paper serves also as a companion to the more detailed [16].

2   The defense of possible worlds

It is interesting to observe that proponents of possible worlds semantics often feel themselves in the position to defend this semantics.[2]

We will start with a list of arguments, given by MARES in [19, p. 23f] in favor of possible worlds semantics for modalities.

> One might wonder what gain is made by translating modal talk into talk about possible worlds. But the virtues of this translation are considerable. First, the semantics is couched in the language of set theory. [...]

> The language of set theory is extensional, unlike the language of modality and, more to the point, it is very well understood mathematically. [...] The translation of modal logic into a set-theoretic semantics has allowed for the proof of many theorems that would have been difficult to prove otherwise.

> Second, the translation of modal logic into possible worlds semantics has given modal logic a compositional semantics. [...]

> Third, possible world semantics has the virtue of providing an intuitive and philosophically satisfying interpretation of modality. [...]

> Fourth, possible world semantics can be used to treat many elements of language other than the operators 'possible' and 'necessarily'. [...] In science, a theory's ability to treat a wide range of phenomena is a good reason to adopt that theory. Possible world semantics has the virtue of being able to treat a wide range of linguistic phenomena.

We think that none of these four points is cogent.

The first two are technical. Of course, set theory is a powerful mathematical tool—but is it the *proper* tool here? Modalities are instances of *intensional* phenomena; set theory is extensional. And although, as MARES noted correctly, set theory is well-understood mathematically (and therefore furnishes us with a lot of powerful theorems), there is no *philosophical* reason why such a reduction from the intensional to the extensional side should be promising.[3] Maybe it is just inadequate. To put it in a drastic picture: Hammers are proven very useful to drive a

---

[2]In contrast, this is not the case for TARSKIan semantics.

[3]Here, we put explicitly the *thesis of extensionality* in question. In [31] this thesis is attributed to CARNAP, [3, p. 950]. However, an inspection of the citation shows that CARNAP himself was quite cautious in his words: "I have the impression that so far nobody has actually refuted the assumption, sometimes called 'thesis of extensionality', which says that every proposition expressible in a non-extensional language is also expressible in a suitable extensional language (in such a way that the two sentences are logically equivalent)." Even if the thesis is yet not refuted, we doubt that there is sufficient evidence to support it, not to mention a *philosophical* argument proving it.

nail into a wall. I can also use a hammer to batter a screw into a wall. But, maybe, it would be better to look for a screwdriver.

Compositionality seems to be a desirable property[4]; but we think it comes for a too high price: the additional ontology. We will come back to the ontological question below.

The third point we just deny: The interpretation of modalities in terms of possible worlds semantics is neither intuitive nor philosophically satisfying. It is not intuitive since it would require an intuition of the different *concrete* possible worlds, an intuition which—in my experience—does not exist. It is philosophically not satisfying since it burdens us with an ontology which "appears from nowhere".

Finally, our criticism reaches also—and in particular—to the fourth point: *Counterfactuals*, for instance, are rather examples for the failure than the success of possible worlds semantics.

One of our main criticisms is the ontological status of possible worlds. It will be in the focus of the main discussion. However, MARES is defending possible worlds semantics only as a technical tool, [19, p. 24]:

> Note that I am not claiming that it is philosophically unproblematic or even intuitive to hold that there are possible worlds other than the one in which we live. Most philosophers have trouble believing in the existence of other worlds, or at least admit that a commitment to them is problematic. I am only claiming that talk about other possible worlds is intelligible and can make modal talk easier to understand.

In this spirit, we received several comments on an earlier version of this paper, saying that possible worlds can be—or even have to be—seen as a way of speaking only, without any commitment to additional ontology. Also, they might be *defined* relative to language, collapsing, e.g., to simple CARNAPian *state descriptions* in the case of propositional logic. They might even be partitioned into equivalence classes depending on which features one is interested in.

Although our main target is the realistic interpretation of possible worlds, one can asked what is actually gained by the alternative views, when we think of possible worlds as *semantics*. In the case where possible worlds are just identified with indices in an abstract Kripke structure, one gets essentially a translation into a formal framework. This framework may have nice structural properties, but, in general, it needs again an interpretation before it can be considered as semantics. Thus, what we actually get is a logical transformation but not a philosophical explanation. For us, that is not much. What is at issue here, is possible worlds as a semantic concept to explain necessity.

---

[4]In fact, it is a reasonable question, why—and how. For a discussion of this question, we like to refer to the work of HODGES, cf. e.g., [13]. One can even ask whether every semantics can be made compositional by allowing higher order concepts, concepts which just take all possible contexts in which a term is used into consideration.

There is, of course, a strong philosophical tradition which interprets possible worlds in a rather realistic way. The most prominent defender of this view is surely DAVID LEWIS [18]. Here, we like to cite a comment on *modal realism* by CHIHARA, [5, p. 3] (see also [2]):

> Philosophers who eschew including possible worlds in their ontology thus seem to be precluded from making use of possible worlds semantics in their account of the truth of modal propositions, in their assessment of the validity and invalidity of modal arguments, and in their explanations of modal principles and modal reasoning. For the Modal Realist, this is too high price to pay for ontological economy.

So far, we could even follow CHIHARA since we have no problems to give up possible worlds semantics (and *a fortiori* modal realism) in the treatment of modalities. What is worrying is the conclusion which CHIHARA presents afterwards:

> Thus, we find Raymond Bradley and Norman Swartz writing: "We feel driven to posit [possible worlds] because we seem unable to make ultimative sense of logic without them."[Footnote: Bradley and Swartz 1979: 64.]

Frankly, we can consider the last statement only as one about BRADLEY and SWARTZ, but not about logic. And we are not alone with this view. FORSTER addresses the problem of possible worlds as basis for logic in drastic words, [8, p. 1f]:

> I once heard a philosophical colleague telling a lecture-theatre full of first year students that an inference was valid if "there was no possible world in which" the antecedent was true and the conclusion was false. Perhaps my readers are not as shocked by this as I am. They should be: the idea of a possible world is not needed for an explanation of the concept of valid argument, since we knew what valid arguments were before we had possible world semantics.

> However the point is not that this is a shocking story, and that people teaching first-year logic ought to know better. The point is not even that this represents an unwarranted intrusion of possible world jargon into an area where it has nothing to offer. The worry is that there are nowadays many other applications of possible world imagery in more complex settings than this (counterfactual, fictions etc etc) and this story raises the possibility that they all may be fully as misconceived as this one was. I fear they are: in fact over the years I have been driven reluctantly to the conclusion that all the uses known to me of possible world semantics beyond formal logic are misconceived, and

that the light that possible world imagery appears to have shed on various philosophical problems in recent years will be seen in the years to come to have been entirely spurious.

This paper should serve to support this view.

Since possible worlds semantics has its origin in modal logic, we will start with a critique of the latter.

3   Against modal logic

Modal logic, as a formal framework[5], pretends to formalize necessity and possibility. In this section we argue that the formalizations of modal logic, as far as they are intended to formalize possible worlds semantics, do not fulfill their purpose.[6] Let us start with monomodal, normal logic. Its key element is the necessitation rule:

$$\frac{\vdash \varphi}{\vdash \Box\varphi}$$

This rule raises the following question (even in the presence of other standard modal axioms containing $\Box$):

> What does modal logic achieve more than mark formulas which are derivable in a background theory?

This question can be answered: It allows one to investigate *nested* modalities. However, we claim that the case of nested modalities is rather uninteresting when we like to investigate modalities; it would be more important to get a theory which would *explain* the (correct) use of modalities at least for the non-nested case. But modal logic does not contribute to this task: Everything which will be necessary is somehow already put in the axioms. By necessitation every axiom (and every derivable formula) turns immediately in a necessary one. Thus, modal logic does not explain necessity, but has it built in.

There is a lot of work concerning alternatives, like *non-normal* and *polymodal*

---

[5]In a broader view, *modal logic* includes, of course, not only the formal systems (usually containing characteristic $\Box$ operators), but also its models, e.g., Kripke frames. Since these *mathematical* models are not directly linked to the question of necessity (at least in the non-technical sense of necessity), they are not considered here. For instance, one of the standard references on the model theory of modal logic, [17], does not even mention "necessity" or "possibility" in the index. In mathematical logic *semantics* is used as a technical term for the models of a formal theory. Here, we explicitly avoid the use of the word "semantics" for the models of formal systems in modal logic, since the term should be reserved for the philosophical notion of semantics, which, in our view, is quite different from the mathematical.

[6]This does not mean that theories like **S4** and **S5** cannot serve for any other purpose.

logics.[7] Since non-normal modal logics[8]—an area to which SHAHID RAHMAN contributed in an important way, cf. e.g., [24]—are not subject to the trivialization critique expressed above, we definitely favor them for a modal approach to necessity; however, the remaining structural properties seem to be too weak to capture a philosophically convincing notion of *necessity*. Polymodal logics contain several □ operators, which, in principle, could serve to formalize different notions of necessity. While they have an interesting model theory, cf. [17], we are not aware of an interpretation in terms of (different?) universes of *possible worlds*. In fact, the way such universes could intersect would be very interesting, but it seems to be rather intractable.

Furthermore, it is worth mentioning that modal logic fails to integrate necessity with actuality (contingent truth). But such an integration is definitely desirable. There are approaches in the literature which add to modal logic an extra *actuality operator*[9]. Although they are an important topic in the philosophical discussion they did not find their way into the standard presentations of modal logic. However, the recent developments of *hybrid logics* which involve actuality operators seem to be steps in the right directions.

Before continuing with our critique of possible worlds semantics we like to include a digression about the virtues of modal logic.

4   The virtue of propositional modal logic

Many propositional modal logics can easily be embedded in predicate logic.[10] So, why do we use modal logic at all, and not its image in predicate logic? WANSING argues against such a view, since the embedding would be only *instrumental*, [27, p. iii]:[11]

> Although established and useful modal logics like **S4** and **S5** may be viewed as fragments of first-order logic, faithful embeddings and 'reductions' are in the first place *instrumental*. Normally, the reducing theory is not really meant to *replace* the reduced theory.

---

[7]Here, we do not consider the approaches using *predicates* instead of *operators* to formalize modal notions, cf. e.g., [22, 1]. Some recent work of HALBACH, LEITGEB, and WELCH shows that this approach can be worked out without resulting in paradoxes; however, as the titles of their papers [9, 10] indicate, they are still related to a possible worlds semantics and therefore subject to our following criticisms.

[8]For non-normal modal logic where the models are given by the more general *neighbourhood semantics* the *congruence rule* is the basic property, cf. e.g., [4].

[9]See, for instance, the list of references in footnote 9 of [30, p. 199]

[10]This holds, for example for the standard systems **S4** and **S5**; some systems would need second order logic, but a switch to this framework does not affect our argument in this section.

[11]WANSING responds in this article to MCCARTHY [20] who questioned the adequacy of modal logic for certain modalities studied in artificial intelligence. There was also an answer to MCCARTY by HALPERN [11], which, however, relates more to the technical rather than the philosophical question of the adequateness of modal logic. This discussion continued with [21] and [12].

This might be true if one starts from modal logic as given theory. However, if one takes possible worlds semantics as a starting point, and if one considers modal logics like **S4** and **S5** "only" as attempts to formalize it, the fragments of first-order logic are definitely more natural. Thus, from the semantic point of view such embeddings are not just instrumental, in contrast, the question arises again why one should consider modal logics at all instead of fragments of predicate logic.

But there is another momentous difference between propositional modal logic and its translated version in predicate logic: Modal logic is *decidable*, while with predicate logic we already entered the world of undecidability.

This is a strong argument in favor of modal logic.

But only, as long as decidability is one of the goals. And, decidability does not mean by any means that this theory contributes anything to the understanding of modalities. It might, of course, be useful for other purposes. Modal logic seems to be, for instance, the adequate framework to formalize *transputers*, a concept of parallel computers fashionable in the early 1990s. Modal logic also provides a lot of technical, mathematical challenging questions which allow deep insights in logical relations, cf. e.g., [17].

But, when we start with decidability, we have to point out that *first order modal logic* is poorly conceived: It will be undecidable from its very beginning. In fact, to investigate how far the border of decidability can be pushed, modal logic has proven to be instrumentally helpful: *Guarded quantifiers*, *multi-modal logics*, and *description logics* which all three are exploring this border can be based on (or at least related to) modal logic. But for first order modal logic one has to ask why one is not using the explicit formalization of modal logic in predicate logic by additional predicates.[12] The fact, that one would be forced to axiomatize these predicates explicitly would probably "solve" the notorious problems of first order modal logic, when one has to postulate the additional axioms.

Thus, we take (propositional) modal logic as a formal system which is interesting from a *mathematical* point of view,[13] but not from a *philosophical* one. And we have to add a warning: Nice *technical* properties should not be taken as arguments for *philosophical* virtues of a theory.

---

[12]One might need second order logic in some cases; second order logic does not have the best reputation as a formal system; however, we think that there is no *philosophical* argument against the use of it. Our arguments do not depend on a specific logical system, so that predicate logic can easily be replaced by any higher order system.

[13]Here, we like to note that modal logic, if taken as starting point, allows for interesting proof-theoretic semantics (i.e., semantics which do not rely on external entities, in particular not on "worlds"), cf. the work of WANSING [26, 28, 29]. This approach differs, however, from our proof-theoretic account described below.

5 Against possible worlds semantics

Our critique of modal logic carries over to a large extent to possible worlds semantics. For instance, necessary truth is usually interpreted as truth in all possible worlds. Given a variety of possible worlds, we therefore can read off what is necessary by looking what is true in all these worlds. But our problem is *how* we get such a variety. As soon as it is given, necessity can be easily explained; but the challenging question is to constitute the variety. Thus, we accuse possible worlds semantics to miss the point: To *explain* necessity it is of no interest to interpret it in a given variety of possible worlds, but the challenge would be to give criteria to find such a variety. This is analogous to the question which would be the non-logical axioms which determine a specific notion of necessity in modal logic.

STALNAKER admits this point in part when he writes [25, p. 333]:

> The possible worlds representation of content and modality should be regarded, not as a proposed solution to the metaphysical problem of the nature of modal truth, but as a framework for articulating and sharpening the problem.

We think that the contribution of possible worlds to articulate and sharpening the problem of modalities is rather limited. In fact, all that remains is the possibility to study some structural properties of nested modalities. This was observed by FØLLESDAL [7, p. 572f] whose citation given at the beginning of this paper reads in a free English translation as:

> All these semantics for modal logic use [...] the notion of possible world as a basic, undefined notion. For this reason one is allowed to consider them only as algebraic structures which clarify the logical interrelations between different modal components, but which do not provide any criteria for what is possible and for what is necessary.

One may object here, that the aim of possible worlds semantics—and modal logic—is to study structural properties of such worlds only, but not to identify them. In the same way, predicate logic studies the logical relations within a model, but without identifying a particular model. This is correct, if we consider modal logic as a purely *logical* enterprise; however, if it wants to contribute to the *philosophical* question to explain necessity—and this question is at issue here—it would have to say something about its worlds.[14] In contrast, predicate logic is philosophically better developed: Although today one hardly claims that its purpose is to explain truth, if needed, one can identify quite detailedly a particular model. Given such a model, the notion of truth is indeed *explained*. And, this model usually serves as *semantics* in a broad philosophical sense.

---

[14] Stalnaker admitted in the citation above that this is not the case.

This leads us to an even deeper philosophical problem with possible worlds semantics: Is it a *semantics*?

To check this, it seems to be sufficient to check whether it fulfills the criteria given in a definition of *semantics*. But, at this point we realize that—despite the fundamental role of semantics in philosophy—there is nothing like a definition of semantics around. Thus, leaving the interesting philosophical question: *What is a semantics?* open,[15] we can only argue on an intuitive level. And on this level, the crucial specification is that semantics should provide *meaning*. This is usually achieved by an interpretation in a realm which is presupposed to be better understood than the formal framework in question.

We doubt that possible worlds semantics can be considered as understood.

Here, we do not mean possible worlds as a theoretical concept, but possible worlds as *semantic objects*. Each of the worlds plays the same role as, e.g., a *model* for predicate logic. As such, it has to be total, i.e., attribute truth values to all sentences. But here, we face the problem of delimitation of the possible worlds. We may ask whether there could be worlds in which bachelors are unmarried, worlds in which there is no number 17,[16] or worlds in which *tertium-non-datur* does not hold. But even for less academic examples, the question remains how the complete world(s) should look like in which tomorrow there is a sea battle. I can imagine the happening of a sea battle without spending a second a thought on the fact whether at the same time a bag of rice is falling down in China, or not, or whether the principle of bivalence does not hold tomorrow for rice bags. To cut the long story short, for us, the meaning of a possibility statement in itself is better understood than the concept of possible world.

The situation is even worse with respect to the attempt to explain counterfactuals in terms of possible worlds.[17] This analysis is based on a completely obscure neighborhood relation on worlds. We are not aware of any concrete example where the notorious *closest possible world* could be defined in a sensible way. Reflecting a little bit on this problem, one should realize that the neighborhood relation is in fact less understood (understandable?) as the counterfactuals themselves. This is the case, because worlds have to take a great deal too many things into account which are obviously irrelevant when we consider a concrete counterfactual statement.

In [19] MARES uses *situations* which are "incomplete" possible worlds, to study relevance in the sense of *relevant logic*. Equipped with a ternary relation (which somehow takes over the role of the accessibility relation) one gets the *Routley-Meyer semantics* for relevant logic. To the surprise of the reader, MARES states [19, p. 28]:

---

[15] Some considerations concerning this question can be found in [?].

[16] See footnote 26 below.

[17] I will not even touch the problem of "impossible" worlds.

In order for the Routley-Meyer semantics to have the same sort of status as a theory of meaning and a theory of truth as the possible world semantics for modal logic, it needs a reasonable interpretation.

As he admits, the *Routley-Meyer semantics* is considered "as an uninterpreted mathematical theory" [19, p. 28]. Thus, it is *not* a semantics. It is used only as an intermediate layer between relevant logic and its (final) semantics. Its use might be justified by the extensionality MARES advocates, but as a semantics it is obviously corrupted. Here the question arises whether the situation is much better for possible worlds semantics: Maybe it is just an interesting mathematical theory, but its role as semantics is as limited as the one of the Routley-Meyer semantics for relevant logic.

6   The proof-theoretic perspective

The last possibility to defend modal logic and possible worlds semantics would be to claim that there is up to today no serious alternative to formalize necessity and possibility. Here, we would like to sketch such an alternative. It starts with the observation that necessity statements are usually *binary*, i.e., instead of

(1) φ is necessary.

we use more often sentences of the form

(2) φ is necessary for ψ.

In fact, we claim that many instances of (1) would have to be extended to instances of (2) if they should not be trivially false.[18] Now, the modal reading of (2) as $\square(\psi \to \varphi)$ is unsatisfactory (although not completely wrong). Not only it is exposed to the problem of logical omniscience—tautologies are necessary for everything; everything is necessary for a contradiction—, but it also converts in some sense the roles of φ and ψ.

Our proposal is to read (2) as

Every proof of ψ uses φ.

Such a reading needs at least a specification of the notion of use of a formula in a proof. It turns out that such a notion is far from being easy to define. However, it is easy to define, if we restrict ourselves to the case that φ is an axiom. And, making a virtue of necessity, we even propose that the choice of φ from a *stipulated* set of axioms is an essential presupposition of a meaningful necessity statement.[19]

---

[18]We think of examples like "I have to go now" standing for "I have to go now to catch the train" or "Club A has to win today" for "Club A has to win today to avoid relegation." For a more detailed discussion of this point, cf. [16, §6.3]. In the same paper we also discuss some cases of non-completable unary necessity [16, §6.4].

[19]For a detailed discussion of this approach, cf. [14] and [16, §6].

With this approach, we abandon the dualism of necessity and possibility. In contrast to necessity, possibility is indeed usually used as a unary relation.

(3) φ is possible.

But there is a straightforward reading of (3) as φ *is independent*, i.e., neither φ nor $\neg\varphi$ is provable. For the natural language use of "possible" it is probably more accurate to replace "neither φ nor $\neg\varphi$ is provable" by "neither φ nor $\neg\varphi$ is *proven*". We suggest to call this form of possibility *epistemic possibility*.[20]

As a consequence of our analysis the *philosophical* status of "modalities" is changed. They are no longer modalities in the original sense of this word;[21] they belong to the same category as provability. And as we can study provability without recourse to a semantics, we can do so for necessity and possibility. But, of course, we can still consider semantics afterwards. And, at this point, possible worlds semantics can enter the stage again—but only as in a "supporting role"; in particular, it does not contribute to the *explanation* of necessity and possibility.

When we have an axiom system from which φ is independent, it allows obviously for different models—at least two: one in which φ holds, one in which $\neg\varphi$ holds. These models can be taken as worlds, and our reading of possibility coincides with the one in terms of possible worlds. In the case of necessity the situation is slightly different, since we consider mainly binary necessity. But the stipulated set of axioms from which the antecedent has to be taken, allows us to take—in a very controlled way—the different models corresponding to the alternatives in the set of axioms as possible worlds.[22]

We do not handle nested modalities, since we think that they are rather rare examples of the use of modalities, if not artificial or academic ones.[23] Instead, we like to look at one example of combined modalities, to exemplify how our approach works.

In a press release from March 23, 2004 I read that a European prime minister declared that "a compromise is not only necessary, but also possible". According to our reading the first part should be completed to "a compromise would be necessary *for something*" (what the concrete "something" is does not further matter).

---

[20] In this reading, actual (and necessary) truth are no longer considered as *possible* (in contrast to the usual formalization of possibility in modal logic). We argue for this on pragmatic grounds: If something is already proven as actually true, why one should still claim its possibility? For a detailed discussion of this approach to possibility, cf. [16, §5].

[21] They do not *modify* a statement. Today, modality is often used in a broader sense, under which our reading still could be subsumed; see, for instance, [23].

[22] For more technical details of this step, cf. [14, 16].

[23] This restriction relates to the nesting of necessity and possibility, but not to the nesting of these modalities with operators for knowledge and belief. While modal logic can, in principle, offer different operators for different modal notions, their interaction seems to be worked out rarely; as already remarked, it remains unclear how the different possible worlds semantics would intersect.

The second part expresses that "a compromise" is not yet excluded by the circum-stances. In this reading it is easy to imagine that the first part would be considered as more likely than the second, such that the use of "not only ... but also" is jus-tified. However, possible worlds semantics, which has normally the inclusion of possibility in necessity built in[24], can hardly contribute to an understanding of this sentence.[25]

Let us finish with an observation how our approach addresses concerns of SHAHID RAHMAN. With a reference to FIELD[26], he advocates a fresh look at a theory of counterpossibles [24, (our emphasis)]:

> These lines [footnote 26] actually express the central motivation for a theory of counterpossibles in formal sciences. Namely, the construc-tion of an alternative system where e.g. *the inter-dependence of some axioms of a given formal system could be studied.*

But, we already have such a system: The formal system itself! And, our analysis of necessity and possibility shows how we can explain these terms on the level of the formal system even better than with reference to some obscure semantics. Of course, you can still put your favorite semantics on top of our analysis—and it might be conceivable that, at the end, one needs always a semantics to have a satisfactory logical approach—but this semantics is *secondary* with respect to the logical relations expressed by necessity and possibility.

BIBLIOGRAPHY

[1] Nicholas Asher and Hans Kamp. Self-reference, attitudes, and paradox. In G. Chierchia et al., editor, *Properties, types and meaning*, volume 1, pages 85–158. Kluwer, 1989.
[2] Bradley, R. and Swartz, N. *Possible Worlds. An Introduction to Logic and Its Philosophy* Blackwell, 1979.
[3] Rudolf Carnap. Replies and systematic expositions. In Paul Arthur Schilpp, editor, *The Philosophy of Rudolf Carnap*, volume XI of *The Library of Living Philosophers*, pages 859–1016. Open Court, 1963.
[4] Brian F. Chellas. *Modal Logic*. Cambridge University Press, 1980.
[5] Charles S. Chiara. *The worlds of possibilty*. Oxford Unviersity Press, 1998.

---

[24]There exist possible worlds semantics in which this is not the case, i.e., where the accessibility relation is not serial. But it does not seem that they are strong competitors in the formalization of necessity.

[25]It would require, in fact, a combination of different modal notions. While polymodal logics may deal with this, a possible worlds semantics for it seems to be lacking; see also footnote 23.

[26]"It is doubtless true that nothing sensible can be said about how things would be different if there were no number 17; that is largely because the antecedent of this counterfactual gives us no hints as to what alternative mathematics is to be regarded as true in the counterfactual situation in question. If one changes the example to 'nothing sensible can be said about how things would be different if the axiom of choice were false', it seems wrong ...: if the axiom of choice were false, the cardinals wouldn't be linearly ordered, the Banach-Tarski theorem would fail and so forth" [6, p. 237].

[6] Hartry Field. Realism, mathematics and modality. In H. Field, editor, *Realism, mathematics and modality*, pages 227–281. Basil Blackwell, 1989.

[7] Dagfinn Føllesdal. Entry *Semantik*. In J. Speck, editor, *Handbuch wissenschaftstheoretischer Begriffe*, volume 3 (R–Z), pages 568–579. Vandenhoek & Ruprecht, 1980.

[8] Thomas Forster. The modal aether. In R. Kahle, editor, *Intensionality*, volume 22 of *Lecture Notes in Logic*, pages 1–19. ASL and AK Peters, 2005.

[9] Volker Halbach, Hannes Leitgeb, and Philip Welch. Possible worlds semantics for modal notions conceived as predicates. *Journal of Philosophical Logic*, 32:179–223, 2003.

[10] Volker Halbach, Hannes Leitgeb, and Philip Welch. Possible worlds semantics for predicates. In R. Kahle, editor, *Intensionality*, volume 22 of *Lecture Notes in Logic*, pages 20–41. ASL and AK Peters, 2005.

[11] Joseph Y. Halpern. On the adequacy of modal logic. *Electronic News Journal on Reasoning about Action and Change. Notes*, 3(3), July–August 1999.

[12] Joseph Y. Halpern. On the adequacy of modal logic, II: A response to McCarthy. *Electronic News Journal on Reasoning about Action and Change. Notes*, 4(1), May 2000.

[13] Wilfrid Hodges. Compositionality is not the problem. *Logic and Logical Philosophy*, 6:7–33, 1998.

[14] Reinhard Kahle. A proof-theoretic view of necessity. *Synthese*, 148(3):659–673, 2006.

[15] Reinhard Kahle. Konstruktivismus und Semantik. In J. Mittelstraß, editor, *Der Konstruktivismus im Ausgang der Philosophie von Wilhelm Kamlah und Paul Lorenzen*, pages 197–212. Mentis, 2008.

[16] Reinhard Kahle. Modalities without worlds. 200x. submitted.

[17] Markus Kracht. *Tools and Techniques in Modal Logic*, volume 142 of *Studies in Logic and the Foundations of Mathematics*. Elsevier, 1999.

[18] David Lewis. *On the Plurality of Worlds*. Blackwell, 1986.

[19] Edwin D. Mares. *Relevant Logic*. Cambridge University Press, 2004.

[20] John McCarthy. Modality Si! Modal Logic, No! *Studia Logica*, 59(1):29–32, 1997.

[21] John McCarthy. Modality for robots – responses to Halpern and Wansing. *Electronic News Journal on Reasoning about Action and Change. Notes*, 3(4), September 1999.

[22] Richard Montague. Syntactical treatments of modality, with corollaries on reflexion principles and finite axiomatizability. *Acta Philosophica Fennica*, 16:153–167, 1963.

[23] W. V. Quine. Three grades of modal involvement. In *Actes du XIème Congrés International de Philosophie*, volume XIV, Volume comlémentaire et communications du Colloque de Logique, pages 65–81. North-Holland and Éditions E. Nauwelaerts, 1953.

[24] Shahid Rahman. A non normal logic for a wonderful world and more. In J. van Benthem et al., editors, *The Age of Alternative Logics*, pages 311–334. Kluwer-Springer, 2006.

[25] R. Stalnaker. Modalities and possible worlds. In J. Kim and E. Sosa, editors, *A Companion to Metaphysics*, pages 333–337. Blackwell, 1995.

[26] Heinrich Wansing. A proof-theoretic proof of functional completeness for many modal and tense logics. In H. Wansing, editor, *Proof Theory of Modal Logic*, pages 123–136. Kluwer, 1996.

[27] Heinrich Wansing. Modality, Of Course! Modal Logic, Si! *Journal of Logic, Language, and Information*, 7:iii–vii, 1998. Editorial.

[28] Heinrich Wansing. The idea of a proof-theoretic semantics. *Studia Logica*, 64:3–20, 2000.

[29] Heinrich Wansing. Logical connectives for constructive modal logic. *Synthese*, 150:459–482, 2006.

[30] Kai Frederick Wehmeier. Modality, mood, and descriptions. In R. Kahle, editor, *Intensionality*, volume 22 of *Lecture Notes in Logic*, pages 187–216. ASL and AK Peters, 2005.

[31] Paul Weingartner. Entry *Extensionalitätsthese*. In J. Speck, editor, *Handbuch wissenschaftstheoretischer Begriffe*, volume 1 (A–F), pages 222–223. Vandenhoek & Ruprecht, 1980.

Predicaments of the Concluding Stage

ERIK C. W. KRABBE

ABSTRACT. Argumentative discussion is successful only if, at the concluding stage, both parties can agree about the result of their enterprise. If they can not, the whole discussion threatens to start all over again. Dialectical ruling should prevent this from happening. The paper investigates whether dialectical rules may enforce a decision one way or the other; either by recognizing some arguments as conclusive or some criticisms as devastating. At the end the pragma-dialectical model appears more successful than even its protagonists have claimed.[1]

1 Introduction

According to common wisdom, it is easier to get into an argument than to get out of it. And it is not even that easy to get an argument started; certainly not when one wants the argument to proceed along sound lines, as stipulated by the pragma-dialectical model of critical discussion [2, 3, 4]. As is commonly known, this model consists of four stages: the confrontation stage, the opening stage, the argumentation stage, and the concluding stage. In an earlier paper [9] I discussed some of the problems that beset the opening stage and make it hard to rationally start an argument. This time I want to concentrate on the concluding stage and the problems that threaten to frustrate the proper ending of an argumentative discussion.

The concluding stage is the one 'in which the parties establish what the result is of an attempt to resolve a difference of opinion' [4, p. 61]. Supposing the discussion to have been centered upon one initial standpoint (thesis), defended by a protagonist and challenged by an antagonist, the questions that need to be answered in this final stage of discussion are the following: Given what happened in the confrontation stage, the opening stage, and the argumentation stage, is the protagonist now obliged to retract his initial standpoint? Is the antagonist obliged to retract her calling into question of the initial standpoint? Or is neither party

[1] Another version of this paper will be published, together with a comment by David M. Godden and a response, in the proceedings of the 2007 OSSA-conference: H. V. Hansen, et al. (eds.), *Dissensus and the Search for Common Ground*, CD-ROM, Windsor, ON: OSSA, 2007. I wish to thank David Godden, as well as the anonymous referee of the present version, for their helpful suggestions.

obliged to retract its original position? If the first question can be answered in the affirmative, the difference of opinion has been resolved in favor of the antagonist; if the second, in favor of the protagonist; otherwise, no resolution has been achieved.

Section 2 will describe some predicaments connected with the concept of a concluding stage of argumentative discussion. Section 3 will investigate whether the idea of a conclusive argument or a devastating criticism can help us to make the concept more definite. A first impression is, that this will not be the case. Section 4 and 5 continue the investigation of conclusiveness in a pragma-dialectical direction, taking it to be a notion that is relative to the discussion at hand. In Section 6 we shall see that, ultimately, there is little about conclusiveness that is conclusive, but that nevertheless, within the model of critical discussion, but contrary to the modesty of the pragma-dialectician's claims, a concluding stage will always result in the resolution of the initial difference of opinion. Some conclusions of this paper are summarized in Section 7.

2   Some predicaments

There is a problem about getting to the concluding stage, a problem about what to do once being in the concluding stage, and also a problem of getting out of the concluding stage.

To start with getting to the concluding stage: if there is a predicament here, this can not literally be a predicament of the concluding stage, but should rather be characterized as a predicament of the argumentation stage, because it is from the argumentation stage that one must enter the concluding stage. When should this transition take place? When neither of the discussants has anything left to say? But that could take an indefinite time. Perhaps someone should, after a reasonable period, propose that the discussion be concluded. But who of the discussants can rightfully claim to be in the position to take the initiative and to say: 'Now we have had enough arguments and comments, let's conclude this session'? And can such an announcement be proclaimed, without any constraints, at any moment of the development of the argumentation stage?

In our salad days, when my brother and I had many arguments, there were no constraints. The concluding stage was announced by a forceful statement of one's opinion and completed by a loud utterance of 'Bang!'. Whoever first remembered that this was the way to end any heated dispute between us, could avail himself of this practical method. For winning the day, just state your final opinion on the issue, followed by 'Bang!'. Later on, we felt this was too easy and one had to utter a more complicated formula: not just 'Bang!', but 'Bang! Stop it! Period!', exactly in that order. This, of course, was much harder to remember in a heated dispute. I think it was my brother who at a certain occasion, after I had performed this little ceremony, and thought my proposition to be safe, continued the argument saying: 'Yes, but we must change this a bit.' This, of course, was intolerable. So, in the end, the formula was extended to: 'Bang! Stop it! Period! No alterations!'. It worked fine. For one thing, this rule did not spoil our arguments, for in the heat of discussion it usually lasted quite some time before anyone remembered about the way we had decided disputes could be ended.

Yet, one feels that in critical discussion one should observe some more sophisticated protocol for entering the concluding stage. The task for the theoretician, then, is to find a set of dialectic rules that prevent indefinite and senseless dilation of the argumentation stage, without giving either discussant the power to curb the other's fundamental rights to bring forward arguments or criticisms.

Second, once the discussants have entered the concluding stage, they are confronted with the problem of what to do next. The predicament here is that, in order to establish the yield of their discussion, they must either take the results of the argumentation stage for granted or make an assessment of the results of the argumentation stage. In the first case, the whole concluding stage would be nugatory, since everything has been established in the argumentation stage, whereas in the second case the attempt to assess the results of the argumentation will involve the use of arguments and hence catapult the discussants back into the same or another argumentation stage.

That it is often not acceptable that people at the concluding stage revert to earlier stages is nicely illustrated by the case of Mrs. Hans, the notorious antagonist figuring in one of the exercises of a pragma-dialectical textbook [5, pp. 33-36]. The exercise presents the case of a discussion at Harrods's department store about whether or not to join a program that would make ex-prisoners available as employees. Mrs. Hans is adamantly opposed to this idea: 'Well, in my view it is sheer madness to employ a bunch of prisoners, and that twenty per cent of my staff members are to be replaced by criminals.' But after some argument from the other side, she seems to be entering the concluding stage when she admits: 'Well, if that's the case, then I can't really say anything more against it.' However, she immediately returns to the confrontation stage, adding: 'But I still can't agree to it.' As the discussion moves towards a positive decision on this issue, she sticks to this attitude of reverting to the confrontation stage, yelling 'No criminals in my department!' and claiming to have insurmountable (but unexplained) objections.

Even if such extremes as those exemplified by Mrs. Hans can be avoided, it is easy to imagine that each concluding stage in which the upshot of an argumentation stage has to be summarized and evaluated, will amply occasion fresh differences of opinion, or revive old ones, leading to more confrontation stages, opening stages, and argumentation stages, that must be ended by concluding stages in which the problem recurs. So, even if all this going back and forth between stages can be satisfactorily regulated, this solution of the second predicament would yield a third predicament: that of how to conclude the concluding stage.

3   Conclusive arguments and devastating criticisms

The predicaments of the concluding stage would not be so threatening if only we had definite, decidable and practicable concepts of what constitutes a conclusive argument for a thesis and of what constitutes a devastating criticism of a thesis. A devastating criticism of a thesis could of course consist of a conclusive argument for the opposite thesis, but it may also amount to a conclusive argument that no conclusive argument for the thesis exists.

Once in possession of such concepts and related decision procedures, discussants could agree to use the following procedure for their concluding stage (assuming there to be just one initial thesis): (1) During the argumentation stage, each discussant may open a concluding stage, but will have to pay a fine if the concluding stage does not lead to a resolution of the difference of opinion. This fine is needed to prevent the discussants from needlessly interrupting the process of argumentation. (2) In the concluding stage the discussants establish the result of their discussion by first making an inventory of all arguments for the thesis that were, during the argumentation stage, presented by the protagonist. They then apply their decision procedure for conclusiveness to each argument. As soon as it has been found that one of the arguments was conclusive, they declare the protagonist to have won the discussion. (3) They also make an inventory of every criticism of the thesis that was, during the argumentation stage, put forward by the antagonist. They then apply their decision procedure for conclusiveness to each criticism. As soon as it has been found that one of the criticisms was conclusive, and therefore devastating, they declare the antagonist to have won the discussion. (4) If it turns out that, during the argumentation stage, neither a conclusive argument for the thesis nor a devastating criticism of the thesis was put forward, the discussion will be declared a draw.

There is one serious drawback to this procedure; that is that, according to Ralph Johnson at least, conclusive arguments do not exist [6, pp. 228-236]. According to Johnson, in order 'to be conclusive, an argument would have to display four properties' (p. 232):

> (C1) Its premises would have to be unimpeachable or uncriticizable. (p. 233)
>
> (C2) The connection between premises and the conclusion would have to be unimpeachable - the strongest possible. (p. 233)
>
> (C3) A conclusive argument is one that can successfully (and rationally) resist every attempt at legitimate criticism. (p. 233)
>
> (C4) The argument would be *regarded* as a conclusive argument. (p. 234)

Johnson argues that no argument 'has satisfied all these conditions' (p. 234).

But are not mathematical proofs the paradigm examples of conclusive arguments? Johnson holds that 'though mathematical proofs are conclusive, they are not arguments and so are not conclusive arguments' (p. 232). But here, to see whether proofs are arguments, we must distinguish between formal and informal proofs [8]. Formal proofs, being purely formal objects or syntactic structures, are indeed not by themselves arguments, though they can be used to express arguments, but then these arguments they express (through some process of interpretation) are themselves at most informal proofs. Informal proofs I hold to be arguments, but then, I must admit, these proofs are seldom if ever conclusive [10]. Even the (informal) proof showing that there is no greatest prime number, may be less conclusive than Johnson seems to assume [6, p. 232], since one could question the underlying logic or deny the possibility of multiplying arbitrarily large numbers. For Johnson, Euclid's proof is conclusive but not an argument, whereas I would hold that it is an argument, and a very strong one, but not in all respects conclusive. A formalization of Euclid's proof may be called conclusive for the system in which it is formalized, but would not be an argument. Neither Johnson nor I have found a conclusive argument in this case.

Again, are there any conclusive arguments? Of course, Johnson will not hold his argument that there are no conclusive arguments to be a conclusive argument. Neither do I. Nevertheless, Johnson's analysis of what it would mean for an argument to be conclusive convincingly shows that conclusive arguments will be extremely rare in argumentative practice and that, consequently, the notion of conclusiveness is otiose; it can certainly not serve as a foundation for the regimentation of the concluding stage. As long as conclusiveness remains a necessary condition for either the protagonist or the antagonist to be declared winner of the discussion (as is the case in the procedures mentioned above), hardly any interesting discussion will ever be won or lost. This means that discussions will generally fail to resolve differences of opinion.

4 Conclusive defense in critical discussion

Given that the notion of conclusiveness does not work, it may seem surprising that the notions of conclusive defense and conclusive attack nevertheless function prominently in the pragma-dialectic model of critical discussion ([4, Ch. 6], esp. Rules 9 and 14). But here those notions are not used in the absolute sense of defenses or attacks that would settle matters to all eternity, but in a sense relative to a particular discussion with particular discussion rules, adopted procedures, and agreed common starting points. The claim is that relative to all these matters (to be settled in the opening stage) defenses and attacks can be conclusive. What constitutes a conclusive defense is given by Rule 9a:

> The protagonist has conclusively defended an initial standpoint or sub-standpoint by means of a complex speech act of argumentation if he has successfully defended both the propositional content called into question by the antagonist and its force of justification or refutation called into question by the antagonist. [4, p. 151]

This rule should be understood bearing the following things in mind. Before the start of the argumentation stage, the initial standpoint[2] has been challenged (called into question) by the antagonist. It is now up to the protagonist to defend his initial standpoint by means of a complex speech act of argumentation, which counts as a provisional defense of the standpoint (Rule 6a, p. 144). There are two ways for the antagonist to react: she may call into question either the 'propositional content' of the argumentation (here to be called: its *premises*) or (in the case of a positive standpoint) its justificatory force (here to be called: its *link*).[3] Both links and premises will be called *inputs* of the arguments in which they figure.[4] Rule 9a is obviously not intended to declare argumentative defenses conclusive when the antagonist has not yet had the opportunity to call into question certain inputs. Rather it presupposes that the antagonist *had her say*, that is, that she had the opportunity to call into question each and every input she wishes (in the spirit of Rule 10, p. 152). The gist of Rule 9a (in the case of a positive standpoint) can now be formulated as follows: If and only if every input of a protagonist's argument that was called into question by the antagonist (where the antagonist had every opportunity to do so) has been successfully defended by the protagonist, has the protagonist conclusively defended his positive standpoint by means of the argument.

To understand what it means to conclusively defend a standpoint, we are thus referred to the notion of a successful defense of the inputs of arguments. For this we must turn to Rules 7a (p. 147) and 8a (p. 150). These rules refer to certain procedures, or tests, that the discussants are supposed to have agreed upon in the opening stage. The *intersubjective identification procedure* can be applied to premises and will check whether a premise is identical to one of the propositions that was, at the opening stage, accepted by both discussants (pp. 145-147).[5] The *intersubjective inference procedure* can be applied to links (in cases where the reasoning has been completely expressed) and tests for deductive validity (p. 148). The *intersubjective testing procedure* can also be applied to links, and checks whether the argumentation scheme that was employed is admissible (according to agreements at the opening stage) and whether it was applied correctly (pp. 149-

---

[2]Following pragma-dialectic theory, I shall distinguish between a (positive or negative) standpoint and the propositional content to which the standpoint refers.

[3]This, of course, corresponds to Arne Næss's dichotomy of tenability and relevance ([11, p. 108ff]; cf. [7]). I shall not discuss negative standpoints.

[4]So, premises are not arguments, but parts of arguments. Arguments are, generally, constellations consisting of premises, conclusions and links. An argument is brought forward by a complex speech act of argumentation.

[5]This procedure is supposed to have been extended (through agreements in the opening stage) by methods that allow for the introduction of fresh information, such as procedures for consulting authoritative sources or for observing the phenomenal world (pp. 146-147). Barth and Krabbe [1, p. 104] mention also computation as one kind of what they call 'material procedures'.

150).[6] When an input has been tested by some procedure that applies to it, and with positive result, it will be called a fixed input. Rules 7a and 8a stipulate that when an input has become fixed it counts as having been successfully defended by the protagonist. For links, this is the whole story, they can only be successfully defended by becoming fixed; that is, by passing either of the two tests for links. But for premises Rule 7a opens up another way: a premise will also count as having been successfully defended if it has been 'accepted by both parties as a result of a sub-discussion in which the protagonist has successfully defended a positive sub-standpoint with regard to this propositional content' (p. 147). This means that the argument may become complex, for premises may be defended by further arguments with links and premises, and these premises again by further arguments, and so on. An argument that is not complex will be called *elementary*.[7]

If we want to use Rule 7a as a definition of 'successfully defended', there is a slight difficulty with it in as far as a notion of successful defense occurs also in the definiens. Not that the definition is circular. It is not circular because in the definiens 'successfully defended' is applied to standpoints, whereas the definiendum is applied to premises. The problem is that we are not told what it means for a standpoint to have been successfully defended.[8] Nevertheless, it seems clear that a standpoint has been defended "successfully" if and only if the protagonist won the critical discussion in which it was defended, that is to say if and only if the antagonist was obliged to retract the calling into question of the standpoint, something that can be the case if and only if the standpoint was defended conclusively (Rule 14, p. 154). Thus, for 'successfully defended' in the definiens, we may read 'conclusively defended'. If this is correct, Rule 7a, in its turn, refers to Rule 9a. Yet, again there is no circularity. Rather Rules 7a, 8a, and 9a together constitute a recursive definition in which the two notions, 'conclusively defended' and 'successfully defended' are simultaneously defined.[9] This definition can, with respect to a possibly complex argument, be formulated as follows:

(1) (basic clause)
Any input[10] that has been fixed counts as having been *successfully defended* (Rules 7a and 8a).

---

[6]For simplicity, I do not describe the role of the *intersubjective explicitization procedure* as a separate component (p. 148-149).

[7]Thus an elementary argument consists just of premises, a link and a conclusion, whereas complex arguments may have intermediate conclusions (conclusions that are also premises). Premises that are not conclusions shall be called *basic premises* and the only conclusion that is not a premise shall be called the *final conclusion* of the argument.

[8]Elsewhere in the model of critical discussion 'successfully defended' always applies to premises (propositional contents) or to links (force of justification). Rule 7a seems to be an exception, and my proposal here is to get rid of this exception by reading 'conclusively' for 'successfully' in the definiens.

[9]Notice that 'successfully defended' is used only for inputs, and that 'conclusively defended' is used only for standpoints.

[10]That is, any input (premise or link) of any of the elementary arguments out of which the (possibly) complex argument consists.

(2) (first inductive clause)
Any positive standpoint[11] defended by an elementary argument all of whose inputs were either not called into question (the antagonist having had every opportunity to do so) or *successfully defended* counts as having been *conclusively defended* (Rule 9a).

(3) (second inductive clause)
Any premise[12] that is the propositional content of a positive standpoint that has been *conclusively defended* counts as *successfully defended* (Rule 7a).

(4) (extremal clause)
No positive standpoint shall count as *conclusively defended* and no input shall count as *successfully defended*, unless this follows from clauses (1) through (3).

The theory can be somewhat simplified if we permit (against Rule 6c, p. 144) a limiting case of critical discussion, where the initial (positive) standpoint is not defended by argument but by an application of the intersubjective identification procedure. That means that the propositional content of the initial standpoint (which content is here called the *final conclusion*, or simply the *conclusion*) might be fixed, and thus might be successfully defended. It is harmless to count the final conclusion as successfully defended also whenever the initial standpoint has been conclusively defended by argument. Let us call each input and also the final conclusion an *element* of the possibly complex argument.[13] In what follows, a defense of an propositional element (a premise or conclusion) is to be understood as a defense of a positive standpoint with regard to that element. It is now possible to unravel the duplex definition given above by first giving a separate recursive definition for 'successfully defended':

a) (basic clause)
Any element that has been fixed counts as having been *successfully defended*.

b) (inductive clause)
Any element defended by an elementary argument all of whose inputs were either not called into question (the antagonist having had every opportunity to do so) or *successfully defended* counts as having been *successfully defended*.

c) (extremal clause)
No element shall count as *successfully defended*, unless this follows from clauses (a) and (b).

In a second step, it may be stipulated that a positive standpoint counts as having been *conclusively defended* if and only if its propositional content counts as having been successfully defended.

The present notion of conclusive defense is very different from Johnson's notion of conclusive argument. None of the necessary conditions discussed by Johnson

---

[11]That is, the initial standpoint or any sub-standpoint.

[12]That is, any premise of any of the elementary arguments out of which the (possibly) complex argument consists.

[13]Thus, arguments consist of elements, each element being either a link or a propositional element, and each propositional element being either a basic premise or an intermediate conclusion or the final conclusion.

[6, p. 232-234], which he plausibly argued never to have been satisfied, applies to the present notion. The premises and the connection between premises and conclusion need not be unimpeachable, rather they must have been either fixed by agreed procedures or settled by further discussion. A conclusive argumentative defense in the present sense need not be immune for legitimate criticism; it is only the present antagonist who sees no further ways of calling inputs of the argument into question. Nor need a conclusive argumentative defense in the present sense be generally regarded as a conclusive argument in Johnson's sense; for its conclusiveness will remain restricted to a specific dialectical situation. The present notion is of course theoretical and idealized, but one can imagine something like it to be exemplified in argumentative practice. Perhaps this notion can support a feasible concluding stage.

5   Conclusive attack in critical discussion

For the most simple type of critical discussion, with only one initial standpoint, one protagonist, and one antagonist, the achievement closest to that of producing a devastating criticism would be that of carrying out a conclusive attack. This notion is defined in Rule 9b:

> The antagonist has conclusively attacked the [complex speech act of argumentation[14]] of the protagonist if he has successfully attacked either the propositional content or the force of justification or refutation of the complex speech act of argumentation. [4, p. 151]

The gist of Rule 9b (in the case of a positive standpoint) can be formulated as follows: If and only if at least one input of a protagonist's argument has been successfully attacked by the antagonist, has the antagonist conclusively attacked the argument presented by the protagonist. To see what it means to successfully attack some input we must turn to Rules 7b (p. 147) and 8b (p. 150). There we learn that, in order to count as successfully attacked, an input should upon attack (calling into question) by the antagonist have failed all tests that were applied to it[15] and, moreover, not have been successfully defended by the protagonist in a subdiscussion. Moreover, it seems reasonable to stipulate that the protagonist must have *had his say*, that is, that he must have had every opportunity to apply tests and to put forward an argumentative defense. But also the antagonist must have had her say about the arguments put forward by the protagonist to defend the attacked input. Otherwise, this input might not have been successfully defended by an argument merely because the antagonist lacked the opportunity to challenge some element (see clause (b) in the definition of 'successfully defended'). In this case it would be premature to say that the attacked input has been attacked conclusively (even if it has failed its tests).

---

[14]The phrase 'complex speech act of argumentation' here replaces the original 'standpoint', which seems inappropriate.

[15]In Rule 8b line 3 I read 'and' instead of 'or'. The rules presuppose that the tests were applied.

As in the preceding section, when discussing successful defense, we may expand also the notion of successful attack so as to apply not only to inputs but also to the conclusion. Further, we may count a (positive) standpoint as conclusively attacked if and only if its propositional content was successfully attacked.[16]

Now it may be shown that, assuming that both parties had their say, that is, that they had every opportunity to put forward their defenses and attacks (including applications of tests), each contested (attacked) element of an argument counts as successfully attacked if and only if it does not count as successfully defended. For, given this assumption, a contested element counts as successfully attacked if and only if (1) it failed all its tests, and (2) no argument for it was presented in which all the contested inputs were successfully defended (which is what successful defense in a sub-discussion amounts to). The conjunction of (1) an (2) is again equivalent to the element's not counting as successfully defended.

The upshot of this exercise is that, as long as it is assumed that both parties had their say, successful attack is not an independent notion but, in the domain of contested elements, just the complement of successful defense. Consequently, the conclusion, being contested, will either count as successfully defended or as successfully attacked and the (positive) initial standpoint will either count as conclusively defended or as conclusively attacked. Does this mean that, once the discussants enter the concluding stage, the resolution of their difference of opinion is guaranteed?

6   The inconclusiveness of conclusiveness

One of the predicaments of the concluding stage was whether to take the results of the argumentation stage for granted, or first to assess them, with the risk of starting the argument all over again. The pragma-dialecticians take the first option: in critical discussion the concluding stage is a rather modest affair. On the basis of what happened in the earlier stages, either the protagonist must retract his initial standpoint, or the antagonist must retract her calling into question of the initial standpoint, or no retractions need to be performed. According to Rule 14, the first speech act is obligatory if and only if the argumentation stage yielded a conclusive attack on the initial standpoint, whereas the second speech act is obligatory if and only if the argumentation stage yielded a conclusive defense of the initial standpoint [4, p. 154].

Only if one of these retractions is performed, does the critical discussion succeed in achieving a resolution of the difference of opinion. From the preceding section it is obvious that this will always be the case when each discussant did have his or her say, that is, if the protagonist had every opportunity to advance

---

[16]Which, because of the result in the next paragraph, can be seen to imply that whenever a standpoint has been conclusively attacked, each complex speech act of argumentation put forward as a defense of the standpoint has been conclusively attacked in the sense of Rule 9b.

arguments and to apply procedures, and the antagonist had every opportunity to call elements into question.

The problem then is how to determine when each discussant has had his or her say. An argument may seem conclusive, but then the antagonist may come up with new doubt and call into question an element that was previously thought to be uncontested. Similarly, a seemingly conclusive attack may be undercut when the protagonist suddenly sees a new possibility for argumentative defense. Thus there is not much conclusiveness about attacks and defenses being or not being conclusive as long as some party can still add some contribution. Since the protagonist can always try a new argumentative defense there is no such thing as an absolutely conclusive attack; the conclusiveness of an attack always depends on the protagonist's having had his say. On the other hand, there can be an absolutely conclusive defense, namely one in which all the links and all the basic premises (premises that are not argumentatively defended) are fixed. For in that case new attacks can nowhere be aimed. But even in that case the conclusiveness need not be everlasting, since the protagonist could still retract some part of the argument (Rule 12, p. 153).

Perhaps the only way, in critical discussion, to determine the moment for both discussants to admit to have had their say and to enter the concluding stage[17] is to let the discussants themselves make the decision. The protocol, which may be started at any moment during the argumentation stage would run as follows:

X: Let us go to the concluding stage!

Y: OK. (If Y refuses X will have to pay a small fine.)

For the concluding stage, it is then simply assumed (stipulated) that each discussant has had his or her say. Moreover, it may be stipulated that the discussants move to the concluding stage, and are supposed to have had their say, as soon as in two consecutive turns both discussants pass.

The effect of these rulings is that (contrary to what is suggested by Rule 14c, p. 154) whenever there is a concluding stage the difference of opinion will always be resolved.[18] This does not mean that critical discussion will always be successful in resolving the difference of opinion, for there is no guarantee that a concluding stage will ever be reached. Moreover, the resolution of a difference of opinion is not itself conclusive, since a discussion may be reopened: 'an argumentative dispute can in principle never be settled once and for all' (p. 138).

---

[17]For each discussant, admitting to have had one's say and expressing one's willingness to go to the concluding stage amount to the same.

[18]This follows from the result of the preceding section that whenever both parties have had their say (as they always have had in the concluding stage) the initial standpoint must either have been conclusively defended or conclusively attacked.

7 Conclusions

For the pragma-dialectical model of critical discussion, the first predicament, that of getting to the concluding stage, was not resolved in the sense that rules were found that could guarantee reaching the concluding stage. Yet, some rules were pointed out that might be helpful in this respect. If one discussant systematically refuses to go to the concluding stage, the other can say 'I pass' to force his interlocutor to either enter the concluding stage or make a further contribution to the argumentation stage. Thus, filibustering can be prevented.

The second predicament, that about what to do in the concluding stage, turns out to present a false dilemma, because, even though the results of the argumentation stage are not re-assessed, the concluding stage of a critical discussion is by no means nugatory. Rather it contains the important feat of ratification of the results and the concomitant retractions.[19]

Given this solution of the second predicament, the third predicament simply disappears.

To sum up, the pragma-dialectical notion of conclusiveness supports a feasible, but modest, concluding stage, but will, happily, not prevent discussions from being reopened and certainly not put issues beyond debate.

BIBLIOGRAPHY

[1] Barth, E.M. & Krabbe, E.C.W. (1982). *From Axiom to Dialogue: A Philosophical Study of Logics and Argumentation.* Berlin & New York: Walter de Gruyter.

[2] Eemeren, F.H. van & Grootendorst, R. (1984). *Speech Acts in Argumentative Discussions: A Theoretical Model for the Analysis of Discussions Directed Towards Solving Conflicts of Opinion.* Dordrecht & Cinnaminson, NJ: Foris.

[3] Eemeren, F.H. van & Grootendorst, R. (1992). *Argumentation, Communication, and Fallacies: A Pragma-Dialectical Perspective.* Hillsdale, NJ: Lawrence Erlbaum.

[4] Eemeren, F.H. van & Grootendorst, R. (2004). *A Systematic Theory of Argumentation: The Pragma-Dialectical Approach.* Cambridge: Cambridge University Press.

[5] Eemeren, F.H. van, Grootendorst, R. & Snoeck Henkemans, A.F. (2002). *Argumentation: Analysis, Evaluation, Presentation.* Mahwah, NJ: Lawrence Erlbaum.

[6] Johnson, R.H. (2000) *Manifest Rationality: A Pragmatic Theory of Argument.* Mahwah, NJ: Lawrence Erlbaum.

[7] Krabbe, E.C.W. (1987). Næss's dichotomy of tenability and relevance. In F.H. van Eemeren, et al. (Eds.), *Argumentation: Across the Lines of Discipline: Proceedings of the Conference on Argumentation 1986* (pp. 307-316). Dordrecht and Providence, RI: Foris.

[8] Krabbe, E.C.W. (1997). Arguments, proofs, and dialogues. In M. Astroh, D. Gerhardus & G. Heinzmann (Eds.), *Dialogisches Handeln: Eine Festschrift für Kuno Lorenz* (pp. 63-75). Heidelberg, Berlin, & Oxford: Spektrum Akademischer Verlag.

[9] Krabbe, E.C.W. (2006). On how to get beyond the opening stage. In. F.H. van Eemeren, et. al. (Eds.), *Proceedings of the Sixth Conference of the International Society for the Study of Argumentation,* (pp. 809-813). Amsterdam: Sic Sat, International Centre for the Study of Argumentation. Also in *Argumentation 21 (3),* 233-242.

---

[19]This was pointed out by the anonymous referee as well as by David Godden in his comment to the OSSA-version of this paper.

[10] Lakatos, I. (1976). *Proofs and Refutations: The Logic of Mathematical Discovery.* J. Worral & E.G. Zahar, (Eds.). Cambridge: Cambridge University Press.

[11] Næss, A. (1966). *Communication and Argument: Elements of Applied Semantics.* A. Hannay (Trans.). Oslo: Universitetsforlaget; London: George Allen & Unwin; Totowe, NJ: Bedminster Press. Translation of: *En del elementære logiske emner*, Oslo, 1947, etc.

Features of Indian Logic

KUNO LORENZ

ABSTRACT. Within an outline of the historical development of Indian Logic focusing on the history of discovering/inventing the concept of logical implication, two points are stressed:
1. The Grammarians and Mīmāṃsakas treat logic as allied with philosophy of language, whereas Naiyāyikas and the Buddhist logicians insist on the interconnectedness of logic and epistemology.
2. The Indian version of the dispute on universals (*apoha-vāda* of the Buddhist logicians versus *sphoṭa-vāda* of the Grammarians) is somewhat different from the Western one, yet without unpleasant effects on the treatment of inference.

Questions of logical reasoning occupy a place as important in Indian philosophy throughout its history as in the Western tradition. This became to be generally recognized in the West quite late, and it is still in danger of being treated condescendingly. It is possible to account for this attitude by paying attention to a number of facts. First of all, logic within the different Indian schools (*darśana*; literally: viewpoint) has never been cut off from its natural connection with philosophy of language or epistemology, and it appears rarely, therefore, as a subject that may be treated just formally. In addition, we face the problem that Indian treatises dealing with problems of logic use, in general, a highly developed technical language which, for a proper understanding, is heavily dependent on an uninterrupted written and oral tradition of commentative argument pro and con assertions in previous expositions. Yet, only rarely whole sequences of commentaries are available at present, not to mention the difficulty of identifying specific problems with their counterparts in the West in case such counterparts exist. Yet, the most influential factors responsible for the widespread lack of attention among logicians to questions of logic in Indian philosophy seem to be the historical circumstances under which Indian philosophy became known in the West. It happened at the turn of the 18th to the 19th century in connection with the discovery of the historical relation among Indo-European languages and the establishment of academic Sanskrit studies. It was a time when research into mythological thinking that was considered to be the starting point for speculative philosophy attracted far more attention than research into first stages of scientific conduct, let alone investigations into the logic of inquiry itself. Logic was treated as being essentially contained in the inherited schemes of syllogistics, hence as a sterile affair which is not worth of serious attention, neither by the working scientist nor by the speculative philosopher. Not before the rise of modern formal logic with G. Frege an adequate appreciation of earlier achievements in logic outside of syllogistics, e.g., by Stoicism or by Scholasticism, was gained. Indian logic suffered the same fate.

Attention to questions of logic is intimately bound to attempts of questioning spiritual authorities. In India we observe it as attention to verbal tools which have to be established in order to know how to conduct disputes over the Vedic tradition. This happened around -500 when Vedic authority became powerfully challenged by Buddhists, Jainists and the even more radical Lokāyatikas – a term usually denoting collectively various groups of materialists, skeptics, and fatalists –, and it occurred once more in the course of many centuries when in defence of the *Veda*, by attempts of an at least partial rational reconstruction of its doctrines, the orthodox *darśanas* develop.

The oldest known term both for the procedure of reasoning and for the theory of reasoning is '*ānvīkṣikī*' (cf. [9]). It appears around -300 in the famous treatise on economics, politics, and administration, the *Arthaśāstra* of Kauṭilya, and it is used there as the common feature of the >rational< schools of his time: Sāṃkhya, Yoga, and Lokāyata, separating them from three other theories, the theory of the three *Vedas* (*Trayī-vidyā*), of economics (*Vārttā-vidyā*), and of politics (*Daṇḍinīti-vidyā*). An even older term used by Buddha and Mahāvīra and their respective followers to characterize the method of dealing with philosophical questions, especially those called '*avyākṛta*' (= unanswerable) – a list of ten in the early Buddhist Sūtra collection Majjhimanikāya contains questions concerning, among others, the finiteness or infinity of the world, and the identity or diversity of soul and body –, is '*vibhajya*' (= splitting [into different cases]). A kind of linguistic analysis serves to transform a question which cannot be answered directly with 'yes' or 'no' into possibly several ones which can be answered or are merely pseudo-questions. In the following centuries this method has been turned into peculiar patterns of argumentation, conspicuously different in the Buddhist Mahāyāna school Mādhyamika and in the later Jainism, in order to find out whether a certain predicate holds of something or not. The antiessentialist Mādhyamikas who allow themselves only noncommittal negative judgments use a tetralemma, called '*catuḥkoṭi*': something is neither (1) so, nor (2) not so, nor (3) both so and not so, nor (4) neither so nor not so. It seems to be derived from a similar fivefold formula of the skeptic Sañjaya. The all-comprehensive Jainas, however, who take every judgment to be affirmable if suitably qualified, use a formula of seven predications, the *saptabhaṅgī*, to show that every object splits into its manners of being given: (1) in a certain sense something is (e.g., a pot is black with respect to state, it is earthen with respect to substance), and (2) in a certain sense something is not (e.g., a pot is not black with respect to substance), and (3) in a certain sense something is and [afterwards] is not, and (4) in a certain sense something is inexpressible [i.e., the predicate is not applicable], and (5) in a certain sense something is and [afterwards] is inexpressible, and (6) in a certain sense something is not and [afterwards] is inexpressible, and (7) in a certain sense something is and [afterwards] is not and [afterwards] is inexpressible. B.K. Matilal (cf. [15]) has convincingly argued that this pattern which has given the Jaina *darśana* the name '*syādvāda*' (= doctrine of it may be [thus or otherwise, depending on the point of view]) should be traced back to a formula with three members of the fatalist Gośāla, who, being a contemporary of Mahāvīra and Buddha like Sañjaya, is the founder of the Ājīvika sect, which says that everything is of triple character, e.g., something living is also non-living and, furthermore, both living and non-living.

Closer scrutiny into the relations of the Lokāyata with Jainism and Buddhism on the one hand (concerning *vibhajya*) and with Sāṃkhya and Yoga on the other hand (concerning *ānvīkṣikī*) reveals that at least two traditions of dealing with methods of rational inquiry should be distinguished. There is a primarily person-oriented *vāda-vidyā* (= doctrine of debate, also called '*tarka-śāstra*') in which rules of argumentation are studied, and a primarily matter-oriented *pramāṇa-śāstra* (= theory of knowledge, also called '*jñāna-vāda*') in which reasons for having a certain knowledge are studied.

Debates (*kathā*) have usually been classified threefold, easily to be mapped on the twofold division into dialectics and eristics in Antiquity: as proper debate or dispute (*vāda*) to find out on what is true, as contention (*jalpa*) to gain fame by winning, and as destructive argument (*vitāṇḍā*) to throw out the other party from the arena by any means without using a thesis of ones own. Hence, also sound argumentations of the Mādhyamikas who only go for refutations, which implies that they consider the law of classical negation duplex negatio affirmat to be invalid, are sometimes called *vitāṇḍā*; in fact, the Mādhyamikas use a kind of intuitionistic strong negation.

The means of knowledge (*pramāṇa*) on the other hand, have been variously classified, but perception (*pratyakṣa*) and inference (*anumāna*) as distinct ways of finding out on what there is are chosen as basic in almost every *darśana*. As early as around -150, in the great subcommentary on Pāṇini's Sanskrit grammar, in the Mahābhāṣya of Patañjali, the role of inference is described in terms which are reminiscent of explications in Stoicism: by inference knowledge of something not perceived or not perceivable is gained through knowledge of something perceived – the latter then being a sign of the former (cf. [21]).

Now, the *vāda-vidyā* becomes the central doctrine of one of the traditional six orthodox *darśanas*, the Nyāya, which developed around the turn of the millenium, and it occupies also an important place within the Buddhist Hīnayāna schools of the same time (famous paradigm cases are: the non-canonical *Milinda-pañhā* around -100 that contains a philosophical disputation between the Greek-Bactrian king Menandros and the Buddhist monk Nāgasena, and: the canonical *Katthāvatthu* around -250 which is the earliest extant treatise that contains discussions about rules of propositional logic as used in debates); both areas belong to the sources for the logical school of Buddhism as initiated by Dignāga (ca. 460-540).

In contrast, the discussion of inference in semiotic terms by the early Grammarians or Vaiyākaraṇas must have stood in close relation with similar conceptions in pre-classical Sāṃkhya, where in distinction to all the other *darśanas*, which treat perception as the first *pramāṇa*, inference occupies the first place among them. From later polemical discussions, especially by Dignāga, we know of an intra-sāṃkhyan transformation of an early doctrine about inference by the Sāṃkhya-teacher Vṛṣagaṇa (around 300) – the reconstruction of the transformation is due to Frauwallner [6] –, which only afterwards got amalgamated with the *vāda*-tradition in Nyāya, partly via its twin-*darśana* Vaiśeṣika. And it is this combination of argumentation theory with theory of knowledge (by inference) which, under the title of a theory of logic (*nyāya-vidyā*), Vātsyāyana (ca. 350-425) in his commentary on the *Nyāya-sūtras*, i.e., the *Nyāya-bhāṣya*, identifies with the *ānvīkṣikī* in Kauṭilya's *Arthaśāstra*. We, thus, get also an explanation of the use of '*pramāṇa-śāstra*' for the Nyāya-darśana as a whole. A reconstruction of the early doctrine of inference faces unsurmountable difficulties, because not even the oldest known Indian commentaries are in accord about the precise meaning of the three kinds of inferences which have come down to us under the titles 'with previous' (*pūrvavat*), 'with the rest' (*śeṣavat*), and 'seen with respect to the common' (*sāmānyato dṛṣṭam*). Among modern readings which use the literal meaning of '*pūrva*' and '*śeṣa*' in grammar, the most convincing one is the identification of a *pūrvavat*-inference with modus ponendo ponens (= if >the previous< holds [then >the following< holds], i.e., by affirming A in $A \rightarrow B$, affirming B; or: A , $A \rightarrow B < B$), of a *śeṣavat*-inference with modus tollendo ponens (= [if one member of an alternative which holds, does not hold] then >the rest< holds, i.e., by negating A in $A \vee B$, affirming B; or: $A \vee B$, $\neg A < B$), and of a *sāmānyato dṛṣṭam*-inference with an inference by analogy, where >the common< acts as the *tertium comparationis*, e.g., if geese are birds then ducks are birds, because both have feathers, provided for the *tertium comparationis* holds: 'having-feathers implies being-bird'. Vṛṣagaṇa transformed this ternary division of inferences into a binary one. He keeps 'seen with respect to the common', yet divides it into *pūrvavat*- and *śeṣavat*-inferences, and adds 'seen with respect to the specific' (*viśeṣato dṛṣṭam*), where the latter is nothing but an instantiation of the former, e.g., in the case of these feathers and this bird. At this stage further treatment of inferences along these lines as well as other more general inquiries into questions of logic took only place in Nyāya and in the logical school of Buddhism. It happened in such a way that argumentation theory and theory of knowledge became intimately interconnected. The once independent inquiries into logical reasoning in Sāṃkhya were, for all we know at present, given up.

This did not happen in the case of the discussions of questions of logic in the school of Grammarians as initiated around -400 by Pāṇini's Sanskrit grammar and which sometimes was treated even as a philosophical system: *Pāṇinīya darśana*. The Grammarians from Kātyāyana onwards, whose *Vārttika* on Pāṇini around -250 had been the subject matter of Patañjali's *Mahābhāṣya*, have treated problems of inference in connection with problems of linguistic representation, mainly of grammatic features, of course. Their work was developed in close interaction with ideas in another at that time gradually evolving orthodox *darśana*, the Mīmāṃsa, whose main concern has been the meticulous interpretation of vedic injunctions (*vidhi*). Semantic analysis together with a careful distinction of object language and metalanguage resulting in explicit metalinguistic rules of interpretation (*paribhāṣā*) for word composition and sentence composition both syntactically and semantically flourishes. Technical innovations abound: there is, e.g., a special particle, '*iti*', which, if added at the end of a text, turns it into a quotation; syntactic and semantic treatment of negation leads to explicit formulations of the rule of contraposition; even the importance of the relative position of negation signs and modal operators is clearly observed (an example relevant for Mīmāṃsakas from the *Mahābhāṣya* is: 'it is obligatory to eat [nothing but] five five-clawed animals' implies 'it is forbidden to eat the others',i.e., $\triangle!(E(x) \rightarrow F(x)) < \neg \triangledown!(E(x) \wedge F(x))$ (cf. [24]).

From the discussion of inference in comparison with perception as tools of knowledge in the *Mahābhāṣya* we learn that it is the semiotic frame of reference which leads to the treatment of implication under the title of connection (*sambhanda*): to make an inference from A to B valid there must exist a connection between the references of the associated terms A* and B*; then A* is called a sign (*liṅga*) for B* (A and A* are by definition related in the following way: A* is satisfied iff A is true). Though it took many centuries that the concept of implication – the term '*vyāpti*' (= pervasion) was eventually used for it, though not everywhere – became clarified, and this happened as a result of the critique launched against the logic of Nyāya by the Buddhist logicians, we already find in the *Mahābhāṣya* the use of the terms '*anvaya*' and '*vyatireka*' for A*<B* and its contraposition ¬B*<¬A*, as it is the case in Buddhist logic but not in the texts of the Naiyāyikas.

The examples show that the Grammarians use a partition of inferences into inferences *pratyakṣato dṛṣṭam* (= seen by referring to perception) and *sāmānyato dṛṣṭam* which obviously reoccurs in Buddhist logic as a partition into two kinds of modus ponens inference: the first one based on causal implications and the second one on conceptual implications. Dharmakīrti (ca. 600-660), the great follower of Dignāga, uses the terms '*kāryānumāna*' (= inference based on effect) and '*svābhāvānumāna*' (= inference based on essence) for these two kinds. The example of an inference *sāmānyato dṛṣṭam* as well as its instantiation *viśeṣato dṛṣṭam* given above may now be combined and rephrased as a *kāryānumāna*, which yields almost the standard form of an Indian syllogism as it is found in the *Nyāya-sūtras* 'here [at the place of the duck] are feathers' therefore 'here is a bird', because having-feathers implies being-bird, i.e., only birds as causes grow feathers as an effect such that feathers become signs for birds; places with geese are supporting examples [for the general implication]. Instead of 'feather' and 'bird' Patañjali uses, e.g., 'leaf' and 'tree', or grammatical features and their meaning.

If it is realized that the doctrines of the Grammarians and the Mīmāṃsakas are characterized by a systematic interrelation of logic and philosophy of language, whereas the doctrines of the Naiyāyikas and the Buddhist logicians are united by their insistence on the interconnectedness of logic and epistemology, it does not come as a surprise when we are confronted with a severe dispute between Mīmāṃsa and Nyāya carried on for almost the whole first millenium of the Christian era about the nature of the relation between word (*śabda*) and object (*artha*). The Mīmāṃsakas, on the one hand, who tried to use the philosophy of language as a means of setting up epistemology, argue for a $\varphi\acute{v}\sigma\epsilon\iota$-theory: since *śabda* – and they refer with this term primarily to the orally preserved *Veda* – is eternal (*nitya*), words refer to their objects by nature, i.e., by a kind of inherent energy (*śakti*). Hence, any verbally represented knowledge is *prima facie* true; if there is a claim to the contrary there should be given reasons for it. Due to their main concern with vedic injunctions their theory contains another interesting feature. It is a kind of reduction principle reminiscent of the Pragmatic Maxim as introduced by C.S. Peirce: every word meaning and every sentence meaning has to be related to actions, e.g., 'this is a rope' gets semantically reduced to 'with this rope it is possible to fasten a cow'. The Naiyāyikas, on the other hand, who tried to justify doctrines pertaining to the philosophy of language epistemologically, argue for a $\theta\acute{\epsilon}\sigma\epsilon\iota$-theory: since *śabda*, being a verbal representation of some knowledge which is essentially independently available by perception and by inference from perception, may appear as the result (*kārya*) of linguistic actions chosen in many different ways, words refer to their objects by convention. Hence, any verbally represented knowledge is a mere claim which has to be investigated as to whether it is true or false.

Both positions, independently of how they are supported by further conceptual constructions, belong to epistemological realism. Their difference may be stated as follows: the Mīmāṃsakas treat objects as universals – *ākṛti*, i.e., form, is what a term refers to –, and the Naiyāyikas treat objects as particulars which has the effect that what a term refers to partakes in general of universality (*sāmānya*) and of individuality (*vyakti*). In the first case language is treated under its type aspect –

the terms are schemata –, and in the second case under its token aspect – the terms are actualizations. Again it is no surprise that Mīmāṃsa is much concerned with norms which serve to ensure invariance of language use, whereas Nyāya pays first of all attention to the factual variability of language use. In fact, every means of knowledge (*pramāṇa*) holds a priori in Mīmāṃsa since its adherents rely on the unconditional validity of vedic revelation (*śruti*; literally: the heard) and, therefore, enforce any effort to preserve its exact wording; the means of knowledge hold a posteriori in Nyāya because there is unconditional validity only of vedic tradition (*smṛti*; literally: the remembered) which by its very existence proves that certain experiences can repeatedly be made.

But in both cases it is essential to be able to infer from what is perceived to what is not perceived or not perceivable. This can be done by means of a theory of inference which received its standard form of a five-membered syllogism (*pañcāvayava vākya*; literally: sentence having five parts) in Nyāya and became a point of reference also for the other *darśanas* when they either use or discuss logical reasoning. Due to the criticism by the Buddhist logicians Dignāga and Dharmakīrti the five-membered form has been restricted to cases of >inference for the sake of others< (*parārthānumāna*); in the case of an >inference for the sake of oneself< (*svārthānumāna*) only the first three or the last three members are necessary, and these members should furthermore be treated as merely conceived term complexes, i.e., as verbal signs in significative function [A*], which are not in need of being uttered by sentences, i.e., as verbal signs in communicative function [A], as in the five-membered case. The paradigm of a five-membered syllogism looks as follows:

assertion (*pratijñā*)	1. [this] mountain has fire
reason (*hetu*)	2. because the smoke
example and counterexample	3. like [it is] in the kitchen
(*udāharaṇa*)	unlike [it is] in the pond
application (*upanaya*)	4. and this [is] so (= this mountain has smoke)
conclusion (*nigamana*)	5. therefore [it is] so (= this mountain has fire)

The mountain is the perceived object of cognition (*pakṣa*; literally: wing). The term '*pakṣa*' originally refers to the place where a thesis in a dispute is put forward; later on '*pakṣa*' together with '*pratipakṣa*' are used for thesis and counterthesis. In the syllogism '*pakṣa*' refers just to the locus where the inference takes place, i.e., the locus [of this] mountain. The smoke is the perceived sign (*liṅga*) used as reason (*hetu*), which is indicated by the ablative case. The fire, finally, is the non-perceived consequence (*sādhya*), and, here again, the term '*sādhya*' originally refers to the whole proposition to be proved, alike '*pakṣa*' in its meaning as thesis.

Of course, the validity of the argumentation is dependent on the validity of the implication between reason and consequence , which originally had only been >perceived< paradigmatically, by example and counterexample. After the general character of implication – in the paradigm case given above: 'wherever [is] smoke there [is] fire' – had been clarified mainly by the criticism of Dignāga and Dharmakīrti, implication soon became the central topic in Indian logic. Several terms for the concept of implication were used, among them '*vyāpti*', which gained common usage through the Mīmāṃsa-teacher Kumārila (ca. 620-680). The *vyāpti* of smoke and fire, i.e., smoke < fire, has to be rendered literally as >smoke pervaded by fire<. Besides '*vyāpti*' also '*avinābhāva*' (= inseparable [connection]) was used for implication, especially by Jaina logicians. With the Nyāya-teacher Udayana (ca. 975-1050) the theory of *vyāpti* began to grow extensively and reached its peak in the *Tattvacintāmaṇi* of Gaṅgeśa (ca. 1300-1360). It was mainly due to the activity of Udayana that Nyāya became united with Vaiśeṣika, thus creating New-Nyāya (Navya Nyāya) or Nyāya-Vaiśeṣika.

In a table with nine entries called '*hetucakra*' (= wheel of reasons) which is found in his *Hetucakra-ḍamaru*, Dignāga was able to give a complete survey of the valid relations between reason and consequence by referring to the possible relations of the syllogistic kind '*a*' (all), '*e*' (no) and 'textiti' (some, but, as always in Indian logic, under exclusion of all), which may hold between either the examples *s*, being *loci* of *sādhya* >like *pakṣa*< (*sapakṣa*), or the counterexamples *v*, being *loci* of the complement of *sādhya* >unlike *pakṣa*< (*vipakṣa*), and the sign *h* (*hetu*):

	vih	*veh*	*vah*
sah	*ai*	<u>*ae*</u>	*aa*
seh	*ei*	*ee*	*ea*
sih	*ii*	<u>*ie*</u>	*ia*

He relies on his critical discussion of Vasubandhu's the Younger (ca. 400-480), his teacher's, *Vādavidhi* where in the context of a treatment of rules for disputations for the first time in Buddhist logic the five-membered syllogism got reduced to the first three members which were called '*sādhana*' (= proof). (Of course, there also occurred elaborate discussions of how to try refutations (*dūṣaṇa*) by finding mistakes in a proof; for this purpose the *vāda-vidyā* did provide techniques of using counterfactuals (*tarka*): if there were A there would be B, but B is not, therefore not A.) Of special importance was Dignāga's investigation into the rule of the >three characteristic features of a sign [as reason]< (*trairūpya liṅgatva*) which had found an unclear formulation in the *Vādavidhi* and is still handled unsatisfactorily by Dignāga's contemporary, the Vaiśeṣika-teacher Praśastapāda (ca.

500-550).

The >wheel of reasons< which supplies the first successful formal treatment of logical inference in Indian logic, is carried by intentions quite alike those of Aristotle in his syllogistics though of a set-up far more alike the one in Stoic logic than in syllogistics. Using the >wheel of reasons< Dignāga states that an argumentation which infers B (e.g., 'at the place of this mountain is fire') from A (e.g., 'at the place of this mountain is smoke') is valid if and only if the following >three characteristic features< are present: (1) the sign occurs in the object (*pakṣadharmatva*; literally: the being-modified of the object by the property [*dharma*]), (2) the sign occurs only where the consequence [*sādhya*] occurs, i.e., in loci alike the object, and (3) the sign is absent where the consequence is absent, i.e., in *loci* unlike the object.

Now, as *loci* alike the object are those of *sādhya*, i.e., those of >similar examples< (*sādharmya dṛṣṭānta* – the Buddhist logicians use '*dṛṣṭānta*' instead of '*udāharaṇa*'), and *loci* unlike the object those of the complement of *sādhya*, i.e., those of >dissimilar examples< (*vaidharmya dṛṣṭānta*), we may restate the three characteristic features in the paradigm case, using '*ιp*' to symbolize the deictic description 'this mountain', as follows: (1) *ιpεh* (= at [the place of] this mountain is smoke), (2) $h < s$ (*anvayī vyāpti*: wherever [is] smoke there [is] fire), and (3) $\neg s < \neg h$ (*vyatirekī vyāpti*: wherever [is] not fire there [is] not smoke).

It was the Buddhist logician Dharmottara (ca. 730-800) who first observed that (2) and (3) should be considered as logically equivalent, because any one of the two clauses expresses the *vyāpti* of smoke and fire. He disregarded Dignāga's insistence on the necessity of all the three features conjunctively. Dignāga's discussion of the insufficiency of just the conjunction of (1) and (3), is a clear indication that he regards the law of double negation : $\neg\neg A < A$ as not being generally valid. This is completely in tune with the primary position of negative concepts in Buddhism in general and with Dignāga's own nominalism in setting up a theory of meaning for general terms by >expulsion [from the complement]< (*apoha*) in particular (cf. [2]). The term '*go*' (= cow), e.g., is defined by the infinite conjunction: 'non-man $\wedge$ non-dog $\wedge$ non-lion $\wedge$ ...' in order to make sure that universal features have to be inferred and cannot be perceived like singulars which alone make up reality. Dharmakīrti intensifies the logical research along these lines in his *Hetubindu*, especially by stressing that the *vyāpti* between the references of two terms must not be understood as merely [general] togetherness (*sāhacarya*) but acts as a kind of rule (*niyama*). He distinguishes three kinds of *vyāpti*. Two of them are kinds of *anvayī vyāpti*, i.e., of $h < s$; the first – in the case when *hetu h* is that which is effected (*kārya*) [only] by *sādhya s* – is the one responsible for causal inference (*kāryānumāna*), the second – in the case when *hetu h* is nothing but the own-being (*svabhāva*) of *sādhya s* (*sādhya-svabhāvatva*), i.e., *s* is part of the concept of *h* – is the one responsible for conceptual inference (*svabhāvānumāna*). The last kind of implication – *vyatirekī vyāpti*: $\neg s < \neg h$ – is a connection of *hetu* and *sādhya ex negativo* which leads to an inference due to non-recognition (*anupalab-dhyānumāna*).

Without further consideration of the problems connected with an explicit incorporation of *apoha-vāda* into a discussion of the >wheel of reasons< one can summarize what it says in the following way: the argumentation for *ιpes* (= at [the place of] this mountain is fire) is valid if and only if both *ιpeh* and $h < s$ hold. Dignāga showed that exactly two of the nine conjunctions of the *hetucakra*, namely '*ae*', i.e., *sah* $\wedge$ *veh*, and '*ie*', i.e., *sih* $\wedge$ *veh*, (they are underlined in the table above) which are each logically equivalent to *has* (= $h < s$), make the inference from *ιpeh* to *ιpes* valid. It is, furthermore, obvious that attempts to read the *hetucakra* as a proof of the syllogistic mood *barbara* with *paksa* as minor term, *sādhya* as major term, and *hetu* as middle term (*pah* $\wedge$ *has* < *pas*) are misguided, because no sentence of the type 'at all mountains there is smoke' occurs in the Indian inference schema.

The Naiyāyikas who did not accept the move of the Buddhist logicians of separating strictly >inference for the sake of oneself< (*svārthānumāna*) from >inference for the sake of others< (*parārthānumāna*), i.e., of keeping the logical question of an inference being valid, because the implication between premiss and conclusion holds formally, apart from the psychological question of being convinced of this validity through a debate about this issue, continued to treat logic, especially questions of inference, in the context of both epistemological and psychological issues. As a famous example, in Gaṅgeśa's *Tattvacintāmani* the inference from *ιpeh* to *ιpes* is treated as a scheme that also includes the case of going from a (perceived) word to the (non-perceived) object it refers to (cf. [12]). Therefore, the knowledge of smoke (*H*: that at the place of this mountain is smoke) is considered to be an efficient cause (*nimitta kārana*, or *karana*) of the knowledge of fire (*S*: that at the place of this mountain is fire). In Navya Nyāya the corresponding relation between such two pieces of knowledge *H* and *S* is called '*vyāpti-jñāna*' (= knowledge of implication); with respect to the standard example it is: *vahni-vyāpyo dhūmaḥ* (= smoke being pervaded by fire). In fact, every piece of knowledge is treated as connected with the cognizing self (*ātman*) by inherence (*samavāya*), which is held to be the same relation as the one between substance (*dravya*) and quality (*guna*), and beOxfordtween a whole (*avayavin*) and one of its parts (*avayava*). The process of producing an effect from a cause is called an operation (*vyāpāra*) which, in the special case of the operating knowledge of implication, is signified by the term '*parāmarṣa*' (= conceiving); with respect to the standard example it is: *vyāpti-viśiṣṭa-paksa-dharmatā-jñāna* (= knowledge of the occurrence of that which is qualified by *vyāpti* [i.e., *vyāpti* of smoke and fire] at the place of the object [i.e., at the mountain]). The cause H together with the *parāmarṣa* yields the effect S, and this may be understood as applying the rule of universal instantiation to $\bigwedge_x(x\epsilon h \rightarrow x\epsilon s)$, thus arriving at $\iota p\epsilon h < \iota p\epsilon s$ whence, in view of H and S being the respective nominalizations of the elementary sentences $\iota p\epsilon h$ and $\iota p\epsilon s$, by applying the rule of modus ponens, we get S from H.

In the old Nyāya H alone was considered to be the cause of S. Hence, after the third member of the five-membered syllogism had been amended by explicitly stating the vyāpti 'wherever [is] smoke there [is] fire' the second member became interpreted as an abbreviation for 'because of the smoke which is pervaded by fire'. This remained the standard interpretation of the classical tradition up to the present where the textbook of logic 'Tarkasaṃgraha' of Annambhaṭṭa (around 1575) is still used in Sanskrit schools as a kind of logical propaedeutic.

Further studies in Navya Nyāya after Gaṅgeśa concerning vyāpti (at Mithilā/Bihār) became increasingly subtle and have led to the construction of a highly technical Sanskrit. It enabled the logician Raghunātha (ca. 1475-1550), who had the intellectual power to start a school of his own (at Navadvīpa/Bengal), to develop, among many other things, a higher order logic of relations which is sufficient to handle logical composition including quantification in all detail. The assertion, e.g., of the adjunction 'the place at which there is fire is a place at which there is water or the place at which there is fire is a place at which there is a mountain' is expressed – as shown by Ingalls (cf. [12], p. 63) – by asserting a certain property of fire: fire ϵ {[where there is] water or at [the place of] a mountain}. Many details still wait for reconstruction by means of modern formal logic.

For the development of Nyāya the work of Dignāga and his followers has been of decisive influence, not only with respect to logic proper concerning the concept of implication (vyāpti) but also with respect to epistemology: From the time of Dignāga onwards Nyāya had to defend its realism against the radical nominalism of Dignāga and his school. Of similar importance for the development of Mīmāṃsa was the elder contemporary of Dignāga, the Grammarian Bhartṛhari (ca. 450-510). He is the founder of a particular branch of the orthodox darśana Advaita-Vedānta which is called Śabdādvaita (= non-duality with respect to word) because it is characterized by the claim that language alone, if taken in its generic aspect, is real. Bhartṛhari draws a radical conclusion from the Mīmāṃsa-doctrine that primitive terms refer to universals: every term irrespective of its level of composition both vertically and horizontally articulates universals. It has the result that one branch of Mīmāṃsa (the Prābhākaras) approaches Buddhism while the other branch (the Bhāṭṭas, i.e., the followers of Kumārila) is driven towards Nyāya.

In Śabdādvaita uttering a sentence does not represent a particular as such and such but is a means – in fact the only one – of making something universal accessible. Reality is linguistic in nature and can be cognized through a particular means of knowledge (pramāṇa), which is called by Bhartṛhari 'pratibhā' (= intuition). By intuition the schematic or generic character of terms – the sphoṭa (= bursting) – on various levels, e.g., on the phonetic or on the semantic one (varṇa-sphoṭa [=

phonem] or *pada-sphoṭa* [= morphem]), is recognized all of a sudden, and only that deserves to be called knowledge of what there is. Perception which is always predicatively determined perception (*savikalpaka pratyakṣa*) – and in this respect Bhartṛhari disagrees with Dignāga who considers pure perception to be predicatively undetermined (*nirvikalpaka pratyakṣa*) – refers to particulars only, and its instantiations being action-tokens on a par with the elements of concrete speech are themselves nothing but particulars; hence, by inference again only particulars can be reached. Neither traditional pramāṇa is fit to make the universal accessible, one needs *pratibhā* for it.

There is agreement between Dignāga and Bhartṛhari in as much as they treat language (*śabda*) and conceptual construction (*kalpanā*) as two sides of the same coin. They differ when calling this coin either '*śūnyatā*' (= emptiness; Dignāga) or '*brahman*' (= '*śabda*' or 'word' in the sense of λόγος'), because for Bhartṛhari, only universals are real, for Dignāga, only singulars, i.e., the actualizations of actions of pure perception. Luckily, the logic of dealing with universals and particulars remains essentially unaffected by this opposition.

BIBLIOGRAPHY

[1] S.S. Barlingay. *A Modern Introduction to Indian Logic.* New Delhi, 1976 (first appeared in 1965).

[2] V. van Bijlert. Apohavāda in Buddhist logic. In M. Dascal/D. Gerhardus/K. Lorenz/G. Meggle (eds.). *Sprachphilosophie. Philosophy of Language. La philosophie du langage. Ein internationales Handbuch zeitgenössischer Forschung 1*, Berlin/New York (1992), 600-609.

[3] R.S.Y. Chi. *Buddhist Formal Logic 1 (A Study of Dignāga's Hetucakra and K'uei-chi's Great Commentary on the Nyāya-praveśa).* London, 1969.

[4] A. Foucher. *Le Compendium des topiques (Tarka Saṃgraha) d'Annambhaṭṭa avec des extraits de trois commentaires indiens (texte et traduction) et un commentaire par A. Foucher.* Paris, 1949.

[5] E. Frauwallner. Vasubandhu's Vādavidhiḥ. In *Wiener Zeitschrift für die Kunde Süd und Ostasiens u. Archiv für indische Philosophie 1* (1957), 104-146.

[6] E. Frauwallner. Die Erkenntnislehre des klassischen Sāṃkhya-Systems. In *Wiener Zeitschrift für die Kunde Süd und Ostasiens u. Archiv für indische Philosophie 2* (1958), 84-139.

[7] E. Frauwallner. Landmarks in the History of Indian Logic. In *Wiener Zeitschrift für die Kunde Süd und Ostasiens u. Archiv für indische Philosophie 5* (1961), 125-148.

[8] D.C. Guha. *Navya Nyāya System of Logic. Some Basic Theories and Techniques.* Varanasi, 1968 (first appeared in Delhi, 1979).

[9] P. Hacker. Ānvīkṣikī. In *Wiener Z. Kunde Süd und Ostasiens u. Archiv ind. Philosophie 2* (1958), 54-83.

[10] R.P. Hayes. Diṅnāga's Views on Reasoning (*Svārthānumāna*). In *Journal of Indian Philosophy 8* (1980), 219-277.

[11] R. Herzberger. *Bhartṛhari and the Buddhists. An Essay in the Development of Fifth and Sixth Century Indian Thought*, Dordrecht, 1986.

[12] D.H.H. Ingalls. *Materials for the Study of Navya-Nyāya Logic.* Cambridge Mass., 1951.

[13] K. Lorenz. *Indische Denker.* München, 1998.

[14] B.K. Matilal. *Epistemology, Logic, and Grammar in Indian Philosophical Analysis.* The Hague/Paris, 1971.

[15] B.K. Matilal. *The Central Philosophy of Jainism (Anekānta-Vāda).* Ahmedabad, 1981.

[16] B.K. Matilal/R.D. Evans (eds.). *Buddhist Logic and Epistemology. Studies in the Buddhist Analysis of Inference and Language.* Dordrecht, 1986.

[17] G. Oberhammer. Ein Beitrag zu den Vāda-Traditionen Indiens. In *Wiener Z. Kunde Süd und Ostasiens u. Archiv ind. Philosophie 7* (1963), 63-103.

[18] T. Patnaik. *Śabda. A Study of Bhartṛhari's Philosophy of Language.* New Delhi, 1994.

[19] K.H. Potter (ed.). *Encyclopedia of Indian Philosophies 2: Indian Metaphysics and Epistemology. The Tradition of Nyāya-Vaiśeṣika up to Gaṅgeśa.* Princeton N.J., 1977.

[20] D.S. Ruegg. The uses of the four positions of the *catuḥkoṭi* and the problem of the description of reality in Mahāyāna Buddhism. In *Journal of Indian Philosophy 5* (1977), 1-72.

[21] H. Scharfe. *Die Logik im Mahābhāṣya.* Berlin, 1961.

[22] E.A. Solomon. *Indian Dialectics: Methods of Philosophical Discussions 1 & 2.* Ahmedabad, 1976/1978.

[23] J.F. Staal. Negation and the Law of Contradiction in Indian Thought. A Comparative Study. In *Bull. School Orient. African Studies 25* (1962), 52-71.

[24] J.F. Staal. Sanskrit Philosophy of Language. In T.A. Sebeok (ed.). *Current Trends in Linguistics 5* (1969) 499-533.

[25] E. Steinkellner. Wirklichkeit und Begriff bei Dharmakīrti. In *Wiener Z. Kunde Südasiens u. Archiv ind. Philosophie 15* (1971), 179-211.

Envelopes and Indifference

GRAHAM PRIEST AND GREG RESTALL

1 The problem

Consider this situation: Here are two envelopes. You have one of them. Each envelope contains some quantity of money, which can be of any positive real magnitude. One contains twice the amount of money that the other contains, but you do not know which one. You can keep the money in your envelope, whose numerical value you do not know at this stage, or you can exchange envelopes and have the money in the other. You wish to maximise your money. What should you do?[1]

Here are three forms of reasoning about this situation, which we shall call FORMS 1, 2 and 3, respectively.

FORM 1 Let n be the minimum of the quantities in the two envelopes. Then there are two possibilities, which we may depict as follows:

	POSSIBILITY 1	POSSIBILITY 2
YOUR ENVELOPE	n	$2n$
OTHER ENVELOPE	$2n$	n

By the principle of indifference, the probability of each possibility is $\frac{1}{2}$. The expected value of keeping your envelope is

$$\frac{1}{2} \times 2n + \frac{1}{2} \times n = \frac{3}{2}n$$

The expected value of switching is

$$\frac{1}{2} \times n + \frac{1}{2} \times 2n = \frac{3}{2}n$$

Conclusion: SWITCHING IS A MATTER OF INDIFFERENCE.[2]

---

[1] We bracket, here, considerations of diminishing returns (you wish to truly *maximise* your monetary return), and the discrete nature of currency. Say, for the purposes of the discussion, the quantity is a cheque made out to you for some positive real quantity, which your bank will deposit into your account.

[2] Some would baulk at calling the quantities computed expectations, since they themselves contain a (random) variable. This is simply a matter of nomenclature. However, we will consider the status of the variables in these expressions in Section 4, below.

FORM 2 Let x be the amount of money in your envelope. Then there are two possibilities, which we may depict as follows:

	POSSIBILITY 1	POSSIBILITY 2
YOUR ENVELOPE	x	x
OTHER ENVELOPE	$2x$	$x/2$

By the principle of indifference, the probability of each possibility is $\frac{1}{2}$. The expected value of keeping your envelope is

$$\frac{1}{2} \times x + \frac{1}{2} \times x = x$$

The expected value of switching is

$$\frac{1}{2} \times 2x + \frac{1}{2} \times \frac{x}{2} = \frac{5}{4}x$$

Conclusion: SWITCH.

FORM 3 Let y be the amount of money in the other envelope. Then there are two possibilities, which we may depict as follows:

	POSSIBILITY 1	POSSIBILITY 2
YOUR ENVELOPE	$2y$	$y/2$
OTHER ENVELOPE	y	y

By the principle of indifference, the probability of each possibility is $\frac{1}{2}$. The expected value of switching is

$$\frac{1}{2} \times y + \frac{1}{2} \times y = y$$

The expected value of keeping is

$$\frac{1}{2} \times 2y + \frac{1}{2} \times \frac{y}{2} = \frac{5}{4}y$$

Conclusion: KEEP.

Prima facie, FORMS 1, 2 and 3 seem equally good as pieces of reasoning. Yet it seems clear that they cannot all be right. What should we say?

2 The solution

In fact, all three answers give you the right solution, in three different circumstances. The relevant reasoning determining what you ought to do to maximise your outcome is *under-determined* by the original description of the situation. The correct way to reason, in the sense of maximising your return given the possibilities — which, after all, is the aim of each kind of reasoning — depends on the process by which the money ends up in the envelopes. For each form of reasoning there are mechanisms such that, if *that* mechanism was employed, the reasoning delivers the correct answer. Here are three examples:

MECHANISM 1 A number, n, is chosen in any way one likes. One of the two envelopes is chosen by the toss of a fair coin, and n is put in that envelope; $2n$ is put in the other.

MECHANISM 2 A number, x, is chosen in any way one likes. That is put in your envelope. Either $2x$ or $x/2$ is then put in the other envelope, depending on the toss of a fair coin.

MECHANISM 3 A number, y, is chosen in any way one likes. This is put in the other envelope. Either $2y$ or $y/2$ is then put in your envelope depending on the toss of a fair coin.

That the three different forms of reasoning are correct for each of the corresponding mechanisms is obvious once one has seen the three possibilities. For example, for Mechanism 1, let us suppose that the number n is chosen with a probability measure P. Since we have not specified any range from which the amount is chosen, the most we can say is that the probability measure defines, for a range of measurable sets S of quantities, the probability $P(n \in S)$ — the probability that the number n is in the set S. So, $P(n \in [0, 1])$ is the probability that the number n is between zero and one, inclusive).[3] Then, since the quantities in the two envelopes are n and $2n$, decided at the toss of a fair coin, the probability that the quantity in *my* envelope is in a set S is $\frac{1}{2}P(n \in S) + \frac{1}{2}P(2n \in S)$, since there are two ways the content of my envelope could be in the set S. One way (with probability $\frac{1}{2}$) is that the amount in my envelope *is* n, and the probability that n is in S is $P(n \in S)$. The other way (also with probability $\frac{1}{2}$) is that the quantity in my envelope is $2n$, and the probability that this is in S is $P(2n \in S) = P(n \in S/2)$, where $S/2$ is the set of all members of S divided by 2. The probability that the quantity in *your* envelope is in set S is $\frac{1}{2}P(2n \in S) + \frac{1}{2}P(n \in S)$, which is the same quantity, so indifference in this is warranted, as the probabilities are identical.

---

[3]We need this delicacy when considering the probabilities, since we cannot in every case define the probability measure on the choice of n by considering the probabilities for the atomic events of each particular choice of n. For example, if the quantity n is chosen uniformly over $[0, 1]$, then $P(n = r)$ is *zero* for each $r \in [0, 1]$. Yet the measure is non-trivial: for example, $P(n \in [0, \frac{1}{3}]) = \frac{1}{3}$.

On the other hand, given Mechanism 2, if the number x (the quantity in *your* envelope) is chosen with measure P' (so $P'(x \in S)$ is the probability that x is in the given set S), then the probability that the quantity in your envelope is in S is simply $P'(x \in S)$, whereas the probability that the quantity in my envelope is in that same set is $\frac{1}{2}P'(2x \in S) + \frac{1}{2}P'(\frac{x}{2} \in S)$, which may diverge significantly from $P'(x \in S)$. If P' is the uniform distribution on $[0, 1]$, than $P'(x \in [0, 1]) = 1$. On the other hand, $\frac{1}{2}P'(2x \in [0, 1]) + \frac{1}{2}P'(\frac{x}{2} \in [0, 1]) = \frac{1}{2} \times \frac{1}{2} + \frac{1}{2} \times 1 = \frac{3}{4}$. The reasoning in the case of Mechanism 3 is similar.

We have specified the state spaces sufficiently to determine enough of the probability measure on each space, in such a way that the probabalistic reasoning is valid. However, if the reader has any doubt about this, intuitions about the scenarios can checked by a series of trials. For example, a sequence of trials is generated employing Mechanism 1. Whether you adopt a policy of keeping or switching or doing either at random, makes no difference in the long run. Similarly, a sequence of trials is employing Mechanism 2: Adopting the policy of switching comes out 5/4 ahead of the policy of keeping in the long term (and changing at random comes out 9/8 ahead). The case is similar for Mechanism 3.

The two envelope paradox is well known.[4] It comes in different versions. The paradigm version is produced by giving reasoning of Form 2 in a context where Mechanism 1 is deployed. This, of course, gives the wrong results. We have just solved this paradox.

3   Bertrand's Paradox

There is a paradox concerning the principle of indifference usually called *Bertrand's Paradox*.[5] This can be put in many well-known ways. Here is one. A train leaves at noon to travel a distance of 300km. It travels at a constant speed of between 100km/h and 300km/h. What is the probability that it arrives before 2pm? We may reason in the following two ways.

1. If the train arrives before 2pm, its velocity must be greater than or equal to 150km/h. Given the range of possible velocities, by the principle of indifference, the probability of this is $\frac{3}{4}$. Hence, the probability is $\frac{3}{4}$.

2. The train must arrive between 1pm and 3pm. 2pm is half way between these two. By the principle of indifference, the train is as likely to arrive before as after. So the probability is $\frac{1}{2}$.

---

[4]The paradox first appeared in the philosophical literature in Cargyle (1992). It continues to generate a substantial literature. See, e.g., Jackson, Menzies and Oppy (1994), Broome (1995), Clark and Shackel (2000), Horgan (2000), Chase (2002). It should be noted that much of the literature appeals to the fact that money is discrete with a minimum or maximum. As is shown by the way that we have set things up, this does not get to the heart of the problem.

[5]See, e.g., Kneale (1952).

The two applications of the principle of indifference seem equally correct, but they result in inconsistent probabilities.

A standard solution to the paradox is to point out that the correct application of the principle of indifference depends upon the mechanism by which the velocity of the train is determined. If, for example, the velocity is determined by setting it to a number between 100 and 300, chosen at random, then reasoning 1 is correct. Suppose, on the other hand, that the velocity is chosen as follows. Choose a number of minutes, n, between 0 and 120, at random. Set the speed of the train to be $300/(1 + n/60)$ (distance/time). Then reasoning 2 is correct. In case it is not *a priori* clear that these are the right ways to reason in the contexts, the matter can be demonstrated by a sequence of appropriately designed trials.

As should now be clear, the paradigm two envelope paradox can be seen as a version of Bertrand's paradox. Perhaps what has prevented it from being seen as such is simply the fact that only one of the ways of applying the principle of indifference is standardly given.

4   Decision theory and designation

So far so good. But where, exactly—it may fairly be asked—does the reasoning in the paradigm two envelope paradox go wrong? The answer is that it depends. It depends on how we conceptualise the designators employed. The reasoning proceeds as follows:

> Let x be the amount of money in your envelope; then there are two possibilities, $\langle x, 2x \rangle$ or $\langle x, x/2 \rangle$ ...

Ask whether 'x' is a rigid designator or simply a definite description. Suppose, first, that it is a rigid designator—say it denotes \$10. Then in the second possibility, that which arises when the envelopes are switched, the amount of money in your envelope is precisely not x, that is, \$10. (It is either \$5 or \$20.) The values involved in the computations of the various expectations (particularly, the second summands) are therefore incorrect.

Suppose, instead that 'x' as it occurs in our reasoning is a definite description, such as 'the amount of money in your envelope', and not a rigid designator. Then it certainly refers to the amount of money in your envelope in possibility 2. But now it refers to a different quantity than it referred to in possibility 1. To go on and compare the values of the expectations computed in this way is therefore a nonsense. This would be like reasoning as follows. The number of sons of the king (in some context) is 4; the number of daughters of the king (in some other context) is 3; hence the king has more sons than daughters. As is clear, both of the kings in question may have more daughters than sons.

Interpreted in one way, then, the reasoning is unsound; interpreted in the other way it is invalid. Similar considerations hold if one applies any of the methods for computing expectations together with a non-corresponding mechanism. To see how the mismatch occurs, consider a circumstance well-suited to the second form of reasoning: the mechanism we have called "mechanism 2". In *this* case, the term 'x' may rigidly refer to the amount of money in your envelope, or it may be a definite description abbreviating 'the number in your envelope'. In either case, the reasoning of form 2 is appropriate because the two possibilities countenanced in that form of reasoning ($\langle x, 2x \rangle$ and $\langle x, x/2 \rangle$) match up precisely with the different outcomes of Mechanism 2, *given that interpretation of the term* 'x'. This cannot be said of Mechanism 1 or Mechanism 3.

The moral of the story is simple: in a computation of expectation, the designation of a variable must refer to the right quantity, and that reference must not vary as the reasoning encompasses different possibilities.

5 Opening the envelope

There are, of course, other versions of the two envelope paradox. In another, one opens the envelope before deciding whether to exchange. Thus, let us suppose again that the money is distributed via Mechanism 1. But suppose that this time we open the envelope and find, say, $10. The expectation of keeping is therefore $10. The contents of the other envelope are either $5 or $20. So the expectation of switching is $12.50. So one should switch. This seems equally paradoxical: the precise amount of money in the envelope seems to provide no significant new information. What is to be said about this? (Note that, in this case, the computation of expectation does not employ variables at all, just numerals—rigid designators.)

Note, for a start, that, relative to certain items of background information, new information provided by opening the envelope can make it rational to switch. Thus, suppose that you know that the minimum amount, n, is an odd integer. Then if you open the envelope and find, e.g., $5, you should change; whereas if you find $10, you should not. If you find $5, then all the possibilities other than $\langle 5, 2.5 \rangle$ and $\langle 5, 10 \rangle$ have zero probability. And if you know that n is an odd integer, the first of these also has zero probability. If one computes the expectation of the two outcomes using this information, the expectation of keeping the envelope is 5, and that of switching is 10. So one should switch.

In general, if one knows a prior probability distribution for the value of n, or at least enough about it, then, by employing Bayes' Law

$$P(h/e) = \frac{P(e/h) \times P(e)}{P(h)}$$

one can compute a posterior probability distribution, given the evidence provided by opening the envelope. The posterior probability distribution thus generated provides the basis for the maximum-expectation computation.

However, if one has no such information, then there is no way one can compute posterior probabilities, and so use these in a computation of expectation. Thus, suppose one knows nothing more than that Mechanism 1 was deployed. One does not know the prior probability distribution, only the following constraint on it: for any n, the probabilities of $\langle n, 2n \rangle$ and $\langle 2n, n \rangle$ are identical (and all other pairs have zero probability). This is sufficient information to compute prior expectations as a function of the value n. Now if one opens the envelope and discovers, say, \$10, the only two possibilities left with non-zero probability are $\langle 10, 5 \rangle$ and $\langle 10, 20 \rangle$. But since one has no information about the prior probabilities of these two possibilities, one cannot compute their posterior probabilities. In particular, one cannot argue that, since there are two possibilities left, each has probability $\frac{1}{2}$. Thus, for example, if the prior probability distribution was such that $\langle 10, 5 \rangle$ and $\langle 5, 10 \rangle$ each had probability $1/2$, whilst everything else had probability 0 (which is consistent with our information), then the posterior probabilities of these two options are 1 and 0, respectively. On the other hand, if it was such that $\langle 10, 20 \rangle$ and $\langle 20, 10 \rangle$ each had probability $1/2$, whilst everything else had probability 0, then the posterior probabilities of these two options are 0 and 1, respectively. To claim that the relevant posterior probabilities are a half each is, therefore, fallacious.

6   Probabilistic ignorance

In the situation we have just considered, we have insufficient information to compute the relevant expectations. The same situation arises, even before opening the envelope, if we have no information concerning the mechanism for distributing the money between envelopes. In this case, the information given so far in the characterisation of the scenario seems to give us no good reason to switch. But how can one justify this view? The answer depends on how one conceptualises rational choice in such situations.

One possibility is to suppose that probability considerations are still relevant to choice. The probabilities in question cannot be objective, of course; not enough about such probabilities is known. So they must be subjective. Now, the prior probability distributions in question are over an infinite space. Since there is no uniform distribution over all the possibilities, there is no one distribution that recommends itself. There are infinitely many equally good distributions consistent with our knowledge. We may nullify any argument to the effect that one should switch or keep based on a probability distribution by pointing out that there are equally good distributions that recommend the opposite. We have already seen this in the case in which we open the envelope. In the case where we do not know the mechanism, all we know about the prior probability distribution is that all the possibilities with non-zero probability are of the form $\langle q_1, q_2 \rangle$, where $q_1 = 2q_2$, or $2q_1 = q_2$. We may therefore nullify any argument to the effect that we ought to keep, based on a certain probability distribution, P, by pointing to its dual, P', obtained by swapping the values of q_1 and q_2. This will recommend the opposite choice. There is therefore nothing to break the symmetry, and so to give ground for anything other than indifference.[6]

The other possibility, one might suppose, is to abandon appeal to probability altogether, in favour of some other principle of decision making. But in this case, there also seems to be nothing to break symmetry in the epistemic situation at hand. Moreover, if any argument for switching or keeping is given, this can again be neutralised by pointing out that there are circumstances (mechanisms, distributions), in which the argument will give us the wrong answer. Again, anything except indifference has no rational ground.

In either case, then, as each case has been sketched, you have no reason to be anything other than indifferent: arguments attempting to justify some difference between switching and keeping get no grip. Does it follow that you *should* be indifferent? Of course not: we have not ruled out any of the infinity of other considerations that might incline you to one envelope rather than another. (One envelope is red and the other is blue and you have promised to accept no blue gifts today. You like the shape of one of the envelopes and it would be a good addition to your collection. And so on.) As the case has been described, we have not ruled out *all* of the possible normative considerations that might apply in the circumstance before you. But when it comes to evaluating those, the probabilities we have discussed will not help. They give you no insight into what to do other than to be indifferent between switching and keeping.[7]

BIBLIOGRAPHY

[1] Broome, J. (1995), 'The Two Envelope Paradox', *Analysis* 5: 6–11.
[2] Cargyle, J. (1992), 'On a Problem about Probability and Decision' *Analysis* 52: 211–216.
[3] Chase, J. (2002), 'The Non-Probabilistic Two Envelope Paradox', *Analysis* 62: 157–160.
[4] Clarke, M. and Shackel, N. (2000), 'The Two Envelope Paradox', *Mind* 109: 415–442.
[5] Horgan, T. (2000), 'The Two Envelope Paradox, Nonstandard Expected Utility, and the Intensionality of Probability', *Noûs* 34: 578–603.
[6] Jackson, F., Menzies, P. and Oppy, G. (1994), 'The Two Envelope "Paradox"', *Analysis* 54: 43-45.
[7] Kneale, W. (1952), *Probability and Induction*, Oxford: Oxford University Press.

---

[6]One might, of course, be a *complete* subjectivist about the matter. Whatever probability distribution does, in fact, reflect your degrees of belief, go with that. If one is *such* a subjectivist, there is nothing more to say.

[7]Thanks go to Bryson Brown and Christine Parker for helpful discussions before we wrote this paper. Versions of the paper itself were given at the Australian National University and the University of Melbourne. Thanks for helpful comments are due to many members of the audiences, including Geoff Brennan, Christian Link, Wlodek Rabinowicz, Laura Schroeter, and especially Bruce Langtry.

Harmony and Modality

STEPHEN READ

ABSTRACT. It is argued that the meaning of the modal connectives must be given inferentially, by the rules for the assertion of formulae containing them, and not semantically by reference to possible worlds. Further, harmony confers transparency on the inferentialist account of meaning, when the introduction-rule specifies both necessary and sufficient conditions for assertion, and the elimination-rule does no more than exhibit the consequences of the meaning so conferred. Hence, harmony is not to be identified with normalization, since the standard modal natural deduction rules, though normalizable, are not in this sense harmonious. Harmonious rules for modality have lately been formulated, using deductive systems.

Shahid Rahman argues in favour of logical pluralism, which I oppose.[1] But what he really favours is an eclecticism of logics which enables different logics to be formulated and compared. That is a project I endorse. It is important to be able to identify the commonality between logics, so that one can discern where they disagree. Only then can one apply appropriate criteria to decide which logic is right. That means that we must distinguish the meanings of the logical connectives from the logical principles which they satisfy. If, as the inferentialist believes, those meanings are given by the rules of inference that may look difficult, if not impossible. The solution lies in distinguishing the rules specific to specific connectives, the operational inferences, from the generic rules, the framework within which the operators work.

1 Inferentialism

Inferentialism about the logical connectives is the claim that their meanings are given by the rules for their use. Robert Brandom places it in opposition to representationalism, the claim that "inferential relations [derive] from the contents of representings" ([5], p. 46). Arthur Prior [15] gave a famous argument to show that inferentialism must be mistaken. Rather, he said, we must first give the meanings of the connectives independently of their use in inference, for only then can we decide whether the rules are valid or not. We cannot just give rules for the

---

[1]See, e.g., [16], §4; [20].

connectives, trusting to faith that they are correct and hope that they will define a meaning.

J.T. Stevenson [23] interpreted Prior as demanding a truth-functional, or more generally, value-functional, explanation of the meaning of each connective. This is clearly unreasonable, as there are many non-truth-functional connectives, forming complex propositions whose truth-value is not determined by any finite set of "truth"-values of their constituents. What one might seek is a possible-worlds semantics for the connectives. Indeed, Richard Routley [21] showed that every logic has a two-valued worlds semantics, so justifying the search for a semantic articulation of the connectives in terms of an indexed map from the proposition's parts to the whole. For each n-place connective, a Routley frame consists of a set of worlds and an $n + 1$-place "accessibility" or "relative possibility" relation, and the connective receives either a universal or an existential truth-condition: for example, for the case of the 1-place necessity ($\square$) and possibility ($\lozenge$) connectives, the familiar conditions:

- $v(\square\alpha, w) = \text{T}$ if $v(\alpha, w') = \text{T}$ for all w' such that Rww';

- $v(\lozenge\alpha, w) = \text{T}$ if $v(\alpha, w') = \text{T}$ for some w' such that Rww'.

Unless one is a modal realist, however, one might worry whether such a world-relative account of the connectives is really independent of their use in inference. The valid rules are all and only those which hold in the canonical model. Consider the canonical model of a modal logic, **S4**, say. The "worlds" of the model are simply prime filters in the Lindenbaum algebra of **S4**. But the Lindenbaum algebra is obtained by dividing the set of formulae into equivalence classes of provably equivalent formulae. And what determines whether formulae are provably equivalent in **S4** are the inference rules of **S4**. So the notion of canonical model is not independent of the specification of the inference rules. Any model will justify a corresponding set of rules.

If one believes in the real existence of possible worlds, one can rebut this scepticism: the canonical model is correct if it describes the possible worlds correctly. If not, the rules to which it is correlated must be wrong. But if there is no independent reality of possible worlds, the canonical model simply is the mathematical model cast in the attractive language of "worlds", corresponding to the particular choice of inference rules. There is no assurance of its correctness independent of the correctness of those inference rules, any more than there was with the Lindenbaum algebra itself.

So too for the ultra-reductionist strategy which identifies worlds with maximally consistent sets of sentences. The notion of consistency depends on the logical inferences which are permitted, and so again cannot constitute the grounds or justification of those rules.

Why might one be sceptical about the reality of possible-worlds talk? Take a sentence like 'Caesar killed Brutus'. The modal realist tells us that there is a possible world, indeed, an infinite class of possible worlds in which Caesar killed Brutus. Each of these worlds differs from the actual world, for in the actual world Brutus killed Caesar and not vice versa. Nonetheless, says the realist, those worlds are like our world in consisting of a maximal collection of states of affairs, the members of the infinite class in question having in common that they each contain the state of affairs of Caesar's killing Brutus. Indeed, even such a reductionist as an advocate of the Combinatorial theory of possibility (see, e.g., [1]) will attribute to each possible world in this class a "recombination" of the constituents of the actual state of affairs in which Brutus killed Caesar constituting the state of affairs of Caesar's killing Brutus - turning the tables on his assassin.

But how is such a recombination produced? Caesar did not kill Brutus, so one has to suppose that, having disassembled the fact that Brutus killed Caesar into its components, Brutus, killing and Caesar, we can somehow recombine them into a different state of affairs. We can certainly think them together. But that only gives us the thought that Caesar killed Brutus, not the corresponding state of affairs. Moreover, we do not want to attribute too much reality to the putative state of affairs of Caesar's killing Brutus, or we might find it made the corresponding proposition, 'Caesar killed Brutus', true, rather than simply possible.

Armstrong ([2], p. 118) replies that states of affairs combine their elements by a non-mereological form of composition. But that is just a name for a problem. How is the actual state of affairs of Brutus' killing Caesar composed? What combines its elements, Brutus, killing and Caesar into the state of affairs of Brutus' killing Caesar? Armstrong ([1], p. 42; [2], p. 30) fears what he calls Bradley's regress argument: it cannot be the addition of a further element, such as instantiation, for we already have Brutus, killing, Caesar and instantiation (since Brutus killed Caesar), but we don't have Caesar killing Brutus. All Armstrong can do is invoke the magic of a non-mereological form of composition, the result of a transcendental argument based on the fact that Brutus killed Caesar. Once conceded, it is open to the modal realist or the combinatorialist to invoke it to combine the elements in a non-actual order, and behold: out of the hat comes the non-actual state of affairs of Caesar's killing Brutus.

But the problem was misconceived before the transcendental argument was invoked. For the answer to the composition question is that killing itself unites Brutus and Caesar into the fact. As a matter simply of fact, killing related Brutus and Caesar as subject and object. Relations are relations only in as much as they relate their terms. As Bradley put it rhetorically in a challenge to Russell in *Mind* in 1911:

"What is the difference between a relation which relates in fact and one which does not relate? ... Is there anything ... in a unity besides its 'constituents', *i.e.* the terms and the relation, and, if there is anything more, in what does this 'more' consist?" ([4], p. 74; and cf. [19])

Certainly, killing might have related Caesar and Brutus as subject and object, and that possibility stands in need of explanation. But it cannot be explained by reference to a real, existing but non-actual state of affairs of Caesar's killing Brutus, for Caesar did not kill Brutus and killing did not relate Caesar to Brutus as subject and object. The modal realist claims that states of affairs are far more numerous than we ordinarily take them to be, claiming that there are states of affairs of Caesar's killing Brutus, of donkeys talking, of Othello's loving Desdemona, of Desdemona's loving Cassio and so on. As evidence, all he can offer is their necessity in explaining modal facts. We might as well explain the roundness of round squares by pointing to their obvious roundness, or Desdemona's loving Othello by pointing to Desdemona. There are no round squares, and Desdemona never existed. Such shadow objects give no explanation and simply confuse with their duplicity and duplication. Fiction, and modality, cannot be explained by simply supposing, or asserting, that what does not exist, exists after all, but in some special, non-actual, way. There is no state of affairs of Caesar's killing Brutus, not even a shadowy, non-actual one, for killing did not relate those Romans as killer and killed. Nor is there a state of affairs of Desdemona's loving Othello - we need some other account of how 'Desdemona loved Othello' seems to be true and 'Desdemona loves Cassio' to be false.

David Lewis' modal realism does not fall to this objection directly, since he does not claim that there is a non-actual state of affairs of Caesar's killing Brutus in order to explain the possibility that Caesar might have killed Brutus. Rather, he claims that there is a state of affairs consisting of Caesar's counterpart killing Brutus' counterpart, non-actual because counterparts are counterfactual. Nonetheless, the explanation still fails, for there is no reason to suppose Caesar and Brutus have such shadowy counterparts, other than the perceived need, on the theory, to postulate them to explain modal facts.

This is not to deny that the trope of possible worlds is useful, as we will see in §4. But modal realism, and combinatorialism, are mistaken theories, and the possible worlds semantics of modal logic is no more than a useful technical device and psychological aid to further our understanding of modality. To show that an inference preserves truth-at-a-world in a suitable class of frames for a modal logic does not show that the inference is sound. Rather, it shows that that class of frames fits that mode of inference.[2] We don't have the independent means of giv-

---

[2]If truth-preservation in a frame were enough, Geach's challenge to the relevance logician to exhibit a model in which $\alpha \vee \beta$ and $\neg\alpha$ were true and β false would suffice to rebut them. But whether Disjunctive syllogism is sound is a much more difficult matter than that - see, e.g., Read [17], ch. 7.

ing meaning to the connectives which Prior demanded. The meaning of the logical connectives, at least in the case of modality, must be given in some other way.

2 Harmony

The meaning of the connectives cannot in general be given a representationalist account, as the argument in §1 shows. So we might seek to give their meaning by the inference rules for their use. However, not just any set of rules suffices to give them meaning, as Prior's example ([15]) of 'tonk' famously revealed. 'tonk' seems to have part of the meaning of 'or', since 'α tonk β' can be inferred from α, while also appearing to have part of the meaning of 'and', since β can be inferred from 'α tonk β'. But worse, since 'α tonk β' can be inferred from α but not from β, 'α tonk β' is in fact equivalent to α, while since β but not α can be inferred from 'α tonk β', 'α tonk β' is equivalent to β. That is, the inference theory of 'tonk' presupposes that there is a proposition which is equivalent both to α and to β, for arbitrary α and β. It is no surprise, therefore, that if inference is transitive, the theory of 'tonk' permits the derivation of any proposition from any other, the absurdity which Prior exposes.

Prior's conclusion is that we need independent evidence that there is a proposition 'α tonk β' with the meaning requisite for the inference rules proposed for it to be sound. But there is a more insightful conclusion present in the observation that 'α tonk β' seems on the one hand, *via* its introduction-rule, to be equivalent to α and on the other, *via* its elimination-rule, to be equivalent to β. It is contained in that clause, 'and not from β': 'α tonk β' can be inferred, in general, from α and not, in general, from β.[3]

In contrast, 'α or β' can be inferred from α or from β. The introduction-rule not only shows what is (severally) sufficient for assertion of the conclusion ('α tonk β' or 'α or β', respectively) but also shows what is (jointly) necessary: not only may 'α or β' be inferred from α or from β, it can be inferred only from one or the other; not only may 'α tonk β' be inferred from α, it may be inferred only from α - not from β.

Of course, 'α or β' (or 'α tonk β' for that matter) can be inferred by Modus Ponens, or Simplification, or any number of other inference rules from more complex formulae not specific to 'or' (or 'tonk'). 'α or β' can also be the conclusion of generic, or structural, rules. What the inference rules for a specific connective make explicit are the logical features of that connective. Hence, what is implicit in the totality of cases of the introduction-rule for a connective is that they exhaust the grounds for assertion of that specific conclusion. Brandom ([5], p. 63) misses this crucial point when he equates the introduction-rule with the set of sufficient conditions for assertion and the elimination-rule with the set of necessary consequences

---

[3]Of course, in special cases, 'α tonk β' can be inferred from β - when $\alpha = \beta$, for example. But inference rules are formal and general.

of that assertion. Each already contains both necessary and sufficient conditions, and it is when these do not agree that problems like those of 'tonk' arise.

The elimination-rule works in the same way. What it says is that from 'α tonk β', β but only β, not α may be inferred; from 'α and β', either α or β may be inferred, but only α or β. The elimination-rule shows not only what is (severally) necessary for the assertion of the premise ('α tonk β', 'α and β', respectively) but also what is (jointly) sufficient: not only may β be inferred from 'α tonk β', but only β may be, not α; not only may either α or β be inferred from 'α and β', but they are both inferable from it. Hence, the theory of 'and' is coherent, for introduction- and elimination-rules agree that both α and β are required for an assertion of 'α and β', as well as sufficient; while the theory of 'tonk' is not coherent, for the introduction-rule says that α is sufficient for assertion of 'α tonk β', while the elimination-rule denies this, and moreover, the introduction-rule says that β is not sufficient for assertion of 'α tonk β', while the elimination-rule says it is.

This requirement of coherence in the rules, in particular, appreciating that what the introduction- and elimination-rules say should be both sufficient and necessary is what seems to have eluded Popper and the other proponents in the 1950s of what Prior dubs the "analytical validity" view of the connectives. But it was not a point which eluded Gentzen, writing some twenty years earlier. His brief comment on the relation between the introduction- and elimination-rules is as acute as it is famous:

> "The introductions represent, as it were, the 'definitions' of the symbols concerned, and the eliminations are no more, in the final analysis, than the consequence of these definitions." ([11], p. 80)

The introduction-rules, by being all and only the rules for asserting a complex formula, already contain all that is needed for formulation of the elimination-rules. If the elimination-rules allow one to infer more than is justified by the introduction-rules, then the meaning conferred by the rules is split between introduction- and elimination-rules, and one possible result is incoherence and inconsistency, as in the case of 'tonk'. If the elimination-rules allow one to infer less than is justified by the introduction-rules, then again, the meaning conferred by the rules is opaquely spread between them: whatever meaning might be promised by the introduction-rule is then restricted and taken back by the elimination-rule. We will consider an example of the latter situation later.

The situation recommended by Gentzen has come to be known as "harmony" (see e.g., [9], p. 455). However, different articulations have been proposed of what this means - normalization, or conservative extension, for example.[4] Actually,

---

[4] See, e.g., [10], pp. 215-20, 246-51.

we should stick closer to Gentzen's original formulation: the introduction- and elimination-rules are in harmony when the elimination-rules do no more than spell out what may be inferred from the assertion of the conclusion of the introduction-rules, given the grounds for its assertion.[5]

That is, we may infer from an assertion all and only what follows from the various grounds for that assertion. Specifically, if the grounds for assertion of δ are Π_i, where $\{\Pi_i : i \in I\}$ is a collection of subproofs, a collection of derivations, degeneratively of formulae, warranting assertion of δ, then the harmonious form of the elimination-rule is

$$\frac{\delta \quad \left\{ \genfrac{}{}{1pt}{0}{(\Pi_i)}{\gamma} \right\}}{\gamma}$$

that is, given an assertion of δ, and (multiple - there may be several cases of the introduction-rule, as in $\vee$I) derivation(s) of γ from the canonical ground(s) for asserting δ, we may infer γ and discharge the assumption of those grounds. Taking Prior's introduction-rule for 'tonk', for example:

$$\frac{\alpha}{\alpha \text{ tonk } \beta} \text{ tonk-I}$$

we obtain the general case of tonk-E, assuming tonk-I to exhaust the grounds for asserting 'α tonk β':

$$(\alpha)$$
$$\vdots$$
$$\frac{\alpha \text{ tonk } \beta \quad \gamma}{\gamma} \text{ tonk-E}$$

In this case, we can simplify this by permuting the derivation of γ from α with the application of the rule, to give the simpler

$$\frac{\alpha \text{ tonk } \beta}{\alpha} \text{ tonk-E}$$
$$\vdots$$
$$\gamma$$

This is, of course, not Prior's tonk-E rule, and does not lead to inconsistency and triviality as did his rule.

---

[5]This account of harmony conflates what, following Dummett, I called harmony and autonomy in [18] §2.1. A connective is "autonomous" if its meaning is (wholly) given by the I-rule; it is "harmonious" if the E-rule does no more than spell out the consequences of the meaning so conferred.

A similar simplification is possible in the case of 'and' ($\wedge$). Given that the sole ground for asserting '$\alpha \wedge \beta$' is derivations of both α and β:

$$\frac{\alpha \quad \beta}{\alpha \wedge \beta} \ \wedge\mathrm{I}$$

it follows that one may infer from '$\alpha \wedge \beta$' whatever one may infer from the grounds for its assertion, that is, both α and β:

$$\underbrace{(\alpha, \beta)}$$

$$\vdots$$

$$\frac{\alpha \wedge \beta \quad \gamma}{\gamma} \ \wedge\mathrm{E}$$

Again, the derivation of γ from α and β may be permuted with the application of $\wedge$E to obtain the more common form of the $\wedge$E-rule:

$$\underbrace{\frac{\alpha \wedge \beta}{\alpha} \quad \frac{\alpha \wedge \beta}{\beta}}$$

$$\vdots$$

$$\gamma$$

But this permutation is not always possible, as the case of $\vee$E shows.[6]

What is important about harmony is that it locates all the meaning-conferring power in the introduction-rule (or alternatively, in the elimination-rule). That does not prevent one introducing inconsistent connectives with harmonious rules. For example, suppose one introduces a zero-place connective $\bullet$ with the rule:

$$(\bullet)$$

$$\vdots$$

$$\frac{\perp}{\bullet} \ \bullet\mathrm{I}$$

That is, from a derivation of absurdity ($\perp$) from $\bullet$, one can infer $\bullet$.[7] Consequently, the elimination-rule will read:

---

[6] Actually, even here permutation is possible, given sufficient constraints on the interpretation of the notion of proof, as Kneale [12] shows for $\vee$ and Copi [6] for $\exists$.

[7] $\bullet$I is not, in Dummett's terminology ([10], p. 257), "pure", in containing another connective, $\perp$. However, as Dummett argues, this is permissible provided it is not circular. First, introduce $\perp$. Then $\neg$ and $\bullet$ can be introduced subsequently, using $\perp$.

$$(\bullet \Rightarrow \bot)$$
$$\vdots$$
$$\frac{\bullet \qquad \gamma}{\gamma} \bullet E$$

Here, $\alpha \Rightarrow \beta$ abbreviates a supposed derivation of β from α. E.g., one might rewrite $\bullet$I as

$$\frac{\bullet \Rightarrow \bot}{\bullet} \bullet I$$

So $\bullet$E says that one may infer from $\bullet$ whatever (*viz* γ) one may infer from supposing one can infer $\bot$ from $\bullet$, which is what $\bullet$I says justifies assertion of $\bullet$. We can see immediately from the I-rule that $\bullet$ is self-contradictory - according to the I-rule, one may assert $\bullet$ if and only if one may infer $\bot$ from it. $\bullet$ is a proof-conditional Liar sentence. However, normalization fails. Replacing

$$\frac{\dfrac{\Pi_1}{\bullet \Rightarrow \bot} \bullet I \quad \dfrac{(\bullet \Rightarrow \bot)}{\Pi_2}}{\gamma} \bullet E \qquad \text{by} \qquad \dfrac{\dfrac{\Pi_1}{\bullet \Rightarrow \bot}}{\dfrac{\Pi_2}{\gamma}}$$

does not eliminate maximum formulae of the form $\bullet$ (those which are both conclusion of an introduction- or $\bot$-rule and at the same time major premise of an elimination-rule) since new maximal formulae of the same degree may occur in the new proof. We can again simplify $\bullet$E by permuting the derivation of γ with the application of the rule to obtain:

$$\frac{\bullet}{\bot} \bullet E$$
$$\vdots$$
$$\gamma$$

Taking familiar, and harmonious, rules for negation:

$$\frac{\alpha \Rightarrow \bot}{\neg \alpha} \neg I \qquad \text{and} \qquad \frac{\neg \alpha \quad \alpha}{\bot} \neg E$$

we can show that $\bullet$ equivalent to its own negation:[8]

---

[8] Substructuralist beancounters who are worried by the double discharge of the assumption in the second proof will realise on reflection that $\bullet$E strictly requires two copies of its premise.

$$\frac{\dfrac{\dfrac{\overset{1}{\bullet}}{\bot}\ \bullet E}{\neg\bullet}\ \neg I(1)}{\bullet \to \neg\bullet}\ \to I(1) \qquad \frac{\dfrac{\dfrac{\overset{2}{\neg\bullet}\quad\overset{3}{\bullet}}{\bot}\ \neg E}{\bullet}\ \bullet I(3)}{\neg\bullet \to \bullet}\ \to I(2)}{\bullet \leftrightarrow \neg\bullet}\ \leftrightarrow I$$

Harmony cannot prevent inconsistency. But it helps to locate and identify that inconsistency. For, taking $\bullet$I as the sole introduction-rule for $\bullet$, we can already see that $\bullet \leftrightarrow \neg\bullet$, for derivation of $\bot$ from $\bullet$ is both enough to assert $\bullet$ (by $\bullet$I) and to deny it (since it entails $\bot$).

Some have wanted not only to claim that harmony prevents inconsistency, which $\bullet$ shows is a mistaken hope, but furthermore, that lack of harmony, if not a source of incoherence and error, at least imports non-logical features and so undermines the inferentialist project. (See, e.g., [10], pp. 246ff.) Certainly, Prior's rules for 'tonk' are incoherent and mistaken, and also inharmonious. But although lack of harmony in Prior's rules serves to obscure the source of that incoherence, it is not itself the source. As the case of $\bullet$ shows, contradictory connectives can be governed by harmonious rules, where the necessary and sufficient conditions for their assertion are given by the I-rules, and the E-rule uses no more than the I-rules justify.

And as we will see in §3, perfectly decent connectives can be governed by inharmonious rules. Such cases are unhelpful, in obscuring the meaning of the connective, but the lack of harmony is not itself a source of incoherence, nor does it mean that the rules do not define the meaning of the connectives as logical.

3 Modality

We now turn to a classic example of such disharmony, the standard rules for '$\Box$' and '$\Diamond$' in modal logic. The pioneers in the development of natural deduction systems for modal logic were Curry ([7], ch. V) and Fitch ([8], ch. 3), both formulating systems of **S4** (variously classical and intuitionistic). Curry's rules for '$\Box$' read:

$$\frac{\alpha}{\Box\alpha}\ \Box I \qquad\qquad \text{and} \qquad\qquad \frac{\Box\alpha}{\alpha}\ \Box E$$

where in $\Box$I every assumption on which α depends must be *modal*, that is, of the form $\Box\beta$ (or $\neg\Diamond\beta$ or $\bot$, for systems containing '$\Diamond$' and/or '$\bot$'). Fitch's system has the same $\Box$E-rule, and essentially the same $\Box$I-rule, where the proof of α is a (what he termed, strict) sub-proof into which only formulae of the form $\Box\beta$ may be reiterated. Fitch adds complementary rules for '$\Diamond$', which Prawitz ([13], ch. VI) brings into a form more comparable to Curry's $\Box$-rules:

$$\frac{\alpha}{\Diamond\alpha}\ \Diamond I \qquad\qquad \text{and} \qquad\qquad \begin{array}{c}(\alpha)\\ \vdots\\ \dfrac{\Diamond\alpha \quad \gamma}{\gamma}\ \Diamond E\end{array}$$

provided that in the case of $\Diamond E$, every assumption on which the minor premise γ depends, apart from α, (the so-called parametric formulae) is modal and γ is *co-modal*, that is, has the form $\Diamond\beta$ (or $\neg\Box\beta$ or $\neg\bot$).

These rules can be strengthened to yield the logic **S5**, by including formulae of the form $\neg\Box\beta$ (equivalently, $\Diamond\beta$) among the modal formulae (and identifying the classes of modal and co-modal formulae). They can also be weakened to yield the system **T** by casting $\Box I$ in the slightly different form:

$$\frac{\alpha}{\Box\alpha}\ \Box I$$

where the conclusion $\Box\alpha$ depends on the formulae $\Box\Gamma$ ($= \Box A_1, \ldots, \Box A_n$) if the premise α depends on Γ ($= A_1, \ldots, A_n$). Finally, **S4** can be weakened to **K4** and **T** can be weakened to **K** by omitting the rule $\Box E$-rule (and taking '$\Diamond$' as a defined symbol), since it is not valid in those logics.

By a clever recasting of the above formulation, preserving the essential nature of the rule-pairs, Prawitz was able to formulate systems of **S4** and **S5** which are normalizable, that is, in which any proof can be replaced by a normal proof, one which contains no "maximum formulae" (as in §2).

Such a normalization result is sometimes identified with harmony. ([10], p. 250) Nonetheless, although normalization can be proved for (a version of) the above rules, from the viewpoint of the account of harmony given in §2, these rules look very unsatisfactory. As formulated above, the logics **T**, **S4** and **S5** differ only in their $\Box I$-rule, sharing $\Box E$ as the elimination-rule; the logics **K**, **K4** and **KB** again differ only in their $\Box I$-rules, having no $\Box E$-rule.

- First puzzle: if logics have different $\Box I$-rules, should we not expect harmony to yield different $\Box E$-rules?

- Secondly, if logics share $\Box I$-rules, should they not be the same logic unless disharmonious? - that is, if one has the $\Box E$-rule, the other not, must not at least one of them be disharmonious?

Consider now the $\Diamond$-rules, and in particular, ask what elimination-rule is justified by $\Diamond I$. From an assertion of $\Diamond\alpha$ we should be able to infer anything we can infer from what justified the assertion of $\Diamond\alpha$. But that was just α itself. That justifies $\Diamond$E as given - but without the restriction on the parametric formulae and on the conclusion γ itself. Indeed, it would warrant inferring α from $\Diamond\alpha$.[9]

But that would collapse modalities. The restriction in $\Diamond$E thus serves to restrict whatever meaning $\Diamond$I might otherwise confer on '$\Diamond$'.

Another indication that the rules are misleading comes from consideration of the proof in **T, S4** or **S5** of an equivalent form of the postulate (K), using '$\Diamond$', ($K_\Diamond$):
$\Box(\alpha \rightarrow \beta) \rightarrow (\Diamond\alpha \rightarrow \Diamond\beta)$

$$
\cfrac{\Diamond\alpha^1 \qquad \cfrac{\cfrac{\cfrac{\Box(\alpha \rightarrow \beta)^2}{\alpha \rightarrow \beta}\ \Box E \qquad \alpha^3}{\beta}\ \rightarrow E}{\Diamond\beta}\ \Diamond I}{\cfrac{\cfrac{\Diamond\beta}{\Diamond\alpha \rightarrow \Diamond\beta}\ \rightarrow I(1)}{\Box(\alpha \rightarrow \beta) \rightarrow (\Diamond\alpha \rightarrow \Diamond\beta)}\ \rightarrow I(2)}\ \Diamond E(3)
$$

The proof uses $\Box$E and $\Diamond$I, in addition to $\Diamond$E. But ($K_\Diamond$) is valid in **K**, and does not depend on the reflexivity of accessibility, and so its proof should not depend on $\Box$E or $\Diamond$I. To formulate the non-reflexive logics, **K, K4** and **KB**, it would be necessary to drop $\Box$E and $\Diamond$I (which are valid only for reflexive logics) and so there would apparently be no introduction-rule for '$\Diamond$'. It is sometimes argued (e.g., [14], p. 243) that $\bot$ has no I-rule, which justifies *Ex Falso Quodlibet*:

$$
\frac{\bot}{\alpha}\ \bot E
$$

as its elimination-rule. We certainly do not want such a rule for '$\Diamond$', so $\Diamond\alpha$ is introduced in **K** and **K4** (as noted above) by definition as $\neg\Box\neg\alpha$. Otherwise, one would at best have conferred no meaning on '$\Diamond$' to justify the elimination-rule. So the rules could not be in harmony.

Finally, consider $\Diamond$I once again. If the rules were in harmony, $\Diamond$I would give the whole meaning of '$\Diamond$'.[10] But even in reflexive logics, α is only one ground for asserting $\Diamond\alpha$. $\Diamond\alpha$ can be true without α also being true. So if $\Diamond$I is the only introduction-rule for '$\Diamond$', part of the meaning of '$\Diamond$' must reside elsewhere, that is, in the elimination-rule.

So again, the rules cannot be in harmony.

---

[9]One spur to the present paper was reflecting on Dummett's difficulties with '$\Diamond$' - see [10] p. 265; cf. [18], p. 129.

[10]More explicitly, in the language of [18], '$\Diamond$' is not autonomous. Part of the meaning of '$\Diamond$' is given by $\Diamond$I, part by $\Diamond$E.

4   Harmony regained

How can we proceed to formulate rules for the normal modal logics which are in harmony - that is, in which the introduction-rule encapsulates the whole meaning of the modal operator - and in which a clear distinction can be made whereby each rule has a form appropriate to each logic? One way to do so is to introduce the notion of a labelled deductive system. (See, e.g., [22], ch. 4, [24]; cf. [16], §3.2.) Whether harmony can be achieved without labels is an open question. But it can certainly be achieved by the use of labels.

In a labelled deductive system, each formula receives a label. The rules can refer to the labels, can change them, and can depend on relations between labels. In fact, in characterizing modal logic, the labels are effectively world-indices, that is, we can read α_i, where 'i' is the label on the wff α, as saying that α is true at the world (with index) i.[11] Then $\Gamma \vdash \alpha$ if there is a derivation of α_0 from $\Gamma' \subseteq \Gamma$ in which every formula in Γ' has the label 0. (0 here can be understood either as an arbitrary index - validity being defined as truth-preservation at an arbitrary world - or as the base, or home, world H of the Kripke semantics.)

Non-modal rules preserve labels in obvious ways - the labels are affected by, and determine the correctness of, the modal rules. In $\wedge$I and $\rightarrow$E, for example, the premises must have the same label. In $\vee$E and $\exists$E, to take another example, the conclusion has the same label as that on the minor premise (which is itself arbitrary).

We need only one explicit relation between labels, which we write $i < j$, and read as saying that (world) j is "possible relative to" (world) i.

Note that such a reading is purely pedagogical. We are setting up a formal system, in which labels and '$<$' are auxiliary symbols whose meaning, if any, is conferred by the rules. If what was said in §1 was right, there are no (other) worlds, and so the reading of $i < j$ is at best a useful metaphor. A similar point is often made about such an expression as

$$\lim_{n\to\infty} a_n,$$

where '∞' does not refer to a value of n, but indicates rather that there is no greatest value of n. One may ask how $\Box$I and $\Box$E, as formulated below, relate to our inferential practice. That is part of a wider question, how any part of the theoretical systematization of logic relates to practice. More needs to be said, but not here. *In nuce*, inferentialism is the thesis that the meaning of all the logical symbols is given by the rules for use, and by nothing else. For $\Box\alpha$ to be true at i is for α to be true at any world j possible relative to, or "accessible from", i.

This motivates the I-rule:

---

[11]But that is, I argued in §1, only a helpful metaphor - helpful, provided it is not taken literally.

$$(i < j)$$

$$\vdots$$

$$\frac{\alpha_j}{\Box\alpha_i} \;\Box\text{I}$$

Here $j \neq i$ and j must not appear in any other assumption on which α_j depends. In other words, $\Box$I combines aspects of $\to$I and $\forall$I, unsurprising given the articulation of '$\Box$' in terms of truth at all (accessible) worlds.

Then $\Box$I warrants, by Gentzen considerations:

$$(i < j \Rightarrow \alpha_j)$$

$$\vdots$$

$$\frac{\Box\alpha_i \qquad \gamma_k}{\gamma_k} \;\Box\text{E}$$

That is, if assuming α true at all worlds j accessible from i allows derivation of γ at k, then so too does assertion of $\Box\alpha_i$. We can simplify this, as in the case of $\wedge$E and $\bullet$E, by inverting the derivation of γ_k with that of α_j, deducing α_j from $i < j$ directly:

$$\frac{\Box\alpha_i \quad i < j}{\alpha_j} \;\Box\text{E}$$

$$\vdots$$

$$\gamma_k$$

These rules suffice to prove (K):

$$\frac{\dfrac{\Box(\alpha \to \beta)_0^1 \quad 0 < 1^2}{\alpha \to \beta_1}\;\Box\text{E} \qquad \dfrac{\Box\alpha_0^3 \quad 0 < 1^2}{\alpha_1}\;\Box\text{E}}{\dfrac{\dfrac{\dfrac{\beta_1}{\Box\beta_0}\;\Box\text{I}(2)}{\Box\alpha \to \Box\beta_0}\;\to\text{I}(3)}{\Box(\alpha \to \beta) \to (\Box\alpha \to \Box\beta)_0}\;\to\text{I}(1)}\;\to\text{E}$$

The fact that $\Box$E has been justified in Gentzen's manner by $\Box$I means that normalization is immediate:

$$\frac{\dfrac{\begin{array}{c}\Pi_1\\ i < j \Rightarrow \alpha_j\end{array}}{\Box\alpha_i}\;\Box\text{I} \qquad \begin{array}{c}(i < j \Rightarrow \alpha_j)\\ \Pi_2\\ \gamma_k\end{array}}{\gamma_k}\;\Box\text{E} \qquad\qquad \text{reduces to} \qquad\qquad \begin{array}{c}\Pi_1\\ i < j \Rightarrow \alpha_j\\ \Pi_2\\ \gamma_k\end{array}$$

eliminating the maximum formula $\Box\alpha_i$.

The rules for '$\Diamond$' are equally straightforward and intuitive, given the interpretation of $\Diamond\alpha$ as indicating the truth of α at some accessible world:

$$\frac{\alpha_j \quad i < j}{\Diamond\alpha_i} \; \Diamond\text{I}$$

$\Diamond$I justifies the E-rule (cf. $\wedge$I and $\wedge$E):

$$(\alpha_j, i < j)$$
$$\vdots$$
$$\frac{\Diamond\alpha_i \qquad \gamma_k}{\gamma_k} \; \Diamond\text{E}$$

where k is any index, $i \neq j \neq k$ and 'j' occurs in no other assumptions on which γ_k depends than those displayed. Note that the restriction on j prevents inversion and simplification of the type available in $\wedge$E and $\Box$E.

We can then prove ($\mathbf{K_\Diamond}$):

$$\cfrac{\Diamond\alpha_0^1 \qquad \cfrac{\cfrac{\cfrac{\Box(\alpha \to \beta)_0^2 \quad 0 < 1^3}{\alpha \to \beta_1} \; \Box\text{E} \qquad \alpha_1^4}{\beta_1} \; \to\text{E} \quad 0 < 1^3}{\Diamond\beta_0} \; \Diamond\text{I}}{\cfrac{\cfrac{\Diamond\beta_0}{\Diamond\alpha \to \Diamond\beta_0} \; \to\text{I}(1)}{\Box(\alpha \to \beta) \to (\Diamond\alpha \to \Diamond\beta)_0} \; \to\text{I}(2)} \; \Diamond\text{E}(3,4)$$

The fact that $\Box$E and $\Diamond$I occur in this proof is no longer a problem, since as now formulated, they are valid in $\mathbf{K}$, and do not connote reflexivity.

These rules suffice for all the theses of the logic $\mathbf{K}$, but they do not yield $\Box\alpha \to \alpha$, or any of the stronger theses of $\mathbf{K4}$, $\mathbf{S4}$ and so on. To strengthen $\mathbf{K}$ to obtain the other normal logics, we need to add structural rules for the manipulation of the index-relation '$<$':

$$\begin{array}{cc}
(i < i) & (i < k) \\
\vdots & \vdots \\
\dfrac{\alpha_j}{\alpha_j}\ T & \dfrac{i < j \quad j < k \quad \alpha_l}{\alpha_l}\ 4
\end{array}$$

$$\begin{array}{cc}
(j < i) & (j < k) \\
\vdots & \vdots \\
\dfrac{i < j \quad \alpha_k}{\alpha_k}\ B & \dfrac{i < j \quad i < k \quad \alpha_m}{\alpha_m}\ 5
\end{array}$$

The system can also be extended to predicate logic, and versions both with varying domains and constant domains (and so validating the Barcan formula) can be
developed by labelling terms in a similar way. (See, e.g., Basin et al. [3].)

We can still be puzzled, however, how the use of auxiliary symbols such as $<$
in the rules for '□' and '◊' gives them the meaning of 'necessarily' and 'possibly'.
The answer lies in the similarity between, say, □I and ∀I:

$$\frac{\alpha(b/a)}{(\forall a)\alpha}\ \forall\mathrm{I}$$

Here the fact that 'b' is free in no assumptions on which $(\forall a)\alpha$ depends means that
α holds regardless of the interpretation of 'a', and so α holds for all a, giving the
necessary universal meaning to '∀'. The similar condition on 'j' in □I means that
α holds for all j. Then we follow out the metaphor, a picture in which worlds are
"possible" relative to one another, and infer that α is true at all worlds "possible"
relative to i, that is, that α is necessarily true at i, i.e., $\square\alpha$ is true at i.

This deductive system provides harmonious modal rules, and provides an answer to the first and second puzzles. The introduction- and elimination-rules
for '□' and '◊' are the same in all the normal modal logics, so we do not have
the situation of the I-rule changing without a change in the elimination-rule, nor
the introduction- or elimination-rule disappearing while the other remains. What
changes are the structural rules governing the assumptions.

The rules by which the logics differ are generic rules, rules governing the auxiliary symbols - the labels and the relation '$<$' - while the operational rules, the
specific rules for the operations, remain constant. What this means is that there
is throughout these logics, these theories of modality, the same sense of necessity
and possibility. What is different is the logic that they satisfy, not the meaning
of '□' and '◊'. The stronger systems permit inferences involving necessity and
possibility which the weaker systems do not.

We can draw an analogy with classical logic and its rivals. We do not want the intuitionist, classicist and relevantist to disagree about the meaning of '→' - that way lies Carnapian tolerance, with a different logic appropriate for each different meaning. (See [20].)

Rather, we want them to agree on what they disagree about, that is, to disagree about the same thing, to attribute different logics to the one connective '→', with a univocal meaning. So too, for modality. Modal tolerance would be the view that each modal system was right for a different sense, or notion of modality.

Sometimes that is appropriate, if one is comparing the structure of consequence in deontic logic with epistemic logic, say. But proper logical disagreement arises when the systems give a different modal theory to the same notions, necessity and possibility. Only then does the question, which modal system is right, become legitimate. We fix that univocal sense of '□' and '◊' by the rules □I and ◊I. The elimination-rules, □E and ◊E, are consequently justified by those senses being univocal and fully specified. What is different, is the logic of the modal notions encapsulated in the structural rules.

5   Conclusion

Unless one believes in the reality of possible-world semantics, it cannot provide an independent and non-circular account of the meaning of the modal terms 'necessary' and 'possible'. Rather, the meaning of these terms must be given inferentially, by laying down rules for their use. Meaning can be usefully made transparent by requiring harmony in the rules of inference wherever possible. For if the elimination-rule is so framed that it is, in Gentzen's phrase, "no more than a consequence" of the meaning conferred by the introduction-rules, then those introduction-rules exhibit the meaning transparently, setting out expressly and fully the conditions for assertion of propositions containing the term in question. We have seen how to express those meanings harmoniously and transparently not only for '∧' and '•', but also for the modal connectives (using auxiliary, uninterpreted labels).

Necessity and possibility have been given a proof-conditional meaning in which the introduction- and elimination-rules lie in harmony, where harmony ensures transparency in the meaning conferred and whose virtue is clarity. There are many suggestions as to the diagnosis of the fallacy in Prior's introduction of 'tonk'. What one seeks is a diagnosis which explains what was mistaken in the inferentialist views that Prior was attacking and helps to improve our understanding of inferentialism. The recognition that the introduction-rules are jointly necessary and the minor premises of the elimination-rule jointly sufficient lets us see how inconsistency can arise from a mismatch in the rules. Harmony can prevent inconsistency arising inadvertently in this way. But it cannot prevent inconsistency *tout court*, since the introduction-rule can be inconsistent in itself, as with '•'. And that is as it should be, for sometimes we wish to study such cases, e.g., in a self-referential paradox.

Harmony does not guarantee normalization nor vice versa, as we have seen in the case of • and of Prawitz' rules for '□' and '◊'. His rules do give the (correct) meaning of '□' and '◊', but they do so opaquely. So harmony should not be identified with normalization. Harmony for the modal rules is achieved when the full meaning of $\Box\alpha$ and $\Diamond\alpha$ is contained in the rules for their assertion. It seems that that cannot be achieved without some auxiliary apparatus. Nothing about the conditions under which α itself can be asserted can tell whether α is necessary or possible. Probably, this is what has inspired the modal sceptics. How can one tell, considering only α, whether α is possible or necessary? Certainly, if α is true it must be possible, and if false, it cannot be necessary. But these are extremal cases. For the general case, we need to resort to some metaphor. But we must be clear that it is a metaphor, and does not represent anything real. The meaning of the modalities is given by the inferential conditions for the assertion of modal formulae.

BIBLIOGRAPHY

[1] D.M. Armstrong, *A Combinatorial Theory of Possibility*. Cambridge: Cambridge U.P. 1989.
[2] D.M. Armstrong, *A World of States of Affairs*. Cambridge: Cambridge U.P. 1997.
[3] D. Basin, S. Matthews and L. Viganò, 'Labelled modal logics: quantifiers', *Journal of Logic, Language and Information* 7, 1998: 237-63.
[4] F.H. Bradley, 'Reply to Mr Russell's Explanations', *Mind* 20, 1911, pp. 74-76.
[5] R.B. Brandom, *Articulating Reasons*. Cambridge, Mass.: Harvard U.P. 2000.
[6] I.M. Copi, *Symbolic Logic*. New York: Macmillan 1954.
[7] H.B. Curry, *A Theory of Formal Deducibility*. Notre Dame: Indiana 1950.
[8] F.B. Fitch, *Symbolic Logic*. New York: The Ronald Press Co. 1952.
[9] M. Dummett, *Frege: Philosophy of Language*. London: Duckworth 1973.
[10] M. Dummett, *Logical Basis of Metaphysics*. London: Duckworth 1991.
[11] G. Gentzen, 'Untersuchungen über das logische Schliessen', in *The Collected Papers of Gerhard Gentzen*, tr. M. Szabo. Amsterdam: North-Holland 1969, pp. 68-131.
[12] W. Kneale, 'The province of logic', in *Contemporary British Philosophy* III, ed. H.D. Lewis. London: George Allen & Unwin 1956, 237-61.
[13] D. Prawitz, *Natural Deduction*. Stockholm: Almqvist & Wiksell 1965.
[14] D. Prawitz, 'Towards a foundation of general proof theory', in *Logic, Methodology and Philosophy of Science* IV, ed. P. Suppes et al. Amsterdam: North-Holland 1973.
[15] A.N. Prior, 'The Runabout Inference-Ticket'. *Analysis* 21, 1960-61: 38-39.
[16] S. Rahman and L. Keiff, 'How to be a dialogician', in *Logic, Thought and Action*, ed. D. Vanderveken. Dordrecht: Kluwer 2004: 359-408.
[17] S. Read, *Relevant Logic*. Oxford: Blackwell 1988.
[18] S. Read, 'Harmony and Autonomy in Classical Logic', *Journal of Philosophical Logic*, 29 (2000), pp. 123-154.
[19] S. Read, 'The Unity of the Fact', *Philosophy* 80, 317-42.
[20] S. Read, 'Monism: the one true logic', in *A Logical Approach to Philosophy: Essays in Memory of Graham Solomon*, ed. David DeVidi and Tim Kenyon. Dordrecht: Springer 2006, pp. 193-209.
[21] R. Routley, 'Universal semantics'. *Journal of Philosophical Logic* 4, 1975, 327-56.
[22] A. Simpson, *The Proof Theory and Semantics of Intuitionistic Modal Logic*. Ph.D. thesis, University of Edinburgh 1994.

[23] J.T. Stevenson, 'Roundabout the Runabout Inference-Ticket'. *Analysis* 21, 1960-61: 124-8.
[24] L. Viganò, *Labelled Non-Classical Logics*. Dordrecht: Reidel 2000.

Contextual Epistemic Logic

MANUEL REBUSCHI AND FRANCK LIHOREAU

ABSTRACT. One of the highlights of recent informal epistemology is its growing theoretical emphasis upon various notions of context. The present paper addresses the connections between knowledge and context within a formal approach. To this end, a "contextual epistemic logic", CEL, is proposed, which consists of an extension of standard S5 epistemic modal logic with appropriate reduction axioms to deal with an extra contextual operator. We describe the axiomatics and supply both a Kripkean and a dialogical semantics for CEL. An illustration of how it may fruitfully be applied to informal epistemological matters is provided.

1 Introduction

The formal approach to knowledge and context that we propose in this paper was originally driven not only by formal logical concerns but also by more informal epistemological concerns. In the last two or three decades, indeed, epistemology has seen two major "turns":

- a "new linguistic turn", as Ludlow [21] calls it, through the increased reliance, in contemporary epistemological debates, upon "evidence" regarding how we ordinarily *talk* about knowledge, most notably as a result of the flourishing discussions about the purported epistemological role of various notions of context and the relative merits of "contextualism" over "scepticism", "anti-scepticism", and "subjectivism", *inter alia*;

- a "logical turn", through the rising conviction that discussions in informal epistemology may benefit from formal epistemology – epistemic logic, formal learning theory, belief revision, and so on –, notably by applying the methods of epistemic logic in order to gain insights into traditional informal epistemological issues. Recent work by van Benthem [3], Hendricks [14, 15] and Hendricks & Pritchard [16], and Stalnaker [32] counts as representative of this trend.

An important part of the background for the present paper consists of the project of taking advantage of the logical turn in order to record some of the main lessons to be drawn from the linguistic turn of epistemology, the acknowledgement of the possible epistemological role of context to start with. This paper, at the junction of informal and formal epistemology, focuses on the question how to introduce the notion of context into epistemic logic. It provides a new logic, contextual epistemic logic (CEL), an extension of standard S5 epistemic modal logic with appropriate reduction axioms to deal with an extra contextual operator.

In section 2 contemporary informal epistemological discussions over the notion of context are briefly exposed, as well as three strategies available to handle context in a formal way. The authors already studied one of them in [20] and the other two seem quite natural, one of which is accounted for in the present paper. The section ends with a presentation of public announcement logic (PAL), which is technically very close to CEL yet very different from it in spirit: after a comparison of the respective objectives of the two logics we mention what must be changed in PAL to reach a logic for contexts.

In section 3 the syntax and Kripke semantics of CEL are given, as well as a complete proof system. In the subsequent section, we present a dialogical version of CEL: after a general presentation of dialogical logic for (multi-)S5 modal systems, we add context relativization and prove its completeness.

Finally, in section 5 we provide a connection between our formal definitions and standard epistemological positions (scepticism and anti-scepticism, contextualism, subjectivism), and we show a few applications of CEL to specifically epistemological questions.

2   Contextual epistemology

2.1   From an informal approach to knowledge and context...

Constitutive of the linguistic turn in epistemology is contextualism. Contextualism can be viewed as an attempt to reconcile scepticism and anti-scepticism, and subjectivism as an alternative to contextualism. All these positions are better thought of as positions about what a satisfactory account for the truth of knowledge sentences like "i knows that φ" or "i does not know that φ" must look like. A way of representing what it takes for a knowledge sentence to be true is by means of an "epistemic relevance set", i.e. a set of "epistemically relevant counter-possibilities" that an agent must be able to rule out, given the evidence available to him, for it to be true that he knows a proposition.

Thus, according to *scepticism* – also called "sceptical invariantism", and defended by Unger [34] for instance – the epistemic standards are so extraordinarily stringent that for any (contingent) proposition, all logical counter-possibilities to this proposition will count as epistemically relevant. That is, for it to be true that an agent i knows a proposition φ, i must rule out the entire set of logical possibilities in which not-φ; this being an unfulfillable task for any i and φ, it is never true that anyone knows anything except, perhaps, necessary truths.[1] On the contrary, for *anti-scepticism* – also called "Moorean invariantism" and defended by Austin [1] and Moore [23] – the standards for the truth of "*i* knows that φ" being those for an ordinarily correct utterance of this sentence, they are lax enough to make (a possibly important) part of our ordinary knowledge claims turn out true: for any proposition, not all logical counter-possibilities to this proposition will count as epistemically relevant. That is, it will be true that *i* knows that φ only if *i* rules out a given (proper) subset of possibilities in which not-φ. Scepticism and anti-scepticism are *absolutist* views about knowledge in the sense that for both of them, whether a logical counter-possibility to a proposition is epistemically relevant or not depends only on what proposition it is that is purportedly known and once this proposition is fixed, the associated epistemic relevance set is not liable to vary.

According to *relativist* views, whether a logical counter-possibility to a proposition is epistemically relevant or not does not only depend on what proposition it is; it also depends on the context, for some sense of "context". For instance, according to *contextualism* – defended by authors like Cohen [7], De Rose [8], Lewis [19] – the context in question will be that of the possible "knowledge ascriber(s)". Although we'd better talk of contextualism in the plural, a common contextualist assumption is that whilst the circumstances of the purported knower are fixed, the epistemic standards, therefore the epistemic relevance set, may vary with features of the context of the knowledge ascriber(s) such as what counter-possibilities they attend to, what propositions they presuppose, what is at stake in their own context, etc. In contrast, *subjectivism* – also called "subject-sensitive invariantism" or "sensitive moderate invariantism" – is defended by authors like Hawthorne [13] and Stanley [33]. This view has it that such factors – viz. attention, interests, presuppositions, stakes, etc. – are not the attention, interests, presuppositions, stakes, etc., of the knowledge ascriber, but those of the "knower" himself. That is, the epistemic standards, therefore the epistemic relevance set, may vary with the context of the purported "knower", even if no change occurs in the circumstances he happens to be in.[2]

---

[1] One might want to be a sceptic as far as contingent truths are concerned while not being a sceptic with respect to necessary truths. Such truths include logical validities as well as analytic truths. Taking the latter into account within a modal-logical framework can be done by adopting meaning postulates.

[2] Absent from the foregoing discussion will be another prominent view on the matter, viz. what may be coined "circumstancialism" – the view, held by Dretske [9, 10] and Nozick [24], that the relevance set depends on the objective situation the subject happens to be in. The reason for not discussing this view here is that one constraint on the formal developments in this paper was to stick to epistemological views whose associated epistemic logics (1) were normal, i.e. incorporating at least the principles of the smallest normal modal logic K – in particular the *Distribution Axiom* and the *Knowledge Generalization Rule* – and thus (2) fully characterized some class of standard Kripke models. This is not the case with circumstancialism which drops the two K-principles just mentioned. However, capturing circumstancialist epistemic logic semantically can be done in several ways compatible with a modal approach, two of the most straightforward being (1) by adding "awareness functions" into the stan-

So, depending on whether one opts for an absolutist or a relativist view about knowledge, one will assume that a notion of context has an epistemological role to play or not. A first incursion into formalizing the possible connections between knowledge and context and the four epistemological views[3] was undertaken in [20].

2.2 ... to formal approaches to reasoning about knowledge and context

The role of context in the above epistemological discussion is to impose a restriction of the relevant set of possible worlds, rather than to go from a possible world to another one. Contexts in a modal formalization thus cannot simply be reduced to standard modalities.

As a consequence, in order to represent contexts in epistemic logic three ways seem available:

1. use non-standard models, i.e. put the context in the metalanguage and evaluate each (standard epistemic) formula relatively to some world and context;

2. use standard models with standard modality, within an extension from basic modal logic;

3. use standard models with a non-standard interpretation for context modalities.

The first option, investigated in [20], requires that the notion of context enter the metalanguage and that the formulas of the object-language be interpreted in "contextual models". The second track would require a kind of hybrid logic to account for the intersection of two accessibility relations. In the remainder of the subsection we briefly discuss the first two approaches before focusing on the third strategy.

---

dard possible worlds models as in [11], (2) or by augmenting such models with "impossible possible worlds" as in [28]. Another important epistemological view is the "assessment sensitivity" (sometimes called "relativist") view, held by MacFarlane [22] and Richard [30] for instance, according to which the relevance set depends neither on the subject's situation or context nor on the attributor's context, but on the context in which a knowledge claim made by an attributor about a subject is being assessed for truth or falsity. For the same reason as before it will not be discussed here, since the formalization of it would require reliance on a semantics allowing formulas to be evaluated not only relative to a world, but also to both a context of attribution and a context of assessment – i.e. a non-standard semantics, perhaps in the vicinity of [6]'s "token semantics". These issues will not be pursued here.

[3] It may appear to some that scepticism is not a respectable language game with a special knowledge operator, but a defective language game with a respectable knowledge operator. This seems intuitively correct, of course. So why bother trying to capture scepticism in a formal framework anyway? Here are two reasons. First, the intuition in question might simply be an effect of a prejudice against scepticism and thus cannot be appealed to in support of a depreciative view of scepticism without begging the question; second, our commitment is to find a formal framework that makes everything "respectable" in each of the epistemological positions it is meant to capture: we want both their respective "language game" and "knowledge operator" to be treated as "respectable".

Contextual models

Formalizing contexts while sticking to the standard syntax of epistemic logic implies the adoption of non-standard models. Contextual models are usual Kripke models for multi-**S5** epistemic languages $\mathcal{M} = \langle W, \{\mathcal{K}_i\}_{i \in I}, V \rangle$ augmented with a pair $\langle C, \mathcal{R} \rangle$, where: $C = \{c_j : j \in J\} \neq \varnothing$ is a set of "contexts", for $I \subseteq J$,[4] and $\mathcal{R} : C \to (W \to \wp(W))$ is a function of "contextual relevance" associating with each context c_j, for each world w, the set $\mathcal{R}(c_j)(w)$ of worlds relevant in that context at that world.

Truth in contextual models is then defined relatively to a world and a context. The clauses are the usual ones for propositional connectives in the sense that contexts do not play any role on them, whereas contexts can modify the evaluation of epistemic operators in one of the following four ways:

$\mathcal{M}, c_i, w \vDash \mathbf{K}_j \varphi$ iff

(1.1)   for all w', if $\mathcal{K}_j ww'$ and $w' \in \mathcal{R}(c_i)(w)$, then $\mathcal{M}, c_i, w' \vDash \varphi$.

(1.2)   for all w', if $\mathcal{K}_j ww'$ and $w' \in \mathcal{R}(c_i)(w)$, then $\mathcal{M}, c_j, w' \vDash \varphi$.

(2.1)   for all w', if $\mathcal{K}_j ww'$ and $w' \in \mathcal{R}(c_j)(w)$, then $\mathcal{M}, c_i, w' \vDash \varphi$.

(2.2)   for all w', if $\mathcal{K}_j ww'$ and $w' \in \mathcal{R}(c_j)(w)$, then $\mathcal{M}, c_j, w' \vDash \varphi$.

The definitions lead to four different logics, i.e. to four interpretations of the epistemic operator $\mathbf{K}_j$.

Taking epistemic accessibility relations to be equivalence relations, a formal characterization of the aforementioned informal views of knowledge can be given through a proper choice among the four cases (1.1)–(2.2) or/and through proper restrictions on the contextual relevance function. The two absolutist views of knowledge – scepticism and anti-scepticism – can thus be associated with case (1.1), differing from each other only in their respective restrictions on $\mathcal{R}$, viz. the restriction that for all i and w, $\mathcal{R}(c_i)(w) = W$ for scepticism – hence a restriction enabling scepticism to drop the condition $w' \in \mathcal{R}(c_i)(w)$ from the truth conditions of epistemic sentences – and the minimal restriction that for all i and w, $\mathcal{R}(c_i)(w) \subsetneq W$ for anti-scepticism. In contrast, the two relativist views of knowledge can be associated with the (–.2) cases, viz. case (1.2) for contextualism and case (2.2) for subjectivism.[5]

---

[4]A context c_i is thus assigned to every agent $i \in I$, but additional contexts may be added for groups of agents, conversations, and so forth.

[5]A possible worry with these formal definitions is that they end up making scepticism look like a limiting case of anti-scepticism and the characterisation of the latter unsatisfactory for anyone interested in a more substantial definition. But first, what we were after was more to contrast absolutist with relativist views than to contrast scepticism with anti-scepticism. Second, the definition we gave of anti-scepticism – viz. "for all i and w, $\mathcal{R}(c_i)(w) \subsetneq W$" – was meant to be broad enough so as to allow for more specific characterisations of one's favourite version of anti-scepticism. This can be done, for instance, by listing all the worlds one thinks should be considered epistemically relevant, that is, included in $\mathcal{R}(c_i)(w)$, or by providing a general criterion for telling the worlds that are epistemically relevant from those that aren't (in terms of "the closest" or "close enough" possible worlds, for

Within this framework one can express *truth-in-a-context* for knowledge claims – an interesting result for contemporary informal epistemology whose interest in the connection between knowledge and context lies primarily in the issue of whether the *truth* of knowledge sentences should be taken as absolute or relative.

Moreover, with contextual models one can account for the logical behaviour of agents of a given epistemological type (sceptical, anti-sceptical, contextualist, or subjectivist) reasoning about the knowledge of other agents regardless of their epistemological type; that is, one can ask such things as *If contextualism is assumed, then if agent 1 knows that agent 2 knows that φ, does agent 1 know that φ?* One can also ask such things as *If a contextualist knows that a subjectivist knows that φ, does the contextualist know that φ?*

However, despite its merits, the contextual models framework is unlikely to win unanimous support from philosophers of language. For consider satisfaction for epistemic formulas of the form $K_j\varphi$ when epistemic operator K_j is taken in its (1.2) and (2.2) interpretations. In these cases, the epistemic operator K_j manipulates the context parameter against which the relevant epistemic formula is evaluated: the truth of $K_j\varphi$ relative to a context c_i (and world w) depends on the truth of the embedded φ relative to *another context* c_j (and world w').

Now, Kaplan [18] famously conjectured that natural language had no such devices as context-shifting operators – which he considered to be "monsters". So, if he is right, there can be no natural language counterpart for K_j in cases (1.2) and (2.2). Schlenker [31] and others have recently challenged Kaplan's conjecture, notably by arguing that natural language allows context-shifting to occur within propositional attitudes. Nevertheless, the existence of "monsters" remains controversial amongst philosophers of language, and the question of context-shifting within attitudes is far from having been settled for that specific kind of attitude reports formed by knowledge claims.

An advantage of the contextual epistemic logic that will be introduced in the present paper is that it is immune to the charge of monstrosity: although something resembling context-shifting takes place in that framework, its language is "monster-free" since its formulas are evaluated against a world parameter only and contexts are of a purely syntactic nature.

Hybrid logic

The semantic intuition lying behind context-relativized attributions of knowledge is that one has to restrict the evaluation to the set of contextually relevant worlds to check whether in this restricted set, the agent knows such and such proposition. If one wants to go back to standard Kripke models and wishes to handle contexts as standard (let us say, S5) modalities $[c_j]$, it is thus required to consider the *intersection* of two accessibility relations: an agent i will be said to know φ (at world

---

instance).

w) relatively to a context c_j – which can be formalized by: $\mathcal{M}, w \vDash \{[c_j]\mathbf{K}_i\}\varphi$ – if and only if $\mathcal{M}, w' \vDash [c_1]\varphi$ at every world w' such that $\mathcal{R}_{j,k}ww'$, where $\mathcal{R}_{j,k}$ is the intersection of the two accessibility relations.

Now it is known that the intersection of two relations cannot be defined within basic modal logics [12]. One has to go beyond basic modal logic to express sound axioms for context-relativized knowledge. Using the hybrid nominals v_x and operators $@_x$, we get the natural axiom schema ($\mathbf{P}_i$ being the dual of $\mathbf{K}_i$): $\{[c_j]\mathbf{K}_i\}\varphi \leftrightarrow ((\mathbf{P}_i v_x \wedge \langle c_k \rangle v_x) \rightarrow @_x[c_l]\varphi)$.

Starting from Blackburn's [5] dialogical version of hybrid logics one could give a natural account of epistemic logic with standard contextual modalities. This is left for another paper.

Model-shifting operators

The way followed in this paper has been paved by recent works about logics of knowledge and communication. The idea, going back to [25], is to add *model-shifting operators* to basic epistemic logics.

Van Benthem et al. [4] recently accounted for model-shifting operators in connection with public announcement. The idea is that the semantic contribution of an announcement, let us say φ, is to eliminate all those epistemic alternatives falsifying φ, so that a new model $\mathcal{M}_{|\varphi}$ is obtained relative to which subsequent formulas are evaluated. In this approach, context modalities are clearly not usual modalities.

In the present paper, we propose to use this Public announcement Logic (PAL) with the required modifications. Before going further into details, let us just add that a frame such as PAL can be combined with some dynamic account of the context operators $[c_i]$. Van Benthem et al.'s paper [4] provides such an account, using standard PDL with an epistemic interpretation. In [20] the authors of the present paper accounted for dynamic aspects of contexts using a formalism based on DRT. Anyway, the way contexts are handled (the dynamic part of the formalism) is relatively independent from their role in epistemic attributions.

2.3   From PAL to CEL

PAL *in a nutshell*

The notion of a public announcement or communication is just that of "a statement made in a conference room in which all agents are present" according to Plaza [25], and that of "an epistemic event where all agents are told simultaneously and transparently that a certain formula holds right now" according to van Benthem *et al.* [4]. The latter intend to model this informal notion by means of a modal operator $[\varphi]$, thus allowing for formulas of the form $[\varphi]\psi$, read intuitively as "ψ holds after the announcement of φ".

The language $\mathcal{L}_{\mathsf{PAL}}$ of PAL, the logic of public announcement proposed by van Benthem *et al.*, results from augmenting the basic epistemic modal language $\mathcal{L}$

with operators of the form $[\varphi]$ where φ is any formula of $\mathcal{L}_{\text{PAL}}$. The semantics for the resulting language $\mathcal{L}_{\text{PAL}}$ is based on standard Kripke models for $\mathcal{L}$, and the definition of $\vDash$ for atoms, negation , conjunction and knowledge operators is as usual with such models. The more "unusual" feature of the semantics of $\mathcal{L}_{\text{PAL}}$ is with the clause of satisfaction for $[\varphi]$, viz.:

$$\mathcal{M}, w \vDash [\varphi]\psi \quad \text{iff} \quad \text{if } \mathcal{M}, w \vDash \varphi \text{ then } \mathcal{M}_{|\varphi}, w \vDash \psi,$$

where $\mathcal{M}_{|\varphi}$ is an "updated model" consisting of a tuple $\left\langle W', \mathcal{K}_1', \ldots, \mathcal{K}_m', \mathcal{V}' \right\rangle$, with:

- $W' = \{v \in W : \mathcal{M}, v \vDash \varphi\}$,
- $\mathcal{K}_i' = \mathcal{K}_i \cap (W' \times W')$ for all agents $i \in \{1, \ldots, m\}$,
- $\mathcal{V}'(p) = \mathcal{V}(p) \cap W'$ for all atomic formulas p.

This appeal to updated models, obtained by restricting a given Kripke model to those worlds where a given formula φ holds, is meant to capture the insight that the epistemic effect of a public announcement of φ at a time t is that at time $t + 1$, all the agents involved will have deleted the possible worlds where they do not know that φ.

Our proposed formalism for reasoning about knowledge and context, although it is strongly inspired by that for reasoning about public announcements, does not appeal to anything like updated models and offers a slightly more complicated semantics for the corresponding language.

Different means for different goals

Let us first compare the objectives of the two formalisms. PAL is specifically designed to account for shifts in common knowledge following public announcements. As a successful announcement is made, its content φ becomes common knowledge and the model is restricted to all and only those possible worlds where φ is true.

The situation is different with CEL: here sentences are presupposed (and as a consequence, known by each agent) as long as they remain implicit; when a presupposition is made explicit – asserted, rejected, questioned. . . – it is removed from the "context". Such a context does not determine the set of epistemic possibilities, but it determines a set of epistemically relevant possibilities.

For instance, in usual (non-sceptical) contexts it is assumed that b, "we are all brains in vats", is false; so $\neg b$ belongs to the context. Now if a provocative sceptic enters the scene and asks whether we are sure that we are not brains in vats, or, better, asserts that we actually are brains in vats, then the context changes: b becomes epistemically relevant. It does not mean that b is true at every accessible world/epistemic possibility, since such a "public announcement" is not necessarily approved by all the agents; but b is now a (counterfactual) possibility that needs to be rejected to grant knowledge.[6] Hence the modifications of contexts according to CEL do not yield any modification of the accessibility relation: they expand or restrict the whole set of possible worlds. Whereas PAL is concerned with changes in the agents' (common) knowledge, CEL aims at accounting for changes of knowledge attributions.

The contrast between the two theories is summarized in the following table:

PAL	CEL
Explicit Public announcement $[\varphi]$	Implicit (Proto-)Context $[c_i]$
Any (complex) formula	Conjunction of literals
What is *a posteriori* known: Set of epistemic possibilities	What is *a priori* assumed: Set of counterfactual possibilities ("epistemically relevant possibilities")
Restriction of the accessibility relation: Removing worlds	Restriction or expansion of the set of relevant possible worlds: Removing or adding worlds
Knowledge shifting	Knowledge attribution shifting

Consequences

PAL is useful to reach CEL for it provides a formalization of a logic with a model-shifting operator. However, a few changes are required to adapt PAL to a logic with contexts:

- Model-shifting operators $[\varphi]$ of PAL are interpreted so that sizes of models are systematically decreasing; in CEL, model-shifting operators $[c_i]$ must allow increasing as well as decreasing sizes of models.

- Whereas the announcements of PAL are obviously linguistic, the contexts of CEL need not be linguistic in nature; what is required is that they be depicted by some linguistic sentence.

- The outcome of a public announcement for the agents' knowledge is straightforward, as is the corresponding axiom schema of PAL. By contrast, several kinds of subtle interactions between context operators and epistemic modalities can be grasped in CEL.

As will be shown in the next section, these slight divergences lead to an important dissimilarity at the semantic level: CEL semantics is much more inelegant than PAL semantics. This will be a good enough reason to go over to the dialogical framework.

---

[6] As a simplification, we consider only literals as presuppositions; it means that an assertion whose content is a complex proposition will modify the context by removing all the literals occurring in it.

3 Contextual epistemic logic

In this section we first introduce the syntax and semantics of CEL, then a sound and complete proof system.

3.1 CEL: Syntax and Kripke semantics

Standard epistemic logic (EL)

The standard formal approach to reasoning about *knowledge* starts with the choice of a basic epistemic modal language. This language $\mathcal{L}_{\mathbf{K}}^m(\mathcal{A}t)$ consists of the set of formulas over a finite or infinite set $\mathcal{A}t = \{p_0, p_1, \ldots\}$ of atomic formulas and a set $J = \{1, \ldots, m\}$ of m agents, given by the following form (F standing for a formula of $\mathcal{L}_{\mathbf{K}}^m(\mathcal{A}t)$):

$$F ::= \mathcal{A}t \mid \neg F \mid (F \wedge F) \mid \mathbf{K}_j(F).$$

This language is then interpreted in Kripke models consisting of tuples $\mathcal{M} = \langle W, \mathcal{R}_1, \ldots, \mathcal{R}_m, \mathcal{V} \rangle$, where: $W \neq \varnothing$ is a set of "worlds"; $\mathcal{R}_j \subseteq W \times W$ is a relation of "epistemic accessibility", for all $j \in J$; $\mathcal{V} : \mathcal{A}t \to \wp(W)$ is a "valuation" mapping each atomic formula onto a set of worlds.

CEL *Syntax*

In order to add contexts to basic epistemic languages, one can choose between two equivalent notations (see [2]). For convenience – especially for the dialogical approach – we will not use the modal contextual operator prefixing formulas $[c_i]\varphi$ but a syntactic relativization of formulas: $(\varphi)^{c_i}$. Moreover, each context c_i is characterized by a conjunction of literals (i.e. atoms and negation of atoms); the "context formula" characterizing a context c_i will be referred to by the same symbol, c_i.[7] So "context formulas" c can be defined as follows:

$$c ::= \mathcal{A}t \mid \neg\mathcal{A}t \mid \top \mid \bot \mid c \wedge c.$$

We thus recapitulate the syntax of epistemic languages with context $\mathcal{L}_{\mathbf{KC}}^m(\mathcal{A}t)$ (G standing for a formula of $\mathcal{L}_{\mathbf{KC}}^m(\mathcal{A}t)$), built upon a set of contexts $C = \{c_i\}_{i \in I}$ ($J \subseteq I$, J being the set of agents):

$$G ::= \mathcal{A}t \mid \neg G \mid (G \wedge G) \mid \mathbf{K}_j(G) \mid (G)^{c_i}.$$

Formulas containing at least one operator $\mathbf{K}_j$ are called *epistemic formulas*; formulas containing at least one subformula of the kind $(\varphi)^{c_i}$ are said to be *context-relativized*, and formulas being not context-relativized are said to be *absolute*. As is immediately seen from the definition, EL is a syntactic fragment of CEL.

---

[7]Contexts need not be *identified* with context formulas; they can be partial models or incomplete possible worlds for instance.

CEL *Kripke semantics*

The contextual epistemic language $\mathcal{L}_{KC}^m(\mathcal{A}t)$ is interpreted relatively to the same Kripke models as $\mathcal{L}_K^m(\mathcal{A}t)$. The satisfaction of atoms, absolute negations and conjunctions is defined as usual. For epistemic operators and contextual relativization, the semantics is a bit more complicated:

$$
\begin{aligned}
&\mathcal{M}, w \vDash p && \text{iff} && w \in \mathcal{V}(p) \quad (p \text{ being atomic})\\
&\mathcal{M}, w \vDash \neg\chi && \text{iff} && \mathcal{M}, w \nvDash \chi\\
&\mathcal{M}, w \vDash \chi \wedge \xi && \text{iff} && \mathcal{M}, w \vDash \chi \text{ and } \mathcal{M}, w \vDash \xi\\
&\mathcal{M}, w \vDash \mathbf{K}_j\varphi && \text{iff} && \text{for every } w' \in W \text{ such that } R_j ww': \mathcal{M}, w' \vDash \varphi\\
&\mathcal{M}, w \vDash (p)^{c_i} && \text{iff} && \mathcal{M}, w \nvDash c_i \text{ or } \mathcal{M}, w \vDash p \quad (p \text{ being atomic})\\
&\mathcal{M}, w \vDash ((\varphi)^{c_k})^{c_i} && \text{iff} && \mathcal{M}, w \nvDash c_i \text{ or } \mathcal{M}, w \vDash (\varphi)^{c_k}\\
&\mathcal{M}, w \vDash (\neg\varphi)^{c_i} && \text{iff} && \mathcal{M}, w \nvDash c_i \text{ or } \mathcal{M}, w \nvDash (\varphi)^{c_i}\\
&\mathcal{M}, w \vDash (\varphi \wedge \psi)^{c_i} && \text{iff} && \mathcal{M}, w \vDash (\varphi)^{c_i} \text{ and } \mathcal{M}, w \vDash (\psi)^{c_i}\\
&\mathcal{M}, w \vDash (\mathbf{K}_j\varphi)^{c_i} && \text{iff} && \mathcal{M}, w \nvDash c_x \text{ or } \mathcal{M}, w \vDash \mathbf{K}_j(\varphi)^{c_y}\\
&&&&& \text{where } \langle x, y \rangle \text{ is chosen in } \{i, j\} \times \{i, j\}
\end{aligned}
$$

There are four different clauses hidden in the last one, depending on the values of x and y. Each choice regiments a specific position about the interaction between knowledge and context. We will refer to these four possibilities as (1.1), (1.2), (2.1), and (2.2), sometimes adding an explicit exponent to the knowledge operator – $\mathbf{K}_j^{1.1}$, $\mathbf{K}_j^{1.2}$, $\mathbf{K}_j^{2.1}$, and $\mathbf{K}_j^{2.2}$ respectively – with the following understanding:

$$
\begin{aligned}
&\mathcal{M}, w \vDash (\mathbf{K}_j^{1.1}\varphi)^{c_i} && \text{iff} && \mathcal{M}, w \nvDash c_i \text{ or } \mathcal{M}, w \vDash \mathbf{K}_j^{1.1}(\varphi)^{c_i}\\
&\mathcal{M}, w \vDash (\mathbf{K}_j^{1.2}\varphi)^{c_i} && \text{iff} && \mathcal{M}, w \nvDash c_i \text{ or } \mathcal{M}, w \vDash \mathbf{K}_j^{1.2}(\varphi)^{c_j}\\
&\mathcal{M}, w \vDash (\mathbf{K}_j^{2.1}\varphi)^{c_j} && \text{iff} && \mathcal{M}, w \nvDash c_j \text{ or } \mathcal{M}, w \vDash \mathbf{K}_j^{2.1}(\varphi)^{c_i}\\
&\mathcal{M}, w \vDash (\mathbf{K}_j^{2.2}\varphi)^{c_i} && \text{iff} && \mathcal{M}, w \nvDash c_j \text{ or } \mathcal{M}, w \vDash \mathbf{K}_j^{2.2}(\varphi)^{c_j}
\end{aligned}
$$

Remarks:
– In the notation $\mathbf{K}_j^{u.v}$, the first superscript u corresponds to the contextual condition, and the second one v to the context according to which the evaluation is to be continued; the possible values of u and v are 1 for the current context, and 2 for the agent.
– It is worth noticing that no restriction of the model $\mathcal{M}$ is required for the evaluation of context-relativized formulas.

According to our definition the context relativization $(\varphi)^{c_i}$ of a non-epistemic formula φ is nothing but a notational variant for the conditional $c_i \rightarrow \varphi$. Since such a relativization has no further impact on epistemic $\mathbf{K}^{1.1}$ formulas, the notion of "context" carried by our formalism will appear to be non-committing for supporters of absolutist conceptions of knowledge. However, this innocent account of context will turn out to be sufficient to model contextualist and subjectivist epistemologies. With CEL context friends and context enemies can thus be put together onto the same neutral field.

3.2 CEL: **Proof system**

Proof systems

There are several proof systems depending on the interaction between epistemic operators and context. The differences are given through axioms of "contextual knowledge". Each proof systems for CEL is that for multi-modal S5 epistemic logic EL plus the following schemas:

$$
\begin{aligned}
\text{Atoms} \quad & \vdash (p)^{c_i} \leftrightarrow (c_i \to p) \\
\text{Contextual negation} \quad & \vdash (\neg\varphi)^{c_i} \leftrightarrow (c_i \to \neg(\varphi)^{c_i}) \\
\text{Contextual conjunction} \quad & \vdash (\varphi \wedge \psi)^{c_i} \leftrightarrow ((\varphi)^{c_i} \wedge (\psi)^{c_i}) \\
\text{Context iteration} \quad & \vdash ((\varphi)^{c_k})^{c_i} \leftrightarrow (c_i \to (\varphi)^{c_k}) \\
\text{Contextual Knowledge } (\langle x, y \rangle \in \{i, j\} \times \{i, j\}) \quad & \vdash (\mathbf{K}_j\varphi)^{c_i} \leftrightarrow (c_x \to \mathbf{K}_j(\varphi)^{c_y})
\end{aligned}
$$

as well as the following rule of inference:

$$
\text{Context generalization} \qquad \text{From} \vdash \varphi, \text{ infer} \vdash (\varphi)^{c_i}.
$$

Like for the semantics of CEL, the schema for Contextual Knowledge can be made more explicit through the following versions:

$$
\begin{aligned}
\text{1.1-Contextual Knowledge} \quad & \vdash (\mathbf{K}_j^{1.1}\varphi)^{c_i} \leftrightarrow (c_i \to \mathbf{K}_j^{1.1}(\varphi)^{c_i}) \\
\text{1.2-Contextual Knowledge} \quad & \vdash (\mathbf{K}_j^{1.2}\varphi)^{c_i} \leftrightarrow (c_i \to \mathbf{K}_j^{1.2}(\varphi)^{c_j}) \\
\text{2.1-Contextual Knowledge} \quad & \vdash (\mathbf{K}_j^{2.1}\varphi)^{c_i} \leftrightarrow (c_j \to \mathbf{K}_j^{2.1}(\varphi)^{c_i}) \\
\text{2.2-Contextual Knowledge} \quad & \vdash (\mathbf{K}_j^{2.2}\varphi)^{c_i} \leftrightarrow (c_j \to \mathbf{K}_j^{2.2}(\varphi)^{c_j})
\end{aligned}
$$

Soundness and Completeness

As CEL axioms are reduction axioms, each CEL-formula φ can be translated into a standard epistemic formula φ' such that for every model $\mathcal{M}$ and world w: $\mathcal{M}, w \vDash \varphi$ iff $\mathcal{M}, w \vDash \varphi'$. completeness of CEL axiomatics relative to the set of formulas valid in the class $\mathcal{M}^{rst}$ of reflexive, symmetric and transitive models is thus inherited from that of usual epistemic logic.

THEOREM 1. *For formulas in $\mathcal{L}_{KC}^m(\mathcal{A}t)$, CEL is a sound and complete axiomatization w.r.t. $\mathcal{M}^{rst}$.*

Proof. From any CEL formula $\varphi \in \mathcal{L}_{KC}^m(\mathcal{A}t)$, one can build a tuple $\langle \varphi_0, \ldots, \varphi_n \rangle$ of CEL formulas and reach a formula $\varphi' \in \mathcal{L}_K^m(\mathcal{A}t)$ such that: $\varphi_0 = \varphi$, $\varphi_n = \varphi'$, and every formula $(\varphi_m \leftrightarrow \varphi_{m+1})$ $(0 \leq m \leq n - 1)$ is an instantiation of an axiom schema. As a consequence:

$$
\vdash_{\text{CEL}} \varphi \text{ iff } \vdash_{\text{CEL}} \varphi_1 \text{ iff } \ldots \text{iff } \vdash_{\text{CEL}} \varphi_{n-1} \text{ iff } \vdash_{\text{CEL}} \varphi'
$$

Since $\varphi' \in \mathcal{L}_K^m(\mathcal{A}t)$, $\vdash_{\text{CEL}} \varphi'$ is equivalent to $\vdash_{\text{EL}} \varphi'$. Now, EL is sound and complete, so:

$$
\vdash_{\text{CEL}} \varphi' \text{ iff } \mathcal{M}^{rst} \vDash \varphi' \qquad \text{(i.e. iff } \varphi' \text{ is valid w.r.t. } \mathcal{M}^{rst})
$$

Hence in order to prove soundness and completeness of CEL, it suffices to prove that for each reduction axiom schema: ψ, one gets: $\mathcal{M}^{rst} \vDash \psi$. This obtains immediately by virtue of the truth definitions.   ∎

4   Dialogical CEL

In this section, we present a dialogical version of CEL. We first introduce dialogical logic for (multi-)S5, then we extend it to context-relativized formulas.

4.1   Dialogical (multi-)modal logic

In a dialogical game, two players argue about a thesis (a formula): The proponent **P** defends it against the attacks of the opponent **O**. For any set of game rules DialΣ associated with some logical theory Σ, we will use the notation DialΣ ⊩ φ to say that there is a winning strategy for the proponent in the dialogical game $G_\Sigma(\varphi)$, i.e. if playing according to the rules of DialΣ, she can defend the formula φ against any attack from the opponent – owning a winning strategy for a game enables a player to win any play of the game. The game rules are defined such that a formula φ is valid or logically true in Σ ($\vDash_\Sigma \varphi$) if and only if DialΣ ⊩ φ.

The rules belong to two categories: particle rules and structural rules. In the remainder of the section, we just give and briefly explain the rules for games reaching multi-S5 valid formulas.[8]

Worlds numbering

The thesis of the dialogue is uttered at a given world w, as well as the subsequent formulas. This world relativization is explicit in dialogue games: we will use labelled formulas of the kind "$w : \varphi$", like in explicit tableau systems. For that purpose, we need a system of world numbering that reflects syntactically the accessibility relation. We will use the following principles, inspired from Fitting numbers for mono-modal logic:

- The initial world is numbered 1. The n immediate successors of w according to the agent j are numbered $wj1, wj2, \ldots, wjn$.

- An immediate successor wju of a world w is said to be of rank $+1$ relative to w, and w is said to be of rank -1 relative to its immediate successors. A successor $wjujv$ of a world w is said to be of rank $+2$ relative to w, etc.

So for instance, a play on a thesis "$1 : \mathbf{K}_i\mathbf{K}_j\varphi$" can reach the following labelled formula: "$1i1j2 : \varphi$."

---

[8] Dialogical modal logics were first introduced in Rahman & Rückert's paper [27]. Readers interested by some more detailed account of both non-modal and modal logics should refer to that paper. For a presentation of game rules close to the present one, see [29].

Particle rules

The meaning of each logical constant is given through a particle rule which determines how to attack and defend a formula whose main connective is the constant in question. The set, **EL^{m}PartRules**, of particle rules for disjunction, conjunction, subjunction, negation, and epistemic operators is recapitulated in the following table:

	Attack	**Defence**
$w : \varphi \vee \psi$	$w : ?$	$w : \varphi$, or $w : \psi$ (The defender chooses)
$w : \varphi \wedge \psi$	$w : ?_L$, or $w : ?_R$ (The attacker chooses)	$w : \varphi$, or $w : \psi$ (respectively)
$w : \varphi \rightarrow \psi$	$w : \varphi$	$w : \psi$
$w : \neg\varphi$	$w : \varphi$	$\otimes$ (No possible defence)
$w : \mathbf{K}_i\varphi$	$w : ?_{\mathbf{K}_i/wiw'}$ (the attacker chooses an available world wiw')	$wiw' : \varphi$
$w : \mathbf{P}_i\varphi$	$w : ?_{\mathbf{P}_i}$	$wiw' : \varphi$ (the available world wiw' being chosen by the defender)

The idea for disjunction is that the proposition $\varphi \vee \psi$, when asserted (at world w) by a player, is challenged by the question *"Which one?"*; the defender has then to choose one of the disjuncts and to defend it against any new attack. The rule is the same for the conjunction $\varphi \wedge \psi$, except that the choice is now made by the attacker: *"Give me the left conjunct ($?_L$)"* or *"Give me the right one ($?_R$)"*, and the defender has to assume the conjunct chosen by his or her challenger. For the conditional $\varphi \rightarrow \psi$, the attacker assumes the antecedent φ and the defender continues with ψ. Negated formulas are attacked by the cancellation of negation, and cannot be defended; the defender in this case can thus only counterattack (if she can).

The particle rules for each epistemic operator $\mathbf{K}_i$ and its dual $\mathbf{P}_i$ ($i \in \{1, \dots, m\}$) enable the players to change the world. They are defined in a way analogous to conjunction and disjunction respectively, regarding the player (challenger or defender) expected to make the relevant choice.

Structural rules

In addition to the particle rules connected to each logical constant, one also needs structural rules to be able to play in such and such a way at the level of the whole game. The first five of them yield games for classical propositional logic, and the last two rules regiment the modal part of epistemic logic.

- (PL-0) **Starting Rule:** The initial formula (the *thesis* of the dialogical game) is asserted by **P** at world 1. Moves are numbered and alternatively uttered by **P** and **O**. Each move after the initial utterance is either an attack or a defence.

- (PL-1) **Winning Rule:** Player **X** wins iff it is **Y**'s turn to play and **Y** cannot perform any move.

- (PL-2) **No Delaying Tactics Rule:** Both players can only perform moves that change the situation.

- (PL-3) **Formal Rule for Atoms:** *At a given world* **P** cannot introduce any new atomic formula; new atomic formulas must be stated by **O** first. Atomic formulas can never be attacked.

- (PL-4c) **Classical Rule:** In any move, each player may attack a complex formula uttered by the other player or defend him/herself against *any attack* (including those that have already been defended).[a]

  A world w is said to be *introduced* by a move in a play when w is first mentioned either through an asserted labeled formula ($w : \varphi$), or through a non-assertive attack ($?_{K_i/w}$).

- (ML-frw) **Formal Rule for Worlds:** **P** *cannot introduce* a new world; new worlds must be introduced by **O** first.

- (ML-S5) **S5 Rule:** **P** can choose any (given) world.[b]

---

[a]This rule can be replaced by the following one to get games for intuitionistic logic:
Intuitionistic Rule: In any move, each player may attack a complex formula uttered by the other player or defend him/herself against *the last attack that has not yet been defended.*
[b]Other structural rules could define other usual modal systems (K, D, T, S4, etc.). See [27].

Now we can build the set of rules for multi-**S5** epistemic logic (**EL**):

$$\text{DialEL} := \textbf{EL}^m \textbf{PartRules}$$
$$\cup \ \{\text{PL-0, PL-1, PL-2, PL-3, PL-4c, ML-frc, ML-S5}\}$$

It is assumed that this dialogical system is sound and complete, i.e.:

$$\text{DialEL} \Vdash \varphi \text{ iff } \vDash_{EL} \varphi.$$

This is shown using strategic tableaus which are similar to usual semantic tableaus, after a reinterpretation of the players' roles.[9]

---

[9]The proofs of soundness and completeness of Dialogical **EL** are non-trivial. They are not explicitly given in [27], even though a halfway point is reached there.

Examples

EXAMPLE 2. Let us consider a substitution instance of the Positive Introspection Property (also known as Axiom **4**): $K_i\varphi \rightarrow K_iK_i\varphi$. As our dialogical rules correspond to **S5**, the proponent is expected to have a winning strategy in the corresponding game.

		O				**P**	
				1 :		$K_ia \rightarrow K_iK_ia$	(0)
(1)	1 :	K_ia	0	1 :		K_iK_ia	(2)
(3)	1 :	$?_{K_i/1i1}$	2	1i1 :	K_ia	(4)
(5)	1i1 :	$?_{K_i/1i1i1}$	4	1i1i1 :	a	(8)
(7)	1i1i1 :	a	1		1 :	$?_{K_i/1i1i1}$	(6)

P wins the play

The numerals within brackets in the external column indicate the moves and the corresponding arguments (here from (0) to (8)); when a move is an attack, the internal array indicates the argument which is under attack; the corresponding defence is written on the same line, even if the move is made later in the play.

Let us comment this particular play in detail. It starts at move (0) with the utterance of the thesis by **P** at world 1 (PL-0). The formula is challenged by **O** at move (1), using the particle rule for implication; at move (2) **P** immediately defends her initial argument. **O** then attacks the epistemic formula at move (3), and using the correlated particle rule as well as (ML-frw), he introduces a new world, 1i1: in dialogical games the opponent is considered as using the best available strategy; he thus jumps from one world to another as much as possible, to prevent the proponent to use his concessions (atomic utterances) at a given world. At move (4) **P** defends her formula using (ML-S5). Then in (5), **O** attacks the epistemic operator introducing once more a new world, 1i1i1. Now the proponent cannot immediately defend her utterance, because it would lead her to utter an atomic formula (a) which has not been previously introduced by **O** at 1i1i1. At move (6), **P** thus counterattacks (1) using the new world introduced by **O**, asking him to utter a at this world (ML-S5); in (7) **O** defends himself uttering a at 1i1i1: the atomic formula is now available for **P**, who can win the play at move (8). As the opponent could not play better – actually, he could not play differently than he did in this play –, this play shows that there is a winning strategy for **P** in the game.

EXAMPLE 3. Now we consider a formula with two epistemic operators: $K_iK_ja \rightarrow (K_ia \wedge K_ja)$. Here it is not enough to consider one play: after move (2), the opponent can choose either the left or the right conjunct. Depending on this choice, the remainder of the play will not be the same. So after checking that the proponent has a winning strategy in plays where **O** chooses the left conjunct, one cannot conclude that she has a winning strategy at all: it must be verified that she can also systematically win against **O** when he chooses the right conjunct.

		O				**P**	
				1 :	$K_i K_j a \rightarrow (K_i a \wedge K_j a)$		(0)
(1)	1 :	$K_i K_j a$	0	1 :	$K_i a \wedge K_j a$		(2)
(3)	1 :	$?_L$	2	1 :	$K_i a$		(4)
(5)	1 :	$?_{K_j/1i1}$	4	1i1 :	a		(10)
(7)	1i1 :	$K_j a$		1 :	$?_{K_i/1i1}$	1	(6)
(9)	1i1 :	a		1i1 :	$?_{K_j/1i1}$	7	(8)

P wins the play

		O				**P**	
				1 :	$K_i K_j a \rightarrow (K_i a \wedge K_j a)$		(0)
(1)	1 :	$K_i K_j a$	0	1 :	$K_i a \wedge K_j a$		(2)
(3*)	1 :	$?_R$	2	1 :	$K_j a$		(4*)
(5*)	1 :	$?_{K_j/1j1}$	4	1j1 :	a		(10*)
(7*)	1 :	$K_j a$		1 :	$?_{K_i/1}$	1	(6*)
(9*)	1j1 :	a		1 :	$?_{K_j/1j1}$	7	(8*)

P wins the play

As expected, there is a winning strategy for the proponent in each case. The formula is thus proved EL valid.

4.2 Adding contextual relativization

Particle rules

The table below gives the particle rules for context-relativized formulas: the rules follow the reduction axioms in a natural way. This new set of particle rules will be referred to as **CEL^m PartRules**.

	Attack	Defence
$w : (p)^{c_i}$ (p being an atom)	$w : c_i$	$w : p$
$w : ((\varphi)^{c_k})^{c_i}$	$w : c_i$	$w : (\varphi)^{c_k}$
$w : (\varphi \vee \psi)^{c_i}$	$w : c_i$	$w : (\varphi)^{c_i} \vee (\psi)^{c_i}$
$w : (\varphi \wedge \psi)^{c_i}$	$w : ?_L$, or $w : ?_R$ (The attacker chooses)	$w : (\varphi)^{c_i}$, or $w : (\psi)^{c_i}$ (respectively)
$w : (\varphi \rightarrow \psi)^{c_i}$	$w : c_i$	$w : (\varphi)^{c_i} \rightarrow (\psi)^{c_i}$
$w : (\neg \varphi)^{c_i}$	$w : c_i$	$w : \neg(\varphi)^{c_i}$
$w : (\mathbf{K}_j \varphi)^{c_i}$	$w : c_x$	$w : \mathbf{K}_j(\varphi)^{c_y}$

Here again, we can give a more explicit version of the last particle rule depending on the kind of the epistemic operator:

	Attack	Defence
$w : (\mathbf{K}_j^{1.1} \varphi)^{c_i}$	$w : c_i$	$w : \mathbf{K}_j^{1.1}(\varphi)^{c_i}$
$w : (\mathbf{K}_j^{1.2} \varphi)^{c_i}$	$w : c_i$	$w : \mathbf{K}_j^{1.2}(\varphi)^{c_j}$
$w : (\mathbf{K}_j^{2.1} \varphi)^{c_i}$	$w : c_j$	$w : \mathbf{K}_j^{2.1}(\varphi)^{c_i}$
$w : (\mathbf{K}_j^{2.2} \varphi)^{c_i}$	$w : c_j$	$w : \mathbf{K}_j^{2.2}(\varphi)^{c_j}$

Structural rule

There is only one structural rule to add to the listing of ML: the rule that states which player can introduce a context c_i, by asserting its characteristic formula c_i. The idea is that a context should not be assumed in any formal proof.

> - (ML-frc) **Formal Rule for contexts: P** *cannot introduce* a new context c in a given world w by playing $w : c$; new contexts must be introduced by **O** first.

The intuition behind such a rule is that a context can be any formula, including atomic ones; so the proponent should have no more power over contexts than she does over atoms.

Now we have the set of rules for **CEL**:

$$\text{DialCEL} := \text{DialEL} \cup \textbf{CEL}^m\textbf{PartRules} \cup \{\text{ML-frc}\}$$

Soundness and completeness will be handled after the following examples.

Examples

In the following two tables, we consider the validity of a contextually modified version of positive introspection, where the consequent is evaluated relative to another context than the antecedent. As easily appears through the games, the upshot depends on the chosen position: the formula is valid according to (2.2), but not valid according to (1.2).

EXAMPLE 4. Using explicit exponents for operators, the formula to be played is the following one: $(\mathbf{K}_i^{1.2}a)^{c_i} \rightarrow (\mathbf{K}_i^{1.2}\mathbf{K}_i^{1.2}a)^{c_j}$.

1.2		O				P	
				1 :	$(\mathbf{K}_i a)^{c_i} \rightarrow (\mathbf{K}_i \mathbf{K}_i a)^{c_j}$		(0)
(1)	1 :	$(\mathbf{K}_i a)^{c_i}$	0	1 :	$(\mathbf{K}_i \mathbf{K}_i a)^{c_j}$		(2)
(3)	1 :	c_j	2	1 :	$\mathbf{K}_i (\mathbf{K}_i a)^{c_i}$		(4)
(5)	1 :	$?_{\mathbf{K}_i/1i1}$	4	1i1 :	$(\mathbf{K}_i a)^{c_i}$		(6)
(7)	1i1 :	c_i	6	1i1 :	$\mathbf{K}_i(a)^{c_i}$		(8)
(9)	1i1 :	$?_{\mathbf{K}/1i1i1}$	8	1i1i1 :	$(a)^{c_i}$		(10)
(11)	1i1i1 :	c_i	10				

O wins the play

After move (11), **P** cannot answer a for it has not been yet introduced by **O** at world $1i1i1$. The only possible solution for **P** would be to attack (1) to force **O** to utter $\mathbf{K}_i a$ at world 1, then to force him to utter a at $1i1i1$. But she cannot, since the opponent never introduced c_i at world 1. So she loses the play.

EXAMPLE 5. Let us now consider the (2.2) version of the same formula: $(\mathbf{K}_i^{2.2}a)^{c_i} \rightarrow (\mathbf{K}_i^{2.2}\mathbf{K}_i^{2.2}a)^{c_j}$.

2.2		O				P		
					1 :	$(\mathbf{K}_i a)^{c_i} \rightarrow (\mathbf{K}_i \mathbf{K}_i a)^{c_j}$		(0)
(1)	1 :	$(\mathbf{K}_i a)^{c_i}$	0		1 :	$(\mathbf{K}_i \mathbf{K}_i a)^{c_j}$		(2)
(3)	1 :	c_i	2		1 :	$\mathbf{K}_i (\mathbf{K}_i a)^{c_i}$		(4)
(5)	1 :	$?_{\mathbf{K}_i/1i1}$	4		1i1 :	$(\mathbf{K}_i a)^{c_i}$		(6)
(7)	1i1 :	c_i	6		1i1 :	$\mathbf{K}_i(a)^{c_i}$		(8)
(9)	1i1 :	$?_{\mathbf{K}_i/1i1i1}$	8		1i1i1 :	$(a)^{c_i}$		(10)
(11)	1i1i1 :	c_i	10		1i1i1 :	a		(18)
(13)	1 :	$\mathbf{K}_i(a)^{c_i}$		1	1 :	c_i		(12)
(15)	1i1i1 :	$(a)^{c_i}$		13	1 :	$?_{\mathbf{K}_i/1i1i1}$		(14)
(17)	1i1i1 :	a		15	1i1i1 :	c_i		(16)

P wins the play

Here the proponent has a winning strategy, thanks to the utterance of c_i by **O** at world 1 in the third move.

Soundness and completeness

THEOREM 6. *Assuming that* DialEL *is sound and complete w.r.t.* $\mathcal{M}^{rst}$, DialCEL *defines a* sound *and* complete *dialogics w.r.t.* $\mathcal{M}^{rst}$, *i.e. for every* CEL *formula* $\varphi \in \mathcal{L}_{KC}^{m}(\mathcal{A}t)$, *the following equivalence holds:*   DialCEL $\Vdash \varphi$ *iff* $\mathcal{M}^{rst} \vDash \varphi$.

The proof is analogous to that of Theorem 1: we use the reduction axioms to translate CEL formulas into EL formulas, and assuming that dialogical logic DialEL is sound and complete for EL, we concentrate on the axiom schemas of CEL.

Proof. From any CEL formula $\varphi \in \mathcal{L}_{KC}^{m}(\mathcal{A}t)$, one can build a tuple $\langle \varphi_0, \ldots, \varphi_n \rangle$ of CEL formulas and reach a formula $\varphi' \in \mathcal{L}_{K}^{m}(\mathcal{A}t)$ such that: $\varphi_0 = \varphi$, $\varphi_n = \varphi'$, and every formula $(\varphi_m \leftrightarrow \varphi_{m+1})$ $(0 \leq m \leq n-1)$ is an instantiation of an axiom schema. As a consequence:

$$\vdash_{\mathsf{CEL}} \varphi \text{ iff } \vdash_{\mathsf{CEL}} \varphi_1 \text{ iff} \ldots \text{iff } \vdash_{\mathsf{CEL}} \varphi_{n-1} \text{ iff } \vdash_{\mathsf{CEL}} \varphi'$$

According to Theorem 1, CEL is complete so:

$$\mathcal{M}^{rst} \vDash \varphi \text{ iff } \mathcal{M}^{rst} \vDash \varphi_1 \text{ iff} \ldots \text{iff } \mathcal{M}^{rst} \vDash \varphi_{n-1} \text{ iff } \mathcal{M}^{rst} \vDash \varphi'$$

Since $\varphi' \in \mathcal{L}_{K}^{m}(\mathcal{A}t)$, as DialEL is sound and complete, we have:

$$\text{DialCEL} \Vdash \varphi' \text{ iff DialEL} \Vdash \varphi' \text{ iff } \mathcal{M}^{rst} \vDash \varphi'.$$

Hence in order to prove soundness and completeness of DialCEL, it suffices to prove that for each reduction axiom schema: $\vdash_{\mathsf{CEL}} \psi \leftrightarrow \psi'$, the following holds: DialCEL $\Vdash \psi \rightarrow \psi'$ and DialCEL $\Vdash \psi' \rightarrow \psi$.

Let us consider the axiom for Negation. We thus have two (kinds of) games.

		O				**P**	
				1 :	$(\neg\varphi)^{c_i} \rightarrow (c_i \rightarrow \neg(\varphi)^{c_i})$	(0)
(1)	1 :	$(\neg\varphi)^{c_i}$	0	1 :	$c_i \rightarrow \neg(\varphi)^{c_i}$	(2)
(3)	1 :	c_i	2	1 :	$\neg(\varphi)^{c_i}$	(4)
(5)	1 :	$(\varphi)^{c_i}$	4		$\otimes$	
(7)	1 :	φ		5	1 :	c_i	(6)
(9)	1 :	$\neg\varphi$		1	1 :	c_i	(8)
		$\otimes$		9	1 :	φ	(10)

O				P		
					1 : $(c_i \rightarrow \neg(\varphi)^{c_i}) \rightarrow (\neg\varphi)^{c_i}$	(0)
(1)	1 :	$c_i \rightarrow \neg(\varphi)^{c_i}$	0		1 : $(\neg\varphi)^{c_i}$	(2)
(3)	1 :	c_i	2		1 : $\neg\varphi$	(4)
(5)	1 :	φ	4		$\otimes$	
(7)	1 :	$\neg(\varphi)^{c_i}$		1	1 : c_i	(6)
	$\otimes$			7	1 : $(\varphi)^{c_i}$	(8)
(9)		c_i	8		1 : φ	(10)

In both plays, the opponent could attack φ after move (10), but as he has already uttered the same formula before, and as the proponent can attack the same argument several times, any strategy deployed by **O** will be turned back as a winning strategy by **P**.

The other implications obtain similarly from the definitions. ∎

5 Epistemological applications

5.1 Epistemological positions formalized

Dialogical CEL provides us with powerful tools for gaining insights into the informal epistemological debate over the context-relativity of knowledge claims presented in section 2. To illustrate this, let us first show how the four epistemological positions alluded to earlier – scepticism, anti-scepticism, contextualism and subjectivism – can be captured within our contextual logico-epistemic framework.

In section 2 the four epistemological positions were introduced in terms of the "ruling out" of "epistemically relevant counter-possibilities", which it is quite natural to understand in terms of S5 epistemic accessibility relations. For if we take it that $\mathcal{R}_i w w'$ iff agent i cannot tell w from w' from all the information available to him at w, we can read "i can rule out w' on the basis of the information he has in w" as "w' is not epistemically accessible for i from w". Moreover, the appropriate epistemic accessibility relations must be equivalence relations, for it is quite natural to think that a possible world is ruled out by a subject as soon as it is not *exactly the same as* the actual world with respect to the totality of the subject's evidence or information, that is, if it differs, were it only in a minimal way, from the actual world in that respect;[10] and *being exactly the same as* is an equivalence relation.

Now, the four epistemological positions are positions for or against the relativity of knowledge claims to a given type of context – relativity to the "knowledge ascriber's" or "attributor's" context for contextualism, to the "knower's" or "knowing

---

[10]Two worlds may differ in numerous respects, and yet be exactly the same with respect to the evidence (conceived of in internalistic terms) at an agent's disposal. For instance, an Evil Genius world would be very different from what we take our world to be, but as the sceptical argument goes, in it we would have exactly the same evidence as we have in our world.

subject's" context for subjectivism, and no relativity to any context for scepticism and anti-scepticism. In order to capture these differences within our contextual logico-epistemic framework, let us opt for the following conventions:

- When dealing with a contextual operator $(\cdot)^{c_i}$, we shall take it that the context c_i stands for the set of formulas that are being presupposed or taken for granted by the agent or group of agents i. So we shall keep in mind that a context c_i may be understood as the result or/and background of a conversation between several agents – what they all take for granted for the purpose of their linguistic interactions –, if i stands for a group of such agents, as well as it may be understood as the result or/and background of an agent's "conversation" with himself – e.g. what he takes for granted for the purpose of his current reflections –, if i is a single agent;

- In general, $(\varphi)^{c_i}$ shall be read as "it follows from context c_i that φ". In particular, even though $(\mathbf{K}_j\varphi)^{c_i}$ can be read as "it follows from context c_i that j knows that φ", it may better be read as "in context c_i, j counts as knowing that φ", that is "given what is taken for granted by i ...". Moreover, when we have a formula of the form $(\mathbf{K}_j\varphi)^{c_i}$, we shall take it that the agent (or group) i is to be called the "attributor" and c_i "attributor i's context", and the agent (or group) j is to be called the "subject". On the other hand, if we need to reflect on a c_x, depending on whether $x = i$ or $x = j$, we shall speak of c_x as of the "attributor's context" or as of the "subject's context".

With these conventions in hand, we can now establish correspondences between the explicitly exponented knowledge operators we distinguished and the different epistemological positions.

Scepticism and anti-scepticism

The two *absolutist* views – scepticism and anti-scepticism – can both be associated with the knowledge operator $\mathbf{K}_j^{1.1}$, whose behaviour with respect to contextual relativization was fully characterized by the following reduction axiom in CEL:

$$\vdash_{\mathsf{CEL}} (\mathbf{K}_j^{1.1}\varphi)^{c_i} \leftrightarrow (c_i \rightarrow \mathbf{K}_j^{1.1}(\varphi)^{c_i})$$

This axiom says that for subject j to count as knowing that φ in attributor i's context c_i, it must follow from this context that j knows that φ holds in that same context. So it is always the very same attributor's context against which an agent will count as knowing something. This is precisely what the two absolutist views about knowledge claims have in common: epistemic standards and relevance sets are not contextually variable but constant matters. The only difference between scepticism and anti-scepticism is that for the former the epistemic standards are too stringent to ever be met while for the latter they are lax enough to be met very often and possibly most of the time. So, the appropriate translation from the informal sceptic/anti-sceptic talk into the formal $\mathbf{K}_j^{1.1}$ talk simply consists in substituting the appropriate contexts – c_{scep} and c_{anti} respectively – for c_i in the axiom above, and in adding the following condition:

$$c_{anti} \rightarrow c_{scep},$$

but not *vice versa*. This is the most minimal way to capture the idea that while c_{anti} excludes all the far-fetched possibilities arising from radical sceptical concerns and encapsulates a rather large set of presuppositions shared by most agents in their ordinary talk about knowledge, c_{scep} excludes all such presuppositions and encapsulates a rather large set of far-fetched possibilities.[11] For instance, the sceptic about contingent truths may not be a sceptic about necessary truths and thus may take the proposition that $2 \times 2 = 4$ into his c_{scep}, just as the anti-sceptic has this in his c_{anti}; but whilst the anti-sceptic will also have the proposition that he is not a victim of an Evil Genius in his c_{anti}, the sceptic will not let this into his c_{scep}. Actually, c_{scep} can be identified with $\top$, which means that context-relativization according to scepticism is no relativization at all.

Contextualism and subjectivism

Contrary to the absolutist views, the two relativist views, according to which epistemic standards and relevance sets are contextually variable, fall under different knowledge operators. Contextualism is the view that the variability in question is a variability according to the attributor's context, not the subject's. It is thus natural to associate contextualism with operator $\mathbf{K}_j^{1.2}$, whose behaviour with respect to contextual relativization was fully characterized by:

$$\vdash_{CEL} (\mathbf{K}_j^{1.2}\varphi)^{c_i} \leftrightarrow (c_i \rightarrow \mathbf{K}_j^{1.2}(\varphi)^{c_j}),$$

for this axiom says that for subject j to count as knowing that φ in attributor i's context c_i, it must follow from *this* context that j knows that φ holds in j's own context, that is, in the context whose attributor is subject j himself. So it is not always the same attributor's context against which something will count as being known. Relativity to the attributor's context is thus encapsulated in knowledge operator $K_j^{1.2}$. In contrast, relativity to the subject's context is encapsulated in knowledge operator $K_j^{2.2}$, since the reduction axiom characterizing its behaviour when contextualized was:

$$\vdash_{CEL} (\mathbf{K}_j^{2.2}\varphi)^{c_i} \leftrightarrow (c_j \rightarrow \mathbf{K}_j^{2.2}(\varphi)^{c_j}),$$

---

[11] What we said in footnote 5 about the definition of anti-scepticism in the contextual model framework holds *mutatis mutandis* in the current framework. One can work with the version of anti-scepticism they favor simply by specifying what, according to their version, can be considered the constant set of epistemically relevant presuppositions, that is, the constant set of literals that constitute c_{anti}.

which says that for subject j to count as knowing that φ in attributor i's context c_i, it must follow from subject j's own context that j knows that φ holds in the same context, that is, subject j's context. We can thus naturally associate the subjectivist view, according to which epistemic standards and relevance sets shift with the subject's context, with operator $\mathbf{K}_j^{2.2}$.

What about the remaining operator?

One may ask which epistemological view could match the $\mathbf{K}^{2.1}$-operator. In our opinion the main interest of this operator does not lie so much in its possible correspondence with a view to be found in the epistemological literature as in the means it would offer us to handle indexicality phenomena, for instance if we wanted to incorporate personal pronouns like 'I' and 'he'. Consider a true utterance of 'He knows that I am here'. To formally account for its truth, we would need a knowledge operator whose logical behavior would match the following equivalence: $(\mathbf{K}_{he} I \ am \ here)^{c_I} \leftrightarrow (c_{he} \rightarrow \mathbf{K}_{he}(I \ am \ here)^{c_I})$, where c_I is the context of the agent who uses 'I' and c_{he} the context of the agent referred to by 'he'. Of course, we would not thereby have accounted for even a bit of the great complexity of natural language indexicality, and it is not our intention to do so at all.

5.2   Knowledge features uncovered

This formal translation of the four epistemological positions can now be exploited within our contextual logico-epistemic framework. In the remainder of this section we will use dialogical CEL to illustrate how differences between the four epistemological positions that have been pointed out in the contemporary philosophical literature on knowledge can be recovered within our formal framework, as well as to illustrate how differences that have not been touched upon in the literature can be discovered through that framework. We will give four such illustrations.

Normality

One thing to note is that the three operators $\mathbf{K}_j^{1.1}$, $\mathbf{K}_j^{1.2}$, $\mathbf{K}_j^{2.2}$, behave in the same manner with respect to the contextualized version of the $\mathbf{K}$ axiom for knowledge, viz.:

$$\vdash_{\mathsf{CEL}} (\mathbf{K}_j^{u.v}(\varphi \rightarrow \psi) \rightarrow (\mathbf{K}_j^{u.v}\varphi \rightarrow \mathbf{K}_j^{u.v}\psi))^{c_i}.$$

For any value of φ, ψ, u and v, this is a theorem of CEL and it is valid with respect to the class $\mathcal{M}^{rst}$ of Kripke models with equivalence accessibility relations. Although it is a trivial result from a logico-epistemic point of view, it is not from an epistemological one. One can tell from the literature that it is important for advocates of the four identified epistemological positions that knowledge closure under known material implication holds. This is crucial to both the famous sceptical argument from ignorance and to the famous Moorean anti-sceptical response to it. Even those who take it that knowledge claims are context-relative admit that closure holds while insisting that it holds only within contexts and not across contexts (see [19] for the contextualist case and [13] for the subjectivist case, for instance).

Factivity (or not)

A more interesting result is that our framework clearly establishes a difference between case (1.1) – absolutists – and cases (1.2) and (2.2) – relativists – with respect to the contextualized version of the T axiom – call it $(T)^c$:

$$(\mathbf{K}_j^{u.v}\varphi \to \varphi)^{c_i}.$$

Indeed, let us consider the following example of CEL-dialogical games for $(T)^c$, where φ is an epistemic formula:

EXAMPLE 7. In what follows, we compare the factivity of contextual knowledge according to 1.1, 1.2 and 2.2. In particular, we will check whether $(\mathbf{K}_j\mathbf{K}_k p \to \mathbf{K}_k p)^{c_i}$ is valid or not.

1.1-version: $(\mathbf{K}_j^{1.1}\mathbf{K}_k^{1.1}p \to \mathbf{K}_k^{1.1}p)^{c_i}$

1.1		**O**				**P**	
					1:	$(\mathbf{K}_j\mathbf{K}_k p \to \mathbf{K}_k p)^{c_i}$	(0)
(1)	1:	c_i	0		1:	$(\mathbf{K}_j\mathbf{K}_k p)^{c_i} \to (\mathbf{K}_k p)^{c_i}$	(2)
(3)	1:	$(\mathbf{K}_j\mathbf{K}_k p)^{c_i}$	2		1:	$(\mathbf{K}_k p)^{c_i}$	(4)
(5)	1:	c_i	4		1:	$\mathbf{K}_k(p)^{c_i}$	(6)
(7)	1:	$?_{\mathbf{K}_k/1k1}$	6		$1k1:$	$(p)^{c_i}$	(8)
(9)	$1k1:$	c_i	8		$1k1:$	p	(20)
(11)	1:	$\mathbf{K}_j(\mathbf{K}_k p)^{c_i}$	3	1:	c_i		(10)
(13)	1:	$(\mathbf{K}_k p)^{c_i}$	11	1:	$?_{\mathbf{K}_j/1}$		(12)
(15)	1:	$\mathbf{K}_k(p)^{c_i}$	13	1:	c_i		(14)
(17)	$1k1:$	$(p)^{c_i}$	15	1:	$?_{\mathbf{K}_k/1k1}$		(16)
(19)	$1k1:$	p	17	$1k1:$	c_i		(18)

P wins the play

1.2-version: $(\mathbf{K}_j^{1.2}\mathbf{K}_k^{1.2}p \to \mathbf{K}_k^{1.2}p)^{c_i}$

1.2	O			P		
				1:	$(K_jK_kp \rightarrow K_kp)^{c_i}$	(0)
(1)	1:	c_i	0	1:	$(K_jK_kp)^{c_i} \rightarrow (K_kp)^{c_i}$	(2)
(3)	1:	$(K_jK_kp)^{c_i}$	2	1:	$(K_kp)^{c_i}$	(4)
(5)	1:	c_i	4	1:	$K_k(p)^{c_k}$	(6)
(7)	1:	$?_{K_k/1k1}$	6	1k1:	$(p)^{c_k}$	(8)
(9)	1k1:	c_k	8			
(11)	1:	$K_j(K_kp)^{c_j}$	3	1:	c_i	(10)
(13)	1:	$(K_kp)^{c_j}$	11	1:	$?_{K_j/1}$	(12)

O wins the play

Short comment: After move (13) **P** would have to utter "$1 : c_j$"; but this context has not been previously introduced by **O**, so she cannot.

$$\text{2.2-version: } (K_j^{2.2}K_k^{2.2}p \rightarrow K_k^{2.2}p)^{c_i}$$

2.2	O			P		
				1:	$(K_jK_kp \rightarrow K_kp)^{c_i}$	(0)
(1)	1:	c_i	0	1:	$(K_jK_kp)^{c_i} \rightarrow (K_kp)^{c_i}$	(2)
(3)	1:	$(K_jK_kp)^{c_i}$	2	1:	$(K_kp)^{c_i}$	(4)
(5)	1:	c_k	4	1:	$K_k(p)^{c_k}$	(6)
(7)	1:	$?_{K_k/1k1}$	6	1k1:	$(p)^{c_k}$	(8)
(9)	1k1:	c_k	8			

O wins the play

Short comment: Here **P** cannot even attack (3): with (2.2), she would have to utter "$1 : c_j$", which has not been introduced by **O**.

What this shows is that absolutist contextual knowledge is always factive while relativist contextual knowledge is not always factive. More specifically, $(T)^c$ holds for relativist knowledge operators $K_j^{1.2}$ and $K_j^{2.2}$ when they bear on "absolute" non-epistemic formulas, but not in the general case.

A precision is required here: There is a possible loss of factivity for contextualism or subjectivism on absolute formulas, but this would be a trivial one like the loss of factivity possibly occurring in standard multi-modal epistemic logic: the formula $K_jp \rightarrow p$ is trivially falsified at any world $w^\star$ lying beyond the scope of the accessibility relation $\mathcal{R}_j$ where p is false. Analogous cases in CEL are formulas like $(K_jp \rightarrow p)^{c_i}$, which are falsified at worlds where c_i is true while c_j and p are false; but of course, c_i is as irrelevant for the agent i's context-relativized knowledge as is the world $w^\star$ for his absolute knowledge. However, the situation is different with epistemic formulas like the ones just evaluated through dialogical games: here the context used to falsify the formulas is perfectly relevant to evaluate the agents' knowledge, as is seen in Figure 1:

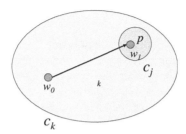

Figure 1. *A counter-model to* $(\mathbf{K}_j^{.2}\mathbf{K}_k^{.2}p \rightarrow \mathbf{K}_k^{.2}p)^{c_i}$

To our knowledge this result that factivity for contextualism and subjectivism is restricted to knowledge of the world (*versus* of other people's knowledge) has never been highlighted in the epistemological literature.[12]

Context-relativized introspection

Another first for epistemological discussions over knowledge and context is the following difference between contextualism and subjectivism: if subjectivism is true, what one knows in one's own context, one also knows that one knows it in anyone else's context; whereas if contextualism is true, what one knows in one's own context, one may not know that one knows it in anyone else's context. This is clear from the dialogical games in examples 4 and 5 in the previous section, which made it explicit that:

- $\nvdash_{\mathrm{CEL}} (\mathbf{K}_i^{1.2}p)^{c_i} \rightarrow (\mathbf{K}_i^{1.2}\mathbf{K}_i^{1.2}p)^{c_j}$
- $\vdash_{\mathrm{CEL}} (\mathbf{K}_i^{2.2}p)^{c_i} \rightarrow (\mathbf{K}_i^{2.2}\mathbf{K}_i^{2.2}p)^{c_j}$

What this means is that for a subjectivist agent i – case (2.2) –, if it follows from the context c_i of which he is the attributor that he knows that p, then it follows from any other attributor j's context that he knows that p; while for a contextualist agent, this is not true. This can be explained informally as follows:

- (Subjectivist case) If subjectivism is true, then for agent i to know that p, he must meet the standards in place in his own context c_i for knowing that p. But he will then *ipso facto* meet the standards for his knowing that he

---

[12]However, see Stanley [33] for a discussion of factivity and related matters in a subjectivist setting.

knows that p. For if he did not know that he knows that p, that would be because he considers it possible that he does not know that p; but he cannot consider this a serious possibility if he already knows that p. And if i counts as knowing that he knows that p in his own context, and since we are dealing with subjectivism, i will count as knowing that he knows that p in any other agent j's context.

- (Contextualist case) If contextualism is true, then the relevant standards for knowing a proposition may vary from one attributor's context to that of another. This being so, agent i may count as knowing that p relative to the context c_i of which he is the attributor, whilst not relative to the context c_j of another attributor j associated with more stringent standards than in c_i. In this case, since one cannot know what is false and since it is false in c_j that i knows that p, it is false in this same c_j that i knows that he knows that p.

It is an advantage of our formal framework that it can capture this informal difference between contextualism and subjectivism.

Mixing agents

Another interesting feature of that framework is that it allows us to reason about knowledge in a group of epistemologically heterogeneous agents and to answer formally such informal questions as "If a contextualist knows that a subjectivist knows this or that, does the contextualist know this or that?".[13] Here we give only one example we find interesting of that feature through the following question: if an absolutist agent knows that a subjectivist agent knows a proposition relative to a context, does the subjectivist know that proposition relative to that context? In our framework this question becomes that of deciding whether the following formula is a theorem of CEL:

$$(\mathbf{K}_j^{1.1}\mathbf{K}_k^{2.2}p)^{c_i} \to (\mathbf{K}_k^{2.2}p)^{c_i}.$$

Now, this question is easily settled by means of the following CEL-dialogical game:

EXAMPLE 8. In the following example, we consider simultaneously agents of different kinds. The upshot is a kind of failure of Axiom T for an absolutist knower.

---

[13] Suppose that agent i is a F-ist (a sceptic, an anti-sceptic, a contextualist, or a subjectivist). If i is coherent with his own theory of knowledge (ascriptions), which he intends to be the only correct one for *any* agent's knowledge, he ought to reason accordingly not only about his own knowledge but also about other agent's knowledge. This amounts to saying *both* that his reasoning about knowledge ought to meet the subjectivist expectations and that he ought to expect other people to meet the same expectations when they reason about knowledge. That is our motivation for talking about sceptical, anti-sceptical, contextualist, and subjectivist *agents* and for asking what they can know about what other types of agents know.

O				P		
	O			**P**		
				$1:$	$(K_j^{1.1} K_k^{2.2} p)^{c_i} \rightarrow (K_k^{2.2} p)^{c_i}$	(0)
(1)	$1:$	$(K_j^{1.1} K_k^{2.2} p)^{c_i}$	0	$1:$	$(K_k^{2.2} p)^{c_i}$	(2)
(3)	$1:$	c_k	2	$1:$	$K_k^{2.2}(p)^{c_k}$	(4)
(5)	$1:$	$?_{K_k/1k1}$	6	$1k1:$	$(p)^{c_k}$	(8)
(9)	$1k1:$	c_k	8			

O wins the play

Short comment: **P** cannot go on and attack (1), for it would require that she utter "$1 : c_i$", which has not been introduced by **O**.

Thus, the formula is not a principle of our contextual epistemic logic. This comes as an amendment to, or better as a complement to what we said earlier about the factivity of absolutist knowledge. For the lesson to be drawn from this is, roughly, that an absolutist agent (a sceptic, an anti-sceptic) may know something without this something being true when this something is about what a subjectivist agent knows.

Incidentally, we may notice that the following formula, differing from the previous one in that the subscripted agent is now the same for each occurrence of a knowledge operator, is not a principle of our contextual epistemic logic either:

$$(K_j^{1.1} K_j^{2.2} p)^{c_i} \rightarrow (K_j^{2.2} p)^{c_i}.$$

Funnily enough, this could be interpreted in terms of "epistemically schizophrenic" agents whose knowledge is compartmented in the sense that they know different things when they are in a subjectivist or in a sceptical mood from what they know when they are in an anti-sceptical or in a contextualist mood, or in the sense that the subjectivist or the sceptical part of them knows different things from what the anti-sceptical or the contextualist part of them knows. Then the lesson to be drawn from the result in question would be that what your sceptic or anti-sceptic compartment knows about your subjectivist compartment, your subjectivist compartment may not know of itself.

6   Conclusion

Our primary goal in this paper was to investigate the relationships between knowledge and context in the formal framework of epistemic modal logic. We thus provided an epistemic logic with context relativization, **PAL**, together with its dialogical semantics, and applied it to epistemological issues.

The subsequent results can be interesting both from a logical and from an epistemological point of view. From the former point of view, the interesting upshot is that the logic of public announcements can be translated into a logic for context.

The interaction between knowledge and context is slightly more subtle than that of knowledge with announcements, but the result is really close to PAL.

From the epistemological point of view, this time, the interesting result is that CEL provides us with a powerful formal tool not only for capturing informal views about knowledge and context, but also for gaining new insights into the debates over their possible interconnections, contributing thereby to the current research program in formal epistemology, at the interface of logic and the philosophy of knowledge.

Let us just add that two typical "Rahmanian issues" came out from this work. First, the dialogical version of PAL and its application to epistemology constitute a new confirmation of the fruitfulness of dialogical logic as a framework to combine different logics. Second, we considered diverging agents mutually reasoning about their respective knowledge; this strongly echoes Shahid's recent work [26] about non-normal logics, classical agents reasoning about intuitionistic ones. Of course, all our agents are normal (and even S5) knowers; anyway as epistemologists, some of them appear to be strange, to say the least.

Acknowledgements The authors wish to thank Berit Brogaard, Bertram Kienzle, Helge Rückert, Tero Tulenheimo and an anonymous referee for their inspiring comments and suggestions about earlier versions of this paper.

BIBLIOGRAPHY

[1] Austin, J., 1946: Other Minds, *Proceedings of the Aristotelian Society, Supplementary Volume* **20**, 148–187.

[2] van Benthem, J., 2000: Information Update as Relativization, ILLC, University of Amsterdam, Available from http://staff.science.uva.nl/~johan/upd=Rel.pdf.

[3] van Benthem, J., 2006: Epistemic Logic and Epistemology: The State of Their Affairs, *Philosophical Studies* **128**: 49–76.

[4] van Benthem, J., J. van Eijck & B. Kooi, 2006: Logics of Communication and Change, *Information and Computation* **204**, 11, 1620–1662.

[5] Blackburn, P., 2001: Modal Logic as Dialogical Logic, *Synthese* **127** (1–2): 57–93.

[6] Bonnay, D. & P. Égré, 2006: A Non-Standard Semantics for Inexact Knowledge with Introspection, in Artemov, S. & R. Parikh (eds.), *Proceedings of the Workshop on Rationality and Knowledge ESSLLI 2006*.

[7] Cohen, S., 2000: Contextualism and Skepticism, in Sosa E. and E. Villanueva (eds.), *Scepticism: Philosophical Issues*, Vol.10, Basil Blackwell: 94–107.

[8] DeRose, K., 1995: Solving the Skeptical Problem, *Philosophical Review* **104**, 1–52.

[9] Dretske, F. I., 1970: Epistemic Operators, *Journal of Philosophy* **67**, 1007–23.

[10] Dretske, F. I., 1981: The Pragmatic Dimension of Knowledge, *Philosophical Studies* **40**, 363–78.

[11] Fagin R. & J. Y. Halpern, 1988: Belief, Awareness, and Limited Reasoning, *Artificial Intelligence* **34**, 39–76.

[12] Gargov, G., S. Passy & T. Tinchev, 1987: Modal Environment for Boolean Speculations, in D. Scordev (ed.), *Mathematical Logic and its Applications*, New York, Plenum Press, 253–263.

[13] Hawthorne, J., 2004: *Knowledge and Lotteries*, Oxford University Press.

[14] Hendricks, V. F., 2004: Hintikka on Epistemological Axiomatizations, *in* Kolak D. & J. Symons (eds.), *Questions, Quantifiers and Quantum physics: Essays on the Philosophy of Jaakko Hintikka*, Springer: 3–34.

[15] Hendricks, V. F., 2006: *Mainstream and Formal Epistemology*, Cambridge University Press.

[16] Hendricks, V. & D. Pritchard (eds.), 2006: *New Waves in Epistemology*, Ashgate.

[17] Hintikka, J., 1962: *Knowledge and Belief: An Introduction to the Logic of the Two Notions*, Cornell University Press.

[18] Kaplan, D., 1989: Demonstratives, *in* J. Almog, J. Perry, & H. Wettstein (eds.), *Themes from Kaplan*: 481–563.

[19] Lewis, D. K., 1996: Elusive Knowledge, *Australasian Journal of Philosophy* **74**, 549–67.

[20] Lihoreau, F. & M. Rebuschi, 2007: Raisonner sur la connaissance en contexte, forthcoming in *Actes du Colloque International 'Computers and Philosophy' 2006*, Editions CNRS.

[21] Ludlow, P., 2005: Contextualism and the New Linguistic Turn In Epistemology, in *Contextualism in Philosophy* (G. Preyer and G. Peter, eds.), Oxford, Oxford University Press.

[22] MacFarlane, J., 2005: The Assessment Sensitivity of Knowledge Attributions, *in* Szabó, T. & J. Hawthorne (eds.), *Oxford Studies in Epistemology 1*, Oxford University Press: 197–233.

[23] Moore, G. E., 1962: *Commonplace Book: 1919–1953*, Allen & Unwin.

[24] Nozick, R., 1981: *Philosophical Explanations*, Oxford University Press.

[25] Plaza, J. A., 1989: Logics of Public Communications, in Emrich, M. *et al.* (eds.), *Proceedings of the 4th International Symposium on Methodologies for Intelligent Systems*, North-Holland: 201–216.

[26] Rahman, S., 2006: Non-Normal Dialogics for a Wonderful World and More, in J. van Benthem *et al*, eds., *The Age of Alternative Logics*, Springer, Dordrecht, 311–334.

[27] Rahman, S. & H. Rückert, 1999: Dialogische Modallogik für T, B, S4 und S5, *Logique et Analyse*, **167-168**, 243–282.

[28] Rantala, V., 1982: Impossible World Semantics and Logical Omniscience, *Acta Philosophica Fennica* **35**, 18–24.

[29] Rebuschi, M., 2005: Implicit vs. Explicit Knowledge in Dialogical Logic, (to appear) *in* O. Majer et al. (eds.), 200?: *Logic, Games and Philosophy: Foundational Perspectives*, Dordrecht, Springer.

[30] Richard, M., 2004: Contextualism and Relativism, *Philosophical Studies* **119**, 215–42.

[31] Schlenker, Ph., 2003: A Plea for Monsters, *Linguistics and Philosophy* **26**: 29–120.

[32] Stalnaker, R., 2006: On the Logics of Knowledge and Belief, *Philosophical Studies* **128**, 169–199.

[33] Stanley, J., 2005: *Knowledge and Practical Interests*, Oxford University Press.

[34] Unger, P., 1971: A Defense of Scepticism, *Philosophical Review* **80**, 198–219.

Common Cause Abduction and the Formation of Theoretical Concepts in Science

GERHARD SCHURZ

ABSTRACT. Abductions are conceived as special patterns of inference to the best explanation whose structure *determines* a particularly *promising* abductive conjecture (conclusion) and thus serves as an *abductive search strategy* (§1). An important distinction is that between *selective abductions*, which choose an optimal candidate from a given multitude of possible explanations, and *creative abductions*, which introduce new theoretical models or concepts (§2). The paper focuses on creative abductions, which are essential for scientific progress, although they are rarely discussed in the literature. It is suggested to demarcate scientifically fruitful abductions from purely speculative abductions by the *criterion of causal unification* (§3). Based on various historical examples it is demonstrated that *common cause abduction from correlated dispositions* is the fundamental abductive operation by which new theoretical concepts are scientifically generated (§4). *Statistical factor analysis* can be regarded as a statistical generalization of common cause abduction (§5). When scientists start to develop theoretical models of their conjectured common (unobservable) causes, common cause abduction turns into what is called *theoretical model abduction* (§6).

1 On the relation between inference to the best explanation and abduction

Harman [21] understood inference to the best explanation (IBE) and abduction as more or less equivalent. Both inferences serve the goal of inferring something about the *unobserved causes* or *explanatory reasons* of the observed events. This was also the understanding of abduction in the mind of the inventor of 'abduction', C.S. Peirce (cf. [47, §5.189]). Nevertheless, I suggest to make a difference here. By an 'inference' I mean a certain (logically explicable) *schematic pattern* which *specifies* the conclusion as a (syntactical) *function* of the premises. In the case of an abductive inference, the premises describe the phenomenon which is in need of explanation, possibly together with background knowledge of a certain form, and the conclusion is an abductive conjecture which is set out to further test operations. However, in the case of IBE no such pattern exists. Rather, the space of possible explanatory hypotheses and their evaluation must already be given in the premises, in order to apply the rule of IBE. In other words, IBE is more an instance of the rule of *rational choice* than an inference operation.

All inferences have a *justificational* (or 'inferential') and a *strategical* (or 'discovery') function, but to a different degree (see also Gabbay/Woods [12, §1.1]). Their justificational function consists in the justification of the conclusion, *conditional* to the justification of the premises. Their strategical function consists in pointing to a most promising conjecture (conclusion) which is set out to further test operations, or in Hintikka's words, which stimulates new questions (Hintikka [22, p. 528]). In abductive inferences the strategical function becomes *crucial*. Different from the situation of induction, in abduction problems we are often confronted with thousands of possible explanatory conjectures (or conclusions) – everyone in the village might be the murderer. The essential function of abductions is their role as *search* strategies which tell us which explanatory conjecture we should set out *first* to further inquiry (cf. [22, p. 528]) – or more generally, which suggest us a *short* and *most promising* (though not necessarily successful) path through the exponentially explosive search space of possible explanatory reasons.

Therefore I suggest in [56] to understand abductions as special patterns of inferences to the best explanation whose structure *determines* a particularly *promising* abductive conjecture (conclusion) and thus serves as an *abductive search strategy* in the space of possible explanatory hypotheses. It is essential for a good search strategy that it leads us to an optimal conjecture not only in a finite but in a *reasonable* time. In this respect, the general rule of IBE fails completely. If you ask which explanatory conjecture you should choose for further investigation among thousands of possible conjectures, the rule IBE just tells us: "Find out which is the best (available) conjecture and then choose it." To see the joke behind, think about someone in a hurry who asks an IBE-philosopher for the right way to the railway station and receives the following answer: "Find out which is the shortest way among all ways between here and the train station which are accessible to you – this is the way you should choose."

In contrast to their strategical function, the justificational function of abductions is minor. Peirce has pointed out that abductive hypotheses are prima facie not even probable, like inductive hypotheses, but merely possible [47, §5.171]. Only upon being confirmed in further tests, an abductive hypothesis may become probable. However, I cannot completely agree with Peirce or other authors (e.g. Hanson [20], Hintikka [22]) who think that abductions are merely a discovery procedure and their justification value is zero. Niiniluoto has pointed out that "abduction as a motive for pursuit cannot always be sharply distinguished from considerations of justification" [40, p. 442]. This paper will confirm Niiniluoto's point for the considered patterns of abduction: their strategical function goes hand in hand with a (weak) justificational value.

2   Selective versus creative abductions

In Schurz [56] a classification of different patterns of abduction is provided, which reaches from fact-abductions, 1st order existential abductions, and law-abductions to theoretical model abductions and 2nd order existential abductions which introduce new concepts. One result of the analysis in [56] is this: the epistemological function and the evaluation criteria of abduction are rather different for different kinds of abduction patterns. An important distinction in this respect is that between selective and creative abductions (cf. also Magnani [35, p. 20]). *Selective* abductions choose the best candidate among a multitude of possible explanations which is determined by the background knowledge, according to some selection strategy. The most frequently discussed pattern of selective abduction is backward reasoning through given causal laws (the double line indicates that the inference is uncertain and non-monotonic):

Known Law: If it rains, the street is wet ($\forall x(Cx \to Ex)$).
Known Evidence: The street is wet (Ea).

Abduced Conjecture: (Probably) It has rained (Ca).

The young Peirce [46] has explicated abduction in this narrow way. Since the background knowledge contains usually many laws of the form "$\forall x(C_i x \to Ex)$", and iteratively "$\forall x(D_j x \to C_i x)$", etc., this kind of abduction leads into backward chaining through known causal from a given explanandum to a set of possible explanatory reasons. Among the possible explanatory reasons one has to select the most plausible one. In this form, abductive inference has been studied in detail in AI research (cf. Paul [44], Josephson & Josephson [27], Flach & Kakas [11]).

Selective abductions remain within the search space determined by the known background laws. They can never create new laws or even new concepts. Exactly this is the task of *creative* abductions. The later Peirce considered scientific abduction as an essentially creative operation by which new laws, theories or concepts are discovered [47, §5.170]. But how and by which kinds of rules should that be possible? For example, can there ever be rules which achieve the following creative abduction?

Known Phenomena: Observable properties of substances
═══════════════════════════════??? ⇓ ??? ═══════════════
Abduced Conjecture: Molecular models of these substances

Peirce kept silent about this question. Until today, most philosophers of science are skeptical whether the scientific discovery of new theories or theoretical concepts follows any abductive rules. Has *Popper* here the last word who has repeatedly argued that only the question of justifying an already *given* theory has a 'logic', while the question of discovering a *new* theory is basically a matter of guessing?

Peirce once remarked that there are sheer myriads of possible hypotheses which would explain the experimental phenomena, and yet scientists have usually managed to find the true hypothesis after only a small number of guesses (cf. [48, §6.5000]). Peirce explained this miraculous ability of human minds by their *abductive instincts* (cf. §5.47, footnote 12, §5.172 and §5.212 of [48]). But I suggest we should not put too much trust in the abductive instincts of humans – for these abductive instincts have produced too many speculative or even irrational pseudo-explanations. The same problem arises for Harman's IBE: if IBE is understood as an inference to the best *available* explanation, then it is unacceptable, because the best available explanation is not always *good enough* to be rationally acceptable (cf. Lipton [34, p. 58]). If a phenomenon is novel and poorly understood, then one's best available explanation is usually a *pure speculation*. For example, in the early animistic word-views of human mankind the best available explanations of natural phenomena such as the sun's path over the sky was that the involved entities (here: the sun) are intentional agents. Such speculative explanations are not acceptable in science, because they do not meet important methodological criteria, which are discussed in §§3-4.

Therefore the Peircean question *returns*: how was it possible for three centuries of scientists, after millennia of idling speculations, to find out so many true theoretical explanations in the astronomically large space of possible explanatory stories? This questions demands for an answer. The answer depends on whether there can exist anything like a 'logic' of discovery. The true observation of Popper and the logical positivists that the justification of a hypothesis is independent from the way it was discovered does not imply that it would not be *desirable* to have in addition *good heuristic rules for discovering new explanatory hypotheses and concepts* – if there only *were* such rules (cf. also Hanson [20]). In the following sections I try to show that there *are* such rules.

Creative abductions are rarely discussed in the literature. The only mechanism which according to my knowledge has been suggested in the abduction literature is *abduction from analogy*. To take Thagard's example ([62, p. 67]), the new concept of a sound wave is achieved by analogical transfer of the already possessed concept of *water waves* to the domain of sound. Analogical abductions are certainly very important in science, but I doubt that they can explain the discovery of all new theoretical models and concepts. How should the concepts of gravitational force, electrical force, atom and molecule, polar and non-polar bonding, acid and base (etc.), be obtained from analogy – analogy with what? Moreover, analogical abductions have led humans more often into error than to the truth – the model of natural phenomena by intentional agents is a nice analogy, but is clearly false from the scientific viewpoint. So there is a need of a more fundamental operation of abduction by which science conjectures new and unobservable kinds of properties or entities or new theoretical models which are not obtained from analogy. I will argue that the structure of this kind of abduction is *hypothetical (common) cause abduction*, and I will elaborate it in the next sections.

3   Hypothetical (common) cause abduction

Hypothetical (common) cause abduction is the essential operation of advanced explanatory reasoning. It starts at the point where the work of empirical induction has already been done. The explanandum of hypothetical (common) cause abduction consists either (a) in one phenomenon or (b) in several mutually *intercorrelated* phenomena (properties or regularities). One abductively conjectures in case (a), that the phenomenon is the effect of a hypothetical (unobservable) cause, and in case (b) that the phenomena are effects of a hypothetical *common* cause. In both cases, the abductive conjecture postulates a *new unobservable entity* (property or kind) together with *new laws* connecting it with the observable properties, without drawing on analogies to concepts with which one is already familiar. In this section I will argue that only case (b) constitutes a scientifically worthwhile abduction, while (a) is a case of pure speculation. Also Salmon ([53, p. 213ff]) has emphasized the importance of finding common cause explanations for the justification of scientific realism. But Salmon did not inform us about the crucial difference between scientific common cause abduction and speculative (cause) abduction. I will argue that the major criterion for this distinction is *causal unification*.

Ockham's razor is a broadly accepted maxim among IBE-theorists: an explanation of observed phenomena should postulate as few unobservable or new entities or properties as possible (cf. Moser [38], p. 97-100, who calls them "gratuitous entities"). After closer inspection this maxim turns into a gradual optimization criterion. For an explanation is the better, the *less* new entities it postulates, and the *more* phenomena it explains (cf. Moser's definition of "decisively better explanations" [38], p. 89). But by introducing sufficiently many 'hidden entities' one can 'explain' anything one wants. Where is the borderline between 'reasonably many' and 'too many' entities postulated for the purpose of explanation? I suggest the following

(CU) *Causal Unification Criterion for Conceptually Creative Abduction*: The introduction of *one* new entity or property merely for the purpose of explaining *one* phenomenon is always speculative and ad hoc. Only if the postulated entity or property explains *many intercorrelated* but *analytically independent* phenomena, and in this sense yields a *causal* or *explanatory unification*, it is a legitimate scientific abduction which is worthwhile to be put under further investigation (cf. also Schurz & Lambert [58, §2.3]).

I first illustrate the criterion by way of examples. The simplest kind of a speculative abduction 'explains' every particular phenomenon by a special 'power' who (or which) has caused this phenomenon as follows (for '$\psi_E x$' read 'a power of kind ψ wanted that E happens to x'):

Speculative Fact Abduction:	*Example*:
Explanandum E: Ea	John got ill.

Conjecture H:	Some power wanted that John gets ill,
$\psi_E a \land \forall x(\psi_E x \rightarrow Ex)$	and whatever this power wants, happens.

This speculative fact-abduction schema has been applied by our human ancestors since the earliest times: all sorts of unexpected events can be explained by assuming one or several God-like power(s). Such pseudo-explanations clearly violate Ockham's razor: they do *not* offer *proper* unification, because for every event (Ea) a special hypothetical 'wish' of God $(\psi_E a)$ has to be postulated (cf. [58, p. 86]). For the same reason, such pseudo-explanations are entirely *post-hoc* and have *no predictive power* at all, because God's unforeseeable decisions can be known only *after* the event has already happened. In §4 it will be shown that there is a systematic connection between causal unification and increase of predictive power. Observe how my analysis differs from Kitcher's analysis [28, p. 528f] who refutes the speculative fact-abduction pattern as a 'spurious' unification because it is not *stringent* enough, in the sense that one may insert any sentence whatsoever for the statement Ea. But according to my suggested criterion (CU), this schema does not provide merely 'non-stringent' or otherwise defective unification – it does not provide unification *at all*.

A *Bayesian* would probably object that there is no *real need* for criterion (CU) – *all* what we need is a good theory of confirmation, and this is *Bayesian confirmation theory*. To this objection I would counter that it is more based on *wishful thinking* than on truth: Bayesian confirmation theory is much *too weak* for demarcating scientifically productive from speculative abductions. Central to Bayesians is the incremental criterion of confirmation, according to which an evidence E confirms a hypothesis H iff H's posterior probability $P(H|E)$ is greater than H's prior probability $P(H)$. It follows from the well-known Bayes-equation $P(H|E) = \frac{P(E|H) \cdot P(H)}{P(E)}$ that E confirms H as long as H's prior probability $P(H)$ is greater zero, and H increases E's probability $(P(E|H) > P(E))$, which is in particular the case if H entails E and $P(E) < 1$. This implies that (almost) *every* speculative abduction would count as confirmed. For example, that God wanted X and whatever God wants, occurs, would be confirmed by the occurrence of the event X. No wonder that philosophers of religion such as Swinburne [61, ch. 6] suggest to confirm religious speculations using this Bayesian criterion. Although these facts are well-known by Bayesians and sometimes even regarded as a success (cf. Earman [10, p. 54]; Howson & Urbach [23, p. 119ff]; Kuipers [31, §2.1.2]), I am inclined to conclude that they imply a *breakdown* of Bayesian incremental confirmation. Bayesians reply to this challenge that they can nevertheless *gradually* distinguish between speculative and scientific explanatory hypotheses by the fact that the *prior* probability of a 'scientific' hypothesis is much higher than that of the 'speculative' one. But prior probabilities are a subjective matter, relative to one's background system of beliefs, and so this Bayesian reply ends up in the unsatisfying position that the difference between science and speculation depends merely on the *subjective prejudices* which are reflected in one's prior probabilities. In contrast, according to criterion (CU) a speculative explanation of an evidence X by a postulated 'X-wish' of God can *never* be regarded as scientifically confirmed by X alone.

A more refined but still speculative abduction schema is the following:

Speculative Law Abduction:	*Example*:
Explanandum E: $\forall x(Fx \rightarrow Dx)$	Opium makes people sleepy (after consuming it).
Conjecture H: $\forall x(Fx \rightarrow \psi_D x)$ $\wedge\, \forall x(\psi_D x \rightarrow Dx)$	Opium has a special power (a 'virtus dormitiva') which causes its capacity to make one sleepy.

Speculative law-abductions of this sort have been common in the explanations of the middle ages: every special effect of a natural *agens* (such as the healing capacity of a certain plant, etc.) was attributed to a special power which God has implanted into nature for human's benefit. The given example of the "virtus dormitiva" had been ironically commentated by Molierè, and many philosophers have used this example as a typical instance of a vacuous pseudo-explanation (cf. Mill [36], Book 5, ch. 7, §2; Ducasse [8], ch. 6, §2). This abduction schema violates Ockham's principle insofar we have already a sufficient cause for the disposition to make one sleepy, namely the natural kind "opium", so that the postulated power amounts simply to a redundant multiplication of causes. More formally, the schema does not offer unification because for every elementary empirical law one has to introduce two elementary hypothetical laws to explain it (cf. [58, p. 87]). Moreover, the abductive conjecture has no predictive power which goes beyond the predictive power of the explained law.

My explication of causal unification – many 'effects' explained by one or just a few 'causes' – requires formal ways to 'count' elementary phenomena, expressed by elementary statements. To be sure, there are some technical difficulties involved in this. Solutions to this problem have been proposed in Schurz [54] and Gemes [13]. The following definition is sufficient for our purpose: a statement S is *elementary* (represents an elementary phenomenon) iff S is not logically equivalent to a non-redundant conjunction of statements each of which is shorter than S. Thereby, the belief system S is represented by those elementary statements S which are *relevant* deductive consequences of K in the sense that no (n-placed) predicate in S is replaceable by an arbitrary other (n-placed) predicate, salva validitate of the entailment $K \vdash S$. However, the following analysis of common cause abduction does *not depend* on this particular proposal; it merely depends on the assumption that a natural method of decomposing the classical consequence class of a belief system into nonredundant sets of *elementary* statements exists.

I do not want to diminish the value of cognitive speculation by my analysis. In fact, cognitive speculations are the predecessor of scientific inquiry. Humans have an inborn instinct to search for causes cf. [60, ch. 3], or in Lipton's words, they are 'obsessed' with the search for explanations ([34, p. 130]). But as it was pointed out in §2, the best available 'explanations' are often not good enough to count as rationally acceptable. The above speculative abduction patterns can be regarded as the *idling* of human's inborn explanatory search activities when applied to events for which a proper explanation is out of reach. In contrast to these empty causal speculations, scientific common cause abductions have usually led to genuine theoretical progress. The leading principle of *causal unification* can be explicated in terms of the following principle (cf. Glymour et al. [15, p. 151]):

(R) *Reichenbach principle*: If two properties or kinds of events are probabilistically dependent, then they are *causally connected* in the sense that either one is a cause of the other (or vice versa), or both are effects of a common cause (where X is a cause of Y iff there leads a directed path of *causal arrows* from X to Y).[1]

Reichenbach's principle does not entail that every phenomenon must have a sufficient cause and, hence, avoids an empty regress of causal speculations – it merely says that all correlations result from causal connections. This principle seems to be the *rationale* which underlies humans' causal instincts. Together with *constraints* on the *causal mechanisms* underlying causal arrows, Reichenbach's principle becomes *empirically non-empty*. The way how Reichenbach's principle leads to common cause abduction is as follows: whenever we encounter several *intercorrelated phenomena*, and – on some reason or other – we can *exclude* that

---

[1] A generalization of (R) is the following principle (M), applying to all triples of variables X, Y, Z: if X and Y are probabilistically dependent given Z, then it holds for the assumed underlying directed acyclic causal graph that either Z is a common effect of X and Y or there exists a causal connection between X and Y which does not go through Z. Thereby, two variables X, Y are called probabilistically dependent (relative to a probability distribution P) iff $P(X = x_i \wedge Y = y_j) \neq P(X = x_i) \cdot P(Y = y_j)$ holds for at least some values x_i and y_j of X and Y, respectively. (M) is equivalent with Glymour's Markov-condition [15, p. 156] and with Pearl's Markov-compatibility. This follows from the fact that a probability distribution over a directed acyclic graph satisfies Pearl's Markov-compatibility iff for all (sets of) nodes X, Y, Z in the graph it holds: X is probabilistically dependent on Y given Z iff there exists an (undirected) path in the graph going from X to Y which is not d-separated by Z (cf. Pearl [45, p. 16-18]). For the equivalence with Glymour's Markov-condition cf. theorem 1.2.7 in Pearl [45, p. 19]. (M) implies Reichenbach's principle (R), and it implies Reichenbach's *screening-off criterion* [51, p. 159], which says that direct causes screen of indirect causes from their effects, and common causes screen off their effects from each other – where Z screens off X from Y iff X and Y are probabilistically dependent, but become independent when conditionalized on Z.

one causes the other(s), then Reichenbach's principle requires that these phenomena must have some (unobservable) common cause which simultaneously explains all of them. In the next section I will show that the most important scientific example of this sort is common cause abduction from *correlated dispositions*: since dispositions cannot cause other dispositions, their correlations must have a common intrinsic cause.

The foremost way of justifying Reichenbach's principle is a kind of *no-miracle argument*: it would be as *unplausible* as a miracle that several properties or kinds of events are persistently *correlated* without that their correlations are the result of a certain causal connection. Reichenbach's principle has been empirically corroborated in *almost every* area of science, in the sense that conjectured common causes have been identified in later stages of inquiry. Only quantum mechanics is the well-known exception. Therefore we treat the Reichenbach-principle not as a *dogma*, but as a *meta-theoretical* principle which *guides* our causal abductions.

In scientific common cause abduction, causality and unification go perfectly hand-in-hand. This is worth emphasizing insofar in the recent philosophy of science literature, causality and unification are frequently set into mutual opposition (cf. de Regt [7]). For example, Barnes [2, p. 265] has put forward the following 'causal' objection against unification: it may well happen that three (kinds of) events E_i ($i = 1, 2, 3$) are caused by three independent causes C_i, and although the corresponding independent explanations do not produce unification, they are certainly not *inferior* as compared to the case when all three events are explainable by one common cause C. What Barnes' example correctly shows is that because not all events have a common cause, the request for unifying explanations cannot always be satisfied. However, Reichenbach's principle allows a very simple analysis of Barnes' example: *either* (1) the three (kinds of) events are probabilistically independent; then they *cannot* have a common cause, or (2) they are probabilistically dependent; then (2.1) either they are related to each other in form of a causal chain, or (2.2) they are effects of a common cause. It is this *latter* case in which an explanation of the three E_i by three distinct C_i is clearly inferior, because, in contrast to the common cause explanation, it cannot explain the correlations between the E_i – it rather shifts this problem into unexplained correlations between the C_i.

4 Strict common cause abduction from correlated dispositions and the discovery of new theoretical kind concepts

In this section I analyze common cause abduction in a simple *deductivistic* setting, which is appropriate when the domain is ruled by *strict* or almost-strict causal laws. Probabilistic generalizations are treated afterwards. Recall the schema of speculative law-abduction, where *one* capacity or disposition D occurring in one (natural) kind F was pseudo-explained by a causal 'power' ψ_D. In this case of a *single* disposition, the postulate of a causal power ψ_D which mediates between

F and D is an unnecessary multiplication of causes. But in the *typical* case of a scientifically productive common cause abduction, we have several (natural) kinds $F_1, \ldots, F_n$ all of which all have a set of characteristic dispositions $D_1, \ldots, D_m$ in common – with the result that all these dispositions are correlated. Given that it is excluded that one disposition can cause another one, then these correlated dispositions must be the common effects of a certain intrinsic structure which is present in all of the kinds $F_1, \ldots, F_n$ as their common cause. For example, the following dispositional properties are common to certain substances such as *iron, copper, tin,* ... (cf. fig. 1): a characteristic glossing, smooth surface, characteristic hardness, elasticity, ductility, high conductivity of heat and of electricity. Already before the era of modern chemistry craftsmen have abduced that their exists a characteristic intrinsic property of substances which is the common cause of all these correlated dispositions, and they have called it *metallic character Mx*.

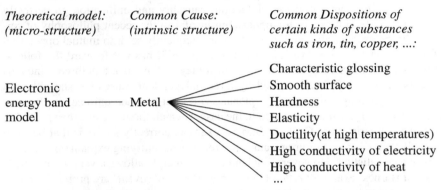

Theoretical model: *Common Cause:*   *Common Dispositions of*
(micro-structure)   *(intrinsic structure)*   *certain kinds of substances*
   such as iron, tin, copper, ...:

Electronic energy band model —— Metal ⟨ Characteristic glossing / Smooth surface / Hardness / Elasticity / Ductility(at high temperatures) / High conductivity of electricity / High conductivity of heat ...

Figure 1. *Common cause abduction of the chemical kind term 'metal'.*

To be sure, the natural kind term *metal* of pre-modern chemistry was theoretically hardly understood. But the introduction of a new (theoretical) natural kind term is the first step in the development of a *new research programme* in the sense of Lakatos [32]. For, the next step then is to construct a *theoretical model* of the postulated kind *metal*, by which one can give an explanation of *how* the structure of a *metal* can cause all these correlated dispositions at once. Especially in *combination* with *atomic (and molecular) hypotheses* the abduced natural kind terms of chemistry became enormously fruitful. In modern chemistry, the molecular microstructure of metals is modeled as a *band* of densely layered electronic energy levels belonging to different nuclei among which the electron can shift easily around, which offers a unifying explanation of all the common dispositions of metals (cf. [41, p. 708ff]).

In the *history of chemistry*, common cause abduction from correlated dispositions was of central importance in the discovery of new (theoretical) kinds of substances. As a second example, consider the 'paradigm' disposition of philosophers: *solubility* in water. Also this disposition does not come in isolation, but is correlated with several further dispositions, such as solubility in ammonium, nonsolubility in oil or benzene, electrolytic conductivity, etc. (see fig. 2). Abduction conjectures an intrinsic property as a common cause, which in early chemistry was called the *hydrophylic* ('water-friendly') character. The corresponding theoretical model of modern chemistry are substances having electrically *polarized* chemical bonds, by which they are solvable in all fluids which have themselves polarized bonds, thereby forming weak electrostatic bondings.

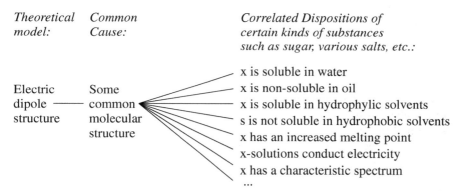

Theoretical model: *Common Cause:*

Correlated Dispositions of certain kinds of substances such as sugar, various salts, etc.:

Electric dipole structure —— Some common molecular structure

x is soluble in water
x is non-soluble in oil
x is soluble in hydrophylic solvents
s is not soluble in hydrophobic solvents
x has an increased melting point
x-solutions conduct electricity
x has a characteristic spectrum
...

Figure 2. *Common cause abduction of the theoretical term "hydrophylic/polar" molecular structure.*

The notion of disposition is discussed rather controversially in the recent literature. According to my understanding of this notion, dispositions are *conditional* (or *functional*) properties. More precisely, that an object x has a (strict) disposition D means that whenever certain initial conditions (or 'stimuli') C are (or would be) satisfied for x, then a certain reaction (or 'response') R of x will (or would) take place. More formally (':$\leftrightarrow$' for 'equivalence by definition'):

(1) $D(x) :\leftrightarrow \forall t \in \Delta(Cxt \rightarrow_n Rxt)$.

Here, $\rightarrow_n$ stands for nomological (or 'counterfactual') implication, and Δ is a more-or-less long temporal interval: if $\Delta = (-\infty, +\infty)$, the disposition is *permanent*, else it is only *temporary*. While (1) expresses a strict disposition, a merely probabilistic disposition is explicated by something like

(2) $D(x) :\leftrightarrow$ (i) $P_{t\in\Delta}(Rxt|Cxt) = $ high $\wedge$ (ii) $P_{t\in\Delta}(Rxt|Cxt) > P_{t\in\Delta}(Rxt)$.

My conditional understanding of dispositions is in accordance with the 'received view' (cf. Carnap [4], §IX-X; Pap [43], p. 44), which has been defended by Prior, Pargetter and Jackson [49]. Dispositional properties are contrasted with *categorial* properties, which are not defined in terms of conditional effects, but in terms of 'occurrent' intrinsic structures or states (in the sense of Earman [9, p. 94]). Dispositional properties in this understanding have categorial properties such as molecular structures as their *causal basis*, but they are *not identical* with them. In particular, since dispositions are '2nd order properties', they can only be the effects of certain (categorial) causes, but cannot themselves act as causes (cf. [49, p. 255]; the same point has been emphasized by Ducasse [8], ch. 6, §2).

In contrast to this view, philosophers such as Quine [50, §§3-4], Armstrong [1, p. 70f] and Mumford [39, p. 205] have argued that dispositions should be identified with categorial and causally effective properties, e.g. with molecular structures etc. There are two main counterarguments against the categorial view of dispositions. The first one is the *multiple realization* argument (cf. [49, p. 253]): the same disposition can be realized by *different* intrinsic structures. For example, a piece of metal and a rubber-band have both the disposition of being *elastic*, although this disposition is caused by very different molecular properties. The second counterargument to the categorial view of dispositions is the situation of correlated dispositions just explained: if several *different* dispositions have the same molecular structure as their common cause, then they *cannot* be identical with this molecular structure because then all of them would be mutually identical, which is counterintuitive.[2]

In conclusion, the categorial view of dispositions is not in accord with the role dispositions play in science: the chemist understands dispositions such as solubility in water clearly in a *conditional* way and separates them from molecular structures which causally explain them. Only in the following *special* situation, the categorial view of dispositions has a rationale behind it: if one has one *isolated* disposition being a conditional property of an (epistemically or ontologically) *primitive* kind, then one may well identify the categorial nature of this kind with this disposition, instead of performing a *speculative* abduction and multiplying causes beyond necessity. As Molnar [37] has pointed out, exactly this situation seems to hold in the case of *elementary particles* (electrons etc.) which are characterized by fundamental dispositions (electric charge etc.) without any further causal explanations for them. So at the fundamental levels of physics there may well be causally ungrounded dispositions. But in all higher levels of science one finds mutually correlated dispositions having a common causal basis – and I argue that this situation gives us a *clear reason* to distinguish between conditionally understood dispositions on the one side and their common causal basis on the other

---

[2]This second counterargument is also a problem for Mumford's "token-identity" view, which would force us to say that this instance of electric conductivity is identical with this instance of elasticity, because both instances are identical with this instance of steel (cf. [39, p. 163]).

side.

A final remark: when I speak of a molecular structure as being the *cause* of a disposition, I understand the notion of 'cause' in a more general sense than the narrow notion of causation between temporally separated events. My usage of 'cause' fits well with ordinary and scientific usage. For the more scrupulous philosopher of causation, let me add that my extended usage of 'cause' is *reducible* to the notion of event-causation as follows: a disposition Dx, being defined as the conditional property $\forall t \in \Delta(Cxt \rightarrow_n Rxt)$, is *caused* by a categorial property Sx iff each *manifestation* of the disposition's reaction, Rxt, is caused by Sx together with the initial conditions Cxt, or formally, iff $\forall x \forall t \in \Delta(Sx \wedge Cxt \rightarrow_n Rxt)$.[3]

The structural pattern of the two examples (figs. 1+2) can be formalized as follows:

Strict Common Cause Abduction (abduced theoretical concept: ψ):

Explanandum E: All kinds $F_1, \ldots, F_n$ have the dispositions $D_1, \ldots, D_m$ in common. $\forall i \in \{1, \ldots, n\} \, \forall j \in \{1, \ldots, m\} : \forall x(F_i x \rightarrow D_j x)$.

Abduced conjecture H: All kinds $F_1, \ldots, F_n$ have a common intrinsic and structural property ψ which is a sufficient [and necessary] cause of all the dispositions $D_1, \ldots, D_m$.
$\forall i \in \{1, \ldots, n\} : \forall x(F_i x \rightarrow \psi x) \wedge \forall j \in \{1, \ldots, m\} : \forall x(\psi x \rightarrow [\leftrightarrow] D_j x)$

The abductive conjecture H logically implies E and it yields a *unification* of $n \cdot m$ empirical (elementary) laws by $n + m$ theoretical (elementary) laws, which is a *polynomial reduction* of elementary laws. H postulates the theoretical property ψx as a merely *sufficient* cause of *all* of the dispositions. If we assume that the dispositions are strictly correlated, then the abductive conjecture even postulates that ψx is both a necessary and sufficient cause of the dispositions (see the version in brackets "[$\leftrightarrow$]"). Note that the given explanandum E would also allow for the possibility that the correlated dispositions $D_1, \ldots, D_m$ have in each kind F_i a *different* common cause ψ_i – but of course, the much *more probable* hypothesis is to assume that they have in all kinds F_i one and the same common cause ψ. For this reason, every single application of this kind of abduction introduces a *new natural kind*: the class of 'ψ-bearers' (e.g. the class of metals, the class of polar substances, etc.).

In conclusion, common cause unification has (at least) three virtues:

---

[3]In this way, also the common-cause-explanation for correlated dispositions $D_1 x, \ldots D_n x$ can be reduced to common cause explanations of the correlated events $R_i x t_i$ *given* $C_i x t_i$, where $t_1, \ldots, t_n$ are different time points in the interval Δ at which the different initial conditions have been realized. The common cause $Sx :\leftrightarrow \forall t \in \Delta : Sxt$ was present during the given temporal interval Δ and thus figures as a common cause of all the conditional events $C_i x t_i$ given $R_i x t_i$: the screening-off relation $P(R_i x t_i \wedge R_j x t_j | C_i x t_i \wedge C_j x t_j \wedge Sx) = P(R_i x t_i | C_i x t_i \wedge Sx) \cdot P(R_j x t_j | C_j x t_j \wedge Sx)$ (recall footnote 1) is satisfied for all pairs of distinct disposition-manifestations $i \neq j \in \{1, \ldots, n\}$.

(1.) The *intrinsic virtue of unification*. Many elementary phenomena (statements) are explained by a few basic principles. Several philosophers, though, are inclined to think that this virtue is merely *instrumentalistic* and, hence, rather weak.

(2.) The virtue of *leading to new predictions*. This may happen in several ways. For example, if for some of the kinds $F_1, \ldots, F_n$, say for F_k, we merely know that it possesses *some* of the dispositions, then the abduced common cause hypothesis predicts that F_k will also possess all the other dispositions. Or, if we know in addition of some independent indicator I for the theoretical property ψ (i.e. $\forall x(Ix \rightarrow \psi x)$), then this knowledge together with the common cause hypothesis predicts I to be an indicator for all of the dispositions D_j. Finally, if ψ is conjectured as being sufficient and necessary for all of the D_j ("[$\leftrightarrow$]"), then this strengthened hypothesis predicts that all the D_j are mutually strictly correlated ($\forall i \neq j \in \{1, \ldots, m\} : D_i x \leftrightarrow D_j x$)). Because of their virtue of producing new predictions, common cause abductions are *independently testable*, in contrast to speculative abductions.

(3.) The virtue of *discovering new (unobservable) kinds or properties* which enlarge our *causal understanding*. This is not only of theoretical, but also of *practical importance*, since knowing a disposition's cause is a necessary step for its technical utilization. Since Reichenbach's causality principle does not hold in every domain (e.g., not in quantum mechanics), there is no guarantee that the hypothetical entities postulated by common cause abduction will always have realistic reference. Nevertheless the following methodological justification can be given: where ever unobservable common causes of observable correlations exist, common cause abduction will find them, while where they don't exist, our efforts to find independent evidence for common causes will fail, and sooner or later we will adopt an instrumentalistic view of our explanatory unification attempts (see §5).

As a further example consider the discovery of *acids*, *bases*, and *salts* which was basically achieved by Glauber in the 17th century. Langley at al. [33, p. 196ff] have given an algorithm for Glauber's discovery which can be reconstructed as the following common cause abduction:

Abduction of the acid-base-salt-system (Glauber):

Phenomena to be explained: There exists a family of substances $F_1, \ldots F_n$ and another family of substances $G_1, \ldots G_m$ which have the following in common:
(i) *Common chemical reactions:*
$\forall i \in \{1, \ldots, n\}$ $\forall j \in \{1, \ldots, m\} : F_i$ and G_j react into a characteristic product (F_i, G_j) and water. (Chemical example: if F_1 is the acid HCl and G_1 is the lime NaOH, then (F_1, G_1) stands for NaCl; etc.).
(ii) *Common monadic properties:*
$\forall i \in \{1, \ldots, n\} : F_i$ has characteristic qualities A (for example, it tastes sour).

$\forall\, j \in \{1,\dots,m\} : G_j$ has characteristic qualities B (for example, it tastes bitter).
$\forall\, i \in \{1,\dots,n\}\ \forall\, j \in \{1,\dots,m\} : (F_i, G_j)$ has qualities C (for example, it tastes salty).

Abduced conjectures (abduced theoretical terms: 'acid', 'base', 'salt'):
(a) $\forall\, i \in \{1,\dots,n\} : F_i$ is an acid.   (b) $\forall\, j \in \{1,\dots,m\} : G_j$ is a lime.
(c) An acid X reacts with a lime Y into the combination (X, Y) and water, where the acid-base-combination (X, Y) is *by definition* called a salt.
(d) An acid has qualities A.   (e) A lime has qualities B.
(f) A salt has qualities C.

The abductive conjectures deductively entail the empirical phenomena (i) and (ii), and thereby they reduce $2 \cdot n \cdot m + n + m$ empirical (elementary) laws to $n + m + 4$ theoretical (elementary) laws.

As a final example we consider *Newtonian physics*. Its fundamental common cause abduction was the abduction of the *sum-of-all-forces* as a common cause for all kinds of accelerations, and the abduction of a universal *gravitational force* as a common cause of the different kinds of movements of bodies in the sky as well as on earth. Thereby, Newton's *qualitative* stipulation of the gravitational force as the counterbalance of the centrifugal force acting on the circulating planets was his *abductive* step, while his *quantitative* calculation of the mathematical form of the gravitational law was a deduction from Kepler's third law plus his abductive conjecture.[4]

Common cause abduction can also be applied to ordinary, non-dispositional properties or (kinds of) events which are correlated. However, in this case one has first to consider more parsimonious causal explanations which do not postulate an unobservable common cause but stipulate one of these events or properties to be the cause of the others. For example, if the three kinds of events F, G, and H (for example, eating a certain poison, having difficulties in breathing and finally dying) are strictly correlated and always occur in form of a temporal chain, then the most parsimonious conjecture is that these event-types form a causal chain. Only in the special case where two (or several) correlated event-types, say F and G, are strongly correlated, but our causal background knowledge tells us that there *cannot* exist a direct causal mechanism which connects them, then a common cause abduction is the most plausible conjecture. An example is the correlation of lightning and thunder: we know by induction from observation that light does not produce sound, and hence, we conjecture that there must exist a common cause of both of them. I call this special case a *missing link common cause abduction*.[5]

---

[4]Glymour [14, p. 203ff] gives illuminating details on Newton's reasoning.

[5]If the correlations between the events are not strict but merely probabilistic, then one may use Reichenbach's *screening off* criterion (recall fn. 1) to distinguish between the case were one of the events E_i causes the other ones from the case where the E_i are effects of a common cause. If the

5 Probabilistic common cause abduction and statistical factor analysis

Statistical factor analysis is an important branch of statistical methodology whose analysis (according to my knowledge) has been neglected by philosophers of science.[6] In this section I want to show that factor analysis is a certain generalization of hypothetical common cause abduction, although sometimes its results may better be interpreted in a purely instrumentalistic way. For this purpose I assume that the parameters are now represented as statistical random variables $X, Y, \ldots$, each of which can take several values $x_i, y_i, \ldots$ (A random variable $X : D \to \mathbb{R}$ assigns to each individual d in the domain D a real-valued number $X(d)$; a dichotomic property F is coded by a binary variable X_F with values 1 and 0.) The variables are assumed to be at least interval-scaled, and the statistical relations between the variables are assumed to be monotonic – only if these conditions are satisfied, the *linearity* assumption of factor analysis yields good approximations.

Let us start from the example of the previous section, where we have n empirically measurable and highly intercorrelated variables $X_1, \ldots, X_n$, i.e. $cor(X_i, X_j)$ = high for all $1 \leqslant i \neq j \leqslant n$. An example would be the scores of test persons in n different intelligence tests. We assume that none of the variables screens off the correlations between any pair of other variables (i.e., $cor(X_i, X_j | X_r) \neq 0$), so that by Reichenbach's principles (recall fn. 1) the abductive conjecture is plausible that these n variables have a common cause, distinct from each of the variables – a theoretical factor, call it F (which is also a real-valued variable). In our example, F would be the theoretical concept of intelligence. Computationally, the abductive conjecture asserts that for each $i \in \{1, \ldots, n\}$, X_i is approximated by a linear function f_i of F, $f_i(F(x)) = a_i \cdot F(x)$, for all individuals x in the given finite sample or domain D (since we assume the variables X_i to be z-standardized, the linear function f_i has no additive term "$+b_i$"). The true X_i-values are scattered around the values predicted by this linear function $f_i(F)$ by a remaining random dispersion s_i; the square s_i^2 is the *remainder variance.* According to standard linear regression technique, the optimal fitting coefficients a_i are computed such as to *minimize* the sum of these remainder variances – which is mathematically equivalent to maximizing the variance of the F-values, $v(F) := \sum_{x \in D} \frac{(F(x) - \mu(F))^2}{|D|}$.[7] Visually speaking, the X_i-values form a stretched cloud of points in an n-dimensional coordinate system, and the F-values form a straight line going through the middle of the cloud such that the squared normal deviations of the points to the straight line are mini-

---

correlations are strict, Reichenbach's screening-off criterion does not work (cf. Otte [42]).

[6]Recently I have discovered an exception, namely Haig [18]. He shares my view of factor analysis as an elaboration of statistical common cause abduction.

[7]This follows from the additivity of the variances: the total variance $v(X_i)$ of a variable X_i equals the sum of the variance of X_i's predicted values $v(a_i \cdot F) = a_i^2 \cdot v(F)$ plus the remainder variance s_i^2 (cf. Bortz [3, p. 233]).

mized.[8]

So far we have described the linear-regression-statistics of the abduction of *one* factor or cause. In factor analysis one takes additionally into account that the mutually intercorrelated variables may have not only one but *several* common causes. For example, the variables may divide into two subgroups with high correlations within each subgroup, but low correlations between the two subgroups. In such a case the abductive conjecture is reasonable that there are two independent common causes F_1 and F_2, each responsible for the variables in one of the two subgroups. In fact, while Spearman had suggested in 1904 one factor being responsible for intelligence, some years later other authors, for example Burt, showed that intelligence is better explained by several independent cognitive factors (cf. [3, p. 618]; [5, p. 8ff]). In the general picture of factor analysis there are given n empirical variables X_i which are explained by $k < n$ theoretical factors (or common causes) F_j as follows:[9]

$$X_1 = a_{11} \cdot F_1 + \ldots + a_{1k} \cdot F_k + s_1.$$
$$\vdots$$
$$X_n = a_{n1} \cdot F_1 + \ldots + a_{nk} \cdot F_k + s_n.$$

This is usually written in a matrix formulation: $\mathbf{X} = \mathbf{F} \cdot \mathbf{A}^t$ ('t' for 'transposed'). While each variable X_i and factor F_j takes different values for the different individuals of the sample or domain, the factor loadings a_{ij} are constant and represent the causal contribution of factor F_j to variable X_i. Given that also the factor variables F_j are been z-standardized, then each *factor loading* a_{ij} expresses the correlation (covariance) between variable X_i and factor F_j, $cor(X_i, F_j)$. Since the variance of each variable X_i equals the sum of the squared factor loadings a_{ij}^2 and the remainder variance s_i^2, each squared factor loading a_{ij}^2 measures the *amount of the variance* of X_i 'explained' (i.e. statistically predicted) by factor F_j and coincides with the so-called *eigenvalue* λ_j of the factor F_j. The sum of all squared factor loadings divided through n measures the percentage of the total variance of the variables which is explained by the factors – this percentage is a direct measure for the explanatory success of the factor-statistical analysis.

The major mathematical technique to find those $k < n$ factors which explain a maximal amount of the total variance is the so-called *principal component analysis*. Instead of any detailed mathematical explanation I confine myself to the

---

[8]The F-axis has a non-standardized length unit given as $\sum_{1 \leqslant i \leqslant n} a_i^2$. This fact and the equation in footnote 7 implies $\sum_{1 \leqslant i \leqslant n} s_i^2 = n - v(F)$, where $\sum_{1 \leqslant i \leqslant n} s_i^2$ is exactly the squared normal deviation mentioned in the text.

[9]For the following cf., e.g., Bortz ([3], ch. 15.1), Kline ([29], ch. 3) or Gorsuch ([16], ch. 6). I only describe the most common method of factor analysis which is used in most computer programs, without discussing the subtle differences between different factor analytic methods.

following remarks. The k factors or axes are determined according to two criteria: (i) they are probabilistically independent (or orthogonal) to each other, and (ii) the amount of explained variance is maximized (i.e., the sum of the remaining variances is minimized). Visually speaking, the first factor F_1 is determined as an axis going through the stretched cloud of points in the n-dimensional coordinate system; then the next factor F_2 is determined as an axis orthogonal to F_1, and so on, until the $k < n$ factor axes are determined by the system of coefficients a_{ij}. Mathematically this is done by rotations of the coordinate system in a way such that the diagonal elements of the $n \times n$ variance-covariance matrix $\mathbf{D}$ are maximized under the constraint of orthogonality; this leads to the matrix equation $(\mathbf{D} - \lambda \cdot \mathbf{I}) \cdot \mathbf{v} = 0$ with unknowns λ and $\mathbf{v}$ ($\mathbf{I}$ is the identity matrix, λ the vector of eigenvalues; and $\mathbf{v}$ is the rotation matrix). The first k eigenvalues λ_j and corresponding rotation vectors v_j obtained from solving this equation determine the system of factor loadings a_{ij}.

The success of an explanation of n variables by $k < n$ factors is the higher, the less the number k compared to n, and the higher the amount of the total variance explained by the k factors. Note that the amount of explained variance goes hand in hand with that amount of the intercorrelation between the variables which is explained by the factors: the amount of the correlation $cor(X_i, X_j)$ explained by all factors is given as $\sum_{1 \leqslant r \leqslant k} a_{ir} \cdot a_{jr}$ (cf. [3, p. 661]; [29, p. 40]). This picture fits perfectly with my account of unification of a given set of empirical variables by a much smaller set of theoretical variables, as explained in §4. While the amount of explained variance of the first factor is usually much greater than one, this amount becomes smaller and smaller when one introduces more and more factors. The trivial limiting case is given when $k = n$, because n empirical variables can always completely be explained by n theoretical factors. According to the Kaiser-Guttman-criterion one should introduce new factors only as long as their amount of explained variance is greater than one (cf. [3, p. 662f], [29, p. 75]). Hence, a theoretical factor is only considered as non-trivial if it explains more than the variance of just one variable and, in this sense, offers a unificatory explanation to at least *some* degree. In other words, the Kaiser-Guttman-criterion is the factor analytic counterpart of my suggested minimal criterion for hypothetical cause abduction (CU).[10]

After the principal component analysis has been performed and the factors have been standardized,[11] the factor axes can be *rotated* without change of the amount

---

[10]The Kaiser-Guttman-criterion is an adequate minimal criterion if one is interested in extracting *non-trivial* common cause factors. A stronger criterion is Catell's *scree test* which in most cases extracts factors whose explained variance is significantly greater than one (cf. [3, p. 662f], [16, p. 167ff], [29, p. 76]). If the intercorrelations between the variables are small and one is interested in representing all correlations by common factors, then one has to extract also factors whose explained variance is smaller than 1 (see [16, p. 163ff]).

[11]This leads to a new coordinate system in which the axes corresponding to the variables are no

Empirical Variables	Factor F_1 Dynamics	Factor F_2 Emot. Value	Factor F_3 Conciseness	E.V. per variable
1: loud-low	**0.84**	-0.08	-0.17	73
2: harmonious-disharm.	-0.26	**0.80**	-0.22	75
3: clear-unclear	0.42	0.03	**-0.86**	91
4: fluent-haltering	0.48	0.45	-0.30	52
5: slow-quick	**-0.86**	0.29	0.07	82
6: articulated-vague	0.28	0.24	**-0.88**	91
7: pleasant-unpleas.	-0.31	**0.86**	-0.21	88
8: active-passive	**0.95**	0.06	-0.23	95
9: strong-weak	**0.67**	**0.66**	-0.17	91
10: deep-high	0.41	**0.80**	0.12	81
11: confident-bashful	**0.69**	0.50	-0.30	81
12: inhibited-free	0.06	**-0.85**	0.27	80
13: quiet-lively	**-0.90**	-0.25	0.03	87
14: hesitating-pressing	**-0.94**	0.06	0.08	90
15: correct-careless	0.01	0.22	**-0.88**	82
16: engaged-tired	**0.93**	0.07	-0.11	88
17: big-little	0.04	**0.94**	0.11	89
18: ugly-nice	0.17	**-0.84**	0.28	80
E.V. per factor:	37	30.4	15.0	Total: 83.3

Figure 3. *18 empirical variables measuring the subjective evaluations of the voices of persons explained by three varimax-factors. E.V. = explained amount of the variance. High loadings in bold. (Taken from [3, p. 672].)*

of explained variance. So the result of a factor analysis is *not unique*. According to the most common *varimax* principle, the factor axes are rotated into a position in which the squared loadings of the factors are, roughly speaking, either very high or very low (cf. [3], p. 665-672, [29], p. 67f). This leads to the effect that the abduced (or 'extracted') factors can most easily be interpreted in terms of certain plus-minus-combinations of the empirical variables. Fig. 3 below offers an example which is taken from Bortz ([3], p. 672):

Prima facie, hypothetical common cause abduction supports a realistic interpretation of the abduced factors. In contrast, for an *instrumentalistic* philosopher of science such as van Fraassen (e.g. 1980), the extracted factors are not taken realistically, and so the factor equations cannot be true in the realistic sense. They can only be more-or-less *empirically adequate*. For the instrumentalist an *abduction pattern* is a useful means of *discovering* an empirically adequate theory – it

---

longer orthogonal to each other; cf. [3, p. 657].

has an important *instrumental* value, but it does not have any *justificational* value. For judgments of empirical adequacy, an abductive inference is not needed – an *epistemic induction* principle is sufficient which infers the future empirical success from its empirical success in the *past*.

In fact, several statisticians tend to interpret the results of a factor analysis cautiously as a merely instrumentalistic means of *data reduction* in the sense of representing a large class of intercorrelated empirical variables by a small class of independent theoretical variables. In spite of this fact I think that the properly intended interpretation of the factors of a factor analysis is their realistic interpretation as common causes, for that is how they are designed. I regard the instrumentalistic perspective as an *important warning* that not every empirically useful theoretical superstructure must correspond to an existing structure of reality (cf. also Cureton & d'Agostino [5], p. 3f). This warning is already entailed by the mentioned fact that the results of a factor analysis are *non-unique modulo rotations* of the standardized factor-axes. Haig [18, p. 319] considers this fact as the factor-analytic counterpart of the general situation of *empirical underdetermination* of theories by empirical evidence in science.

6   Theoretical model abduction

Every successful fundamental common cause abduction in the natural sciences has been a *germ* for a new theoretical research programme in the sense of Lakatos [32], in which scientists attempt to develop theoretical and quantitative *models* for their conjectured common causes. For example, chemical kind concepts get replaced by molecular models (recall §4), or qualitative force concepts by quantitative equations. In this way, fundamental common cause abduction turns gradually into what I call *theoretical model abduction*. The capacity of producing *novel predictions* is significantly enhanced by this transformation. This concluding section of my paper provides an analysis of theoretical model abduction. Theoretical model abduction takes also place in the social sciences, for example as emerging from the results of a factor analysis. However, theories in social sciences are formulated usually only in a qualitative form, without precise equations from which quantitative predictions could be derived. I don't know whether this is only due to complexity reasons, or to reasons of principle – but anyhow, it seems to me that this fact reflects a major contemporary difference between theories of natural and social sciences.

In this section we consider theoretical model abductions in more detail. Their explanandum is a well-confirmed empirical phenomenon expressed by an empirical law – for example, the phenomenon that wood swims in water but a stone sinks in it. The abduction is driven by an *already established* scientific theory which is usually quantitatively formulated. The theory itself is the historical outcome of a scientific common cause abduction which identifies the qualitative kind of causes in terms of which the phenomenon has to be explained. The abductive task consists in *finding theoretical (initial and boundary) conditions* which describe the causes of the phenomenon in the theoretical language and which allow the mathematical derivation of the phenomenon from the theory. Formally, these theoretical conditions are expressed by factual or lawlike statements, but their semantic content corresponds to what one typically calls a *theoretical model* for a particular kind of phenomenon within an already *given* theory. Note also that with my notion of a 'model' I do not imply a particular kind of formalization of models: they can be represented by *statements* as well as by *set-theoretical* models (which in turn are characterized by statements of a set-theoretical meta-language).

As an example, consider Archimedes' famous explanation of the phenomenon of buoyancy. Here one searches for a theoretical explanation of the fact that certain substances like stones or metals sink in water while others like wood or ice swim on water, *solely in terms of mechanical and gravitational effects*. Archimedes' ingenious abductive conjecture was that the amount of water which is supplanted by the swimming or sinking body tends to lift the body upwards, with a force f_W which equals the weight of the supplanted water (see fig. 4). If this force is greater than the weight of the body (f_B) the body will swim, otherwise it will sink. Since the volume of supplanted water equals the volume of the part of the body which is under water, and since the weight is proportional to the mass of a body, it follows that the body will sink exactly if its density (mass per volume) is greater than the density of water.

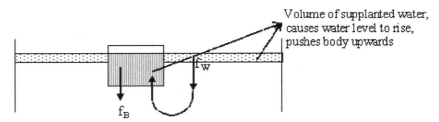

Volume of supplanted water, causes water level to rise, pushes body upwards

Figure 4. *Theoretical conditions which allow the mechanical derivation of the law of buoyancy.*

The example shows clearly that this kind of abduction is tantamount to the formation of a *theoretical model* for a given kind of lawlike phenomenon within a *given* theory. This situation is very different from *selective* factual abductions: one does *not* face here the problem of a huge multitude of possible theoretical models or conjectures. Rather, the given theory *constrains* the space of possible causes to a small class of basic parameters (or generalized 'forces') by which the theory models the domain of phenomena which it intends to explain. In the Archimedean case, the given theory presupposes that the ultimate causes are only contact forces and gravitational forces – other ultimate causes such as intrinsic swimming capacities of bodies or invisible water creatures etc. are excluded. Therefore, the real difficulty of theoretical model-abduction does not consist in the elimination of possible explanations – this elimination is already achieved by the given theory – but in finding just *one* plausible theoretical model which allows the derivation of the phenomenon to be explained. If such a theoretical model is found, this is usually celebrated as a great scientific success.

Theoretical model-abduction is the typical theoretical activity of *normal* science in the sense of Kuhn [30], that is, the activity of extending a given theory core (or paradigm) to new application cases, rather than changing a theory core or creating a new one. If the governing theory is *classical physics*, then examples of theoretical model abductions come into hundreds, and physics text books are full of them. Examples are the theoretical models underlying

(1.) the trajectories (paths) of rigid bodies in the constant gravitational field of the earth (free fall, parabolic path of ballistic objects, gravitational pendulum, etc.);

(2.) the trajectories of cosmological objects in position-dependent gravitational fields (the elliptic orbits of planets – Kepler's laws, the moon's orbit around the earth and the lunar tides, inter-planet perturbations, etc.);

(3.) the behaviour of solid, fluid or gaseous macroscopical objects viewed as systems of more-or-less coupled mechanical atoms (the modeling of pressure, friction, viscosity, the thermodynamic explanation of heat and temperature, etc.); and finally

(4.) the explanation of electromagnetic phenomena by incorporating electromagnetic forces into classical physics [19, §3].

While for other kinds of abductions we can provided a *general* formal pattern by which one can *generate* a most promising explanatory hypothesis, we cannot provide such a general pattern for theoretical model abduction because here all depends on *what theory* we are in. But if the theory is specified, then such patterns can often be provided: they are very similar to what Kitcher [28, p. 517] has called a schematic explanatory argument, except that the explanandum is now given and the particular explanatory premises have to be found within the framework of the given theory. For example, the abduction pattern of Newtonian particle mechanics would be something like the following:

Explanandum: a kinematical process involving (a) some *moving* particles whose position, velocity and acceleration at a variable time *t* is an *empirical function* of their initial conditions, and (b) certain objects defining constant boundary conditions (e.g. a rigid plane on which a ball is rolling, or a large object which exerts a gravitational force, or a spring with Hooke force, etc.).

Generate the abduced conjecture as follows: (i) specify for each particle its mass and all non-neglectible forces acting on it in dependence on the boundary conditions and on the particle's position at the given time; (ii) insert these specifications into Newton's 2nd axiom (which says that for each particle *x* and time *t*, sum-of-all-forces-on-*x*-at-*t* = mass-of-*x* times acceleration-of-*x*-at-*t*); (iii) try to solve the resulting system of *differential equations*; and finally (iv) check whether the resulting time-dependent trajectories fit the empirical function mentioned in the explanandum – if yes, the conjecture is preliminarily confirmed; if no, then search for (perturbing) boundary conditions and/or forces which may have been overlooked.

Theoretical model abduction can also be found in 'higher' sciences which are working with explicitly formulated theories. In *chemistry*, the explanations of the atomic component ratios (the chemical gross formulae) by a three-dimensional molecular structure are the results of theoretical model abductions; the given theory here is the periodic table plus Lewis' octet rule for forming chemical bonds. Theoretical model abductions take also place in *evolutionary theory*. For example, the reconstruction of phylogenetic trees of descendance from phenotypic similarities (and other empirical data) is a typical abduction process. The basic evolution-theoretical premise here is that different biological species descend from common biological ancestors from which they have split apart by discriminative mutation and selection processes. The alternative abductive conjectures about trees of descendance explaining given phenotypic similarities can be evaluated by probability considerations. Assume three species S_1, S_2, and S_3 where both S_1 and S_2 but not S_3 have a *new* property F – in Sober's example, S_1 = sparrows, S_2 = robins, S_3 = crocs, and F = having wings ([59], p. 174-176). Then the tree of descendance T_1, in which the common ancestor A first splits into S_3 and the common ancestor of S_1 and S_2 which has already F, requires only one mutation-driven change of non-F into F, while the alternative tree of descendance T_2 in which A first splits into S_1 and a common F-less ancestor of S_2 and S_3 requires two such mutations (see fig. 5).

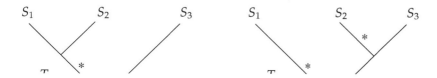

So, probabilistically T_1 is favored as against T_2. There are some well-known examples were closeness of species due to common descent does *not* go hand in hand with closeness in terms of phenotypic similarities (e.g., *birds, crocs* and *lizards*): examples of this sort are recognized because there are several mutually *independent* kinds of evidences which the tree of descendance must *simultaneously* explain, in particular (i) phenotypic similarities, (ii) molecular similarities, and (iii) fossil record (cf. Ridley [52], ch. 17).

An example of qualitative model-abduction in the area of humanities is *interpretation* (an illuminating analysis is found in Gabbay & Woods [12], §4.1). The explanandum of interpretations are the utterances, written text, or the behaviour of given persons (speakers, authors, or agents). The abduced models are conjectures about the beliefs and intentions of the given persons. The general background theory is formed by certain parts of (so-called) folk psychology, in particular the general premise of all rational explanations of actions, namely, that normally or ceteris paribus, persons act in a way which is suited to fulfill their goals given their beliefs about the given circumstances (cf. Schurz [55], §1). More specific background assumptions are hermeneutic rationality presumptions (Davidson [6]), Grice's maxims of communicative cooperation [17], and common contextual knowledge. Interpretative abductions may both be selective or creative: in the case of interpretations, the question whether there will be many possible interpretations (and the difficulty will be their elimination), or whether it will be hard to find just one coherent interpretation, depends crucially on *what* the speaker says and *how* (s)he says it.

As a final example I discuss the theoretical model of cultural development which Inglehart has developed based on his factor-analytic analysis of the data of the *Word Value Survey (WVS) project* (see http://www.worldvaluessurvey.org/). In this project, social scientists have developed and successively improved detailed questionnaires, by which people in more than 65 countries are asked questions concerning their *attitudes* (e.g. towards religion, authority, labor, family, political organization, personal self realization, gender roles, homosexuality, liberty etc.) as well as concerning their *lifestyle* (married or single, how much time they spend for work, family, clubs, etc.).

Each answer to a question corresponds to one *empirical variable*. In order to interpret the more than 250 variables, Inglehart has performed a factor analysis. In several independent WVS-studies or 'waves' (1990, 1995, 2000, 2005), two stable (statistically independent) factors have been obtained which explain more than 50% of the total variance of the empirical variables. The factor analysis has been performed with nations as empirical units, i.e., with the mean values for each country. The two obtained factors have led Inglehart to the following theoretical interpretation of two major causes of the cultural characteristics of nations:

Factor 1: tradition-religious versus secular-rational orientation. This factor

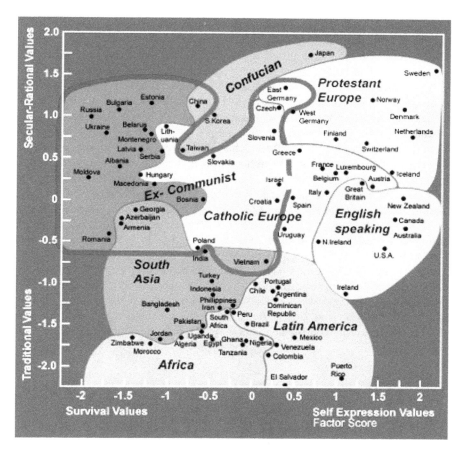

Figure 6. *The cultural world-map. Most recent data from 2000 and 2005. Source:* http://www.worldvaluessurvey.org/

correlates strongly with: religious vs. rational-secular value-orientation, importance of family bonds vs. individual freedom, national pride high vs. low, respect vis-à-vis state authority high vs. low, birth rate high vs. low, estimation of labor high vs. low. The transition on the 1st factor axis from traditional-religious to rational-secular values is theoretically interpreted as the social transition from *agraic* to *industrial* civilizations – the so-called *process of modernization*.

Factor 2: survival values versus self-expression values. This factor correlates strongly with: low vs. high economic standards of life, importance of existential security & work vs. pleasure & quality of life, low vs. high appreciation of gender equality, rejection vs. acceptance of homosexuality, intolerance vs. tolerance towards foreign immigrants, low vs. high interpersonal trust, acceptance vs. non-acceptance of authoritarian regimes, devaluation vs. appreciation of democracy. The transition on the 2nd factor axis from survival to self-expression values is theoretically interpreted as the transition from industrial to post-industrial societies with a high prosperity level and a dense social service infrastructure – the so-called *process of post-modernization.*

In his earlier books (cf. [24]), Inglehart defended a version of modernization theory in which more-or-less all nations follow in their development the same major modernization trend which goes from agrarian to industrial to post-industrial societies. Because of the WVS-data, Inglehart later weakens his view and recognizes a large amount of *culture-specific path-dependence* in this development (cf. Inglehart & Baker [25, p. 21]). Fig. 6 shows the most recent cultural world map of nations based on the WVS-data of 2000 and 2005 (they include more nations than 1995).

Fig. 6 and related results have led to a variety of interesting questions and discoveries, which cannot be discussed here (for details cf. [25], [26], [57]). In the perspective of this paper, Inglehart's work shows how the results of factor analysis in social science can lead to the abduction of fruitful theoretical models even in so complex areas as the world-wide cultural development and cultural divergence of nations.

BIBLIOGRAPHY

[1] Armstrong, D.M.: "Dispositions as Causes", *Analysis* 30, 1997, 23-26.
[2] Barnes, E.: "Inference to the Loveliest Explanation", *Synthese* 103, 1995, 251-277.
[3] Bortz, J.: *Lehrbuch der Statistik*, Springer, Berlin 1985 (6th ed. 2005).
[4] Carnap, R.: "The Methodological Character of Theoretical Concepts", in: Feigl, H./ Scriven, M. (eds.), *Minnesota Studies in the Philosophy of Science Vol. I*, Univ. of Minnesota Press, Minneapolis, 1956, 38-76.
[5] Cureton, E., and D'Agostino, R.: *Factor Analysis: An Applied Approach*, Lawrence Erlbaum Ass., Hillsdale, N.J. 1983.
[6] Davidson, D.: *Inquiries into Truth and Interpretation*, Oxford Univ. Press 1984.
[7] De Regt, H.W.: "Wesley Salmon's Complementarity Thesis: Causality and Unificationism Reconciled?", International Studies in the Philosophy of Science 20/2, 2006, 129-147.
[8] Ducasse, J.: *A Critical Examination of the Belief in a Life after Death*, Charles and Thomas, Springfield 1974.
[9] Earman, J.: *A Primer on Determinism*, Reidel, Dordrecht 1986.
[10] Earman, J. (ed.): *Bayes or Bust?*, MIT Press, Cambridge/MA 1992.
[11] Flach, P. and Kakas, A. (eds.): *Abduction and Induction*, Kluwer, Dordrecht 2000.
[12] Gabbay, D., and Woods, J.: *The Reach of Abduction: Insight and Trial (A Practical Logic of Cognitive Systems Vol. 2)*, North-Holland, Amsterdam 2005.
[13] Gemes, K.: "Hypothetico-Deductivism, Content, and the Natural Axiomatization of Theories", *Philosophy of Science*, 54, 1993, 477-487.
[14] Glymour, C.: *Theory and Evidence*, Princeton Univ. Press, Princeton 1981.
[15] Glymour, C., Spirtes, P., and Scheines, R.: "Causal Inference", *Erkenntnis* 35, 1991, 151-189.

[16] Gorsuch, R.: *Factor Analysis*, Lawrence Erlbaum Ass., Hillsdale, N.J. 1983 (2nd ed.).
[17] Grice, H. P.: "Logic and Conversation", in P. Cole, J. Morgan (eds.), *Syntax and Semantics* Vol. 3, New York 1975, 41-58.
[18] Haig, B.: "Exploratory Factor Analysis, Theory Generation, and Scientific Method", *Multivariate Behavioral Research* 40(3), 2005, 303-329.
[19] Halonen, I., and Hintikka, J.: "Towards a Theory of the Process of Explanation", *Synthese* 143/1-2, 2005, 5-61.
[20] Hanson, N.R.: "Is There a Logic of Discovery?", in: H. Feigl, G. Maxwell (eds.), *Current Issues in the Philosophy of Science* Holt, Rinehart and Winston, New York, 1996, 20-35.
[21] Harman, G. H.: "The Inference to the Best Explanation", *Philosophical Review* 74, 1965, 173-228.
[22] Hintikka, J. (1998): "What is Abduction? The Fundamental Problem of Contemporary Epistemology", *Transactions of the Charles Sanders Peirce Society*, Vol. XXXIV, No. 3, 1998, 503 - 533.
[23] Howson, C., and Urbach, P.: *Scientific Reasoning: The Bayesian Approach*, Open Court, Chicago 1996 (2nd ed.).
[24] Inglehart, R.: *Cultural Shift in Advanced Industrial Society*, Princeton Univ. Press, Princeton 1990.
[25] Inglehart, R., and Baker, W.: "Modernization, Cultural Change, and the Persistence of Traditional Values", *American Sociological Review* 65, 2000, 19-51.
[26] Norris, P., and Inglehart, R.: *Sacred and Secular. Religion and Politics Worldwide*,Cambridge Univ. Press New York 2004.
[27] Josephson, J. and Josephson, S. (eds.): *Abductive Inference*, Cambridge Univ. Press, New York 1994.
[28] Kitcher, P.: "Explanatory Unification", *Philosophy of Science* 48, 1981, 507-531.
[29] Kline, P.: *An Easy Guide to Factor Analysis*, Routledge, London 1994.
[30] Kuhn, T.S.: *The Structure of Scientific Revolutions*, Chicago Univ. Press, Chicago 1962.
[31] Kuipers, T. A. F.: *From Instrumentalism to Constructive Realism*, Kluwer, Dordrecht 2000.
[32] Lakatos, I.: "Falsification and the Methodology of Scientific Research Programmes", in: I. Lakatos, A. Musgrave (eds.), *Criticism and the Growth of Knowledge*, Cambridge Univ. Press, Cambridge, 1970, 91-195.
[33] Langley, P., et al.: *Scientific Discovery. Computational Explorations of the Creative Process*, MIT Press, Cambridge/Mass 1987.
[34] Lipton, P.: *Inference to the Best Explanation*, Routledge, London 1991.
[35] Magnani, L.: *Abduction, Reason, and Science*, Kluwer, Dordrecht 2001.
[36] Mill, J. St.: *System of Logic*, Parker, Son, and Bourn, London 1865 (6th ed.).
[37] Molnar, G.: "Are Dispositions Reducible?", *Philosophical Quarterly* 49, 1999, 1-19.
[38] Moser, P.K.: *Knowledge and Evidence*, Cambridge Univ. Press, Cambridge 1989.
[39] Mumford, S.: *Dispositions*, Oxford Univ. Press, Oxford 1998.
[40] Niiniluoto, I.: "Defending Abduction", *Philosophy of Science (Proceedings)* 66, 1999, S436 - S451.
[41] Octoby, D.W. et al.: *Modern Chemistry*, Saunders College Publ., Orlando 1999.
[42] Otte, R.: "A Critique of Suppes' Theory of Probabilistic Causality", *Synthese* 48, 1981, 167-189.
[43] Pap, A.: "Disposition Concepts and Extensional Logic", in: Tuomela, R. (ed.): *Dispositions*, Reidel, Dordrecht, 1978, 27-54.
[44] Paul, G.: "Approaches to Abductive Reasoning", *Artificial Intelligence Review* 7, 1993, 109-152.
[45] Pearl, J.: *Causality*, Cambridge Univ. Press, Cambridge 2000.
[46] Peirce, C.S.): "Deduction, Induction, and Hypothesis" (1878), in *Collected Papers* 2.619 - 2.644.
[47] Peirce, C. S.: "Lectures on Pragmatism" (1903), in *Collected Papers* CP 5.14 - 5.212.
[48] Peirce, C. S.: *Collected Papers*, ed. by C. Hartshorne, P. Weiss, Harvard Univ. Press, Cambridge/Mass, 1931 - 1935.
[49] Prior, E.W., Pargetter, R. and Jackson, F.: "Three Theses about Dispositions", *American Philosophical Quarterly* 19, 1982, 251-257.
[50] Quine, W.v.O.: *Roots of Reference*, La Salle, Open Court 1974.
[51] Reichenbach, H.: *The Direction of Time*, Univ. of California Press, Berkeley 1956.
[52] Ridley, M.: *Evolution*. Oxford: Blackwell Scientific Publications 1993 .
[53] Salmon, W.: *Scientific Explanation and the Causal Structure of the World*, Princeton Univ. Press, Princeton 1984.
[54] Schurz, G.: "Relevant Deduction", *Erkenntnis* 35, 1991, 391-437.

[55] Schurz, G.: "What Is 'Normal'? An Evolution-Theoretic Foundation of Normic Laws", Philosophy of Science 28, 2001, 476-97.

[56] Schurz, G.: "Patterns of Abduction", *Synthese* 164, 2008, 201-234.

[57] Schurz, G.: "Clash of Civilizations? An Evolution-Theoretic and Empirical Investigation of Huntington's Theses", to appear in: C. Kanzian, E. Runggaldier (eds.), *Cultures. Conflict-Analysis-Diagnosis*, Ontos Verlag, Frankfurt 2007.

[58] Schurz, G., and Lambert, K.: "Outline of a Theory of Scientific Understanding", *Synthese* 101/1, 1994, 65-120.

[59] Sober, E.: *Philosophy of Biology*, Westview Press, Boulder 1993.

[60] Sperber, D. et. al (eds.): *Causal Cognition*, Clarendon Press, Oxford 1995.

[61] Swinburne, R.: *The Existence of God*, Clarendon Press, Oxford 1979 (revised 2nd ed. 2004).

[62] Thagard, P.: Common cause!)Computational Philosophy of Science, MIT Press, Cambridge/MA 1988.

Abduction, Unification, and Bayesian Confirmation: Comment on Schurz

ILKKA NIINILUOTO

Gerhard Schurz's paper "Common Cause Abduction and the Formation of Theoretical Concepts in Science" is an excellent account of several important aspects of abductive inference. He makes a useful distinction between selective abduction (i.e., the choice of the optimal one among given potential explanations) and creative abduction (i.e., inferences involving the discovery of new explanatory concepts and hypotheses). He emphasizes the strategic role of abductive inferences in the search of promising conjectures, but admits that abduction has at least weak justificational value as well. Schurz argues further that scientifically worthwhile creative abductions can be distinguished from illegitimate speculations by employing the criterion of causal unification. The nature of common cause abduction and theoretical model abduction is illustrated by well-chosen examples from chemistry, statistical factor analysis, and evolutionary theory.

In this comment, I shall concentrate on one basic issue in Schurz's analysis, viz. his claim about the weak prospects of the Bayesian confirmation theory to motivate the causal unification criterion.

1 Consilience of inductions

Schurz's basic argument is as follows. If a hypothesis H explains only one phenomenon E, then it is ad hoc and remains "a case of pure speculation". A scientifically interesting hypothesis H, worthy of further test operations, has to explain several mutually intercorrelated but analytically independent phenomena, thereby showing that these phenomena are effects of a hypothetical common cause.

This idea is historically related to the classical exposition of the method of hypothesis by William Whewell in *The Philosophy of the Inductive Sciences* in 1840. According to Whewell, a scientific hypothesis discovered by induction to account for phenomena should be independently testable:

> "The hypotheses which we accept ought to explain phenomena which we have observed. But they ought to do more than this: our hypothesis ought to *fortel* phenomena which have not yet been observed; – at least *of the same kind* as those which the hypothesis was invented to explain.
> ... We have here spoken of the predictions of facts of the same kind as those from which our rule was collected. But the evidence in favour of our induction is of a much higher and more forcible

character when it enables us to explain and determine cases of *a kind different* from those which were contemplated in the formation of our hypothesis." (Whewell [12, 62-65])

Similar requirements were put forward by Charles S. Peirce (*CP* 2.738) and Karl Popper [9, 241]. (Cf. Niiniluoto [4, 37-38])

Whewell argued further that the strongest sign of scientific progress occurs with "the consilience of inductions", where "an induction, obtained from one class of facts, coincides with an induction, obtained from another class" (cf. [8, 228]). In consilience, inductions "jump together": two separate generalizations are found to be consequences of the same comprehensive theory. His paradigm example was Newton's theory of gravity which explains both Galileo's and Kepler's laws.

Whewell regarded consilience as "a test of the theory in which it occurs", or "a criterion of reality, which has never yet been produced in favour of falsehood" (see [11]). Whewell seems to propose here an optimistic meta-induction from the history of science, claiming that the Peircean truth-frequency of consilience equals one. But this is clearly an exaggeration: a fallibilist has to admit that theoretical unification in science has lead to mistakes as well. How strong confirming power does such causal unification really have? How can we measure the strength of such confirmation? These questions lead us back to Schurz's account of abduction.

2   Relative simplicity

Schurz himself mentions one proposal: "An explanation is the better, the *less* new entities it postulates, and the *more* phenomena it explains." This idea, which has been prominent in recent discussions of explanatory unification, was formulated by Eino Kaila in 1935 by his notion of *relative simplicity* (see [5, 182]). The relative simplicity of theory H is the ratio between the explanatory power of H (the multitude of empirical data E derivable from H) and the complexity of H (the number of logically independent basic assumptions of H). Kaila suggested that relative simplicity would define the "inductive probability" of H on E, *if* it could be exactly defined and measured.

3   Qualitative abductive confirmation

Another approach is to analyse confirmation by qualitative or structural principles. Inspired by Carl G. Hempel's work, Howard Smokler [10] proposed that the following principles of *Converse Entailment* and *Converse Consequence* are satisfied by "abductive inference":

(CE)   If hypothesis H logically entails evidence E, then E confirms H.

(CC)   If K logically entails evidence H and E confirms H, then E confirms K.

One way of guaranteeing these principles is to define confirmation simply as *inverse deduction*:

(ID)   E confirms H iff H entails E.

To avoid trivial cases, it has to be assumed in (ID) that H is consistent and E is non-tautologous. More plausible versions of these principles – call them CE*, CC*, and ID* – are obtained by replacing deductive entailment by the stronger condition of deductive explanation. Thus, *abductive confirmation* can be defined by

(ID*)   E abductively confirms H iff H deductively explains E.

Then confirmation defined by (ID*) satisfies (CE*), but not generally (CC*). (See [8, 227].)

4 Bayesian confirmation

The *Bayesian* approach analyses inferences in terms of epistemic probabilities. Probability $P(H/E)$ expresses a rational degree of belief in the truth of hypothesis H given evidence E. According to Bayes's Theorem, this posterior probability is a function of the prior probability $P(H)$ of H and the likelihood $P(E/H)$ of H relative to E:

(1) $P(H/E) = P(H) \cdot P(E/H)/P(E).$

The principles (CE) and (CE*) receive immediately a Bayesian justification, if an epistemic probability measure P is available for the language including H and E, and if confirmation is defined by the *Positive Relevance* criterion:

(PR)   E confirms H iff $P(H/E) > P(H).$

If now H entails E, we have $P(E/H) = 1$. Hence, by (1),

(2)
If H logically entails E, and if $P(H) > 0$ and $P(E) < 1$, then $P(H/E) > P(H).$

More generally, as positive relevance is a symmetric relation, it is sufficient for the (PR)-confirmation of H by E that H is positively relevant to E. If inductive explanation is defined by the positive relevance condition, i.e., by requiring that $P(E/H) > P(E)$ (see [8]), then we have the general result:

(3)
   If hypothesis H deductively or inductively explains data E, then E confirms H.

The same result about indirect confirmation holds of course for successful predictions E from a hypothetical theory H. Note that these results are valid for all epistemic probability measures P. Note also that positive relevance (PR) does not generally satisfy the controversial principle (CC). The strength of the quantitative Bayesian treatment is thus its ability to account for the most central qualitative principle (CE) (and (CE^*)) of abductive confirmation.

Schurz complains that probabilistic incremental confirmation allows "(almost) *every* speculative abduction to be confirmed", so that result (2) in fact implies "a breakdown" of the Bayesian approach. In my view, the prospects of the Bayesian framework are much more promising (cf. [6, 7]).

First, it should be admitted that positive relevance is a weak criterion of confirmation: the increase of probability may be small, and several incompatible hypotheses may receive confirmation by the same evidence. This helps to understand the point repeated by Schurz: in some cases the best available explanation is not *good enough* to be acceptable. Indeed, a hypothetical explanation H (PR)-confirmed by E need not be rationally *acceptable* on E.

Secondly, for the purpose of comparing rival explanations, it is useful to introduce quantitative degrees of confirmation and explanatory power (cf. Kaila's notion discussed above). In a more detailed Bayesian analysis, if degrees of confirmation are defined by the difference measure

(4) $$conf(H/E) = P(H/E) - P(H),$$

then we can see that evidence gives strongest support to a *minimal explanation*, i.e., only to the part of an explanatory hypothesis H that is indispensable for the deductive explanation of the evidence E. So irrelevant additions to an explanatory theory do not improve its status with respect to abduction.

Thirdly, Bayesian comparisons of rival explanations depend also on their prior probabilities. Such comparisons need not rely only on "subjective matters", as Schurz states, since such explanatory theories may have objective logical relations between each other (e.g., logically stronger theories have lower prior probabilities than weaker ones).

Fourthly, the Bayesian approach can be generalized in interesting ways (see [7]). Instead of posterior probability, we may estimate the degree of truthlikeness of a theory on the basis of its success in explanation and prediction. This treatment can be applied to cases, where a hypothesis H has the probability zero, or H is known to be a false idealization.

Fifthly, for the evaluation of Schurz's thesis, it is important to ask whether the Bayesian approach is able justify the criterion of causal unification. In other words, is it possible to prove that a hypothesis which explains many phenomena has better confirmation than a hypothesis which explains only one phenomenon? Can the Whewellian idea of "consilience of inductions" be justified in Bayesian terms?

A recent result of Wayne C. Myrwold [3] gives light to these issues. Suppose that theory H achieves inductive systematization between empirical propositions E and E' (cf. [8]). This means that E and E' are probabilistically independent, but they are informationally relevant to each other given H:

(5) $P(E'/E) = P(E')$ and $P(E'/E\&H) > P(E'/H)$.

Myrwold proposes to measure the degree of *unification* of E and E' achieved by H by means of the difference

(6) $log[P(E'/E\&H)/P(E'/H)] - log[P(E'/E)/P(E')]$.

Then he argues, by applying again the logarithmic ratio measure of confirmation (i.e., a logarithmic variant of (4)), that the degree of confirmation of H by $E\&E'$ can be divided into three additive parts: confirmation of H by E, confirmation of H by E', and the degree of unification of E and E' achieved by H.

Essentially the same result is proved by Timothy McGrew [2, 562]. Assume that H_1 and H_2 are equally good explanations of E and E', i.e., $P(H_1) = P(H_2)$, $P(E/H_1) = P(E/H_2)$, and $P(E'/H_1) = P(E'/H_2)$. Assume further that H_1 achieves inductive systematization between E and E' in the sense of (5), but E and E' are independent conditional on H_2, i.e., $P(E/E'\&H_2) = P(E/H_2)$. Then a direct calculation by Bayes's Theorem (1) shows that $P(H_1/E\&E') > P(H_2/E\&E')$. Hence, the theory H_1 which unifies E and E' gains more confirmation by evidence $E\&E'$ than the theory H_2 without such unifying power.

I conclude with a simple and straightforward argument about the Whewellian situation where theory H deductively explains two independent phenomena E and E'. Note that in this case Myrwold's degree of unification (6) is zero, since $P(E'/E\&H) = P(E'/H) = 1$. (For another similar example, see [1].)

Let $E_1, E_2, \ldots, E_n$ be repeated occurrences of the phenomenon E, and let $E(n)$ be their conjunction. Suppose that H logically entails $E(n)$, so that $P(E(n)/H) = 1$. Then by (2) evidence $E(n)$ confirms H. But repetitions of the same kind of evidence give diminishing returns, since by inductive learning the (n+1)st occurrence of E is more probable than its earlier occurrences: $P(E_{n+1}/E(n)) > P(E_{n+1}) = P(E_1)$. Now suppose further that H logically entails another phenomenon E' which is probabilistically independent of E, i.e., $P(E'/H) = 1$ and $P(E'/E) = P(E')$. It follows from these conditions that

$$P(H/E(n)\&E_{n+1}) < P(H/E(n)\&E')$$
$$\text{iff } P(E(n)\&E') < P(E(n)\&E_{n+1})$$
$$\text{iff } P(E'/E(n)) < P(E_{n+1}/E(n))$$
$$\text{iff } P(E') < P(E_{n+1}/E(n)).$$

The last condition holds under quite broad conditions, e.g., if the initial probabilities $P(E)$ and $P(E')$ are about the same magnitude. Hence, relative to $E(n)$, new evidence E' gives more confirmation to H than old evidence E_{n+1}. This argument thus proves that successful explanation of the new kind of phenomenon E' gives more confirmation to theory H than the repetition of the old kind of phenomenon E.

BIBLIOGRAPHY

[1] Marc Lange. Bayesianism and unification. *Philosophy of Science*, 71:205–215, 2004.
[2] Timothy J. McGrew. Confirmation, heuristics, and explanatory reasoning. *The British Journal for the Philosophy of Science*, 54:553–567, 2003.
[3] Wayne C. Myrwold. A Bayesian account of the virtue of unification. *Philosophy of Science*, 70:399–423, 2003.
[4] Ilkka Niiniluoto. *Is Science Progressive?* D. Reidel, Dordrecht, 1984.
[5] Ilkka Niiniluoto. *Critical Scientific Realism*. Oxford University Press, Oxford, 1999.
[6] Ilkka Niiniluoto. Defending abduction. *Philosophy of Science*, 66:436–451, 1999.
[7] Ilkka Niiniluoto. Truth-seeking by abduction. In F. Stadler, editor, *Induction and Deduction in the Sciences*, pages 57–82. Dordrecht, 2005.
[8] Ilkka Niiniluoto and Raimo Tuomela. *Theoretical Concepts and Hypothetico-Inductive Inference*. D. Reidel, Dordrecht, 1973.
[9] Karl R. Popper. *Conjectures and Refutations: The Growth of Scientific Knowledge*. Routledge and Kegan Paul, London, 1963.
[10] Howard Smokler. Conflicting conceptions of confirmation. *The Journal of Philosophy*, 65:300–312, 1968.
[11] Laura J. Snyder. Confirmation for a modest realism. *Philosophy of Science*, 72:839–849, 2005.
[12] William Whewell. *The Philosophy of Inductive Sciences, Founded upon their History*. John W. Parker and Sons, London, second edition, 1847.

Reply to Ilkka Niiniluoto

GERHARD SCHURZ

ABSTRACT. A detailed analysis of the refined Bayesian criterion for unification proposed by Niiniluoto and Myrwold shows that it faces similar problems as those which are mentioned in §3 of my paper.

Ilkka Niiniluoto's note 'Abduction, Unification, and Bayesian Confirmation' is a very useful comment to my paper 'Common Cause Abduction and the Formation of Theoretical Concepts in Science'. Niiniluoto illuminates the close relationship between my notion of causal unification and Whewell's notion of consilience of inductions. In particular, he puts my skeptical remarks about the ability of Bayesian confirmation accounts to capture causal unification under closer examination. Niiniluoto's discussion of more fine-grained Bayesian criteria are a welcome supplement to my admittedly very brief remarks on Bayesian confirmation contained in §3 of my paper. In the following reply I want to show that a deeper analysis of the refined Bayesian criteria which are defended by Niiniluoto and have been suggested by Myrwold [2] may strengthen the skeptical remarks of my paper. My criticism of Myrwold's account is partly based on the criticism of Lange [1, §2], who is quoted by Niiniluoto, although Niiniluoto does not pick up Lange's criticism of Myrwold's account.

In §3 of my paper I ask the question under which conditions one is scientifically justified to explain some given empirical phenomena E_i by a hypothesis H which introduces new theoretical concepts and hence postulates new unobservable entities and/or properties. I suggest a criterion of causal unification (CU) according to which the postulated entity or property has to explain *many intercorrelated* but analytically independent phenomena, and in this sense yields a causal or explanatory unification. In contrast, the introduction of *one* new theoretical postulate merely for the purpose of explaining *one* empirical phenomenon is a case of pure speculation – an information-less pseudo-explanation, whose ex-post construction is always possible. I argue that the Bayesian incremental confirmation criterion cannot demarcate scientifically worthwhile hypotheses from pure speculations because according to this criterion, every hypothesis H which deductively entails an empirical phenomenon E is confirmed by E, provided only that $P(H) > 0$ and $P(E) < 1$. For example, the hypothesis $H :=$ 'God wants E and whatever God wants, happens' is confirmed by E.

Niiniluoto replies that my criterion of unification – or at least that part of it which is independent from causality considerations – can also be justified in Bayesian terms, based on the proposal of Myrwold [2, p. 410]. According to Myrwold a hypothesis H unifies two (or several) empirical phenomena E_1 and E_2 iff E_1 and E_2 are probabilistically independent but become dependent conditional on H. Myrwold defines the (Bayesian) degree of unification of $\{E_1, E_2\}$ achieved by H, $U(E_1, E_2; H)$, as

$$U(E_1, E_2; H) := log_2(\tfrac{P(E_1|E_2 \wedge H)}{P(E_1|H)}) - log_2(\tfrac{P(E_1|E_2)}{P(E_1)})$$

(see Niiniluoto's equation 6), where the Bayesian probability measure is relative to some given background knowledge B. Myrwold proves that the degree of confirmation of H by $E_1 \wedge E_2$ as explicated by the logarithmic ratio measure is the sum of the confirmation of H provided by E_1 plus that of H provided by E_2 plus the degree of unification $U(E_1, E_2; H)$. He concludes that unification is a *part* of Bayesian confirmation. However, as Lange [1, p. 208] has pointed out, $U(E_1, E_2; H)$ is not an adequate measure of unification because it cannot distinguish unification from pseudo-unification. To see this, simply note that for any two given independent phenomena E_1 and E_2, the hypothesis $H := E_2 \rightarrow E_1$ will unify $\{E_1, E_2\}$ with a high degree (since $P(E_1|E_2 \wedge H) = 1$, $P(E_1|H)$ is small, and $\frac{P(E_1|E_2)}{P(E_1)} = 1$). For example, according to Myrwold's measure, the two phenomena 'grass is green' and 'snow is white' are unified by the hypothesis 'if snow is white then grass is green'. But clearly, $H := E_2 \rightarrow E_1$ does not unify $\{E_1, E_2\}$ in my sense of unification which is explicated in detail in Schurz and Lambert [3]. This 'unification' merely replaces the set of phenomena $\{E_1, E_2\}$ by the set $\{H, E_2\}$ (where $H, E_2 \models E_1$). No reduction of the number of unexplained elementary (empirical or theoretical) phenomena is achieved, and hence this is not a case of unification in my sense of this notion. In the view of Schurz and Lambert [3], the mistake in Myrwold's account lies in the fact that it considers only the unification *gain* of the hypothesis *on* the explained phenomena, but it ignores the unification *cost* of the introduction of the hypothesis itself. In order to distinguish genuine unification from speculative pseudo-unification, this cost must be taken into account.

This part of my critique was technical. Now I turn to the intended applications of Myrwold's account. Myrwold's example of the Copernican versus the Ptolemaic explanation of the correlations between the retrograde motions and the period-relations of the planets is very similar to my examples of common cause explanations of correlated dispositions in §4 of my paper. We have the planets as our domain of objects, and we have two empirically discovered dispositional properties of them: disposition D_1 of a planet is its so-called synodic period, which reflects the relation of its period to the sun-earth-period, and disposition D_2 expresses the phase of its maximal retrograde motion in relation to the earth. The two kinds of dispositions are observed to be strictly correlated. As Myrwold explains in all detail [2, §2.1], the simple Ptolemaic cosmological model H_P cannot explain these correlations. The Copernican cosmological model H_C explains these correlations by the common cause that all planets, including the earth, are revolving around the sun, with different angular velocities. Thus, the Copernican theory (or model) H_C together with D_1 strictly entails D_2 (for every planet x; within a given degree of approximation). Hence $P(D_2|D_1 \wedge H_C) = 1 > P(D_2|H_C)$, and so, H_C unifies $\{D_1, D_2\}$ according to Myrwold. In contrast, the simple Ptolemaic theory H_P cannot derive D_1 from D_2 and, thus, does not yield unification in Myrwold's sense.

However, as Myrwold himself remarks later on, the full-fledged Ptolemaic theory, call it H_P^+, included two additional theoretical postulates (concerning the relative positions of sun, earth, planet, and the center of the planet's epicycle) which just entail that the two dispositions D_1 and D_2 are strictly correlated. These two ad hoc postulates provide no common cause explanation; they rather stipulate unexplained correlations on the theoretical level which 'save' the empirically observed correlations. Myrwold, however, faces the consequence that on his account the full-fledged Ptolemaic theory H_P^+ achieves exactly the same 'unification' as the Copernican theory [2, p. 415f]. This problem illustrates the weakness of Myrwold's unification account on a deeper level. The theory H_P^+ is obviously a conjunction of H_P with additional theoretical adhoc postulates, call them H_{cor}, which are at least as complex as the claim that the dispositions D_1 and D_2 are correlated. No unification in my sense of this notion is provided here. But since Myrwold does not analyze the compositional structure of the hypothesis H_P^+, he cannot distinguish between H_C and H_P^+ in terms of unification. In §3 of my paper I say that a Bayesian will probably deal with such a problem in terms of comparing the prior probabilities of the competing hypotheses. Exactly this is what Myrwold does. He argues that (a) the prior probability of H_C and that of the simple Ptolemaic hypothesis H_P should be approximately equal. Since the full-fledged Ptolemaic theory H_P^+ contains the additional postulate H_{cor} as compared to H_P, Myrwold further argues (b) that the prior probability of H_P^+ is significantly smaller than that of H_P. From these considerations Myrwold then derives that the confirmation of H_C by $D_1 \wedge D_2$ is greater than that of H_P^+ by $D_1 \wedge D_2$. Myrwold's claim (b) is certainly correct. But his claim (a) – to repeat the point in §3 of my paper – is clearly a subjective matter, which does not have an objective justification.

In conclusion, my skeptical remarks about Bayesian accounts seem to be good also for Myrwold's measure of unification. Given the flexibility of Bayesian accounts one may, of course, uphold the hope that a satisfying Bayesian account of unification will once be found. What I doubt is that a satisfying account of unification can be given without means for analyzing the compositional structure of the hypotheses whose confirmation is to be measured. I agree with Niiniluoto's points in his last paragraph – namely that the Bayesian account can at least say the following: (a) every new evidence E_{n+1} for a hypothesis H (which is not entailed by the other evidences) increases the confirmation of H; (b) if E_{n+1} is inductively supported by $E_1, \ldots, E_n$, the increase in confirmation of H by E_{n+1} gives diminishing returns, and (c) if E_{n+1} is independent of $E_1, \ldots, E_n$, E_{n+1} conveys more confirmation to H than in case (b). These observations are certainly useful, but unfortunately they do not tell us how to distinguish pure speculations from scientific hypotheses.

BIBLIOGRAPHY

[1] Lange, M.: "Bayesianism and Unification", *Philosophy of Science* 71, 2004, 205-215.
[2] Myrwold, W.C.: "A Bayesian Account of the Virtue of Unification", *Philosophy of Science* 70, 2003, 399-423.
[3] Schurz, G., and Lambert, K.: "Outline of a Theory of Scientific Understanding", *Synthese* 101/1, 1994, 65-120.

A Novel Paradox?

GÖRAN SUNDHOLM

The naming of paradoxes is usually a straight-forward matter: they are named after their inventors. Thus we have the paradoxes of Epimenides *(Liar)*, (Zermelo-) Russell,[1], Burali-Forti[2] Cantor,[3] Richard,[4] Grelling,[5] Mirimanoff,[6] Berry, [7] Curry,[8] Kleene-Rosser,[9] Hempel,[10] Girard, [11] ...., and of course, in recent years, of Fitch [12] . The scholarly grapewine [13] now tells that the paradox of Fitch really is that of Church, which is awkward, since there is already a function-theoretic paradox known by the name of Church. [14] The paradox of Fitch/Church is sometimes also referred to as that of *knowability*, much in the same way that the Liar is sometimes referred to as the Paradox of *Truth*. [15] Personally, even though I

---

[1]Cf. B. Rang and W. Thomas, 'Zermelo's discovery of the "Russell paradox"', *Historia Mathematica* 8 (1981), 15-22.

[2]C. Burali-Forti, 'Una questione sui numeri transfiniti', *Rendic. Palermo* 11 (1897), 154-164.

[3]G. Cantor, 'Letter to Richard Dedekind', in: *Gesammelte Abhandlungen mathematischen und philosophischen Inhalts*, Springer, Berlin, 1932, 443-447.

[4]J. Richard, 'Les principes des mathématiques et le problème des ensembles', *Revue générale des sciences pures et appliqués*, 16 (1905), 541-543.

[5]K. Grelling and L. Nelson, 'Bemerkungen zu den Paradoxien von Russell und Burali-Forti', *Abhandlungen der Friesschen Schule*, N.S. 2 (1907/8), 301-324.

[6]D. Mirimanoff, 'Les antinomies de Russell et de Burali-Forti et le problème fondamental de la théorie des ensembles', *L'Enseignement Mathématique*, 19 (1917), 37-52.

[7]Cf. B. Russell, 'Mathematical Logic based on the Theory of Types', §1, Section 10, footnote 3, *American Journal of Mathematics*, 20 (1908), 222-252.

[8]H. B. Curry, 'The inconsistency of certain formal logics', *The Journal of Symbolic Logic* 7 (1942), 115-117.

[9]S. C. Kleene and B. Rosser, 'The inconsistency of certain formal logics', *Annals of Mathematics*, (2) 36 (1935), 630-636.

[10]C. G. Hempel, 'A Note on the Paradoxes of Confirmation', *Mind* 55 (1946), 79-82.

[11]Cf. P. Martin-Löf, 'An Intuitionistic Theory of Types', in: G. Sambin and J. Smith (eds.), *Twenty-Five Years of Constructive Type Theory*, Oxford University Press, 1998 (reprinted version of an unpublished report from 1972), 127-172.

[12]F. Fitch, 'A Logical Analysis of Some Value Concepts', *The Journal of Symbolic Logic* 28 (1963), 135-142.

[13]With Joe Salerno as its agent.

[14]So called by Georg Kreisel in his monumental survey article 'Mathematical Logic', in: Thomas Saaty (ed.), *Lectures on modern mathematics*, volume 3, Wiley, New York, 1965, 95-195, §§2.151, 2.152, at pp. 124-125. Church's harvest of named concepts is quite astounding: one (Undecidability) *Theorem*, one *Thesis*, and now *two* paradoxes!

[15]D. Edgington, 'The Paradox of Knowability', *Mind* 94 (1985), 557-568; B. Brogaard and J.

admit of constructivist tendencies, I have never felt in the slightest threatened by the Church/Fitch reasoning. To my mind, its rendering of the knowability of truth simply does no justice to what is involved in constructivist knowability. Thus, I do not wish to squander the venerable term of *knowability* of Fitch's (Church's) reasoning – a paradox most certainly does not smell as sweet by any other name! – but instead it is my purpose to offer Shahid a proper *Paradox of Knowability* for his 52th birthday, in the hope that it shall not trouble him to the extent that the Liar troubled Philetos from the (present holiday-paradise) island of Kos. Here goes. We consider:

(*)   This sentence is unknowable.

(1)   Assume that (*) can be known.   Assumption

Then

(2)   (*) is true

whence

(3)   This sentence cannot be known.   From (*) and (2); T-Schema

(4)   Contradiction   From (1) and (3)

Therefore

(5)   (*) cannot be known.   From (1) and (2),
                                                 without assumptions

That is

(6)   This sentence is unknowable.   (*) is demonstrated on no assumptions

Therefore

(7)   (*) is known.   From (7), what is demonstrated is known

(8)   CONTRADICTION   From (5) and (7) on no assumptions

Remarks

(i) I have used an indexical formulation analogous to the "This sentence is false" version of the Epimenides. This can be circumvented using substitution

---

Salerno, 'Fitch's Paradox of Knowability', http://plato.stanford.edu/entries/fitch-paradox/.

in order to construct a fix-point, after the fashion of Quine's *Mathematical Logic*.[16] We then have a sentence S such that

$$S =_{def} S \text{ cannot be known,}$$

which can be used to carry out the reasoning without the above anaphoric self-reference.

(ii) From a constructivist point of view this is very close to the Liar since, in some suitable version or other, truth is nothing but knowability.

(iii) Only the delicate question now remains: by what name should one call this novel(?) paradox...?

BIBLIOGRAPHY

[1] Cf. B. Rang and W. Thomas: 'Zermelo's discovery of the "Russell paradox"', *Historia Mathematica* 8 (1981), 15-22.

[2] C. Burali-Forti: 'Una questione sui numeri transfiniti', *Rendic. Palermo* 11 (1897), 154-164.

[3] G. Cantor: 'Letter to Richard Dedekind', in: *Gesammelte Abhandlungen mathematischen und philosophischen Inhalts*, Springer, Berlin, 1932, 443-447

[4] J. Richard: 'Les principes des mathématiques et le problème des ensembles', *Revue générale des sciences pures et appliqués* 16 (1905), 541-543

[5] K. Grelling and L. Nelson: 'Bemerkungen zu den Paradoxien von Russell und Burali-Forti', *Abhandlungen der Friesschen Schule*, N.S. 2 (1907/8), 301-324

[6] D. Mirimanoff: 'Les antinomies de Russell et de Burali-Forti et le problème fondamental de la théorie des ensembles', *L'Enseignement Mathématique* 19 (1917), 37-52

[7] Cf. B. Russell: 'Mathematical Logic based on the Theory of Types', §1, Section 10, footnote 3, *American Journal of Mathematics* 20 (1908), 222-252

[8] H. B. Curry: 'The inconsistency of certain formal logics', *The Journal of Symbolic Logic*, 7 (1942), 115-117

[9] S. C. Kleene and B. Rosser: 'The inconsistency of certain formal logics', *Annals of Mathematics*, (2) 36 (1935), 630-636

[10] C. G. Hempel: 'A Note on the Paradoxes of Confirmation', *Mind*, 55 (1946), 79-82

[11] Cf. P. Martin-Löf: 'An Intuitionistic Theory of Types', in: G. Sambin and J. Smith (eds.), *Twenty-Five Years of Constructive Type Theory*, Oxford University Press, 1998 (reprinted version of an unpublished report from 1972), 127-172.

[12] F. Fitch: 'A Logical analysis of Some Value Concepts', *The Journal of Symbolic Logic*, 28 (1963), 135-142

[13] D. Edgington: 'The Paradox of Knowability', *Mind*, 94 (1985), 557-568; B. Brogaard and J. Salerno, 'Fitch's Paradox of Knowability', http://plato.stanford.edu/entries/fitch-paradox/.

[14] W. Stegmüller: *Das Wahrheitsproblem und die Idee der Semantik*, Harvard UP, 1940, Springer Verlag Wien, 1957

---

[16]Harvard UP, 1940, 59. W. Stegmüller, *Das Wahrheitsproblem und die Idee der Semantik*, Springer Verlag Wien, 1957, p. 31 ff, considers the (substitution-theoretic) construction in great detail.

Inquiry and Functionalism

JOHN SYMONS

Functionalism rests on the claim that mental states are multiply realizable. One way to see them as multiply realizable is to understand mental states as causal roles rather than particular physical structures. Thus, functionalism can be contrasted with metaphysical accounts which treat mental states as instances of mental substance. Instead of puzzling over the relationship between mental and physical kinds, functionalists understand our talk of minds as a way of describing the functions of bodies. In addition to its metaphysical content, functionalism has been connected with more or less concrete empirical hypotheses about psychological states and the prospects for psychology as a scientific theory. The connection between the metaphysics of mental states and the status of psychology as a theory is straightforward. It is a small step from a functional characterization of psychological states to a similar treatment of psychological types and laws. A type is functional if all instances of the type are functional and laws are functional if they relate some functional type with something else in a law-like way. (See [8, p. 533]) Psychological theories can be understood as functional if one agrees that all psychological types are functional and that the laws of psychology relate those types. When applied to psychological theories, functionalist arguments are widely seen as providing a bulwark against scientific reductionism.

Functionalists who argue for non-reductive physicalism often combine their metaphysical arguments with defenses of favored psychological taxonomies. For example, the fate of functionalism is often mistakenly conflated with those of representationalism and computationalism. For the past three decades, functionalism has played a central role in both the metaphysics of mind and development of scientific theories of mind. This paper examines the relationship between functionalism as a metaphysical position and the possibility of progressive inquiry into the nature of mental states. Few philosophers would wish to give up on the possibility that we might be wrong about at least some features of our mental life. However, functionalism by itself, does not have the resources to explain how we might correct our errors or about how progress in the science of mind is possible.

One might argue that it is not the job of functionalism to account for scientific progress; functionalism is a metaphysical doctrine, not a contribution to philosophy of science or epistemology. I agree. However, there are significant risks for the functionalist in distinguishing her metaphysical and scientific claims about the mind too sharply. Strikingly, for example, functionalism is not, strictly speaking, a theory of mind. As Brie Gertler points out, functionalism per se neither entails nor explains a distinction between mental and nonmental properties [2, p. 77]. Both Gertler and Alvin Goldman [3, p. 24] note that functionalism does not possess the resources to mark the mental/nonmental distinction. The continuity of the mental and the nonmental is the great virtue of functionalism, but it is also, potentially, the source of its irrelevance. Few functionalists would wish to give up on the connection between the metaphysics of mind and our ordinary or even our scientific psychology. It would be odd if there were no relationship between our views on the nature of mental life and our metaphysics of mind. Might functionalism have nothing to say about the nature of mind?

Functionalism would be a strikingly barren doctrine if it could be shown to make no difference with respect to how we ought to conduct of our inquiry into the nature of mind and, perhaps more importantly, for philosophers, if it could be shown to tell us nothing about the mind itself. Could it be that functionalism per se can't even tell us whether mental states are multiply realizable?

At this point, a careful student of philosophy of mind might tell us that we have gone too far and that the answer to our questions will depend on the type of functionalism we are addressing. Different kinds of functionalism will vary with respect to the kinds of inquiry they take to be authoritative. For instance, common-sense or analytic or conceptual functionalists take folk views on psychology as primary. On views of this kind, the way the folk characterizes the causal role of some mental state defines the nature of that mental state. To use a term in a manner which differs from the way the folk use the term is to misuse the term. Pain is what pain does and if we don't know what pain's role in the causal economy of nature is, we ask the folk. We can find no better account of pain than the folk account of what causes pains and of what pains cause. By contrast, a range of competing forms of functionalism, call them, scientific functionalism for convenience, will see folk accounts as subject to modification in light of evidence from scientific inquiry.

In this paper a minimal version of functionalism is outlined and separated off from the various empirical and other commitments which are often associated with functionalism in the philosophy of mind. Because of its failure to accommodate theory change, this minimal form of functionalism is shown to become virtually irrelevant to our most pressing philosophical concerns.

1   Finished theories

In 'Theories' (1929), Frank Ramsey provides a technique whereby 'theoretical concepts' like is a proton or is a center of gravity can be treated as existentially bound variables rather than predicates. By 'theoretical concept' he means those which are indispensable for the generalizations and predictions of a theory but whose members are unobservable or otherwise philosophically problematic. Treating such concepts as existentially bound variables specifies the role of theoretical terms via the system of relationships defined by the structure of the theory [7, p. 212–236]. Following David Lewis [4], functionalists have understood Ramsey sentences as a way of resolving, at least in principle, worries about the ontological status of psychological predicates. Given some psychological theory, the Ramsey sentence can serve as a way of providing definitions for mental terms that do not themselves include mental terms. Metaphorically speaking, we can say that they serve to provide non-question begging definitions of mental terms by treating them as locations in the network provided by a theory. The virtues of the Ramsey-Lewis strategy to psychological predicates extend only to what I will call minimal functionalism. Minimal functionalism is the view that we can give a Ramsey sentence for some finished psychological theory and that the psychological predicates are multiply realizable.

Even given multiple realizability, minimal functionalism is compatible with virtually any scientific treatment of mental life. As Robert Richardson has argued, functionalist commitments to multiple realizability are compatible with reductionist explanations of psychological predicates [8]. Likewise, converting psychological predicates into relative positions in a theory about the causal structure of things has no serious implications for the direction of psychological inquiry, insofar as such inquiry is a matter of discovering the appropriate causal structure or theory.

As discussed below, Ramsey elimination does not make any significant difference in the development of a scientific theory of mind since it assumes the existence of a theory that is both finished and true. Given some scientific theory T where T ranges over unobservable properties $A_1 \dots A_n$ and observable properties $O_1 \dots O_n$, it is also possible to introduce reference to individuals $a_1 \dots a_n$.

(1.1) $$T(A_1 \dots A_n, O_1 \dots O_n)$$

The ascription of some unobservable property (say the property of being a neutron) to some individual or region of space-time a can be carried out via a sentence containing a higher-order existential quantifier along the following lines:

(1.2) $$(\exists A_1) \dots (\exists A_n)[T(A_1 \dots A_n, O_1 \dots O_n) \& A_i a)]$$

This definition characterizes unobservable theoretical terms based solely on existential quantification, observables and the structure provided by the theory. If we understand our theory T as providing a unique ordering of properties, then reference for problematic terms, things like neutrons, beliefs and market forces can be fixed via their relationships with one another and with the observable phenomena described by the relevant theory. The structure of relationships between the elements of a theory is presented by the theory T and to say that some individual has some property can be converted into a claim about relative placement within the structure described by T, in this case that a has the i_{th} of $A_1 \ldots A_n$.

Despite its virtues, the minimal, Ramsey-Lewis version of functionalism tells us nothing about how one might settle on a causal structure appropriate to particular explanations. Instead, Ramsey's definition assumes T's ordering without saying anything about what it is, or how one might decide between alternatives. Of course, Ramsey's approach was not originally intended to answer such questions and so this defect does not matter for his purposes. His goal was to settle the meaningfulness of theoretical terms in an established theory. As such, recourse to Ramsey-Lewis definitions is a way of dissolving a certain kind of ontological concern with respect to theoretical terms in some well-established theory via the causal structure it specifies.

Philosophically interesting theories of psychological states go well beyond the minimal metaphysical doctrine of causal-role functionalism that we can extract from Ramsey and Lewis. While they may be couched in the metaphysical framework of functionalism, theories that are worth arguing about are likely to concern the nature of the functions which comprise intelligent behavior and the manner in which they are organized. For instance, the generalizations of computational or representational theories might be understood to be true of human minds insofar as they make the claim that there must be particular systems of embodied representations responsible for thought, perception, or action.

It is frequently noted that computational functionalism requires an additional non-trivial set of commitments over and above the simple claim that mental states are functional states. This, of course, is no great surprise. However, it is also worth noting the status of these additional representationalist assumptions in philosophical argument relative to minimal functionalism. For example, the additional assumptions that would turn a minimal functionalist into a computational functionalist might include for instance the notion that mental states are representations, that they are interpretable only from within a language of thought, that they are governed by algorithms, etc. Each of the additional computational assumptions is itself a matter for either scientific investigation or arguments from intuition or common sense. Insofar as they go well beyond minimal functionalism, these additional claims do not inherit the strengths of that view.

2   Functionalists as scientists

Consider a concept like memory. The minimal functionalist assumes that the possessors of some true psychological theory will know what memory is and that as a functional concept it has a variety of possible instantiations. However, we have, at best a very rough idea of what memory is and in the history of recent science psychologists have disagreed over the appropriate taxonomies and definition of memory. Apparently, inquiry in the brain and behavioral sciences involves entities and patterns we either do not yet understand, or that we understand only partly via strategies that are subject to revision or rejection.

This relatively straightforward point illuminates reductionist criticisms of functionalism of the kind presented by John Bickle. Bickle claims that multiple realizability has been shown to be false by the discovery that the molecular mechanisms for memory consolidation are conserved throughout evolutionary history [1]. Clearly Bickle is not arguing that some other mechanism could not possibly form the basis of memory. His claim is directed specifically towards an empirical reading of Putnam's original formulation of multiple realizability and concerns the nature of the concept of memory as instantiated in the actual world. He is not arguing that memory or any other trait might not have been instantiated in a machine made of rubber bands or some other structure in some possible world; just that it has not been instantiated in any other form to the best of our knowledge and that an empirical argument for multiple realizability has no legs to stand on.

Without addressing the specifics of Bickle's argument, consider whether and where such arguments can confute minimal functionalism. Clearly, insofar as the minimal functionalist is invoking a Ramsey sentence, he will be invoking the Ramsey sentence of some true psychological theory. To refer to such a sentence does not require perfect knowledge of it. Therefore, the minimal functionalist is off the hook on that score and can cancel his subscription to Nature. But, what about multiple realizability? Since one can imagine any particular function being performed by a variety of physical structures, it is possible to think of functions as having a kind of independence from the structures that perform them. In terms of raw logical possibility this is clearly correct.

No conceptual barrier blocks the possibility that my current psychological functions could be installed on my computer's hard drive or in a wedge of Swiss cheese. Of course, many empirical constraints stand in the way of thinking cheese, but these should be of no concern to the functionalist. Given a thin enough account of the concept of thought and a rich enough interpretation of what cheese could be up to, it is logically possible that cheese could think. This is what Putnam meant by his claim that as far as our study of mental life is concerned "we could be made out of Swiss cheese and it wouldn't matter." [6, p. 291]

Since functionalism is a view concerning the metaphysical status of minds and not a recipe for scientific progress, the strangeness or implausibility of the Swiss cheese mind is not, by itself, a reason to abandon functionalism. For the functionalist, particular realizations of a mental property, whether carbon-, silicon- or dairy-based, are an accidental matter. This is because, according to the functionalist, minds possess no physical features essentially. Functions are simply relative locations within some true theoretical structure. If a functionalist believes that minds can only have a causal role via some physical instantiation, then perhaps having some physical property is an essential feature of mental entities. However, even this is already a step beyond the core doctrine of functionalism since it draws on a set of beliefs concerning the nature of causal power and entails a physicalist ontology. Minimal functionalism per se is not wedded to any one view of what it is to be a cause nor is it committed to a single ontology.

Returning to Bickle's arguments, we find that they can only be effective when aimed at a view with more content than minimal functionalism. Without offering some kind of additional, non-functional evidence (empirical, intuitive, or whatever) a minimal functionalist view of memory, for instance, will not go beyond the claim that 'Memory is what memory does'. If anyone ever thought that there is some extra memoryness that makes memory what it is, over and above what memories do, functionalists stand ready to correct them. As we have seen, in this basic form, philosophical functionalism is a virtually unobjectionable doctrine. While being unobjectionable is a virtue, it is also the source of functionalism's one major weakness - its irrelevance. What we could call the 'modal refuge' defense against empirically based arguments is powerful but very costly. One must be willing to forego resting one's view on any specific claims concerning what psychological phenomena are in the actual world. That is, the functionalist must detach his metaphysical claim about the multiple realizability of a concept from any empirically verifiable features of that concept. Empirical claims must be minimized so as to ensure that the view does not depend on some fact about the actual world that could be shown to be false. This is why the functionalist risks irrelevance.

Nobody wants to be irrelevant and so functionalists will still wish to say that claims that they make concerning, for example, the essential character of memory are applicable, insofar as memory as we find it here in the actual world must also possess the essential properties of memory. However, here the burden falls on functionalists to provide an account of how any proffered list of essential properties for a particular psychological concept is relevant to that concept without thereby being open to correction and without thereby sacrificing functionalism's principal metaphysical advantage.

3   Conclusion: Inquiry and the revision of our psychological taxonomy

At this point it is clear that the minimal version of functionalism is a rather sterile doctrine and I recognize that no philosopher would argue for it as outlined so far. What I would suggest is that when self-described functionalists take sides in disputes concerning the status of folk psychology or the relevance of empirical results, their arguments draw from sources beyond the scope of functionalism proper. common-sense or analytical functionalism, teleological functionalism, machine functionalism, empirical functionalism; the list of functionalisms is long and the differences between them is what most philosophers of mind earn their keep arguing about. However, contention over these positions derives from their prefixes. Arguments from empirical functionalists are, at bottom, empirical; arguments from common-sense or analytical functionalists are derived from the intuitions we claim to have about our own minds, etc. Prefixes to functionalism allow the functionalist to avoid triviality and to participate in interesting debates.

Whatever merit various forms of functionalism might have in debates over the scientific investigation of mental life, they should not be understood to inherit the strengths or invulnerabilities of the minimal and I would suggest irrelevant "handsome is as handsome does" functionalism. The functionalist qua minimal or non-prefixed functionalist is in no position to help us understand much of anything. In order to support or attack a view on for example the status of folk psychology or the role of neuroscience in philosophy of mind they must resort to non-functionalist premises.

If we settle apriori on a particular psychological taxonomy and decide that it is not subject to revision, then clearly minimal functionalism would suffice as a theory of mind. However, few of us would wish to deny the possibility of scientific progress in psychology completely. This is true even for those who regard common-sense or intuition as the source for our psychological taxonomy. Even introspection can be improved or refined by further introspective inquiry.

If inquiry is contributing to the development of improved scientific explanation, we will add to our understanding of the explanandum. As such the inquiring agent will modify its characterization of the target of scientific explanation. Obviously this change is unrelated to the question of whether the actual object under investigation is being modified in any significant way by the process of inquiry. Unlike objects understood from a metaphysical realist perspective, an explanandum for a project of scientific inquiry is determined to a large extent by the interests and knowledge of the community of inquirers. While, the most minimal kind of change that explananda will undergo over the course of scientific progress results from simply knowing more about the explanandum, there are other, more dramatic ways that it can change and in the history of science it sometimes happens that the explanandum changes so dramatically that a conservative critic may argue that a modern explanation or theory has nothing to do with the original explanandum. Under ordinary circumstances, and as long as we can track some kind of continuous discourse, and as long as each twist and turn in the scientific process had some reasonable justification, the conservative objection is not a relevant obstacle to inquiry.

As inquirers learn more about something, it is likely to appear differently to them. It is beyond the scope of this paper to provide an historical study of some particular case of theory change, but there are many obvious examples of this kind of shifting explanandum in the history of science. We are not wedded to a fixed characterization of the nature of the mental insofar as we would not want to mistakenly latch onto a particular psychological taxonomy prematurely.

Acknowledgments

Shahid Rahman will understand why I chose this paper for a volume in his honor. His brilliant, but gentle insights and criticism have helped me to organize my thinking about the relationship between logic, metaphysics and philosophy of mind. This paper is an unlikely product of his influence, but I think he will welcome it as a familiar spirit. I am very fortunate to have such a generous and deep collaborator, colleague and friend.

With respect to this little paper, I am grateful to Jorge Valadez for his careful critical reading and commentary on a previous version of this paper at the New Mexico-West Texas Philosophical Society. Philosophers at the Society for Exact Philosophy provided exactly the right kinds of criticism and I know they will remain unsatisfied by this version of the paper. This paper also benefited from the comments of Clifford Hill and Tim Cleveland.

BIBLIOGRAPHY

[1] John Bickle. *Philosophy and Neuroscience: A Ruthlessly Reductive Approach*. Kluwer, Dordrecht, 2003.

[2] Brie Gertler. Functionalism's methodological predicament. *Southern Journal of Philosophy*, 38(1), 2000.

[3] Alvin I. Goldman. The psychology of folk psychology. *The Behavioral and Brain Sciences*, 16:15–28, 1993.

[4] David Lewis. Psychophysical and theoretical identifications. *Australasian Journal of Philosophy*, 50:249–58, 1972. Reprinted in [5], references are to reprint.

[5] David Lewis. *Papers in Metaphysics and Epistemology*. Cambridge University Press, Cambridge, 1999.

[6] Hilary Putnam. *Philosophical Papers*. Cambridge University Press, New York, 1975.

[7] Franck Plumpton Ramsey. *The Foundations of Mathematics and other Logical Essays*. Routledge & Keegan Paul, London, 1931.

[8] Robert C. Richardson. Functionalism and reductionism. *Philosophy of Science*, 46:553–558, 1979.

What is it Like to be Formal?

CHRISTIAN THIEL

In the third act of *The Taming of the Shrew*, Lucentio teases Hortensio in front of Bianca (whom they are both courting) by asking him: "Are you so formal, sir?"[1] It is easy to see that Shakespeare's wordplay on "formal" is here quite different from that intended in the title of the present paper. A modern ambiguity was pointed to by the knowing smile of an English native speaker upon reading my title, perhaps at the idea that a German would be the right kind of person to exemplify what it is like to be formal.

With words like "formal", "form", or "schema", each has an unusually broad spectrum of meaning in common usage as well as in the language of the Arts and Sciences, and the spectra extend from one of these areas deep into the other. Even where we succeed in bringing this undesirable abundance under control by terminological regulation, we remain dependent on a lot of more informal and even historically earlier uses which often guide our terminological decisions. So let me begin my survey and my delimitations here.

In expressions like "formal science", "formal system" or "formal language" we make reference to the notion of form, the designations of which in modern Indoeuropean languages as well as in Latin or its Greek equivalents $\mu\rho\rho\varphi\dot{\eta}$ and $\epsilon\hat{\iota}\delta\sigma\varsigma$ show an impressive variety of meanings. As scientists, we are inclined to overlook an every-day use of "form" that is not grounded in abstraction due to some relation of equiformity or, in Greek, isomorphism. We refer straightforwardly, as did the ancient Greeks, to shapes like a circle, a sphere, a straight line or a right angle (the designated rôle of the right angle seems even to be programmed neurally), and we need no abstraction to realize the decrease or increase of the values of a function in a graph. In some sense, therefore, there are cases of a direct perception and understanding of (some) forms. But if we wish to go beyond identification of and successful communication about shapes, and aim at a precise description, this will not suffice. We have to forget about forms in the sense of shapes, and about forms in the sense of moulds for giving shape to something else (like a bell or a pudding having a form is often the same as having been formed). We have also to forget about phrases like "water in the form of vapour", and about being in good form

---

[1] Wells/Taylor, 45 a 59; Tieck translates this as "Seid Ihr so pünktlich?", with a meaning of "pünktlich" which is obsolete today.

or in good shape. Instead, we have to focus our attention on form in the sense of pattern, style, or order (e.g. of words, or of sounds combining to form a melody), that is to say, on form as opposed to content, like colour or some material that *has* the form.

This is the origin, as I take it, of the use of *schematic letters* in mathematics and formal logic. This is essentially different from giving letters as quasi-names to geometrical figures and their elements, a procedure which we find in Euclid but which must have been customary a long time before. Here, we can denote by A, B, C three points on a line, three points on a circle, or the three vertices of a triangle; g and h may be two arbitrary lines, two parallel lines or two perpendicular lines. We could have interchanged the letters at the beginning without affecting the validity of our reasoning about the figure; the letters, being used as names or quasi-names, do not exhibit any pattern, scheme, order or form. A, B and C might be used for the vertices of an equilateral triangle, of an isosceles triangle, of a right-angled triangle, or some other triangle, but we could have permuted them, or used other letters instead. Rarely do we give some letter a designated meaning, as e.g. to M as the centre of a circle.

If I am not mistaken, a different approach first appeared with Aristotle's description of the propositions to be joined to yield a syllogism. The procedure is semi-formal: the four kinds of proposition which are admitted are "P applies to all S", "P applies to some S", "P does not apply to any S", and "P does not apply to some S". "P" and "S" are schematic letters, standing for terms where P is the major term, the predicate, and S the minor term, the subject. The "form" of these propositions is P–S, but they differ in the relation between the two terms. The different relations are *not* distinguished by different schematic letters in Aristotle; their designation by a, i, e and o was introduced much later. Nevertheless, the four kinds of proposition admitted for a syllogism are called the "standard forms".

The letters are not wholly arbitrary, as appears when we combine three propositions of standard form to make up a syllogism. A first proposition of this kind has the two terms P and M (the "middle term") and is called the major premise; a second proposition has the terms M and S and is called the minor premise, and a third proposition has the terms P and S, in this order, and is called the conclusion of the syllogism. P is the predicate in the conclusion, and S the subject in the conclusion, but in each of the premises M may be the predicate or the subject, and P or S (respectively) will take the remaining part in the proposition. Corresponding to the position of M in either premise, we get four different arrangements of P, M and S that make up a syllogism, and these arrangements are exhibited by the position of the letters P, M and S, independently of the particular relation between the terms in each proposition. Aristotle called the four resulting arrangements the four syllogistic "schemata", which was translated into Latin as *"figurae"* and is the origin of our syllogistic "figures".

Obviously, Aristotle's doctrine of the syllogism is a semi-formal theory: it is material with respect to the propositions admitted, i.e. to the relations later named a, i, e, and o, but it is formal with respect to the terms entering into the propositions of a syllogism. Any term capable of being predicated meaningfully of another one may replace P, M, or S in a syllogistic schema, and in this sense the letters "P", "M", and "S" are schematic letters.

Aristotle also reflects on the form of elementary propositions when he analyzes the meaning of predication, but he does not use schematic letters for predicates and nominators. Likewise, he talks about inference schemata beside syllogistic schemata, but he does not systematically use schematic letters for propositions.[2] The attractiveness of such a device seems to have dawned upon the Stoic logicians although they did not use schematic letters either, but employed ordinals as a surrogate, e.g. when they codified the detachment rule in the form[3]

If the first, then the second

But the first [holds]

Therefore the second.

Schematic letters have often been called "variables", but variables in the proper sense serve different purposes.[4] If we can trust Cajori (which is not always a good idea), "indeterminate" quantities had "unknown" quantities as their forerunners.[5] These, however, also have a different task, e.g. in an equation $x + 2 = 0$ or $x^2 = 1$. They mark or indicate an "empty" place here, too, but it is a place kept free only provisionally because of our temporary ignorance, an ignorance that will hopefully be removed by finding a "solution" of the equation by suitable manipulation of its parts, including the "unknowns". And of course, they have a different purpose from schematic letters.

---

[2]Bocheński pointed out long ago that Aristotle in the *Prior Analytics* (I. XV, 84a6 ff.) seems to ascribe possibility and impossibility to propositions A and B, using "A" and "B" as schematic letters for propositions. The matter is, however, not beyond doubt since immediately before this passage, Aristotle talks about some A's possible applying to something, so that A's "being" may be understood not as the circumstance that A holds, but that the predicate A pertains to some object. An additional problem is posed by the fact that Aristotle (op. cit., 34a24) considers the case "that A represents the premises and B the conclusion", where the plural of the premises makes it doubtful that "A" stands schematically for a single proposition.

[3]See, e.g., Sextus Empiricus, *Adversus mathematicos* VIII, 227 (*Opera* ed. H. Mutschmann, vol. II, B.G. Teubner: Leipzig 1914, repr. 1984, 154; English translation in I.M. Bocheński, *A History of Formal Logic*, University of Notre Dame Press: Notre Dame, Ind. 1961, repr. Chelsea Publishing Company: New York 1970, 125, quotation 21.22).

[4]See below. Non-schematic letters in a stricter sense are mathematical proper names like "i", "e", and "π", or binding operators like Frege's "β" in, e.g., "$M_\beta(\varphi(\beta))$", although this "β" may be replaced by some other letter like "γ" serving the same purpose.

[5]Cf. Florian Cajori, *A History of Mathematical Notations*, vol. II, The Open Court: La Salle, Ill. 1929 (repr. together with vol. I, Dover: New York 1993), 1 ff.

Historically, the transition to variables proper took place in a hazy period. Leonardo of Pisa, alias Fibonacci (c. 1200), may have been the first to use them, but I am not sure they really lost their character as "unknowns" with him.[6] Vieta (François Viète, 1540–1604) is the first to aim at the expression of the "general equation" of a certain degree, and at general methods and strategies for their solution, and to develop for this purpose what may be regarded as the earliest form of a "symbolic algebra". He complements the *logistica numerosa* (elementary arithmetic dealing with numerals) by a *logistica speciosa* which is an "algebra of letters", dealing with "species" or "forms" ("*quae per species seu rerum formas exhibetur*"[7]). The manipulation of the letters, however, follows only the rules of an *algebra speciosa* explicitly developed by Vieta for this purpose. While ancient and hellenistic algebraists had used, at most, abbreviations for numbers, numerical operations or (with Diophantos) the "unknown", Vieta denotes the various unknowns of an algebraic problem by capital vowels A, E, I, O, Y, and the "known" quantities (parameters) by consonants B, C, D, ... Even though Vieta did not invent further devices like exponentiation (he writes *latus* A for A, *quadratum* A for A^2, *cubo-cubo-cubus* A for A^9),[8] he reaches in his *Ars Analytica* a spectacular new survey of algorithmic and systematic relations in arithmetic, in the theory of equations and in general algebra, and so successfully in the scientific community that even in the 19th century we often find "algebra" and "analysis" used synonymously.

The trick of using variables or schematic letters to exhibit the form or construction of an expression led to a novel approach to *functions*. Although the term dates back to Leibniz and the Bernoullis,[9] a reflection on the general notion did not commence before the 19th century. In our context, the most fruitful approach is that of Frege who used to present the idea by inviting his readers to reflect on expressions like

$$3 + 5 \cdot 3 = 6$$
$$1 + 5 \cdot 1 = 6$$
$$4 + 5 \cdot 4 = 6$$

---

[6]The situation is difficult to assess (see Cajori, op. cit.). Fibonacci's presumable contemporary Jordanus Nemorarius has also been credited with the introduction of schematic letters proper, and indeed he uses letters instead of "arbitrary" numbers (in the sense of numbers taken merely as an example); but as he sometimes changes an assigned letter in the course of a calculation or transformation procedure, the claim is doubtful.

[7]Franciscus Vieta, *Opera*, ed. Franciscus A. Schooten (Elzevier: Leiden 1646, repr. Georg Olms: Hildesheim/New York 1970), 4.

[8]Op. cit. (fn. 7), 6.

[9]Leibniz uses the term, in a new mathematical sense (but different from our modern sense) in 1692 (G.W. Leibniz, *Mathematische Schriften*, ed. C.I. Gerhardt, vol. V, Halle 1858, repr. Georg Olms: Hildesheim 1962, 306), and it is used often in his subsequent correspondence with Jacob and Johann Bernoulli (Leibniz is pleased, in his letter to Johann Bernoulli of July 29, 1698, that Johann uses the term in Leibniz's sense).

and to "discover" their common form

$$\xi + 5 \cdot \xi = 6$$

which represents a function, in this case a propositional function or sentential form. Leibniz, who had used the word "function" in the calculus, i.e. in analysis, had also given the word "calculus" a new meaning which has survived to the present day. One often reads that Leibniz introduced the concept of a calculus in the modern sense as used, e.g., in the expression "propositional calculus". Although there are good reasons for saying so, the matter is not altogether clear, since Leibniz in many places used the term "calculus" simply to denote a calculation procedure, perhaps even in the passage normally quoted for the more "progressive" notion.[10]

The central idea is that used by many authors in their characterization of the universality of "formal languages", viz. the idea that they admit of different interpretations. Indeed, the reflection on form took a big jump forward with this idea which is quite clear in Leibniz's statement that his logical calculi admit of a geometrical interpretation (or rather an extensional interpretation and an intensional one), an arithmetical interpretation and, presumably, many others.[11]

George Boole re-invented this approach or feature, at first without knowledge of his predecessor, and the correlation of a calculus or formal system with its interpretations has held a prominent place in mathematical logic ever since. Volker Peckhaus has pointed out[12] that a similar relation was propagated by Ernst Schröder who conceived of "number systems" as formal systems with many possible interpretations, numerically, logically, by whole finite classes of formulae etc. This point of view was perhaps stimulated by the so-called extensions of number systems in the early 19th century, but I cannot enter into that here.

I must, however, touch on the effect of the slightly later invention of non-Euclidean geometries. The procedure of replacing one or more of Euclid's axioms by others contradicting them, or of simply removing them without replacement, lay completely in line with Leibniz, Boole (and later Schröder) who studied interrelations of formal systems and their models. E.g., a three-axiom incidence "geometry" could be interpreted by the "concrete" objects of Euclidean geometry, but as well by finite systems of natural numbers, of pairs of such numbers, or even

---

[10]Cf. Leibniz, *Fundamenta calculi ratiocinatoris*, Akad. VI, iv, part A (Akademie-Verlag: Berlin 1999), 917–922, quotation 921; also (with German translation) in id., *Die Grundlagen des logischen Kalküls*, ed. Franz Schupp (Felix Meiner: Hamburg 2000), 22–23.

[11]Cf., e.g., Leibniz, *Specimen calculi coincidentium et inexistentium*, Akad. VI, iv, part A (Akademie-Verlag: Berlin 1999), 830–845, 832; also (with German translation) in: id., *Die Grundlagen des logischen Kalküls*, op. cit. fn.10, 122–123.

[12]See Volker Peckhaus, *Schröder's Logic*, in: Dov Gabbay / John Woods (eds.), *Handbook of the History of Logic. Volume 3. The Rise of Modern Logic: From Leibniz to Frege* (Elsevier/North Holland: Amsterdam etc. 2004), 557– 609, especially 567.

pairs of meaningless letters.[13] Our philosophy of mathematics has inherited from this period the notion and the problems of the so-called "intended interpretation". More important, however, was the enlivening of studies in the *axiomatic method* and axiomatization of contentual mathematical theories, and a temporary boom of "formal theories" of contentual mathematical disciplines (or rather their domains of objects).

Let me take a look at an historically somewhat later step and say a few words on "formal theories". The main motivation was the understanding that systems admitting different interpretations must themselves be purely *formal* systems, and that the theorems of a mathematical discipline could be deduced *formally* from the formal system alone without having to decide in favour of a particular interpretation. It seemed advantageous to get rid of the necessity to justify the decision, and the necessity to face up to philosophical interrogation afterwards.

A "formal arithmetic" was declared superior to the traditional one, and a plurality of differing "arithmetics" was not even excluded. Eduard Heine and Johannes Thomae produced "formal theories" of arithmetic and analysis, an idea that Frege attacked in his *Die Grundlagen der Arithmetik* in 1884, in "Über formale Theorien der Arithmetik" in 1885, and in the second volume of *Grundgesetze der Arithmetik* in 1903.[14] There he gave a brilliant description of the aim of formal theories of arithmetic (and, conceivably, logic) and of the means that would be necessary for their establishment, a description superior indeed to the expositions of their original proponents. These passages, especially § 90, are an admirable piece of elementary and non-formal proof theory.

Yet, they helped to pave the way to the rather unhappy Frege-Hilbert controversy. Hilbert had published his *Grundlagen der Geometrie* in 1899, and a correspondence on the aims and the feasibility of the whole endeavour followed.[15] Frege and Hilbert did not reach an agreement, and most of the current secondary literature on the debate seems to me to miss the point of it.

Apparently, the issue is between the traditional and the Hilbertian notion of axiom. Frege sided with the tradition, regarding axioms as contentual true propositions and letting himself be pushed as far as denying the acceptability of any non-Euclidean geometry for just that reason (but not out of ignorance, as some have claimed). Hilbert, on the other side, insisted that in mathematics concepts

---

[13]See the examples given in my *Philosophie und Mathematik. Eine Einführung in ihre Wechselwirkungen und in die Philosophie der Mathematik*, Wissenschaftliche Buchgesellschaft: Darmstadt 1995, 262 ff.

[14]Gottlob Frege, *Grundgesetze der Arithmetik, begriffsschriftlich abgeleitet*, vol. II, Hermann Pohle: Jena 1903 (last reprinted, together with vol. I, by Georg Olms: Hildesheim/Zürich/New York 1998, with the subtitle ridiculously distorted into "begriffsgeschichtlich abgeleitet"), §§ 86 ff., pp. 96 ff.

[15]Cf. Gottlob Frege, *Wissenschaftlicher Briefwechsel*, ed. Gottfried Gabriel et al., Felix Meiner: Hamburg 1976, 55–80; English translation in Gottlob Frege, *Philosophical and Mathematical Correspondence*, ed. Gottfried Gabriel et al., (abridged), Basil Blackwell: Oxford 1980, 31–52.

could be defined only by "mutual determination", i.e. by the unlucky "implicit definition" by which in an axiom system the fundamental notions involved were to define each other.

Frege, never a good combatant in scientific controversies, failed to convince Hilbert that what was really being defined was – in Frege's more advanced terminology – a concept of the second order; in the pertinent case, we do not define points, lines and planes etc., but the second order concept of a *Euclidean geometry*. While thinking (mistakenly) that he had defined the basic notions of Euclidean geometry, Hilbert had (correctly) defined the *structure* of a Euclidean geometry itself. To show this today, one may apply a method which Frege had already conceived in 1884 in the *Grundlagen*, without realizing it in 1899 (in my opinion) that it could also be used to introduce the second order concept of a geometry in an impeccable manner. As the method also contributes to an analysis of a whole type of (seemingly) elementary propositions, let us take a short look at it.

What is the object of our talk when we speak of the concept unicorn, the number 7, the direction of the straight line g, the set of rational numbers etc.? Are we referring to concepts, numbers, directions and sets as inhabitants of some "third realm", as "ideal-world phenomena"? Frege, though firmly believing in such a third world, did not think that we acquire knowledge about its components and states-of-affairs by direct inspection. He saw that in a proposition not about the word "unicorn", but about the concept unicorn, we want to say that our proposition is to hold likewise for every word synonymous to the word "unicorn" according to some accepted system of linguistic rules. If "Nothing falls under the concept unicorn" is true, then "Nothing falls under the concept Einhorn" is also true. Propositions about "abstract objects" like concepts, classes, numbers etc. are propositions about representative objects (named in the proposition before us), which are invariant against replacement of the representative or representing object by an equivalent one with regard to a clearly specified equivalence relation. If "αB" is the compound of an abstractor "α" and an object's name "B",

$$A(\alpha B) \text{ means}$$

$$\bigwedge_X (X \sim B \rightarrow A(X))$$

This is the understanding of abstraction (called "modern abstraction" by Ignacio Angelelli) that was first envisaged by Frege in his *Grundlagen*, stated more clearly and less philosophically by Hermann Weyl some thirty years later, and analyzed in the nineteen-sixties by Paul Lorenzen.[16] Note that I have not resorted

---

[16]Cf. Gottlob Frege, *Die Grundlagen der Arithmetik*, Wilhelm Koebner: Breslau 1884 (repr. in the *Centenarausgabe*, ed. Christian Thiel, Felix Meiner: Hamburg 1986), § 65 (p. 77 and 74, respectively;

to equivalence *classes* – classes or sets may also be introduced by this method of abstraction.

The importance for our reflections on the development of the notion of "form" and the term "formal" resides in the fact that taking *interderivability* (with or without additional definitions) as the equivalence relation between axiom systems, we can formulate propositions about corresponding abstract objects. These abstract objects are nothing other than that which has been called the *structures* represented (*dargestellt*) by each of the axiom systems from which we departed.[17] Obviously, this is a late reconstruction, an explication of *intentions* present at the turn from the 19th century to the 20th, and therefore totally a matter of *heritage*, not influencing the historical development *then*.

Frege was also the first to gain complete insight into the interplay of quantifiers, strings of quantifiers, and quantified variables. For a long time past, mathematicians had used letters not only schematically to exhibit the form of a term or a proposition, but also for expressing the universal validity of a formula, e.g. of $(a + b)^2 = a^2 + 2ab + b^2$, tacitly quantifying (universally) over a and b. They had to get along, however, with verbal circumlocutions like "For all positive numbers, there is ...", or "Some functions are discontinuous at all of their argument places", where existence is involved. Frege, in his *Begriffsschrift* of 1879, was the first to introduce in a strictly formal way, viz. within a calculus of logic, not only quantifiers (universal and existential, and of the first and of the second degree) but also "scopes" for them to indicate how "far" into the adjoining expression their binding force extends. Beside the use of variables for the adequate expression of functional relations, this is the best known and most characteristic use of variables today, and Michael Dummett has aptly called this step of Frege's "the deepest single technical advance ever made in logic".[18] It made it possible to express in strict manner, concepts like continuity and uniform continuity by applying four consecutive quantifiers and stating that the second property logically implies the first.[19] I can only mention in passing Frege's useful distinction of schematic, free,

---

see also my introduction to the 1986 edition, p. XXXIX f.); Paul Lorenzen, *Gleichheit und Abstraktion*, Ratio 4 (1962), 77–81 of the German edition, 85–90 of the English edition (English translation also in Paul Lorenzen, *Constructive Philosophy*, University of Massachusetts Press: Amherst 1987, 71–77; Ignacio Angelelli, *Frege and Abstraction*, in: Paul Weingartner / Christine Pühringer (eds.), *Philosophy of Science, History of Science. A Selection of Contributed Papers of the 7th International Congress of Logic, Methodology and Philosophy of Science Salzburg, 1983* (Anton Hain: Meisenheim am Glan 1984; *Philosophia Naturalis* 21, issue 2–4), 453–471.

[17]For details see Friedrich Kambartel, *Zur Rede von "formal" und "Form" in sprachanalytischer Absicht, neue hefte für philosophie, Heft 1: Phänomenologie und Sprachanalyse* (Vandenhoeck and Ruprecht: Göttingen 1971), 51–67, and chapter 12 ("Der Strukturbegriff in der Mathematik", 261–272) of my *Philosophie und Mathematik* quoted in footnote 13.

[18]Michael Dummett, *Frege: Philosophy of Language* (Duckworth: London 1973, ²1981), XV.

[19]This example found its way into many textbooks of logic, cf., e.g., Alonzo Church. *Introduction to Mathematical Logic, vol. I* (Princeton University Press: Princeton, N.J. 1956, 43 (fn. 102).

and bound variables, and auxiliary letters as e.g. mere placeholders or the "free" and the quasi-bound letters like $\xi, \zeta, ..., \varphi, ...$ and β and γ in function names of the first order and of higher orders, respectively. No emphasis is needed to bring it home that the sorting of classes of expressions in the decision problem according to their prefix type, and the derivation of the prefix forms of given expressions of quantificational logic would never have come about without the formative means introduced by Frege in 1879.

Let me finally take a glance at the actual development at the turn of the 19th to the 20th century. Frege's *Grundgesetze* had been established on the basis of syntactical rules of a preciseness unknown before, an exactness that enabled Russell to construct a formal inconsistency in the *Grundgesetze* system – the Zermelo-Russell antinomy, so called because Zermelo had independently found it slightly before Russell (although not by syntactical analysis), but left it unpublished.

But that antinomy gave rise to the great progress in logic and foundations which Frege had prophesied in his answer to Russell's letter telling him about the paradox.[20] More and much better methods of syntactical analysis were developed (beside semantical ones, of course). I mention only Hilbert's proof theoretical methods, Schönfinkel's and Curry's combinatorial logic, Hilbert and Bernays and their rules of substitution, Carnap's *Abriß der Logistik* and *Logische Syntax der Sprache,* and Gödel with the concept of a formal system for which I use the German term *"Vollformalismus"* following Lorenzen's *Metamathematik.*[21] Carnap described *formal theories* as having reference only to the kind and order of symbols (not to their meaning or reference). "Formal" has, in this way, once more become opposed to "contentual" or "material".

Parallel to this, formal theories in the modern sense have sometimes been explained as formalized theories, presented in a form from which "their construction [*Aufbau*] appears without any doubt in all its relevant details."[22]

Today, *formal* sciences are even more often characterized by *formalization* than were Carnap's *Formalwissenschaften*, which were opposed to *Realwissenschaften* which comprised natural sciences, the Humanities, and psychology.[23] By Carnap's *Formalwissenschaften*, whose domain had been narrowed down by elimi-

---

[20]Frege to Russell, June 22, 1902, see *Wissenschaftlicher Briefwechsel* (quoted in footnote 15), pp. 213 and 132, respectively.

[21]Paul Lorenzen, *Metamathematik*, Bibliographisches Institut: Mannheim 1962, B.I. Wissenschaftsverlag: Mannheim [2]1980.

[22]"[...] in einer Form [...], aus der ihr Aufbau in allen relevanten Einzelheiten zweifelsfrei hervorgeht" (Peter Hinst, "Formalisierung". In: *Handbuch philosophischer Grundbegriffe*, ed. Hermann Krings / Hans-Michael Baumgartner / Christoph Wild, Kösel: München 1973, vol. 1, 465–472, quoted from p. 465).

[23]Cf. Rudolf Carnap, *Der logische Aufbau der Welt*, Weltkreis-Verlag: Berlin-Schlachtensee 1928, Felix Meiner: Hamburg 1974 (English as *The Logical Structure of the World*, in: id., *The Logical Structure of the World. Pseudoproblems in Philosophy*, University of California Press: Berkeley, Calif./ Routledge and Kegan Paul: London 1967, 1–300).

nating e.g. grammar (still included by Lichtenfels in 1842),[24] we now understand the disciplines the theorems of which are logically valid, or formally-analytic (if they have reference to definitions or predicator rules in addition to logical rules) or, in constructivistic philosophy of science, formally-synthetic as in mathematics (excluding geometry and kinematics). In spite of the variety of "formal" concepts commonly used in modern formal sciences and their philosophies, and of the knowledge available today of the most important steps in the history of the concept "formal", it is obvious that the story (and the history) of this important notion still has to be written.

BIBLIOGRAPHY

[1] Ignacio Angelelli, *Frege and Abstraction*, in: Paul Weingartner/Christine Pühringer (eds.), *Philosophy of Science, History of Science. A Selection of Contributed Papers of the 7th International Congress of Logic, Methodology and Philosophy of Science Salzburg, 1983*, Anton Hain: Meisenheim am Glan 1984 (also in *Philosophia Naturalis* 21, issue 2–4)

[2] Aristotle, *Prior Analytics* in W. D. Ross, *Aristotle's Prior and Posterior Analytics: A Revised Text with Introduction and Commentary*, Oxford, Oxford University Press Academic Monograph Reprints: 2000

[3] Florian Cajori, *A History of Mathematical Notations*, vol. II The Open Court: La Salle, Ill. 1929 (repr. together with vol. I, Dover: New York 1993)

[4] Rudolf Carnap, *Der logische Aufbau der Welt*, Weltkreis-Verlag: Berlin-Schlachtensee 1928, and reprinted Felix Meiner: Hamburg 1974 (English as *The Logical Structure of the World*, in: id., *The Logical Structure of the World. Pseudoproblems in Philosophy*, University of California Press: Berkeley, Calif./Routledge and Kegan Paul: London 1967)

[5] Alonzo Church, *Introduction to Mathematical Logic, vol. I*, Princeton University Press: Princeton, N.J. 1956

[6] Michael Dummett, *Frege: Philosophy of Language*, Duckworth: London 1973

[7] Gottlob Frege , *Die Grundlagen der Arithmetik*, Wilhelm Koebner: Breslau, 1884. (repr. in the *Centenarausgabe*, ed. Christian Thiel, Felix Meiner: Hamburg 1986)

[8] Gottlob Frege, *Grundgesetze der Arithmetik, begriffsschriftlich abgeleitet*, vol. II, Hermann Pohle: Jena 1903 (last reprinted, together with vol. I, by Georg Olms: Hildesheim/Zürich/New York 1998)

[9] Gottlob Frege, *Wissenschaftlicher Briefwechsel*, ed. Gottfried Gabriel et al. Felix Meiner: Hamburg 1976. English translation in Gottlob Frege, *Philosophical and Mathematical Correspondence*, ed. Gottfried Gabriel et al., (abridged), Basil Blackwell: Oxford 1980

[10] Peter Hinst, "Formalisierung", In: *Handbuch philosophischer Grundbegriffe*, vol. 1, ed. Hermann Krings, Hans-Michael Baumgartner and Christoph Wild, Kösel: München 1973

[11] Friedrich Kambartel, *Zur Rede von "formal" und "Form" in sprachanalytischer Absicht, neue hefte für philosophie, Heft 1: Phänomenologie und Sprachanalyse*, Vandenhoeck and Ruprecht: Göttingen 1971

[12] G. W. Leibniz, *Mathematische Schriften*, ed. C.I. Gerhardt, vol. V, Halle 1858 (repr. Georg Olms: Hildesheim 1962)

[13] G. W. Leibniz, *Fundamenta calculi ratiocinatoris*, Akademie-Verlag: Berlin 1999 also (with German translation) in *Die Grundlagen des logischen Kalküls*, ed. Franz Schupp (Felix Meiner: Hamburg 2000)

---

[24]Cf. Ritter Johann Lichtenfels, *Lehrbuch der Logik*, J.G. Heubner: Wien 1842, 5 (almost literally repeated in the author's *Lehrbuch zur Einleitung in die Philosophie*, Wilhelm Braumüller: Wien 1850, 136).

[14] G. W. Leibniz, *Specimen calculi coincidentium et inexistentium*, Akademie-Verlag: Berlin 1999); also (with German translation) in: id., *Die Grundlagen des logischen Kalküls*, ed. Franz Schupp (Felix Meiner: Hamburg 2000)

[15] Ritter Johann Lichtenfels, *Lehrbuch der Logik*, J.G. Heubner: Wien 1842

[16] Paul Lorenzen, "Gleichheit und Abstraktion", *Ratio* 4 (1962), 77–81 of the German edition, 85–90 of the English edition (English translation also in Paul Lorenzen, *Constructive Philosophy*, University of Massachusetts Press: Amherst 1987, 71–77)

[17] Paul Lorenzen, *Metamathematik*, Bibliographisches Institut: Mannheim 1962 (reprinted in B.I. Wissenschaftsverlag: Mannheim 1980)

[18] Volker Peckhaus, *Schröder's Logic*, in Dov Gabbay/John Woods (eds.), *Handbook of the History of Logic. Volume 3. The Rise of Modern Logic: From Leibniz to Frege*, Elsevier/North Holland: Amsterdam 2004

[19] Sextus Empiricus, *Adversus mathematicos*, in *Opera*, ed. H. Mutschmann, vol. II, B.G. Teubner: Leipzig 1914, repr. 1984. English translation in I.M. Bocheński, *A History of Formal Logic*, University of Notre Dame Press: Notre Dame, Ind. 1961, repr. Chelsea Publishing Company: New York 1970

[20] William Shakespeare, *The Complete Works (Oxford Shakespeare)*, S. W. Wells, G. Taylor et al., Oxford, Oxford University Press: 2005 (sec. ed.)

[21] Christian Thiel, *Philosophie und Mathematik. Eine Einführung in ihre Wechselwirkungen und in die Philosophie der Mathematik*, Wissenschaftliche Buchgesellschaft: Darmstadt 1995

[22] Franciscus Vieta, *Opera*, ed. Franciscus A. Schooten, Elzevier: Leiden 1646 (repr. Georg Olms: Hildesheim/New York 1970)

Propositional Logics for Three

Tero Tulenheimo and Yde Venema

ABSTRACT. Semantics of propositional logic can be formulated in terms of 2-player games of perfect information. In the present paper the question is posed what would a generalization of propositional logic to a 3-player setting look like. Two formulations of such a '3-player propositional logic' are given, denoted PL_0^3 and PL^3. An overview of some metalogical properties of these logics is provided.

Semantics of classical propositional logic is typically given by laying down recursive rules which compute the truth-values of complex formulas from valuations that specify the truth-values of propositional atoms. Alternatively, the very same truth-conditions can be captured by defining semantics in terms of games of perfect information between two players (say *Eloise*, *Abelard*), with the property that a formula φ is true (false) under a valuation V in the usual sense if and only if there is a winning strategy for *Eloise* (*Abelard*) in the associated game $G(\varphi, V)$.[1] Conjunction and disjunction are interpreted by choices (between conjuncts and disjuncts) made by the two players, while negation is interpreted in terms of role switch (from 'Verifier' to 'Falsifier' and *vice versa*).

Classical propositional logic can, then, be seen as a logic for choice and role switching in a 2-player setting. In the present paper we wish to ask what 'logics for choice and role switching' would look like in multi-player settings.[2] In particular, we take first steps in exploring how to generalize propositional logic to the case where there are three players. The main goal of this paper is conceptual: to see how such a generalization can be carried out. The success of the proposed generalizations can, of course, only be assessed by reference to their formal properties, which is why we take up various related technical issues. The framework of the

---

[1] The original definition stems from Hintikka's [10]. Game-theoretic ideas were systematically applied in 20th century logic before Hintikka, in dialogical logic (starting with Lorenzen's [11, 12]). However, dialogic deals primarily with uninterpreted formulas and proof theory, while Hintikka's approach, influenced by Henkin's [9], is model-theoretic.

[2] Just before sending off the final version of the paper to the editors, we became aware of two other publications on multi-player logic. In [1, 2], Abramsky develops a compositional semantics of such logics in terms of multiple concurrent strategies (formalized as closure operators on certain concrete domains). It will be of obvious interest to further investigate this connection.

present paper can be approached from many perspectives: besides technical developments, one can analyze the use of game-theoretical notions for logical purposes and *vice versa*; also the philosophical significance of the emerging framework can be discussed. Given that the ground being covered is previously unexplored, we consider it as rewarding to ask questions on several fronts, and to follow in our discussion more than one lead from more than one viewpoint.

Two formulations of a '3-player propositional logic' will be presented, denoted PL_0^3 and PL^3. (Whether these are *stricto sensu* logics or not, it turns out that they can be studied as if they were.) Concerning these logics, we ask: Which are the 'semantic attributes' corresponding to the truth-values *true* and *false*? In which form, if any, do analogues of the *law of excluded middle*, *law of double negation*, *negation normal form*, or *conjunctive* and *disjunctive normal forms* emerge? What is the computational complexity of determining the semantic attribute of a formula relative to a valuation in the 3-player setting? Regarding one of the formulations, PL^3, some remarks are further made concerning the existence of a tableau-based proof system; 32 related decision problems are furthermore solved. Interestingly, many properties that fall together in classical propositional logic — typically due to the determinacy of the corresponding games — turn out to be distinguished in the multi-player setting.

1   Propositional logic or PL^2

Throughout the present paper, **prop** will be a countable set of propositional atoms. Formulas of *propositional logic* are generated by the grammar

$$\varphi ::= p \mid (\varphi \vee \varphi) \mid (\varphi \wedge \varphi) \mid \neg\varphi,$$

with $p \in$ **prop**. Propositional logic will be denoted by PL^2. The basic semantic notion of PL^2 is, of course, that of a valuation. *Valuations* for PL^2 are functions $V :$ **prop** $\longrightarrow \{true, false\}$; they provide a distribution of truth-values over the propositional atoms considered. Well-known recursive clauses compute the truth-values of complex formulas, relative to a valuation V, from the truth-values that V gives to the atoms. We write $V \models \varphi$ to indicate that φ is true under the valuation V, whereby $V \not\models \varphi$ will indicate that φ is false under V. The truth-values *true* and *false* will occasionally be referred to as *semantic attributes* of propositional formulas.

There is an alternative way to define the semantics of PL^2, employing tools from game theory. Let us associate, with every formula φ and valuation V, a 2-player game $G(\varphi, V)$ of perfect information (between *Eloise* and *Abelard*). The following game rules determine the set of all (partial) *plays* of game $G(\varphi, V)$. The players are associated roles. In the beginning, *Eloise* occupies the role of 'Verifier' and *Abelard* that of 'Falsifier'; in the course of a play, the roles may get switched.

- If $\varphi := p$, a play of the game has come to an end. If $V(p) = true$, the player whose current role is 'Verifier' wins and the one whose current role is 'Falsifier' loses; otherwise 'Falsifier' wins and 'Verifier' loses.

- If $\varphi := (\psi \vee \chi)$, then 'Verifier' chooses a disjunct $\theta \in \{\psi, \chi\}$ and the play of the game continues as $G(\theta, V)$.

- If $\varphi := (\psi \wedge \chi)$, then 'Falsifier' chooses a conjunct $\theta \in \{\psi, \chi\}$ and the play of the game continues as $G(\theta, V)$.

- If $\varphi := \neg\psi$, the players switch their roles ('Verifier' assumes the role of 'Falsifier', and *vice versa*), and the play continues as $G(\psi, V)$.

If $\mathcal{P}$ is one of the players, a *strategy* for $\mathcal{P}$ is any function that specifies a move for $\mathcal{P}$ corresponding to each partial play at which it is $\mathcal{P}$'s turn to move, depending on the opponent's earlier moves. A strategy for $\mathcal{P}$ is *winning*, if it leads to a play won by $\mathcal{P}$ against any sequence of the opponent's moves. The usual propositional semantics is captured by the above games:

FACT 1. Let φ be a formula of PL^2 and V a valuation. Then: $V \models \varphi$ iff there is a winning strategy for *Eloise* in $G(\varphi, V)$; and $V \not\models \varphi$ iff there is a winning strategy for *Abelard* in $G(\varphi, V)$.   □

2   From two to three — basic ideas

In n-player logic games, there will be n roles in addition to n players. This generalizes the 2-player case where there are two players (*Eloise*, *Abelard*) and two roles ('Verifier', 'Falsifier'). The roles are bijectively distributed to the players at each stage in a play of a game. Conceptually, players and their roles must be kept apart. In order to be able to say who is having which role at a given stage of a play, one cannot simply identify players with their roles. The same player may assume different roles during a play. Various semantically crucial notions will be defined by reference to the *initial role distribution*: the roles the players have when the playing of the game begins.

When generalizing PL^2 to multi-player settings, there are two mutually independent parameters that admit of variation: (1) the payoff function, and (2) the interpretation of negation symbols. Two generalizations of PL^2 to the 3-player setting will be presented, to be denoted PL_0^3 and PL^3. The former will be technically somewhat simpler to deal with. It will retain the binary character of the payoff function: each play is won by some players and lost by the others. Further, for each pair (i, j) of distinct roles there is a negation symbol $\sim_{ij}$, interpreted in terms of a transposition of the roles i and j. By contrast, in PL^3 payoffs are defined in terms of rankings of the players. In this respect it represents a more straightforward generalization of PL^2 than PL_0^3 does. On the other hand, the treatment of negations is more complicated: the syntax of PL^3 provides two negation symbols, which are interpreted by functions mapping role distributions to role distributions — instead of being interpreted simply by permutations of roles.

It should be noted that the parameters (1) and (2) could be instantiated in further ways. In particular, it might be of interest to combine the definition of payoffs as rankings with the interpretation of negations in terms of transpositions. The investigations into the 'logics' PL_0^3 and PL^3 presented in this paper are best viewed as case studies.

The most central question in the generalization of classical propositional logic is what happens to negation. Games defining the semantics of PL^2 are *determined*: in each game either *Eloise* or *Abelard* has a winning strategy.[3] Hence in PL^2, the truth of $\neg\varphi$ under V can be equivalently characterized in one of the two ways: (*a*) there is a winning strategy for 'Falsifier' in $G(\varphi, V)$; and (*b*) there is no winning strategy for 'Verifier' in $G(\varphi, V)$. In the former case negation is defined in terms of role shift, in the latter case by the absence of a winning strategy. In multi-player settings no analogous equivalence holds. Precisely because the two characterizations of negation in PL^2 — (*a*) and (*b*) — are equivalent, the classical framework as such does not dictate which characterization we should take as the model of our generalization. In connection with PL_0^3 and PL^3, we will continue to interpret negations in terms of changing roles. There is, admittedly, a rather strong pretheoretical tendency to construe negation in terms of 'complementation' or 'absence'. Therefore one might wish to think of the negations in the logics PL_0^3 and PL^3 as *contrarieties* rather than negations proper.

2.1 First formulation: logic PL_0^3

Syntax. Formulas of PL_0^3 are generated by the grammar

$$\varphi ::= p \mid (\varphi \vee_i \varphi) \mid \sim_{ij}\varphi,$$

where $i, j \in \{0, 1, 2\}$, $i \neq j$, and $p \in \mathbf{prop}$. Intuitively, there are three players. The numbers $0, 1, 2$ stand for the roles that the players may have. The connective $\vee_i$ is interpreted by the player whose current role is i. Note that the syntax of PL_0^3 involves 3 disjunction signs and 6 negation signs. Whenever no confusion threatens, brackets may be dropped.

Valuations. Valuations of the logic PL_0^3 assign to propositional atoms subsets of the set $\{0, 1, 2\}$ of all roles, viz. they are functions $V : \mathbf{prop} \longrightarrow Pow(\{0, 1, 2\})$. Intuitively, those players whose roles at the end of a play are in the set $V(p)$ *all win* the play, and the rest lose the play.

---

[3]This follows from the Gale-Stewart theorem [8]. The theorem saying that all two-player zero-sum perfect information games of finite length are determined, is often termed 'Zermelo's theorem'. However, the result is not due to Zermelo. For details, see [13].

If *arbitrary* subsets of $\{0, 1, 2\}$ are allowed as values of V, it may happen that all players win a play, or that no player does. Not to deviate from the 2-player setting in our generalization, we should ban $\varnothing$ and $\{0, 1, 2\}$ as possible values of a valuation function. In what follows, valuations V of PL_0^3 meeting the extra requirement that neither $\varnothing$ nor $\{0, 1, 2\}$ lies in the image of V, will be termed *restricted valuations*.[4] The counterpart in PL^2 of arbitrary PL_0^3-valuations would be the generalized valuations allowing any of the sets $\varnothing$, $\{true\}$, $\{false\}$ and $\{true, false\}$ as possible values. A 4-valued propositional logic evaluated precisely relative to such valuations was introduced by Belnap in [3]. (See also Dunn's article [7].)

Game rules. The games are played by three players: *Alice*, *Bob* and *Cecile* (in short: *a*, *b*, *c*). Relative to valuations $V : \textbf{prop} \longrightarrow Pow(\{0, 1, 2\})$, the semantics of PL_0^3-formulas is specified by means of 3-player games $G(\varphi, V)$. To introduce these games, we first define something a bit more general, namely 3-player games $G(\varphi, V, \rho)$, where the extra input ρ is a bijection $\{0, 1, 2\} \longrightarrow \{a, b, c\}$, i.e., a distribution of roles to the players. If ρ is a role distribution, let ρ_{ij} be its transposition satisfying: $\rho_{ij}(i) = \rho(j)$ and $\rho_{ij}(j) = \rho(i)$ and $\rho_{ij}(k) = \rho(k)$ for $k \notin \{i, j\}$. With every formula φ of PL^3, valuation V and role distribution ρ, a 3-player game $G(\varphi, V, \rho)$ of perfect information between a, b and c is introduced. The game rules are these:

- If $\varphi \in \textbf{prop}$, a play of the game has come to an end. Those players whose roles are in the set $V(\varphi)$ win the play, the others lose it. That is, a player $\mathcal{P}$ is one of the winners of the play iff $\rho^{-1}(\mathcal{P}) \in V(\varphi)$.[5]

- Let $i \in \{0, 1, 2\}$. If $\varphi = (\psi \vee_i \chi)$, then the player $\rho(i)$ chooses $\theta \in \{\psi, \chi\}$, and the play goes on as $G(\theta, V, \rho)$.

- Let $i, j \in \{0, 1, 2\}$ and $i \neq j$. If $\varphi = \sim_{ij} \psi$, the play continues as $G(\psi, V, \rho_{ij})$.

By stipulation the game $G(\varphi, V)$ equals the game $G(\varphi, V, \rho_0)$, where ρ_0 is specified by putting $\rho_0(0) = a$, $\rho_0(1) = b$, and $\rho_0(2) = c$. This particular role distribution ρ_0 will be referred to as the 'standard initial role distribution'.

Strategies. Any sequence of moves made according to the game rules is a *partial play*. A *play* is a partial play at which no player is to move. If $\mathcal{P}$ is any of the players, a *strategy* for $\mathcal{P}$ is any function providing a choice for $\mathcal{P}$ at any partial play at which it is his or her turn to move; the choice may depend on the moves made by $\mathcal{P}$'s opponents earlier in the course of the relevant partial play. A strategy for $\mathcal{P}$ is *winning*, if against any sequence of moves by his or her opponents, it leads to a play won by $\mathcal{P}$.

---

[4] If $f : A \longrightarrow B$ is a function, its *image*, denoted $Im(f)$, is the set $\{f(a) : a \in A\}$.

[5] Were payoffs taken to be numbers, the games being defined would *not* in general be constant-sum games. If winning corresponds to 1 and losing to 0, the sum of the players' payoffs may be any integer m with $0 \leq m \leq 3$.

Semantic attributes. For each set S of players, exactly the members of S might have a winning strategy in a game $G(\psi, V)$. Each subset S of $\{a, b, c\}$ constitutes a *semantic attribute* such that φ has by definition the attribute S relative to V, if the set of players having a winning strategy in $G(\varphi, V)$ is S. There are, then, 8 semantic attributes. We will write $|\varphi, V| = S$ to indicate that the semantic attribute of φ relative to V is S.

REMARK 2. Semantic attributes were just defined in terms of players. Arguably a more intrinsic definition would be in terms of roles. However, throughout this paper we will think of formulas as evaluated starting with the standard initial role distribution, assigning to *Alice* the role 0, to *Bob* the role 1, and to *Cecile* the role 2. We could indeed leave the semantics neutral with respect to the initial role distribution, and let semantic attributes to be sets of roles rather than sets of players. The players corresponding to these attributes would then vary with the particular initial role distribution. We stay with a fixed initial role distribution for clarity of exposition.   □

Observe that directly by the semantics, for any formula φ, any valuation V, and any distinct roles i and j, we have: $|\sim_{ij}\varphi, V| = |\sim_{ji}\varphi, V|$. Therefore the logic PL_0^3 involves 'really' only three negations.

Let us think of the 8 attributes. There is a one-one correspondence between sets of roles and sets of players via the standard initial role distribution ρ_0. Given a set S of players, we write $\rho_0^{-1}(S)$ for the set $\{\rho_0^{-1}(\mathcal{P}) : \mathcal{P} \in S\}$. Similarly, given a set R of roles, we write $\rho_0(R)$ for $\{\rho_0(i) : i \in R\}$. Observe that trivially, each attribute can appear as an attribute of an atomic formula. If S is a set of players and $V(p) = \rho_0^{-1}(S)$, then $|p, V| = S$. Next, note that even if $\varnothing$ were not allowed in the image of a valuation function — as when considering only restricted valuations — *still* there would be formulas having $\varnothing$ as their semantic attribute relative to some valuation. To see this, let $V(p) = \{1\}$, $V(q) = \{2\}$, and consider determining the value $|p \vee_0 q, V|$. Now no player has a winning strategy in $G(p \vee_0 q, V)$. For, no matter which disjunct *Alice* chooses, she herself loses. On the other hand, choosing 'left' she will prevent *Cecile* from winning, and choosing 'right' she will prevent *Bob* from winning. Thus $|p \vee_0 q, V| = \varnothing$.

By contrast, no similar fact holds for the other extreme. If V is a valuation whose image does not involve the full set $\{0, 1, 2\}$, no formula can have $\{a, b, c\}$ as its semantic value relative to V. For, suppose that $\{0, 1, 2\} \notin Im(V)$, but still all players have a winning strategy in $G(\varphi, V)$. Let f, g and h be winning strategies of *Alice*, *Bob* resp. *Cecile*. These strategies determine a certain play of $G(\varphi, V)$. Let p be the atom reached at the end of the play. Since the play is determined by the three winning strategies, it is won by all players, i.e., $V(p) = \{0, 1, 2\}$. This is a contradiction.

It is noteworthy that when arbitrary valuations are employed, the very same 8 semantic attributes are available for all formulas, both atomic and complex. But if restricted valuations are used — valuations whose image excludes both $\varnothing$ and the full set $\{0, 1, 2\}$ — then there are 6 semantic attributes available for atomic formulas, but 7 attributes for complex formulas.

Every formula $\varphi \in PL_0^3$ determines a map, call it $|\varphi|$, from restricted valuations to semantic attributes, namely $|\varphi| : V \mapsto S$, with $S = |\varphi, V|$. On the other hand, as noted above, any semantic attribute except $\{a, b, c\}$ is realizable by a formula of PL^3 under restricted valuations. Let **prop** be a finite set of propositional atoms, and f a map from restricted valuations on **prop** to realizable semantic attributes. An important systematic question related to PL_0^3 then is whether there always is a formula φ of PL_0^3 such that $f = |\varphi|$. This issue of functional completeness is left as an open question.

Case of n players. It would be straightforward to generalize PL_0^3 to the case of an arbitrary finite number n of players. This would involve having in the syntax n disjunction symbols and $n \cdot (n - 1)$ negation symbols. The semantics would require n roles in addition to the n players. The disjunction symbol $\vee_i$ would be interpreted by the player having the role i, and the negation symbol $\sim_{ij}$ by the transposition of the roles i and j. Semantic attributes would simply be subsets of the set of all players. Hence there would be 2^n distinct semantic attributes.

2.2 Second formulation: logic PL^3

In PL_0^3, certain features of the 2-player framework were preserved that admit of generalization. For one thing, payoffs in games for PL_0^3 are simply *win* and *loss* — the relevant difference with respect to PL^2 is just that several players may receive a given payoff. For another thing, the negations of PL_0^3 are interpreted by means of transpositions — just like the negation of PL^2. One might consider interpreting negation symbols of a 3-player logic by permutations of the three roles, not in general merely switching two roles at a time. Such permutations are arbitrary bijections of type $\{0, 1, 2\} \longrightarrow \{0, 1, 2\}$. However, in PL^3 we will go one step further, and interpret negation symbols by bijections taking role distributions as arguments, and yielding role distributions as values: bijections of type $\mathbf{P} \longrightarrow \mathbf{P}$, where $\mathbf{P}$ is the set of all role distributions (the set of all bijections in the set $\{a, b, c\}^{\{0,1,2\}}$). In PL^3, negations are hence interpreted by *permutations of role distributions* rather than by permutations of the set of roles. Such a 'higher-order' interpretation of negation symbols has obvious drawbacks. There are $6! = 720$ such bijections, so which ones should we consider? Should all these functions be expressible by means of those we introduce explicitly? In connection with PL^3 a more modest approach is adopted: we introduce 2 negation symbols, and simply content ourselves with being able to express the 6 permutations of the set of roles in terms of these 2 negation symbols — interpreted by means of permutations of role distributions.

Syntax. Formulas of PL^3 are generated by the following grammar:

$$\varphi ::= p \mid (\varphi \vee_i \varphi) \mid \neg\varphi \mid {\sim}\varphi,$$

with $i \in \{0,1,2\}$ and $p \in$ **prop**. Again, there are intuitively three players, each of whom occupies one of the roles $0,1,2$. And as before, the connective $\vee_i$ is interpreted by the player whose current role is i. To the two negation signs, $\neg$ and $\sim$, two permutations of role distributions will correspond.

Valuations. Valuations of the logic PL^3 are functions from propositional atoms to role distributions, i.e., functions of type **prop** $\longrightarrow$ **P**. Let us write $V(p) = (r, r', r'')$ to indicate that $V(p)(r) = a$, $V(p)(r') = b$, and $V(p)(r'') = c$. Intuitively, the value $V(p)$ serves to rank the players in a linear order with respect to the propositional atom p. If $V(p) = (r, r', r'')$, this means that relative to the atom p, Alice has the role r, Bob the role r', and Cecile the role r''. The numerical values of r, r' and r'' then determine a ranking among the three players. Such a numerical value is termed the *rank* of the player. The best rank is 0, the next best 1, and the worst 2. If, e.g., $V(p) = (2,0,1)$, then relative to p Alice gets the worst rank, Bob the best rank and Cecile the next best rank.[6]

Negations. Negation of PL^2 turns truths into falsehoods and *vice versa*; game-theoretically this negation is interpreted by a transposition acting on a pair of roles. transpositions are a (representative) special case of permutations of a finite set. The generalization to be considered next involves interpreting the two negations of PL^3 by permutations of *role distributions*. The 'higher order' permutations to be considered are $\pi_\neg$ and $\pi_\sim$:

$\pi_\neg(0,1,2) = (1,2,0)$	$\pi_\neg(0,2,1) = (1,0,2)$
$\pi_\neg(1,2,0) = (2,0,1)$	$\pi_\neg(1,0,2) = (2,1,0)$
$\pi_\neg(2,0,1) = (0,1,2)$	$\pi_\neg(2,1,0) = (0,2,1)$

$\pi_\sim(0,1,2) = (0,2,1)$	$\pi_\sim(0,2,1) = (0,1,2)$
$\pi_\sim(1,2,0) = (1,0,2)$	$\pi_\sim(1,0,2) = (1,2,0)$
$\pi_\sim(2,0,1) = (2,1,0)$	$\pi_\sim(2,1,0) = (2,0,1)$

FIGURE 1

The two negations $\neg$ and $\sim$ will, then, correspond to two different ways in which distributions of roles to the players are changed: in one case according to $\pi_\neg$, in the other according to $\pi_\sim$. It can be noted that $\pi_\neg$ is definable as a permutation of roles, mapping the role 0 to 1, the role 1 to 2, and the role 2 to 0. By contrast, $\pi_\sim$ cannot be defined simply by reference to roles, but is genuinely a permutation of entire role distributions. As a matter of fact, all permutations of roles can now be defined by means of compositions of these two permutations of role distributions. (Certainly not all of the 720 permutations of role distributions are definable in

---

[6]For games with rankings as payoffs, see [4, 5].

terms of $\pi_\neg$ and $\pi_\sim$, but then again, this is not posed as a *desideratum* in our case study of PL^3.)

Game rules. With every formula φ of PL^3, valuation $V : \mathbf{prop} \longrightarrow \mathbf{P}$ and role distribution $\rho : \{0,1,2\} \longrightarrow \{a,b,c\}$, a 3-player game $G(\varphi, V, \rho)$ of perfect information between *Alice*, *Bob* and *Cecile* is introduced.[7] The game rules are as follows:

- If $\varphi \in \mathbf{prop}$, a play of the game has come to an end. If $V(\varphi) = (r, r', r'')$, then r determines the payoff for *Alice*, r' for *Bob*, and r'' for *Cecile*. Role 0 yields payoff $\mathfrak{g}$, role 1 payoff $\mathfrak{s}$, and role 2 payoff $\mathfrak{b}$. (Mnemonics: $\mathfrak{g}$ for 'gold', $\mathfrak{s}$ for 'silver' and $\mathfrak{b}$ for 'bronze'.)

- Let $i \in \{0,1,2\}$. If $\varphi = (\psi \vee_i \chi)$, then the player $\rho(i)$ chooses $\theta \in \{\psi, \chi\}$, and the play goes on as $G(\theta, V, \rho)$.

- If $\varphi = \neg\psi$, then the play continues as $G(\psi, V, \rho')$ with $\rho' = \pi_\neg(\rho)$.

- If $\varphi = \sim\psi$, then the play continues as $G(\psi, V, \rho')$ with $\rho' = \pi_\sim(\rho)$.

We stipulate that the game $G(\varphi, V)$ equals the game $G(\varphi, V, \rho_0)$, where ρ_0 is the standard initial role distribution, defined as in *Subsection 2.1*.

Strategies. The notions of (partial) play and strategy are defined as with PL_0^3. Since in PL^3 a play has more than two possible outcomes (there are $3! = 6$ possible outcomes), unqualified talk of winning strategies would not make sense. This fact motivates the following definitions: a $\mathfrak{g}$-*strategy* for player $\mathcal{P}$ is a strategy for $\mathcal{P}$ leading to the payoff $\mathfrak{g}$ for $\mathcal{P}$, against any sequence of moves by $\mathcal{P}$'s opponents; and an $\mathfrak{s}$-*strategy* for $\mathcal{P}$ is a strategy for $\mathcal{P}$ which is not a $\mathfrak{g}$-strategy, and which leads *at least* to the payoff $\mathfrak{s}$ for $\mathcal{P}$, against any sequence of moves by $\mathcal{P}$'s opponents. If *Alice*, say, has an $\mathfrak{s}$-strategy, she *cannot* use it to gain $\mathfrak{g}$ against *all* sequences of moves by *Bob* and *Cecile*. Yet, if she follows her $\mathfrak{s}$-strategy, she may obtain $\mathfrak{g}$ against *some* moves by them, and whenever she does not, she gains the payoff $\mathfrak{s}$.

Semantic attributes. When extending the semantics from atomic to complex formulas, there are the following questions to consider: *Is there a $\mathfrak{g}$-strategy for one of the players? Is there an $\mathfrak{s}$-strategy for (at least) one of the players?* It can happen that no player has a $\mathfrak{g}$-strategy, and it can happen that exactly one has. Further, it can happen that no player has an $\mathfrak{s}$-strategy, and it can happen that one player has or that two players have an $\mathfrak{s}$-strategy. If no player has a y-strategy (for $y \in \{\mathfrak{g}, \mathfrak{s}\}$) in a game, the game is said to be *non-determined* with respect to having a y-strategy; if more than one player has a y-strategy (for $y = \mathfrak{s}$), the game is *over-determined* with respect to having a y-strategy. Schematically, we are interested

---

[7]Unlike games for PL_0^3, games for PL^3 would indeed be constant-sum games, should we define payoffs as numbers.

in global properties of games $G(\varphi, V)$ represented by the following 16 pairs ('?' stands for non-determinacy, '$!_{\mathcal{P},\mathcal{P'}}$' for overdeterminacy due to players' $\mathcal{P}$ and $\mathcal{P'}$ both having an s-strategy):

(a,b)	(a,c)	$(a,?)$		
(b,a)	(b,c)	$(b,?)$		$(?, !_{a,b})$
(c,a)	(c,b)	$(c,?)$		$(?, !_{a,c})$
$(?,a)$	$(?,b)$	$(?,c)$	$(?,?)$	$(?, !_{b,c})$

FIGURE 2

The first member x of a pair (x, y) indicates whether one of the players has a g-strategy, and if one of them has, it also indicates who does. The second member y indicates whether some of the players have an s-strategy, and if at least one of them has, it indicates which one does or which ones do.

We distinguish between 16 semantic attributes of a formula: each pair P listed in Figure 2 constitutes a *semantic attribute* such that φ has by definition the attribute P relative to a valuation V, if the players' status in terms of having or lacking a g-strategy *resp.* an s-strategy in game $G(\varphi, V)$ is as specified by the pair P. Hence for instance the attribute (a, b) corresponds to *Alice*'s having a g-strategy and *Bob*'s (and only *Bob*'s) having an s-strategy; while $(?, !_{b,c})$ corresponds to no player's having a g-strategy and both of the players' *Bob* and *Cecile* having an s-strategy. We write $|\varphi, V| = P$ to indicate that the semantic attribute of φ relative to V is P.

Like with PL_0^3, also in connection with PL^3 the semantics might be left neutral with respect to the initial role distribution, and we could define the semantic attributes by reference to roles rather than players. However, we believe it to serve clarity of exposition to refer to players and stay with the standard initial role distribution (cf. Remark 2).

REMARK 3. Semantic attributes in logics PL^2, PL_0^3 and PL^3 can be viewed from a unifying perspective using the game-theoretic notion of *security level* (guaranteed minimum payoff);[8] for this notion see e.g. [5]. In each case semantic attributes can be considered as maps representing the security levels of the players. With PL^2, the only possible maps are $\{(Eloise, 1), (Abelard, 0)\}$ and $\{(Eloise, 0), (Abelard, 1)\}$ — since the corresponding games are determined. Semantic attributes S of PL_0^3 give rise to maps f_S of type $\{a, b, c\} \longrightarrow \{0, 1\}$, where $f_S(\mathcal{P}) = 1$ iff $\mathcal{P} \in S$. Finally, in PL^3 semantic attributes P induce maps $f_P : \{a, b, c\} \longrightarrow \{g, s, b\}$, where the value $f_P(\mathcal{P})$ indicates the optimal minimum rank that player $\mathcal{P}$ can guarantee by a suitable choice of strategy. We leave systematic investigation of the use of security levels and related game-theoretic notions in connection with game-theoretically defined logics for future research.   □

---

[8]We are indebted to an anonymous referee for this observation.

Let us now check in detail that all the 16 pairs of Figure 2 can indeed occur as semantic attributes of a formula. Let us begin by considering g-strategies. The number of players having a g-strategy in a game $G(\varphi, V)$ can be zero or one. (Evidently there are no games where more than one player has a g-strategy.)

EXAMPLE 4 (No one has a g-strategy). Put $V(p) = (2, 0, 1)$, $V(q) = (0, 2, 1)$, and $V(r) = (1, 2, 0)$; and let $\varphi := (p \vee_0 (q \vee_1 r))$. No player has a g-strategy in $G(\varphi, V)$: *Alice* gets second position if *Bob* chooses 'right' for $\vee_1$ (*Alice* having first herself chosen 'right' for $\vee_0$). *Cecile* gets second position if *Alice* chooses 'left' for $\vee_0$. And *Bob* gets only the last position if *Alice* chooses 'right' for $\vee_0$, no matter what *Bob* himself chooses for $\vee_1$. □

EXAMPLE 5 (Exactly one player has a g-strategy). Let $V(p) = (0, 1, 2)$. Then trivially *Alice* has a g-strategy in game $G(p, V)$. □

Let us proceed to think of s-strategies. The number of players having an s-strategy in a game $G(\varphi, V)$ can be zero, one or two. (Two players can have an s-strategy only if no player has a g-strategy.)

EXAMPLE 6 (No one has a g-strategy, nor an s-strategy). Consider the formula $\varphi := ((p \vee_1 q) \vee_0 (r \vee_2 s))$. Let $V(p) = (0, 1, 2)$, $V(q) = (2, 1, 0)$, $V(r) = (0, 2, 1)$, $V(s) = (2, 0, 1)$. In the left 0-disjunct *Bob* (i.e., the player responsible for moving) becomes second no matter which 1-disjunct he chooses, but he can decide whether *Alice* becomes first or third; and in the right 0-disjunct *Cecile* (i.e., the player responsible for moving) becomes second no matter which 2-disjunct she chooses, but she can decide whether *Alice* becomes first or third. It is *Alice* who decides whether the play proceeds to the left or to the right 0-disjunct. Hence in $G(\varphi, V)$ no player has a g-strategy. Actually no player even has an s-strategy. For, by what noted above, *Alice* cannot exclude the possibility that she becomes third. And jointly *Alice* and *Bob* can guarantee that *Cecile* becomes third, and likewise jointly *Alice* and *Cecile* can guarantee that *Bob* becomes third. □

EXAMPLE 7 (Someone has a g-strategy; no one has an s-strategy). Consider the formula $\varphi := (p \vee_0 q)$, and the valuation V with $V(p) = (0, 1, 2)$ and $V(q) = (0, 2, 1)$. *Alice* has a g-strategy in $G(\varphi, V)$, but neither *Bob* nor *Cecile* can guarantee second position: it is up to *Alice* to decide. □

Trivially there are games where some player has a g-strategy and another one has an s-strategy. E.g., in the game of Example 5, *Alice* has a g-strategy and *Bob* an s-strategy. Let us look at games where no one has a g-strategy.

EXAMPLE 8 (No one has a g-strategy; exactly one player has an s-startegy). Consider the formula $\varphi := (p \vee_0 (q \vee_1 r))$, and let $V(p) = (2, 1, 0)$, $V(q) = (2, 0, 1)$, $V(r) = (0, 1, 2)$. Clearly no player has a g-strategy in $G(\varphi, V)$: *Alice* cannot force that all plays of the game end with the atom r; *Bob* cannot force that all plays end with q; and *Cecile* cannot force that all plays end with p. On the other hand, *Bob* has an s-strategy: it consists of doing nothing if *Alice* picks out the left 0-disjunct, and choosing either of the 1-disjuncts if *Alice* picks out the right 0-disjunct. By contrast, *Cecile* cannot guarantee becoming second (since *Alice* can choose 'right' for $\vee_0$ and *Bob* can continue by picking out 'right' for $\vee_1$). And *Alice* cannot exclude ending up third (for if she chooses 'right' for $\vee_0$, *Bob* can choose 'left' for $\vee_1$). Hence *Bob* alone has an s-strategy in $G(\varphi, V)$.   □

EXAMPLE 9 (No one has a g-strategy; two players have an s-strategy). A case in point is the formula $\varphi := (p \vee_0 q)$ and the valuation V such that $V(p) = (2, 0, 1)$ and $V(q) = (2, 1, 0)$. No player has a g-strategy in $G(\varphi, V)$; but both *Bob* and *Cecile* have a way of guaranteeing (without any personal effort, for that matter) that they receive at least the payoff s. By contrast, *Alice* ends inevitably last no matter how the play goes.   □

The following fact holds trivially:

FACT 10. (a) It cannot happen that each of the three players has either a g-strategy or an s-strategy. (b) It cannot happen that one of the players has a g-strategy, and both remaining players have an s-strategy.   □

The above considerations show that indeed each of the 16 pairs listed in Figure 2 determines a possible semantic attribute of a PL^3-formula.

Like in the case of PL_0^3, also in connection with PL^3 the question of functional completeness naturally arises. Every PL^3-formula φ determines a map — let us call it $|\varphi|$ — from valuations to semantic attributes, namely $|\varphi| : V \mapsto P$, where $P = |\varphi, V|$. Let, then, **prop** be a finite set of propositional atoms, and f any map from valuations on **prop** to the 16 semantic attributes. The question then is whether there always is a formula φ of PL^3 such that $f = |\varphi|$. We leave this as an open question.

Case of n players. Could PL^3 be naturally generalized to the case of n players? Insofar as the payoff function is considered, the answer is affirmative. There would be $n!$ possible rankings of the n players. The semantic attribute of a formula relative to a valuation would be defined by specifying for each player his or her optimal minimum rank in the correlated game. It would be straightforward to define the semantics of disjunction symbols. By contrast, it is less evident how to effect a generalization with respect to negations. If $\mathbf{P}_n$ is the set of all distributions of the n roles to the n players, the requirement would be to choose a minimal set of bijections of type $\mathbf{P}_n \longrightarrow \mathbf{P}_n$ which would suffice for defining all bijections of type $n \longrightarrow n$. The fact that the relevant set of permutations of role distributions is not uniquely determined can be seen as diminishing the theoretical interest of PL^3. Another theoretically problematic feature of PL^3 and its generalizations is that if we are willing to climb to the level of permutations of role distributions, why should not all permutations of role distributions be studied within the logic? In the n-player case there are $(n!)!$ of them.

3 Basic features of PL_0^3 or PL^3

Let L be one of the logics PL_0^3 or PL^3. Valuations for L will be called L-*valuations*. The possible semantic attributes of L-formulas will be called L-*attributes*. In what follows, L-formulas φ and ψ are said to be *logically equivalent*, in symbols $\varphi \Longleftrightarrow \psi$, if for all L-valuations V and L-attributes A, we have that $|\varphi, V| = A$ iff $|\psi, V| = A$. Formulas $\varphi, \psi \in L$ are *incompatible*, if there is no L-valuation V such that $|\varphi, V| = |\psi, V|$. Let us proceed to make some basic observations about the behavior of negations, literals, and the definability of semantic attributes in the logics PL_0^3 and PL^3.

3.1 Behavior of negations

In PL^2, the negation $\neg$ obeys the law of double negation: φ and $\neg\neg\varphi$ are logically equivalent, for any formula φ. Let us consider what types of laws hold for iterations of negation symbols in PL_0^3 and PL^3.

Let us begin with PL_0^3. Let $\varphi \in PL_0^3$ be arbitrary. We have already seen that whenever i, j are distinct roles, the formulas $\sim_{ij}\varphi$ and $\sim_{ji}\varphi$ are logically equivalent. Furthermore, it is easily observed that the following *laws of double negation* hold in PL_0^3: $\sim_{ij}\sim_{ij}\varphi \Longleftrightarrow \varphi$.

What about the interaction of different negation signs? To begin with, it is readily checked that $\sim_{01}\sim_{02}\varphi \Longleftrightarrow \sim_{02}\sim_{12}\varphi \Longleftrightarrow \sim_{12}\sim_{01}\varphi$, and $\sim_{01}\sim_{12}\varphi \Longleftrightarrow \sim_{02}\sim_{01}\varphi \Longleftrightarrow \sim_{12}\sim_{02}\varphi$.

Obviously all six role distributions $\rho \in \{a, b, c\}^{\{0,1,2\}}$ can be expressed in terms of the three negations $\sim_{01}$, $\sim_{02}$, and $\sim_{12}$. That is, for any such ρ there is a string $\bar{n}$ of length at most 2, formed in the alphabet $\{\sim_{01}, \sim_{02}, \sim_{12}\}$, such that if $\pi_{\bar{n}} \in \{0, 1, 2\}^{\{0,1,2\}}$ is the permutation corresponding to the string of negations $\bar{n}$, then $\rho = (\rho_0 \circ \pi_{\bar{n}})$, where ρ_0 is the standard initial role distribution.[9] Identifying role distributions with the triples of roles they give to *Alice, Bob* and *Cecile* in this order, we may note that for the role distributions $(0, 1, 2)$, $(0, 2, 1)$, $(1, 0, 2)$, $(1, 2, 0)$, $(2, 1, 0)$ and $(2, 0, 1)$, the corresponding strings $\bar{n}$ are:[10] empty string; $\sim_{12}$; $\sim_{01}$; $\sim_{01}\sim_{02}$; $\sim_{02}$; and $\sim_{01}\sim_{12}$. It should be observed that from the viewpoint of expressive power, not all negations $\sim_{01}$, $\sim_{02}$ and $\sim_{12}$ are actually *needed* for expressing all role distributions. For instance $\sim_{12}$ is definable in terms of $\sim_{01}$ and $\sim_{02}$: namely, $\sim_{12}\varphi \Longleftrightarrow \sim_{01}\sim_{02}\sim_{01}\varphi$, for any formula φ.

Negations of the logic PL^3 behave rather differently from the case of PL_0^3. Let $\varphi \in PL^3$ be arbitrary. It is straightforward to see that $\sim$ obeys the *law of double negation*, $\sim\sim\varphi \Longleftrightarrow \varphi$, while $\neg$ obeys the *law of triple negation*: $\neg\neg\neg\varphi \Longleftrightarrow \varphi$. Further, the relative order of the different negation symbols obeys the *law of*

---

[9]If $f : A \longrightarrow B$ and $g : B \longrightarrow C$ are functions, the composite function $(g \circ f) : A \longrightarrow C$ satisfies: $(g \circ f)(a) = g(f(a))$, for any $a \in A$.

[10]Instead of the combination $\sim_{01}\sim_{02}$, we might just as well use either $\sim_{02}\sim_{12}$ or $\sim_{12}\sim_{01}$; and in place of $\sim_{01}\sim_{12}$ either $\sim_{02}\sim_{01}$ or $\sim_{12}\sim_{02}$ could be used.

order invariance: $\sim\neg\varphi \iff \neg\sim\varphi$. (That is, $\pi_\neg$ and $\pi_\sim$ commute.) By contrast, the following six formulas are pairwise incompatible: φ, $\neg\varphi$, $\neg\neg\varphi$, $\sim\varphi$, $\neg\sim\varphi$, $\neg\neg\sim\varphi$. Note that on the other hand, by the *law of order invariance* we have: $\sim\neg\neg\varphi \iff \neg\sim\neg\varphi \iff \neg\neg\sim\varphi$. Inspecting the definitions of the permutations $\pi_\neg$ and $\pi_\sim$ (given in Fig. 1), it is evident that the two negations $\neg$ and $\sim$ are *not* interdefinable.

3.2   Literals

In PL^2, there are for each propositional atom p precisely two mutually incompatible literals, namely p and $\neg p$. Let L be either of the logics PL^3_0 or PL^3. If $p \in$ **prop**, a formula ℓ of L is a *p-literal*, if ℓ is syntactically built from p using the negation symbols of L only. A set Λ of p-literals is *basic*, if (*i*) any two elements in Λ are pairwise incompatible, and (*ii*) every p-literal is logically equivalent to some element of Λ. (Recall the definition of incompatibility from the beginning of the present section.) Extending this terminology to PL^2, for instance sets $\{p, \neg p\}$ and $\{\neg\neg p, \neg p\}$ are both basic sets of p-literals, while the sets $\{\neg p\}$ and $\{p, \neg p, \neg\neg p\}$ are not. A formula of L is a *literal*, if it is a p-literal for some atom p. What interests us in connection with L is determining whether *some* basic set of p-literals exists, for any given atom p.

Consider PL^3_0 first. By observations made in *Subsection* 3.1, in order to find out whether PL^3_0 admits of basic sets of p-literals in the first place, it suffices to check whether $X := \{p, \sim_{01} p, \sim_{02} p, \sim_{12} p, \sim_{01}\sim_{02} p, \sim_{01}\sim_{12} p\}$ is such a set. Let S be either $\{a, b, c\}$ or $\varnothing$. Now if $V(p) = \rho_0^{-1}(S)$, then actually the semantic attribute of *all* formulas of the set X is S, relative to V. It follows that relative to arbitrary valuations, there simply exists no basic set of p-literals at all.

What about considering restricted valuations only — valuations whose value on an atom is neither $\varnothing$ nor $\{0, 1, 2\}$? Under this assumption, if $V(p) = R$ and $S = \rho_0(R)$, we can find distinct roles i, j such that $\{i, j\}$ coincides either with R or with $\{0, 1, 2\} \setminus R$. In both cases $|p, V| = |\sim_{ij} p, V| = S$. It follows that the p-literals p, $\sim_{01} p$, $\sim_{02} p$ and $\sim_{12} p$ are *not* pairwise incompatible in the sense that under no valuation no two of them would have the same semantic attribute. Actually, under any restricted valuation exactly two of them have the same attribute. Hence no basic set of p-literals exists, even when attention is confined to restricted valuations. With respect to literals, PL^3_0 behaves, then, very differently from PL^2.

By contrast, PL^3 comes closer to PL^2 in its behavior with respect to literals. The p-literals p, $\neg p$, $\neg\neg p$, $\sim p$, $\neg\sim p$ and $\neg\neg\sim p$ are indeed pairwise incompatible, and moreover any further p-literal is equivalent to one of them. Therefore these formulas form a basic set of p-literals in PL^3.

3.3 Capturing semantic attributes

In PL^2, both truth-values can be *captured* by a formula — or are *definable* — in the sense that there is a formula $\top$ of PL^2 whose truth-value is *true* under all valuations, and likewise there is a formula $\bot$ of PL^2 whose truth-value is *false* under all valuations. For instance $(p \vee \neg p)$ is such a formula $\top$, and $(p \wedge \neg p)$ such a formula $\bot$. Letting L be one of the logics PL^3_0 or PL^3, let us now consider the question whether all L-attributes can be similarly captured within L, that is, whether for each L-attribute A there is a formula φ_A of L such that $|\varphi_A, V| = A$, for all L-valuations V.

Let us begin by considering the logic PL^3_0. First we may notice that relative to arbitrary valuations such capturing is *not* possible in PL^3_0. This is a corollary to the following lemma:

LEMMA 11. *Let R be either $\{0, 1, 2\}$ or $\varnothing$; and let V be a valuation mapping all propositional atoms to the set R. Then all formulas of PL^3_0 receive, relative to V, the semantic attribute $S = \rho_0(R)$.*

Proof. By assumption the claim holds for atoms. Assuming inductively that the claim holds for formulas φ, ψ, it immediately follows that it also holds for formulas $(\varphi \vee_i \psi)$ and $\sim_{ij} \varphi$. ∎

If $\{a, b, c\} \neq S \neq \varnothing$, then by Lemma 11 the semantic attribute S cannot be captured by any formula of PL^3_0. For, relative to a valuation as in the statement of the lemma, any formula will receive a semantic attribute other than S. What about turning attention to restricted valuations, then?

Consider the attribute $\{a\}$. Evidently under any restricted valuation V, *Alice* has a winning strategy in one of the games $G(p, V)$, $G(\sim_{01} p, V)$ and $G(\sim_{02} p, V)$. Now in order for a formula to have the semantic attribute $\{a\}$, it is required, not just that *Alice* has a winning strategy, but also that neither of the other two players has one. Actually the formula

$$(p \vee_0 \sim_{12} p) \vee_0 (\sim_{01} p \vee_0 \sim_{01} \sim_{02} p) \vee_0 (\sim_{02} p \vee_0 \sim_{01} \sim_{12} p)$$

captures the attribute $\{a\}$ relative to restricted valuations. To see this, let V be any such valuation. If $0 \in V(p)$, then already by her first choice *Alice* can guarantee that she will win: by choosing the leftmost disjunct. What is more, by her next choice she may prevent any of the other two players from winning. Namely, since V is a restricted valuation, either 1 or 2 falls outside of $V(p)$. Hence neither *Bob* nor *Cecile* has a winning strategy in both games $G(p, V)$ and $G(\sim_{12} p, V)$, and consequently neither of them has a winning strategy for $(p \vee_0 \sim_{12} p)$, since it is *Alice* who chooses the disjunct. Similarly, if $1 \in V(p)$, *Alice* may choose the middle disjunct, and if $2 \in V(p)$, the rightmost disjunct, being able to make sure by her remaining move that neither of the other players can also win the play.

The attributes $\{b\}$ and $\{c\}$ are captured similarly. On the other hand, by Lemma 12 the attributes $\{a, b\}$, $\{a, c\}$ and $\{b, c\}$ are *not* definable.

LEMMA 12. *Suppose* $|V(p)| = 1$ *for all propositional atoms* p.[11] *Then no formula* φ *of* PL_0^3 *has relative to* V *an attribute* S, *for any* S *with* $|S| = 2$.

Proof. Suppose V satisfies the premise of the lemma. Then the claim holds for atomic formulas. Go on to assume inductively that the claim holds for formulas φ, ψ, and let i, j be any two roles. Consider the formula $(\varphi \vee_i \psi)$. By inductive hypothesis, in both games $G(\varphi, V)$ and $G(\psi, V)$ at most one player has a winning strategy. Clearly the number of players with a winning strategy in game $G(\varphi \vee_i \psi, V)$ can be at most $\max\{n_\varphi, n_\psi\}$, where $n_\theta \leq 1$ is the number of players having a winning strategy in $G(\theta, V)$, with $\theta \in \{\varphi, \psi\}$. Further, given the inductive hypothesis, evidently at most one player has a winning strategy in $G(\sim_{ij}\varphi, V)$. ∎

As witnessed by the fact that the singleton attributes S can be captured relative to restricted valuations, the 'mirror image' of Lemma 12 does *not* hold: it *can* very well happen that a formula has the semantic attribute S for a singleton S, even relative to a valuation V with $|V(p)| = 2$, for all atoms p. Generally, the number of players having a winning strategy for a complex formula can very well be smaller than the number of players having a winning strategy for its components.

Let us, then, take a look at the logic PL^3. As regards capacity to capture semantic attributes, PL^3 turns out to differ from PL_0^3. Actually, each of the 16 semantic attributes of PL^3 can be captured.

First think of the attribute (a, b). Divide the p-literals into three 'cells': $\{p, \sim p\}$, $\{\neg p, \neg \sim p\}$, $\{\neg \neg p, \neg \neg \sim p\}$, and observe that under any valuation, there is a cell such that *Alice* receives the highest rank relative to both formulas of that cell. For one of the formulas in that cell, it is *Bob* who becomes second in the ranking, while for the other, *Cecile* becomes second. So the formula $(p \vee_1 \sim p) \vee_0 (\neg p \vee_1 \neg \sim p) \vee_0 (\neg \neg p \vee_1 \neg \neg \sim p)$ receives under any valuation the attribute (a, b). For, *Alice* may make sure that she becomes first, whereafter *Bob* can guarantee that he becomes second. The attributes P with $P \in \{(a, c), (b, a), (b, c), (c, a), (c, b)\}$ are captured similarly.

Let us, then, consider the case of $(a, ?)$. This attribute is simply captured by the formula $(p \vee_0 \neg p \vee_0 \neg \neg p \vee_0 \sim p \vee_0 \neg \sim p \vee_0 \neg \neg \sim p)$. Namely, under any valuation, there are two 0-disjuncts which *Alice* can choose so that she herself reaches the highest rank. In one of them *Bob* becomes second, while in the other *Cecile* becomes second. *Bob* and *Cecile* have no control over which 0-disjunct *Alice* chooses. The attributes $(b, ?)$ and $(c, ?)$ are captured similarly.

For the remaining attributes, we state the relevant formulas; the reader is invited to check in detail that indeed they fit the bill. Let $\Lambda := \{p, \neg p, \neg \neg p, \sim p, \neg \sim p,$

---

[11] If X is a set, $|X|$ stands for its cardinality.

$\neg\neg\sim p\}$, and define formulas C_1 and C_2 by setting:

$$C_1 := \bigvee_{\ell\in\Lambda}(\ell \vee_1 \neg\ell) \text{ and } C_2 := \bigvee_{\ell\in\Lambda}(\sim\ell \vee_1 \neg\ell).$$

The formula $(C_1 \vee_0 C_2)$ captures $(?, a)$; while $(?, b)$ and $(?, c)$ are captured similarly. Further, $(?, ?)$ is captured by the formula $(D_1 \vee_0 D_2)$, where

$$D_1 := \bigvee_{\ell\in\Lambda}\!{}_1\, \ell \text{ and } D_2 := \bigvee_{\ell\in\Lambda}\!{}_2\, \ell.$$

Finally, the attribute $(?, !_{a,c})$ is captured by

$$[(p \vee_2 \sim p) \vee_1 (\neg p \vee_2 \neg\sim p)] \vee_0 [(\neg p \vee_2 \neg\sim p) \vee_1 (\neg\neg p \vee_2 \neg\neg\sim p)]$$
$$\vee_0 [(\neg\neg p \vee_2 \neg\neg\sim p) \vee_1 (p \vee_2 \sim p)],$$

the attributes $(?, !_{a,b})$ and $(?, !_{b,c})$ being captured similarly.

3.4 The notion of consequence

In connection with PL^2, logical consequence is defined in terms of truth-preservation: ψ is a logical consequence of φ, if all valuations making φ true, make ψ true as well. The game-theoretic content of this condition is as follows: for every valuation V, if there is a winning strategy for *Eloise* in $G(\varphi, V)$, there also is one for her in $G(\psi, V)$. A corresponding consequence relation for *Abelard* is defined by the requirement that for all valuations V, whenever there is a winning strategy for *Abelard* in $G(\varphi, V)$, there is one for him in $G(\psi, V)$. The relation defined by this condition is the converse of the relation of logical consequence, and is characterized by falsity-preservation.

The notion of consequence relation is naturally generalized to the multi-player setting by introducing one consequence relation for *each player*. Let $\mathcal{P}$ be one of the three players. It is said that ψ is a *$\mathcal{P}$-consequence of φ in PL_0^3*, provided that all valuations V satisfy: if $\mathcal{P}$ has a winning strategy in $G(\varphi, V)$, then $\mathcal{P}$ also has one in $G(\psi, V)$. This condition can be otherwise expressed thus: for all valuations V, if $\mathcal{P} \in |\varphi, V|$, then $\mathcal{P} \in |\psi, V|$.

In games corresponding to PL^3, the payoffs are in terms of rankings. Accordingly, what is of interest to a given player $\mathcal{P}$ on the level of strategies is the optimal minimum rank that $\mathcal{P}$ can guarantee by a suitable choice of strategy. In *Section 2* the relevant notions of strategy were conceptualized in terms of the notions of g-strategy and s-strategy. (Alternatively, we might speak of security levels g and s, respectively, as noted in Remark 3.) Now ψ is said to be a *$\mathcal{P}$-consequence of φ in PL^3*, if for all valuations V, the optimal minimum rank that $\mathcal{P}$ can guarantee in $G(\psi, V)$ is the same or better than the optimal minimum rank that $\mathcal{P}$ can guarantee in $G(\varphi, V)$. In terms of semantic attributes, this condition means the following. Suppose that $|\varphi, V| = (x, y)$ and $|\psi, V| = (x', y')$. Then if $x = \mathcal{P}$, also

$x' = \mathcal{P}$, while if y marks $\mathcal{P}$ as a player with an s-strategy, then either $x' = \mathcal{P}$ or y' marks $\mathcal{P}$ as a player with an s-strategy.

From the viewpoint of PL^2 one might equally well think of a generalization where *each semantic attribute* would induce its own consequence relation. We believe, however, that the generalization in terms of players is the fruitful one. In particular, when consequence relations for PL_0^3 and PL^3 are defined as above, they seem to have good algebraic counterparts in terms of semi-lattices. Systematically studying the consequence relations is left for future research.

4 Further features

The logic PL^2 satisfies the *law of excluded middle*: for any valuation V and any formula φ, we have $V \models (\varphi \vee \neg\varphi)$. Furthermore, PL^2 is subject to the *semantic principle of bivalence*: relative to any valuation, any formula φ is either true or false.[12] Let us take a look whether the logics PL_0^3 and PL^3 admit of analogous logical laws and semantic principles. We also take up the question whether analogues to De Morgan's laws and distributive laws can be formulated in PL_0^3 and PL^3.

The following definitions will be needed subsequently. Let L be any of the logics PL^2, PL_0^3 or PL^3. If φ is a formula of L, let us agree to write $L(\varphi)$ for the class of L-formulas that can be formed from (any number of tokens of) the formula φ using the operators of L. Hence for instance $\sim_{01}(p \vee_0 q) \in L(p \vee_0 q)$, but $p \notin L(p \vee_0 q)$.

DEFINITION 13 (Uniform characterization; characterizability). If $\chi \in L(p)$, write f_χ for the syntactic transformation of type $L \longrightarrow L$ defined by the following condition: for all $\varphi \in L$,

$$f_\chi(\varphi) = \chi[p/\varphi],$$

where $\chi[p/\varphi]$ stands for the result of substituting everywhere φ for p in χ. Let, then, A_0 be a fixed L-attribute, and A an arbitrary L-attribute. It is said that a formula $\chi \in L(p)$ *uniformly characterizes* A in terms of A_0, if for all formulas $\psi \in L$ and L-valuations V,

$$|\psi, V| = A \text{ iff } |f_\chi(\psi), V| = A_0.$$

It is merely said that A *can be characterized* in terms A_0, if for all $\psi \in L$, there is $\chi_\psi \in L$ such that for all V: $|\psi, V| = A$ iff $|\chi_\psi, V| = A_0$.   □

Whenever a formula $\chi \in L(p)$ uniformly characterizes A in terms of A_0, the attribute A can of course be characterized in terms of A_0, but the converse does not hold. Crucially, in connection with the stronger notion, the *form* of the formula

---

[12] As stressed notably by Dummett (see, e.g., [6]), logical laws must be distinguished from semantic principles. From the perspective of the present paper, what is crucial about logical laws is that they are expressed in terms of a designated truth-value (*true*).

$f_\chi(\psi)$ does *not* depend on the particular input ψ: the form of $f_\chi(\psi)$ is that of χ, modulo ψ being substituted everywhere for p. With the weaker notion, again, there is no guarantee of syntactic similarity between formulas χ_ψ corresponding to distinct formulas ψ.

4.1 Analogues of the law of excluded middle

In PL^2, the truth-value *true* is captured by the formula $(p \lor \neg p)$. As it happens, the atom p can be replaced by an arbitrary formula φ, and the result will continue to capture *true*. This fact alone shows that the law of excluded middle holds for PL^2. In the case of a logic subject to the semantic principle of bivalence, the law of excluded middle can be understood as providing at the object language level an exhaustive list of the semantic values that an arbitrary formula φ may have. Expressing the different truth-values on the object language level is possible in PL^2, since falsity of a formula can be expressed by truth of another formula. Let us consider whether similar phenomena occur in the 3-player setting.

Logic PL_0^3. A necessary condition for the existence of an analogue of the law of excluded middle in PL_0^3 would be that any semantic attribute S of any given formula χ could be expressed in terms of some *designated* attribute S_0 of some other formula ψ_S^χ, i.e., that for all valuations V, we had $|\chi, V| = S$ iff $|\psi_S^\chi, V| = S_0$. Let us take $\{a\}$ as the designated attribute. The choice is to some extent arbitrary. In particular, one might ask why the designated attribute should be a singleton rather than a pair. To this a possible reply would be that the former are better-behaved from the logic-internal viewpoint: singleton attributes are definable in PL_0^3, while two-element attributes are not (see Lemma 12). Let us now turn attention to propositional atoms, and show that indeed for each atom p there is a formula φ_p which lists all the six semantic attributes that p may have relative to *restricted valuations*. (By Lemma 11 we know such a result cannot hold with respect to arbitrary valuations.) We first prove a lemma.

LEMMA 14. *Let V be an arbitrary restricted valuation. Define formulas $\psi_{\{a,b\}}$, $\psi_{\{a,c\}}$ and $\psi_{\{b,c\}}$ as follows:*

$$\psi_{\{a,b\}} := (p \lor_0 {\sim_{12}} p) \lor_2 ({\sim_{01}} p \lor_1 {\sim_{12}} p),$$
$$\psi_{\{a,c\}} := (p \lor_0 {\sim_{12}} p) \lor_1 ({\sim_{02}} p \lor_2 {\sim_{12}} p),$$
$$\psi_{\{b,c\}} := ({\sim_{01}} p \lor_1 {\sim_{02}} p) \lor_0 ({\sim_{01}} p \lor_2 {\sim_{02}} p).$$

For all sets $S \in \{\{a, b\}, \{a, c\}, \{b, c\}\}$: $|p, V| = S$ iff $|\psi_S, V| = \{a\}$.

Proof. Let us check the case $S := \{a, b\}$; the other cases can be proven similarly. Let V be a restricted valuation. Suppose first that $|\psi_{\{a,b\}}, V| = \{a\}$. Hence *Alice* has a winning strategy in both games $G({\sim_{01}} p, V)$ and $G({\sim_{12}} p, V)$. Therefore $0, 1 \in V(p)$, and since V is restricted, in fact $V(p) = \{0, 1\}$. So $|p, V| = \{a, b\}$. Conversely, suppose $|p, V| = \{a, b\}$. Thus *Alice* has a winning strategy in all games

$G(\sim_{01} p, V)$, $G(\sim_{12} p, V)$ and $G(p, V)$, and therefore in $G(\psi_{\{a,b\}}, V)$. What is more, *Bob* does not have a winning strategy in the game corresponding to $\psi_{\{a,b\}}$ (since *Cecile* may choose 'left' after which *Alice* may choose 'right'); and *Cecile* has no winning strategy either, because if she chooses 'left', *Alice* may choose 'left' to prevent *Cecile* from winning; and if she chooses 'right', *Bob* may choose 'left', with the consequence that *Cecile* loses. Hence indeed $|\psi_{\{a,b\}}, V| = \{a\}$.   ∎

THEOREM 15 (Atomic law of excluded seventh). *Let $S_1, \dots, S_6$ be a list of all subsets of $\{a, b, c\}$ except $\{a, b, c\}$ and $\varnothing$. There is a formula $\varphi_p := (\psi_{S_1} \vee_0 \dots \vee_0 \psi_{S_6})$ of PL_0^3 with p as its only atom, such that for all restricted valuations V,*

$$|\varphi_p, V| = \{a\}, \quad \text{and} \quad |\psi_{S_i}, V| = \{a\} \;\; \text{iff} \;\; |p, V| = S_i \; (1 \le i \le 6).$$

Proof. Let $\psi_{\{a\}} := p$, $\psi_{\{b\}} := \sim_{01} p$, and $\psi_{\{c\}} := \sim_{02} p$. Further, let $\psi_{\{a,b\}}$, $\psi_{\{a,c\}}$ and $\psi_{\{b,c\}}$ be as in the statement of Lemma 14. Then we may take φ_p to be the formula $(\psi_{\{a\}} \vee_0 \psi_{\{b\}} \vee_0 \psi_{\{c\}} \vee_0 \psi_{\{a,b\}} \vee_0 \psi_{\{a,c\}} \vee_0 \psi_{\{b,c\}})$.   ∎

The next thing to ask is whether the atomic law of excluded seventh can be generalized so as to apply to arbitrary formulas. Let us first formulate the question more precisely. To begin with, note that the law of excluded middle of PL^2 has the following general format: for any formula φ of PL^2, the formula $(f_p(\varphi) \vee f_{\neg p}(\varphi))$ is true under any valuation. Observe that p serves to uniformly characterize the truth-value *true* in terms of *true*, while $\neg p$ uniformly characterizes *false* in terms of *true*. (Recall the notion of uniform characterizability from Definition 13.) Looking for an analogue to the law of excluded middle in PL_0^3, we must first of all ask whether each of the relevant semantic attributes of PL_0^3 can be uniformly characterized in terms of a fixed attribute. But which *are* the relevant attributes?

As already noted at the end of *Subsection* 2.1, relative to restricted valuations the number of possible semantic attributes increases from 6 to 7 when considering arbitrary formulas instead of literals: there are complex formulas for which no player has a winning strategy, i.e., which have the attribute $\varnothing$ relative to suitable restricted valuations. It is not difficult to check that each of the $PL_0^3(p)$-formulas $\psi_{\{a\}}$, $\psi_{\{b\}}$, $\psi_{\{c\}}$, $\psi_{\{a,b\}}$, $\psi_{\{a,c\}}$, $\psi_{\{b,c\}}$ referred to in the proof of Theorem 15 uniformly characterizes the corresponding semantic attribute in terms of the attribute $\{a\}$. But is there a formula uniformly characterizing the attribute $\varnothing$ in terms of $\{a\}$? That is, can we find a formula χ such that $|\psi, V| = \varnothing$ iff $|f_\chi(\psi), V| = \{a\}$? By the argument given for Lemma 11, it is immediate that the answer is negative: if the attribute of ψ is $\varnothing$ relative to a valuation V, so will be the attribute of any formula of the class $PL_0^3(\psi)$.

We may conclude that a natural generalization of the law of excluded middle, applicable to arbitrary formulas, is *not* possible for PL_0^3. As a semantic principle we still have the principle of *7-valence*: any formula of PL_0^3 has, relative to restricted valuations, one of seven semantic attributes.[13]

So the attribute $\varnothing$ cannot be *uniformly* characterized in terms of $\{a\}$. Might it still not be possible to simply characterize $\varnothing$ in terms of $\{a\}$? We leave settling this issue for future research. The following example however shows that at least for some formulas ψ, a suitable formula χ_ψ can be found such that for all restricted valuations V: $|\psi, V| = \varnothing$ iff $|\chi_\psi, V| = \{a\}$.

EXAMPLE 16. Consider the formula $(p \vee_0 q)$, letting V be a restricted valuation. Evidently $|p \vee_0 q, V| = \varnothing$ iff: $[V(p) = \{1\}$ and $V(q) = \{2\}]$ or $[V(p) = \{2\}$ and $V(q) = \{1\}]$. Let, then, φ be the following formula:

$$(\sim_{01} p \vee_1 \sim_{02} q) \vee_0 (\sim_{01} q \vee_1 \sim_{02} p).$$

It is easy to check that $|\varphi, V| = \{a\}$ iff $|p \vee_0 q, V| = \varnothing$.   □

Logic PL^3. Let us turn to PL^3. Does PL^3 have an analogue of the law of excluded middle? Again, an analogue is found when attention is restricted to propositional atoms. Let us take (a, b) as the designated attribute. (It has a clearly better claim on being a designated attribute than any attribute involving non-determinacy or overdeterminacy.) Let P_i denote the i-th pair in the list (a, b), (a, c), (b, a), (b, c), (c, a), (c, b); and let ψ_{P_i} stand for the i-th formula in the list p, $\sim p$, $\neg\neg\sim p$, $\neg\neg p$, $\neg p$, $\neg\neg\sim p$. It is straightforward to check that $|\psi_{P_i}, V| = (a, b)$ iff $|p, V| = P_i$, for any valuation V $(1 \le i \le 6)$.

THEOREM 17 (Atomic law of excluded seventh). *There is a formula $\varphi_p :=$ $(\psi_{P_1} \vee_1 \psi_{P_2}) \vee_0 (\psi_{P_3} \vee_1 \psi_{P_4}) \vee_0 (\psi_{P_5} \vee_1 \psi_{P_6})$ of PL^3 with p as its only atom, such that for all valuations V,*

$$|\varphi_p, V| = (a, b), \quad \text{and} \quad |\psi_{P_i}, V| = (a, b) \quad \text{iff} \quad |p, V| = P_i \ (1 \le i \le 6).$$

Proof. Put $\varphi_p := (p \vee_1 \sim p) \vee_0 (\neg\neg\sim p \vee_1 \neg\neg p) \vee_0 (\neg p \vee_1 \neg\sim p)$.   ∎

When formulating the atomic law of excluded seventh for PL^3, two types of disjunction symbol are needed — unlike in the cases of PL^2 or PL_0^3. This is due to the 'combinatorial' nature of payoffs in games correlated with PL^3-formulas: not only must we express that one of the players has a g-strategy, but we also must express that another player has an s-strategy.

As in the case of PL_0^3, also in connection with PL^3 the next thing to ask is whether a variant of the atomic law of excluded seventh can be formulated which is applicable to arbitrary formulas. Recall that in PL^3 there are 16 semantic attributes

---

[13]Dummett [6, p. xix] remarks: "[W]hile acceptance of the semantic principle normally entails acceptance of the corresponding logical law, the converse does not hold." It may be of some interest to observe that by these criteria, PL_0^3 offers an instance of abnormality: while the semantic principle of 7-valence holds, the corresponding logical law, the law of excluded eighth, does not. (The latter cannot even be formulated.)

to consider: 10 attributes in addition to the 6 possible attributes P_i $(1 \leq i \leq 6)$ of atomic formulas. It is not difficult to check that the above $PL^3(p)$-formula ψ_{P_i} uniformly characterizes the semantic attribute P_i in terms of the attribute (a, b) (with $1 \leq i \leq 6$). To see what happens with the further attributes, let us first prove a result that can be compared to Lemma 11, the case $R := \varnothing$. It should be noted, however, that whereas that lemma applies to all formulas of PL_0^3, the present result only pertains to complex PL^3-formulas capable of non-determinacy.

LEMMA 18. *Let (x, y) be any attribute such that $\{x, y\} \cap \{?\} \neq \varnothing$. Suppose ψ is a formula of PL^3 and V is a valuation such that $|\psi, V| = (x, y)$. Then for all formulas $\chi \in PL^3(\psi)$, we have that there is a pair (x', y') such that $|\chi, V| = (x', y')$, where: $(x' = ?, if x = ?)$ and $(y' = ?, if y = ?)$.*

Proof. By the premise of the lemma the claim holds for ψ. Assume inductively that it holds for $\theta, \theta' \in PL^3(\psi)$. No switching of roles can yield any of the players a z-strategy (with $z \in \{\mathtt{g}, \mathtt{s}\}$) in $G(\neg\theta, V)$ or in $G(\sim \theta, V)$, if no player has a z-strategy in $G(\theta, V)$. Similarly no player can have a z-strategy in $G(\theta \vee_i \theta', V)$ without having one either in $G(\theta, V)$ or in $G(\theta', V)$.   ∎

It follows from Lemma 18 that whenever at least one of x, y equals ?, no formula can uniformly characterize the attribute (x, y) in terms of the attribute (a, b). Consequently *only* the attributes P_i $(1 \leq i \leq 6)$ can be uniformly characterized. Therefore we may conclude that a natural generalization of the law of excluded middle, applicable to arbitrary formulas, is *not* possible for PL^3 — just as we saw that it is not possible for PL_0^3. On the other hand, PL^3 is subject to the semantic principle of *16-valence*.[14]

Not all semantic attributes can, then, be uniformly characterized in terms of the attribute (a, b). Settling the issue whether all semantic attributes of PL^3 could, however, be characterized in the weak sense in terms of the attribute (a, b) is left for another occasion. Let us conclude the present considerations by looking at a particular semantic attribute involving non-determinacy, namely $(?, a)$. We may observe that actually there are formulas ψ for which another formula χ_ψ can be found such that $|\psi, V| = (?, a)$ iff $|\chi_\psi, V| = (a, b)$, for all valuations V.

EXAMPLE 19. Consider the formula $(p \vee_0 q)$, letting V be an arbitrary valuation. Let, then, φ be the following formula:

$$(\neg\neg\sim p \ \vee_2 \ \neg q) \ \vee_0 \ (\neg\neg\sim q \ \vee_2 \ \neg p).$$

Now $|\varphi, V| = (a, b)$ iff $|p \vee_0 q, V| = (?, a)$. Namely, $|\varphi, V| = (a, b)$ iff: either $|\neg\neg\sim p, V| = |\neg q, V| = (a, b)$, or $|\neg\neg\sim q, V| = |\neg p, V| = (a, b)$. This condition, again, is equivalent to requiring that either $V(p) = (1, 0, 2)$ and $V(q) = (2, 0, 1)$,

---

[14]Hence also PL^3 is 'abnormal', in the way PL_0^3 was observed to be in footnote 13.

or else $V(q) = (1, 0, 2)$ and $V(p) = (2, 0, 1)$, which is simply equivalent to the condition $|p \vee_0 q, V| = (?, a)$.   □

We have now seen that in the logics PL_0^3 and PL^3, those semantic attributes that are possible attributes of *propositional atoms* can be uniformly characterized in terms of the fixed attribute $\{a\}$ *resp.* (a, b). By contrast, in both cases all attributes that only emerge in connection with complex formulas defy *uniform* characterization. This fact is perhaps made understandable by the negative character of those attributes: each of them involves the failure of all players to have a strategy of a certain type. The particular reason *why* in each relevant case this failure takes place is *not* a straightforward matter of logical form. The 'explanation' of the failure is sensitive to the syntax of the formula in a less robust way, in a way that precludes the possibility of uniform characterizability.

An important closure property that can be formulated in terms of a designated attribute A_0 is closure under 'complementary negation'. If L is one of the logics PL_0^3 or PL^3, it may be asked whether for every formula φ of L there is a formula $neg(\varphi)$ of L such that every valuation V satisfies: $|neg(\varphi), V| = A_0$ iff $|\varphi, V| \neq A_0$. We conjecture that for neither logic PL_0^3 or PL^3 can an attribute A_0 be so chosen as to admit of this property; settling the issue is left for another occasion.

4.2 Normal forms

De Morgan's laws for PL^2 enable transforming formulas into *negation normal form*, and distributive laws allow putting formulas, already in negation normal form, further into *disjunctive* and *conjunctive normal forms*. All these normal forms admit of analogues in the 3-player setting.

Let us first consider the logic PL_0^3. If i, j are distinct roles and ψ, χ any formulas, the following De Morgan's laws hold:

$$\sim_{ij}(\psi \vee_i \chi) \Longleftrightarrow (\sim_{ij}\psi \vee_j \sim_{ij}\chi)$$
$$\sim_{ij}(\psi \vee_j \chi) \Longleftrightarrow (\sim_{ij}\psi \vee_i \sim_{ij}\chi)$$
$$\sim_{ij}(\psi \vee_k \chi) \Longleftrightarrow (\sim_{ij}\psi \vee_k \sim_{ij}\chi) \quad \text{if } i \neq k \neq j$$

By successive applications of these laws, together with the laws of double negation, any formula can be brought to an equivalent form where negation symbols only appear in literals, and the literals furthermore contain at most two negation signs. Such a formula is said to be in *negation normal form*, and can be formed from literals of the forms p, $\sim_{01} p$, $\sim_{02} p$, $\sim_{12} p$, $\sim_{01}\sim_{02} p$, $\sim_{01}\sim_{12} p$ using binary connectives $\vee_0$, $\vee_1$ and $\vee_2$ only.

Consider, then, the logic PL^3. As already noted in *Subsection 2.2*, the map $\pi_\neg$ is in effect a permutation of roles. For any argument $(r, r', r'') \in \mathbf{P}$, it does the same: *replaces 0 by 1, 1 by 2, and 2 by 0.* Define an operation $\ominus$ by putting $0 \ominus 1 = 2$, $1 \ominus 1 = 0$, and $2 \ominus 1 = 1$. The rule for driving $\neg$ deeper is this: if i is any role, then $\neg(\varphi \vee_i \psi) \Longleftrightarrow (\neg\varphi \vee_{i \ominus 1} \neg\psi)$.

The behavior of $\sim$ is trickier: $\pi_\sim$ is genuinely a permutation of role distributions, as opposed to a permutation of roles. The rule for driving the negation $\sim$ deeper is sensitive to the relative location of the occurrence of $\sim$ within the larger formula considered. To describe the effect of the map $\pi_\sim$, we must take into account *who* is having which role, or to be more precise, we must know which role *Alice* is having. What $\pi_\sim$ does is this: it keeps *Alice*'s role intact, no matter which role she has, while the roles of *Bob* and *Cecile* are interchanged. Since $\sim$ precisely does not affect *Alice*'s role, her role relative to a subformula token $\sim \psi$ only depends on the number of occurrences of the negation $\neg$ to which $\sim \psi$ is subordinate in the relevant larger formula χ.

Let us write $n[\chi, \theta]$ for the number of occurrences of $\neg$ to which a given subformula token θ is subordinate in a formula χ. Define, then, $a[\chi, \theta]$ as the unique number $m \in \{0, 1, 2\}$ such that

$$n[\chi, \theta] \equiv m \pmod 3.$$

It is immediate that (relative to the standard initial role distribution) *Alice*'s role at a subformula token θ in χ is $a[\chi, \theta]$.

If i is a role and $\sim(\varphi \vee_i \psi)$ is a subformula token in χ, put

$$i^* := \begin{cases} i, \text{ if } i = a[\chi, (\varphi \vee_i \psi)] \\ j, \text{ with } i \neq j \neq a[\chi, (\varphi \vee_i \psi)], \text{ otherwise} \end{cases}$$

If it is, e.g., *Bob* who has the role i at $(\varphi \vee_i \psi)$, then i^* is *Bob*'s role at the negated subformula token $\sim(\varphi \vee_i \psi)$. The following rule allows pushing $\sim$ deeper in a formula: if $\sim(\varphi \vee_i \psi)$ is a subformula token in a formula χ, it may be replaced in χ by the formula $(\sim\varphi \vee_{i^*} \sim\psi)$, and the resulting formula will be logically equivalent to χ. Observe that since the result of pushing an occurrence of $\sim$ deeper in a formula χ is sensitive to the relative location of this very occurrence within χ, the rule under consideration actually must be formulated for subformula tokens relative to a given larger formula.

By applying the given rules for driving the negation signs deeper, together with the laws of double negation, triple negation and order invariance, one can produce out of any PL^3-formula a logically equivalent PL^3-formula in which the negation signs $\neg$ and $\sim$ appear on the atomic level only, and which can be built from literals of the forms $p, \neg p, \neg\neg p, \sim p, \neg\sim p, \neg\neg\sim p$ using only the binary connectives $\vee_0$, $\vee_1$ and $\vee_2$. Such a formula of PL^3 is said to be in *negation normal form*.

Let L be either of the logics PL_0^3 or PL^3. Having seen that L admits of a negation normal form, let us proceed to ask whether it allows for analogues of disjunctive and conjunctive normal forms of PL^2. To begin with, straightforward distribution laws hold in L for any distinct roles i and j:

$$\varphi \vee_i (\psi \vee_j \chi) \Longleftrightarrow (\varphi \vee_i \psi) \vee_j (\varphi \vee_i \chi).$$

Provided that (i, j, k) is a triple of pairwise distinct roles, a formula is said to be in (i, j, k) *normal form*, if it has the form

$$\bigvee_i^{x \in I} \bigvee_j^{y \in J_x} \bigvee_k^{z \in K_{xy}} \theta_{xyz},$$

where I, the J_x and the K_{xy} all are finite sets of natural numbers, and θ_{def} is a literal of L, for any assignment of suitable values d, e, f to the variables x, y, z, respectively. For all six permutations (i, j, k) of the set $\{0, 1, 2\}$ and any formula φ of L, there is a logically equivalent formula $\psi_\varphi \in L$ which is in (i, j, k) normal form. This follows immediately from the distribution laws together with the De Morgan's laws.

5 Computing semantic attributes

Model-checking PL^2 can be done in linear time: there is an algorithm which, given a formula φ and valuation V as inputs, decides whether the relation $V \models \varphi$ holds or not; and the number of computation steps the algorithm uses — computation time — is bounded by a linear function of the size of φ. (The size of a formula is the number of tokens of symbols it contains.) It turns out that in this respect, no additional difficulties arise from moving to the logics PL_0^3 and PL^3.

THEOREM 20. *Let L be either PL_0^3 or PL^3. There is an algorithm such that given any L-formula φ and any L-valuation V, the algorithm computes the attribute $|\varphi, V|$ in computation time linear in the size of φ.*

Proof. First note that each L-formula φ has a uniquely determined *syntactic tree*. The root of the tree is the formula φ itself, and its leaves are propositional atoms. Any node $(\psi \vee_i \chi)$ has two immediate successors, namely ψ and χ; and any node $\sim_{ij} \psi$ (case $L := PL_0^3$) *resp.* any node $\neg\psi$ or $\sim\psi$ (case $L := PL^3$) has ψ as its unique immediate successor. Fix, then, a formula $\varphi \in L$ and an L-valuation V. The attribute $|\varphi, V|$ may be determined by labeling the nodes of the syntactic tree of φ. First label each leaf p with the attribute $V(p)$. Then work the way towards the root of the tree by computing the semantic attribute of each complex node from its simpler parts: the attributes of ψ and χ determine in a straightforward way the attribute of $(\psi \vee_i \chi)$, and the attribute of ψ determines the attribute of the result of applying any of the relevant negations to ψ. In this way the formula φ receives a label; this label is the attribute $|\varphi, V|$. Since the number of nodes cannot exceed the size of φ, the algorithm just described runs in time linear in the size of the input formula φ. ∎

As pointed out in the proof of Theorem 20, it is straightforward to formulate recursive rules that determine the semantic attribute of an arbitrary formula of PL_0^3 or PL^3 from the attributes of its atomic components. That is to say, nothing prevents from defining the semantics of these 3-player logics by laying down a set of recursive semantic clauses — in perfect analogy to the usual recursive rules defining the semantics of PL^2. It should, however, be noted that in the case of PL_0^3 and PL^3, the motivation of such recursive rules comes from their relation to the game-theoretically defined semantics.

6   Issues of satisfiability and validity[15]

Let us take a look at some possible further developments in the context of 3-player propositional logic. More specifically, let us see how various issues related to satisfiability and validity can be studied in connection with PL^3.

6.1   Tableaus

What would *proof theory* of PL^3 look like? Let us restrict attention to formulas in negation normal form. A proof system for such formulas can be defined in terms of a set of tableau rules.[16] Our tableaus will deal with 'signed formulas'. There are two basic signs T and F as usual, but they appear relativized both to a player (a, b, or c) and to a feature (g for 'having a g-strategy', s for 'having an s-strategy'): the 12 signs $T_a^g, T_b^g, T_c^g, F_a^g, F_b^g, F_c^g, T_a^s, T_b^s, T_c^s, F_a^s, F_b^s, F_c^s$ are introduced. A *signed formula* is any expression $S_x^y \varphi$ with $S \in \{T, F\}$, $y \in \{g, s\}$ and $x \in \{a, b, c\}$, where φ is a formula of PL^3. A set B of signed formulas is *realizable*, if there is a valuation V such that any signed formula $S_x^y \varphi \in B$ satisfies: player x has ($S := T$) resp. lacks ($S := F$) a y-strategy in game $G(\varphi, V)$. Then V is said to *realize* the set B. From the fact that V realizes B, it does *not* follow that all formulas φ with $S_x^y \varphi \in B$ (for some S, y, x) have the same semantic attribute relative to V. It only follows that the requirements induced by the signed formulas in B can all be simultaneously satisfied by one and the same valuation.

No single sign S_x^y will correspond to a semantic attribute of PL^3. On the other hand, the attributes of PL^3-formulas can be analyzed by reference to the players: an attribute is expressed by specifying for each player his or her optimal minimum rank in the relevant game. It is precisely for dealing with such an analysis that the 12 signs are introduced. An example may clarify the idea. A formula φ receives the attribute (a, b) under all valuations, if for all valuations V, it is impossible for a to fail having a g-strategy in $G(\varphi, V)$, for b to fail having an s-strategy in $G(\varphi, V)$, and for c to have an s-strategy in $G(\varphi, V)$. The idea will be that in order to check whether these three conditions are met, it suffices to see whether the tableaus for the following three signed formulas are 'closed': $F_a^g \varphi, F_b^s \varphi, T_c^s \varphi$. Note that there is no obvious way in which to dispense with the signs F_x^y in favor of the two negations of PL^3 and the signs T_x^y. The tableau for $F_a^g \varphi$, for instance, will have an 'open' branch iff there is a valuation V such that a fails to have a g-strategy in $G(\varphi, V)$. It is far from evident that there are $y \in \{g, s\}$ and $x \in \{a, b, c\}$ such that for all φ there

---

[15] *Sect.* 6 describes continuing work by one of the authors (TT), cf. [15].

[16] For a classical presentation of tableau systems, see [14].

is φ' satisfying: a tableau for $T_x^y \varphi'$ has an 'open' branch iff a tableau for $F_a^g \varphi$ has one. It is an open question whether PL^3 enjoys such a player-relative property of 'closure under complementation'; we conjecture that it does not.

There will be two types of tableau rules: those that allow extending a branch, and those that allow closing a branch, or schematically:

In the left rule, we have $S \in \{T, F\}$, $y \in \{g, s\}$, $x \in \{a, b, c\}$, and $\mathfrak{B}$ is a set of sets of signed formulas. In the right rule, B is a set of signed formulas and X is a specific additional symbol. The rules of the latter kind are termed 'closing rules'. Rules of the former kind will be so chosen that they can be proven to satisfy the following two properties: (P1) The singleton set $\{S_x^y \varphi\}$ is realized by a valuation V *only if* one of the branches $B \in \mathfrak{B}$ is realized by V; and (P2) If some of the sets $B \in \mathfrak{B}$ is realized by a valuation V, then $\{S_x^y \varphi\}$ is realized by V. Each closing rule, in turn, will be chosen so that (Q1) the set B that triggers closing a branch is not realized by any valuation. Furthermore, the totality of all closing rules is so selected that (Q2) if B' is a set of signed formulas not realized by any valuation, it has a subset B to which some closing rule can be applied.

Familiarity with basic notions related to tableaus for PL^2 is assumed (maximal/closed/open branch, extending a branch); the reader may consult e.g. [14] for details. These notions are straightforwardly extended to the case of PL^3, cf. [15].

Tableau rules. Let us introduce a tableau system, to be denoted TS. Its tableau rules are divided into three groups: (1) 'g-rules', (2) 's-rules', and (3) closing rules. Here are the g-rules for the binary connective $\vee_0$:

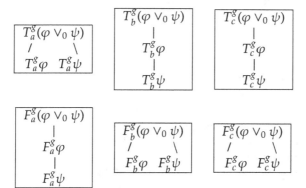

g-rules for $\vee_1$ ($\vee_2$) are entirely analogous: the only branching T-rule is the rule for T_b^g (*resp.* T_c^g), and the only non-branching F-rule is the one for F_b^g (*resp.* F_c^g). The s-rules for $\vee_0$ are as follows:

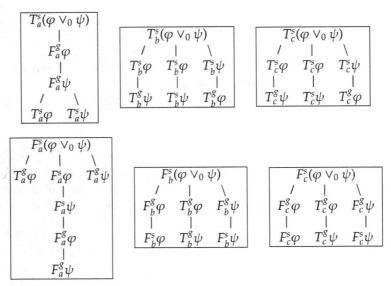

s-rules for $\vee_1$ ($\vee_2$) are again completely analogous. Now it can be shown that both g-rules and s-rules satisfy the properties (P1) and (P2) mentioned in the beginning of the present subsection.

Within the confines of this article, we cannot formulate a full collection of closing rules with properties (Q1) and (Q2). Therefore we content us with an entirely 'non-constructive' proof to the effect that a suitable set of closing rules *exists*. Let ℓ be a fixed literal, and let $\{\ell_1, \ldots, \ell_6\}$ be a basic set of p-literals, for a fixed atom p. Define $B_\ell := \{S_x^y \ell : S \in \{T, F\}, y \in \{g, s\} \text{ and } x \in \{a, b, c\}\}$, and write B_p for the set of sets of signed formulas obtained by assigning one of the 12 possible signs to all members of some non-empty subset of $\{\ell_1, \ldots, \ell_6\}$. Then the size of the set $B_\ell \cup B_p$ is 4,095+4,826,808 = 4,830,903. *A fortiori*, the number of closing rules needed for meeting condition (Q2) cannot exceed 4,830,903. Hence there unavoidably exists a set of closing rules such that each rule individually satisfies (Q1), and they jointly satisfy (Q2). As a matter of fact — luckily enough! — we can do with only a handful of closing rules; for details, cf. [15].

We may, then, be convinced that a suitable set TS of tableau rules can be formulated. What are these rules applied to? As an *input* of a tableau we consider (finite) sets of signed formulas. If C is a set of signed formulas, a *tableau for C* is the result of applying the tableau rules of TS to formulas of the set C, and to the formulas thereby recursively generated, until all branches produced are maximal, i.e., cannot be further extended by the tableau rules. A tableau for a single signed formula $S_x^y \varphi$ is by definition a tableau for the set $\{S_x^y \varphi\}$. There are in general several tableaus for a given set of signed formulas. The differences between these tableaus are, however, immaterial for our purposes: If τ, τ' are both tableaus for

C, then τ is closed if and only if τ' is closed; and τ has an open maximal branch if and only if τ' has an open maximal branch.

6.2 Collection of sound and complete proof systems

If P is one of the 16 semantic attributes, a formula is *P-valid*, if it has the attribute P under all valuations. Dually, a formula is *P-satisfiable*, if it has the attribute P under some valuation. For every attribute P and every formula φ, there are sets $C_{P,\varphi}^{VAL}$ and $C_{P,\varphi}^{SAT}$ — each consisting of three signed formulas — with the following properties: φ is P-valid iff each signed formula in the set $C_{P,\varphi}^{VAL}$ individually has a closed tableau; and φ is P-satisfiable iff there is a closed tableau for the set $C_{P,\varphi}^{SAT}$. For instance, for the attribute $P_{(a,b)}$ the corresponding sets are $C_{(a,b),\varphi}^{VAL} :=$ $\{F_a^g\varphi, F_b^s\varphi, T_c^s\varphi\}$ and $C_{(a,b),\varphi}^{SAT} := \{T_a^g\varphi, T_b^s\varphi, F_c^s\varphi\}$.

A *P-proof* of a formula φ is by definition a triple of closed tableaus, one for each signed formula in the set $C_{P,\varphi}^{VAL}$. It will be said that our tableau system TS is *P-sound*, if actually every formula with a P-proof is P-valid. And TS is *P-complete*, if every P-valid formula has a P-proof.

THEOREM 21 (*P*-soundness, *P*-completeness). *For each of the 16 semantic attributes, P, TS is P-sound and P-complete.*

Proof. Let an attribute P and a formula φ be given. For P-soundness, it suffices to prove that if φ is *not* P-valid, then a tableau for one of the signed formulas in $C_{P,\varphi}^{VAL}$ is *not* closed. This can be done thanks to the properties (P1) and (Q1) of TS. For P-completeness, it is enough to show that if a tableau for one of the signed formulas in $C_{P,\varphi}^{VAL}$ is *not* closed, then φ is *not* P-valid. This, again, can be done by virtue of the properties (P2) and (Q2) of TS. For details, cf. [15]. ∎

6.3 Decidability issues

If P is one of the 16 attributes, the *P-satisfiability problem* (or *P-SAT*) is by definition the problem of deciding whether a given formula of PL^3 is P-satisfiable. Similarly, the *P-validity problem* (or *P-VAL*) is the problem of deciding whether a given formula is P-valid. Actually, each of these 32 decision problems is decidable. What is more, the 16 satisfiability problems are decidable in **NP** and the 16 validity problems in **coNP**.

THEOREM 22. *Let P be any PL^3-attribute. Then P-VAL is in **coNP** and P-SAT is in **NP**.*

Proof. Consider the attribute (a, b); the claim can be similarly proven for the rest of the attributes. Let φ be any formula of PL^3. To decide whether φ is (a, b)-valid, first non-deterministically guess a valuation V. (The guess will yield a counterexample to the claim that φ is (a, b)-valid, if such a counterexample exists.) Then apply the polynomial time algorithm described in *Section* 5 to compute the attribute $|\varphi, V|$. If $|\varphi, V| = (a, b)$, then φ is (a, b)-valid, otherwise not. We have just described an algorithm that solves (a, b)-VAL and runs in **coNP**. Consider, then, (a, b)-SAT. Given a formula φ, non-deterministically guess a valuation V. (The guess will yield a witness to the claim that φ is (a, b)-satisfiable, if such a witness exists.) Then apply the polytime algorithm of *Section* 5 to determine whether $|\varphi, V| = (a, b)$. If yes, φ is (a, b)-satisfiable, otherwise not. The algorithm just described solves (a, b)-SAT and runs in **NP**.   ∎

It is possible to prove, at least for some semantic attributes, that the corresponding validity problem is **coNP**-*hard* and the corresponding satisfiability problem **NP**-*hard*.[17] In those cases, then, we have **NP**-complete and **coNP**-complete decision problems about the logic PL^3.

7   Concluding remarks

What has been accomplished in the present paper? Looking at the outcome, a minimalist answer would be that we have introduced two 'systems' of the general form $(\mathcal{L}, \mathcal{V}, A)$, where $\mathcal{L}$ is a set of 'formulas', $\mathcal{V}$ is a set of 'valuations', and $A : \mathcal{L} \times \mathcal{V} \longrightarrow \mathbb{N}$ is a function, mapping pairs of 'formulas' and 'valuations' to encodings of what we called 'semantic attributes'.

For each of the two systems, PL_0^3 and PL^3, the corresponding function A was, as a matter of fact, determined by reference to certain 3-player games. These games were obtained by generalizing in certain respects 2-player evaluation games of classical propositional logic, PL^2. Given this background, the two systems were thought of as 'logics' in an extended sense of the word. As evidenced by the body of the present paper, viewing PL_0^3 and PL^3 as logics is at least heuristically justified — virtually any question that can be asked of PL^2 can be formulated in connection with the two systems. Studying the systems PL_0^3 and PL^3 from a logical viewpoint may even be seen as throwing light on PL^2 itself. For, properties of PL^2 can be classified according to whether they do or do not survive the transition to the 3-player setting. The former properties can be regarded as 'robust'. Applying these criteria, for example model-checking in linear time is a robust property of PL^2 (*Sect.* 5), while uniform characterizability of semantic attributes is not (*Sect.* 4).

Various specific questions suggest themselves for future work. Algebraic perspective on multi-player logics may turn out to be of considerable interest. It is natural to study consequence relations proper to these logics in this connection. Further, multi-player logics offer novel ways of introducing game-theoretical notions into logical contexts — a case in point is the notion of security level, referred

---

[17]In [15] it is proven that whenever $P \in \{(a, b), (a, c), (b, a), (b, c), (c, a), (c, b)\}$, P-SAT is **NP**-hard and P-VAL is **coNP**-hard. The proof is based on showing that the satisfiability problem of PL^2-formulas in conjunctive normal form can be reduced in polynomial time to P-SAT, and that the validity problem of PL^2-formulas in disjunctive normal form has a polytime reduction to P-VAL.

to in Remark 3. Indeed, the framework of multi-player logic games calls for a systematic study of correspondences between theorems about multi-player logics and game-theoretic principles.

Many-valued logics and n-player logic games could be compared. In particular, one can ask for which many-valued logic PL_0^3 resp. PL^3 provides an alternative, game-theoretical semantics. (The precise formulation of this question is dependent on how consequence relations are defined for these 3-player logics.) Also, it might be of interest to study the variant of game-theoretical semantics for PL^2 with 4 possible payoffs: both players win, no player wins, and one of the two wins. The resulting logic could then be compared with Belnap's and Dunn's 4-valued logics [3, 7].

Acknowledgments

The first author wishes to thank Arcady Blinov for the thought-provoking suggestion (personal communication, 2002) that the credibility of a theory of meaning involving intrinsically game-theoretic conceptualizations — such as game-theoretical semantics — would be undermined by the possibility of generalizations — such as 3-player semantic games — which by themselves have nothing to do with theorizing about meaning.

The authors are thankful to the two anonymous referees, who helped to improve the paper by their most useful suggestions and criticisms.

The research of the first author was carried out within the project "Modalities, Games and Independence in Logic" funded by the Academy of Finland. The research of the second author has been made possible by a grant of the Netherlands Organization for Scientific Research (NWO) on the vici project "Algebra and Coalgebra".

BIBLIOGRAPHY

[1] S. Abramsky. Socially responsive, environmentally friendly logic. In T. Aho & A.-V. Pietarinen, editors, *Truth and Games*, pages 17–45. Acta Philosophica Fennica, 2006.

[2] S. Abramsky. A compositional game semantics for multi-agent logics of partial information. In J. van Benthem, D. Gabbay & B. Löwe, editors, *Interactive Logic. Selected Papers from the 7th Augustus de Morgan Workshop, London*, pages 11–47, Amsterdam University Press, 2007.

[3] N. Belnap. A useful four-valued logic. In G. Epstein and J. M. Dunn, editors, *Modern Uses of Multiple-Valued Logic*, pages 8–37. Reidel, 1977.

[4] F. Brandt, F. Fischer, and Y. Shoham. On strictly competitive multi-player games. In Y. Gil and R. Mooney, editors, *Proc. of the 21st National Conf. on Artificial Intelligence (AAAI)*, pages 605–612. AAAI Press, 2006.

[5] F. Brandt, F. Fischer, P. Harrenstein, and Y. Shoham. A game-theoretic analysis of strictly competitive multiagent scenarios. In M. Veloso, editor, *Proc. of the 20th Int. Joint Conf. on Artificial Intelligence (IJCAI)*, pages 1199–1206. AAAI Press, 2007.

[6] M. Dummett. *Truth and Other Enigmas*. Duckworth, 1978.

[7] J. M. Dunn. Partiality and its dual. *Studia Logica*, 66(1):5–40, 2000.

[8]   D. Gale and F. M. Stewart. Infinite games with perfect information. In H. W. Kuhn and A. W. Tucker, editors, *Contributions to the Theory of Games II*, Annals of Mathematics Studies 28, pages 245–266. Princeton University Press, 1953.

[9]   L. Henkin. Some remarks on infinitely long formulas. In *Infinitistic Methods*, pages 167–183. Pergamon Press, 1961.

[10] J. Hintikka. Language-games for quantifiers. In *Americal Philosophical Quarterly Monograph Series 2: Studies in Logical Theory*, pages 46–72. Blackwell, 1968.

[11] P. Lorenzen. Logik und Agon. In *Atti del XII Congresso Internazionale di Filosofia (Venezia, 1958)*, pages 187–94. Sansoni, 1960.

[12] P. Lorenzen. Ein dialogisches Konstruktivitätskriterium. In *Infinitistic Methods*, pages 193–200. Pergamon Press, 1961.

[13] U. Schwalbe and P. Walker. Zermelo and the Early History of Game Theory. *Games and Economic Behavior* 34: 123–37, 2001.

[14] R. M. Smullyan. *First-Order Logic*. Dover, 1995 (first appeared in 1968).

[15] T. Tulenheimo. Satisfiability and validity from a 3-player perspective. Unpublished manuscript, 2007.

Elements for a Rhetoric of Mathematics: How Proofs can be Convincing

JEAN PAUL VAN BENDEGEM

1 Rhetorics, old and new, and mathematics: never the twain shall meet

Both rhetorics and mathematics are ancient, elaborate, and still active fields of study and cover a time span of two millennia. That, at least, they have undisputedly in common. However, in the domain of mathematics one will search in vain for traces, positive or negative, of rhetorics, and in the domain of rhetorics, although the relation between mathematics and rhetorics is often discussed, the standard claim is to deny that they are intimately related or intertwined. Moreover, things have hardly changed over two millennia. Let me present two examples, slightly biased I freely admit, one belonging to the so-called old rhetoric and one belonging to the equally so-called new rhetoric.

My first example is the famous book by Quintilian, the Roman orator, *Institutes of Oratory*.[1] (Of course, numerous other choices were possible, but it is generally accepted that Quintilian's book should be included in any overview of rhetoric.) First it is worth noting that mathematics is not intensely discussed throughout the book. And, secondly, what is being said is quite meagre. The most interesting passage is to be found in chapter ten of the first book, where two statements are put forward:

> (a) Mathematics, in this case basically arithmetic and geometry, is important in the court room to avoid making silly mistakes. Making a calculating error creates a very bad impression and insufficient geometrical knowledge can lead to painful and costly mistakes.[2] Here mathemat-

---

[1] I have used an on-line edition, to be found at:
http://honeyl.public.iastate.edu/quintilian/index.html.

[2] These are the actual wordings of Quintilian (paragraphs 39-41): "**39**. Geometry often, moreover, by demonstration, proves what is apparently true to be false. This is also done with respect to numbers, by means of certain figures which they call *pseudographs*, and at which we were accustomed to play when we were boys. But there are other questions of a higher nature. For who would not believe the asserter of the following proposition: "Of whatever places the boundary lines measure the same length, of those places the areas also, which are contained by those lines, must necessarily be equal?" **40**. But this proposition is fallacious, for it makes a vast difference what figure the boundary lines may form, and historians, who have thought that the dimensions of islands are sufficiently indicated by the space traversed in sailing round them, have been justly censured by geometricians. **41**. For the nearer to

ics is reduced to required background knowledge and, in that sense, there is surely no rhetoric of mathematics, save perhaps the rather modest rule: "Keep the number of errors as small as possible."

(b) Occasionally mathematics does occur within an oration itself. What happens then? The following quote is quite clear on the matter: "Besides, of all proofs, the strongest are what are called geometrical demonstrations, and what does oratory make its object more indisputably than proof?" (Book 1, chapter 10, paragraph 38, second sentence) In other words, mathematical proof is the highest standard to aim for.

Both (a) and (b) show the special status of mathematics. If it intervenes, it does so with certainty, with conviction, with not-being-questioned. Obviously no room for rhetorical considerations within the mathematical realm. Quite the contrary: whenever some mathematical argument enters into the discourse, for a brief moment all rhetoric can be suspended and for a short time, clarity and transparancy reign.

My second example comes from the work of Chaïm Perelman and Lucie Olbrechts-Tyteca, who are credited with the idea of the "new rhetoric". It is true that, in an essential way, new approaches and conceptions of rhetoric are explored, but, strangely enough, when it comes down to mathematics (including also logic in their case), one must unfortunately notice that mathematics keeps its special status, witness the following quote:

"Le langage artificiel des mathématiciens fournit, depuis des siècles, à beaucoup de bons esprits, un idéal de clarté et d'univocité que les langues naturelles, moins élaborées, devraient s'efforcer d'imiter. Toute ambiguÃ⁻té, toute obscurité, toute confusion sont, dans cette perspective, considérées comme des imperfections, éliminables non seulement en principe, mais encore en fait. L'univocité et la précision de ses termes feraient du langage scientifique l'instrument le meilleur pour les fonctions de démonstration et de vérification, et ce sont ces caractères que l'on voudrait imposer à tout langage."[3] (Perelman and Olbrechts-Tyteca [10], pp. 174-175)

It is quite remarkable that both authors seem to differ so little from Quintilian. At first sight, the label "new" does not seem to apply to the role mathematics can play in a rhetorical setting. However, it is perhaps necessary to explain why Perelman and Olbrechts-Tyteca remain so close to Quintilian. As it happens, it is an ingredient of their strategy to establish rhetoric and argumentation theory as independent fields of research. Or, in other words, by breaking the ties between everyday reasoning and arguing and mathematical reasoning as an ideal to strive for,

---

perfection any figure is, the greater is its capacity, and if the boundary line, accordingly, shall form a circle, which of all plane figures is the most perfect, it will embrace a larger area than if it shall form a square of equal circumference. Squares, again, contain more than triangles of equal circuit, and triangles themselves contain more when their sides are equal than when they are unequal."

[3]"The artificial language of mathematics presents, since centuries, to many fine minds, an ideal of clarity and unicity that natural languages, less developed, should do their best to imitate. All ambiguity, all opacity, all confusion are, from this perspective, seen as imperfections, to be eliminated not merely in principle, but also in fact. The unicity and the precision of its terms turn the scientific language into the best instrument for demonstrations and verifications, and these are the characteristics one would like to see imposed on any language." (my translation)

it follows that argumentation and rhetoric have to be studied on their own without reference to mathematics. Ironically, this strategy implies that mathematics now becomes something different altogether, an extremely particular language game, practised by a happy few in their own special environments.

As the title of this paper indicates, I want to defend the thesis in this paper that (a) a rhetoric of mathematics is possible, and (b) it is necessary (or at least desirable) to formulate such a rhetoric in order to understand what happens when mathematical proofs turn up in everyday life. The focus will be on part (a), which will be elaborated in sections 2, 3 and 4. In section 5 I will briefly deal with part (b) of the thesis: what interests me more specifically is the problem why correct mathematical arguments often tend to loose their force, when they end up in non-mathematical contexts.

2   A mathematical proof to start

Imagine one is presented with the following problem. Today is day 1 and you receive 1 euro. Tomorrow, day 2, you receive 1/2 euro. Day 3, you get 1/3 euro (we make a slight abstraction here, assuming that the sum of 0,33333... euro actually exists). Generally speaking, on the n-th day, $1/n$ euro is given to you. Assume finally that this process continues until the end of times, in this case infinitely far away. How rich will you be? I guess that most, if not all non-mathematicians will wonder if the term rich should not be put between quotation marks, as this is not a successful road towards wealth. And, yet, as mathematicians know, at the end of times, one will be unavoidably filthy rich, since one will have an infinite amount of money.

Let the mathematician step forward to explain what is going on and this is the most likely story she will present to a non-mathematical audience. The question that is being asked here is: what is the sum of

$$S = 1 + 1/2 + 1/3 + 1/4 + 1/5 + ... + 1/n + ...,$$

because that is the total amount you will receive in the end.
Now look at the following curious phenomenon:

$1 + 1/2 > 1/2$

$1/3 + 1/4 > 1/4 + 1/4 = 1/2$

$1/5 + 1/6 + 1/7 + 1/8 > 1/8 + 1/8 + 1/8 + 1/8 = 1/2$

$1/9 + 1/10 + 1/11 + 1/12 + 1/13 + 1/14 + 1/15 + 1/16 >$
$1/16 + 1/16 + 1/16 + 1/16 + 1/16 + 1/16 + 1/16 + 1/16 = 1/2$

and so on.
In other words, by a correct grouping of the terms in S, one sees that

$$S > 1/2 + 1/2 + 1/2 + \ldots,$$

counted an infinite number of times, hence S = infinite.

No doubt, all mathematicians will end the proof with QED. But has the statement actually been proved? Let me rephrase that question in a more precise manner: is this proof convincing for everyone who goes through it, whereby "everyone" includes mathematicians and non-mathematicians[4] alike? The answer seems to be "no".

Without any doubt the mathematician will remark that roughly the proof looks OK, but that it needs cleaning-up a bit here and there. To name but one sloppy element in the proof: there are conditions to be satisfied when grouping terms in a series (which happen to be satisfied in this particular case), that should be mentioned in a decent version of the proof. It is clear, however, that the mathematician will be convinced by this proof. The mere fact that he or she requires a "clean-up" indicates that there is something to be cleaned-up in the first place and what else can that be but a proof?

How about the non-mathematician? There are in a straightforward way several reasons one can imagine why this proof is not convincing at all (sidestepping the simple lack of elementary mathematical competencies such as numbers, addition, fractions, comparing numbers, and so forth):

(a) There is an obvious contradiction or, at least, conflict between what the theorem says and our relevant intuitions: how is it possible to become infinitely rich by getting less every day. It simply does not make any sense.[5]

(b) Reasonings involving infinities connect very badly, if at all, with everyday reasoning. This is reflected in the fact, sustained by personal teaching experience, spanning 25 years and on average about 800 students per year, that, if one is clever enough, one can prove almost anything one likes.[6] Given this bad connection, it follows that one does not have straightforward standards

---

[4]The negative description "non-mathematician" is meant as neutral as possible. I object to the use of the expression "layperson", as it carries too many religious connotations, including a hierarchical relation between layperson and expert. This being said, "non-mathematician" is not a success either.

[5]This observation is supported by the fact that in mathematics courses the harmonic series is often presented as a fine case to show that, given a series $S = a_1 + a_2 + \ldots + a_n + \ldots$, S need not be convergent if the condition $a_n \to 0$, for n going to infinity, is satisfied. So, it is assumed that students might think otherwise.

[6]A typical reaction is that the student spontaneously presents a clearly invalid argument that is believed to correspond to a correct form of reasoning, such as: "I fit in my pyjamas, my pyjamas fit in my suitcase, hence I fit in my suitcase", hence demonstrating that everything can be proven correct (being of a friendly disposition, one does not point out immediately the self-refuting nature of this line of reasoning). As it happens, it does take time to convince a student that formal logic is the tool par excellence to show why the argument is indeed invalid. One has to suppose that the relation "fits in" is transitive, which is obviously not the case.

to judge and evaluate such reasonings, hence they appear to a large extent to be arbitrary, hence the distrust.

(c) Related to the previous point, is the observation that the proof does not explain why the terms of the series S are grouped in this particular way. (Note, of course, that the mathematician will reply that a proof is not supposed to do such a thing, which is correct, but is differently perceived by the non-mathematician.) One suspects a form of "backwards reasoning": start from what you want to obtain and construct the proof so that that result is obtained. Put differently, one is not convinced that there could not be a different grouping that shows S to be convergent.[7] In a curious way, the mathematician is here perceived corresponding to the standard, hence heavily biased image of the sophist: prepared to defend any position whatsoever (if payment is alright, that is).

These three reasons give a first insight as to why this beautiful mathematical proof ceases to be convincing in particular settings. It is my firm belief that the mentioned reasons are merely the tip of the often-quoted iceberg, and it is my claim that the best way to describe the iceberg is by reference to a rhetoric of mathematics.

3 Implicitly shared practices and characteristics of style

Let us have a second look at the above proof on a more detailed level. It is clear that a set of practices, conventions and customs need to be shared:[8]

(a) A symbol introduced in the course of a proof, maintains its meaning throughout the proof. At the same time, it is perfectly acceptable, if within a proof a subproof is started, to use the same symbol, yet with a different meaning. In short, a subtle play of signs, symbols and meanings.

(b) Steps in a proof that are considered evident are contracted into a single step. Let me illustrate this point straightaway: have another look at the proof and notice that I used the obviously non-mathematical expression "and so on". To replace this expression by a (more) correct proof would make the proof quite elaborate and complex, thereby less transparant – quite paradoxical, in fact! – and therefore it is accepted as part and parcel of mathematical practice. The extreme variations on this theme are, of course, "trivial" and "proof left to the reader".

---

[7]In a way this is of course correct: for all we know, the theory of convergence and divergence of series may well turn out to be inconsistent and then obviously anything is provable. That, however, is not what the non-mathematician has in mind.

[8]The details mentioned here are well-known. One of the first formulations I saw, was Karl Menger [7], but these considerations have been presented on countless occasions.

(c) Related to (b), is the fact that for every mathematical domain there exists a set of statements, although not axioms, which can be used without proof, because this set is supposed to be shared background knowledge. Or, if challenged to produce a proof, one sees immediately how that should be done. If, e.g., one is working in number theory, then there is no need at all to separately prove that the sum of two odd numbers is even (although a full proof is not trivial at all[9]).

On the basis of these few characteristics, it seems quite acceptable to conclude that mathematical texts, proofs in particular, are very specific texts, aimed at a very specific public, namely mathematicians (and often a subset of this set), and have specific properties that are not reducible to purely formal aspects. This last statement is rather easy to check. Take any mathematical manual or textbook. The first chapter comes very close to the formal(ist) ideal: all steps in proofs are explicitly mentioned, everything is presented as transparantly as possible. Here only formal characteristics are at work. But as the number of the chapter increases, so does the distance between real mathematical text and formal ideal. The reason why this happens is indeed straightforward, as already mentioned: if all steps are written out in full, proof length increases exponentially.[10] So one is simply forced to "summarize" proofs. But do note that there is not a unique way of achieving this, so choices have to be made. In addition, these summarized parts need to be "glued" together, so these binding texts become crucially important.

A useful comparison is a cookery book. Imagine such a book meant to be helpful for the fully uninitiated. Imagine all the actions that need to be mentioned

---

[9]Suppose you are challenged to produce a proof. What would it look like? Something like this, I imagine: an odd number can be written as $2n + 1$. So, given $2n + 1$ and $2m + 1$, their *sum* $= (2n + 1) + (2m + 1) = 2(n + m) + 2 = 2(n + m + 1)$, which is an even number. However a critical spirit will not be satisfied because she is left with the following questions: (a) why does an odd number always have the form $2n + 1$?, (b) why is it alright to move from $(2n+1) + (2m+1)$ to $2(n+m+1)$? There is a very subtle play going on with brackets, associativity and commutativity: $(2n + 1) + (2m + 1) = 2n + (1 + (2m + 1)) = 2n + (1 + (1 + 2m)) = 2n + ((1 + 1) + 2m) = 2n + (2 + 2m) = 2n + (2m + 2) = \underline{(2n + 2m) + 2} = 2(n + m) + 2 = 2(n + m + 1)$. Note that in the summary-proof above only the underlined statement has been mentioned, (c) why, if an even number is represented as $2k$, is it correct to conclude that $2(n + m + 1)$ is also an even number? After all, I have to be certain that, for every term $n + m + 1$, there is a k, such that $k = n + m + 1$. Not trivial at all, because this question is related to the closure of addition. Perhaps one is inclined to think "much ado about nothing", but there exists plenty of alternative arithmetical structures where one of these properties does not hold and the proof does not carry through. So the critical spirit is quite right to be critical.

[10]A rather amusing but at the same time entirely convincing example: the Bourbaki attempt to rewrite mathematics in a clear and explicit formal manner. Nevertheless they had to use summarized proofs when things became sufficiently complex. At one particular moment in the treatises, the number 1 is defined and the authors mention the fact that this definition could be written out in full, but then it would consist of a couple of tens of thousands symbols. In the curious paper [6] the author performs the actual calculation for the number of symbols needed and the answer is (exactly!): 4.523.659.424.929, a number a bit above ten thousand.

explicitly and in full detail (just imagine having to deal with the existing variety of cookers). Instead real cookery books provide summaries ("Light a fire" applies to all cookers). What determines its quality is precisely the balance between what is left out and what remains. Otherwise a single instruction should do the trick: "Prepare dish". The comparison is indeed useful as often mathematical proofs are seen as recipes. So what goes for recipes, holds for proofs.

The thoughts formulated here, are anything but refreshingly new. They are to be found in many places, among others in the philosophy of the Signific Movement, a group having its peak moment in the first quarter of the twentieth century. There are a number of parallels to be drawn with the logical empiricism of the Wiener Kreis, but just as many differences, e.g., in the idea of Gerrit Mannoury, one of the central figures, that what language is, how language functions, should also be applicable to the mathematical language. This creates a continuum with ordinary, everyday language at one side and mathematical and logical languages at the other. The movement from one side to the other requires "language-steps":[11]

> "All language means, used by members of a society, can be divided up in language steps, that differ gradually from one another. This occurs on the basis of regularities related to the use and, above all, to the interconnections and resulting stability of the meaning of signs and speech acts. With every step these regularities become more important; in parallel run a stepwise social differentiation and stability of meaning. The resulting language steps serve specific communicative aims and the formulation of specific contents; furthermore they shape the social relations between individuals, who, while communicating, use a particular language step." ([13], p. 352, my translation)

At first sight what is put forward here seems relatively trivial, at least what the quote says is generally accepted in socio-linguistics today. After all, is more being said than the statement that ordinary language is spoken by everyone, whereas mathematical language is only used by a very select subgroup in society (and simultaneously serves as a means of identification)? True, if that were the only thing the quote above claims. However, there is the additional aspect of the stability of meaning.

Stability of meaning implies increased clarity and therefore a decreased ambiguity of meaning. In its turn this implies that, if we had the thought of striving

---

[11]I quote from the excellent historical survey, Schmitz [13], as it still stands on its own as a presentation of Signific thinking. Apart from Mannoury, most prominent members were the author and philosopher Frederik van Eeden and the mathematician-mystic L.E.J. Brouwer (at least for some time). Mannoury corresponded with Otto Neurath, one of the core members of the Wiener Kreis. This also establishes an immediate connection with the work of Shahid Rahman, see, e.g., the introduction to the first volume in the LEUS series, [11].

towards a full mutual understanding, we would have to eliminate all ambiguities, but that implies that we end up in a higher language step and that finally implies social differentiation. Conclusion: perfect mutual understanding can only occur in very select, very small groups, preferably isolated from the rest of society. It is obvious that such a view goes in the opposite direction from any attempt to spread a disambiguous language worldwide and to promote it as a new "universal" language. Any such project is considered doomed right from the outset from the perspective of the Significs.[12]

The next obvious question is what guarantees this meaning stability and my proposal is: the intricate network of mainly implicit, occasionally explicit practices that ensure that a particular text is recognized, read and interpreted as a mathematical proof. It is crucial to note that these practices have often little to do with formal correctness – in most cases there is little or no doubt about the proof being a proof, i.e., it shows that the mathematical statement is indeed the case – but rather with power of persuasion, i.e., how "well" it is done. If a name needs to be found, "style" seems the best choice. And, if style can be considered as an essential ingredient of a rhetoric, then we are well on our way to develop a rhetoric of mathematics. What kind of rhetoric I have in mind, will be dealt with at the end of this paper. Let me, for the moment, strengthen my case.

4   Additional evidence

In this part of the paper I will present four examples that, I hope, will show what further elements are needed to formulate a full-fledged theory of a rhetoric of mathematics. It is too early at this stage to present such a theory itself, as too much work remains to be done.

First example   Mathematicians themselves make a clear distinction between "ordinary" proofs and "excellent" proofs. In its turn this is closely related to "ugly" and "beautiful" proofs. Excellence implies beauty. A perfect example is [1]. The book's title, *Proofs from the Book*, is itself a reference to a statement of the Hungarian mathematician Paul Erdös. The Book refers to a book kept by God himself, containing all perfect proofs. What are the properties of a perfect proof? They must be surprising, clever, ingenious, and, preferably, concise. In addition they require a "grasping time", i.e., the proof must be such that the reader needs quite a bit of time to grasp the proof, but, once one has grasped it, one must experience something more than merely the idea that the proof is correct. One is impressed and illuminated. From a rhetorical point of view, it is important to note that the excellence-beauty connection (often) goes together with an emotional experience. Both from my own experience and mathematical literature, I can testify that it is

---

[12]I refer to my [16] for more details on the failed attempts to find such a universal language.

indeed the case that a mathematical proof can be "thrilling" and "touching".[13]

As an example take the proposition that the number of primes is infinite. We all know the standard proof:

Suppose to the contrary that the number of primes is finite and let these be $2, 3, 5, ..., p$, where p is the last prime number. Form the number $N = 2 \cdot 3 \cdot 5 ... p + 1$. There are two possibilities: either N is a prime itself and then we have a prime number larger than p, hence the list is incomplete. Or N is not a prime number, but then it is composed of prime factors, say some factor q. Necessarily q must differ from the given prime numbers as N is not divisible by any of them. So once again the list is incomplete. As each possibility leads to the same result, we must conclude that the list is incomplete.[14] *QED*

Here is a proof that is generally considered to be more excellent, more beautiful. It connects the problem of the number of primes with the divergence of the harmonic series, $1 + 1/2 + 1/3 + 1/4 + 1/5 + ... + 1/n + ...$, presented at the beginning of this paper. I will suppose that the theorem that any number n can be uniquely written as a product of prime factors is known. Suppose once again that the number of primes is finite: $2, 3, 5, ..., p$. An arbitrary number n will consists of a product of primes selected from this list, raised to some power. So, $6 = 2 \cdot 3, 24 = 2^3 \cdot 3, 36 = 2^2 \cdot 3^2$, and so on.

Now look at these expressions:

$$1 + 1/2 + 1/2^2 + ... + 1/2^n + ...$$
and $$1 + 1/3 + 1/3^2 + ... + 1/3^n + ...,$$
$$...,$$
up to $$1 + 1/p + 1/p^2 + ... + 1/p^n + ...$$

Multiply all these expressions. What do you get? A general term will consist of a multiplication, whereby one element is selected of each expression, in short, a multiplication of prime factors raised to some power. That simply means that a general term has the form $1/n$, for an arbitrary number, so the product of all these expressions is equal to the harmonic series! In short:

$$(1 + 1/2 + 1/2^2 + ... + 1/2^n + ...)(1 + 1/3 + 1/3^2 + ... + 1/3^n + ...)...(1 + 1/p + 1/p^2 + ... + 1/p^n + ...) = 1 + 1/2 + 1/3 + 1/4 + 1/5 + ... + 1/n + ...$$

The next step relies on a well-known result, viz., the proof that an infinite geometric series, starting with 1 and with factor r has a sum $1/(1 - r)$. This in itself

---

[13]This is probably one of the reasons why mathematics is so often compared to poetry. Whether such comparisons are apt or not, requires a separate paper.

[14]Note that this proof has been written down by the author of this paper, who has been and is strongly influenced by formal logic. Hence it is important (for me) to stress the formal structure of the proof: a dilemma of the form $p \vee \sim p, p \supset q, \sim p \supset q \vdash q$. See the difference with this randomly chosen proof from [2]: "There exist infinitely many primes, for if $p_1, ..., p_n$ is any finite set of primes then $p_1 ... p_n + 1$ is divisible by a prime different from $p_1, ..., p_n$." (p. 4) No sign of dilemma or excluded third being used!

is a nice problem giving rise to beautiful proofs. A classic way is to note that, if
$R = 1 + r + r^2 + ... + r^n + ...$, then $rR = r + r^2 + ... + r^n + ...$, so $R - rR = 1$, so
$R \cdot (1 - r) = 1$, hence $R = 1/(1 - r)$.

This means that:

$$1 + 1/2 + 1/2^2 + ... + 1/2^n + ... = 1/(1 - 1/2) = 2 \qquad \text{(n.b. } r = 1/2),$$
and $\quad 1 + 1/3 + 1/3^2 + ... + 1/3^n + ... = 1/(1 - 1/3) = 3/2 \qquad \text{(n.b. } r = 1/3),$
...,
and $\quad 1 + 1/p + 1/p^2 + ... + 1/p^n + ... = 1/(1 - 1/p) = p/(p - 1) \quad \text{(n.b. } r = 1/p).$

And now for the "finale":

Since each of the expressions corresponds to a finite number and the finite multiplication of such numbers is itself finite, it follows that the harmonic series must
have a finite sum, but it does not, hence the number of primes must be infinite.
QED

The superiority of this proof is clearly related to the unexpected connection
between, on the one hand, the infinity of the primes and, on the other hand, the
divergence of the harmonic series. So, if someone asks why the harmonic series
diverges, one can answer quite seriously "because there are an infinite number of
primes". And vice versa, so these two statements are equivalent. In this sense it
generates a surprising and deep insight in the structure of, in this case, number
theory.

Second example In the second half of the previous century computer programs
have in diverse ways made their entry in the search for proofs. Whether or not
this is to be considered a success, it has its place in present-day mathematical
practice. One of the well-known uses is to have the computer program check a
finite, but large set of cases for a specific property. Often mathematicians reduce
an infinite set equivalently to a finite set, i.e., if all members of the finite set have
a certain property, so does the infinite set. Checking the set of finite cases thereby
guarantees the correctness of the proof.[15] It is a separate issue, not to be dealt with
here, whether such proofs are reliable or not, a discussion still going on within
mathematics and the philosophy of mathematics. The main problem is, of course,
how to check a program.

Another well-known use, not to be treated in this paper, is the role of the computer as a source of inspiration. Imagine a geometer investigating topological properties of geometrical surfaces, it is obvious that she can learn many things (within
certain boundaries of reliability, of course) from even approximate visualisations.
This would lead me straightaway not merely into a discussion about computer
drawings, but into the more general topic of the use, relevance and necessity of
drawings and diagrams.

What I do want to talk about here, is the use of the computer for the *verification*
of mathematical proofs.

---

[15] A classic in this connection is the proof of the four colour theorem. See [18].

It is a well-known recent phenomenon in mathematics that the length of proofs continues to rise, even taking into account the contracted steps and the summaries.[16] Consequently, the error rate rises equally so. It seems more than reasonable to call in the help of computers to verify such proofs. This raises the following, interesting and solvable problem: the proofs need to be reformulated into a standard format, suitable for automatic treatment. The problem is solvable because, to a certain extent, the mathematician's behaviour can be copied. The basic proofs need to be spelt out in detail, but a memory store can be built-up so that statements already proven, need only be mentioned, much in the same way we do it. Yet a clear distinction remains between human and computer proofs. Barendregt and Wiedijk [3] actually use the labels *romantic* for human and *cool* or *cold* for computer proofs. My suggestion would be that the computer proofs are cold because all rhetorical elements have been eliminated through the standardization process. There is, quite literally, no more exciting story to be told. No wonder the authors claim that computer proofs are often incomprehensible. Precisely because every detail is spelt out, the oversight is lost and insight reduced to zero. As mentioned before, quite an intriguing paradox!

Let me give an example of a very romantic proof for the irrationality of $\sqrt{2}$. The key to the proof is the following drawing:[17]

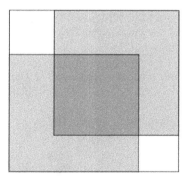

How does this drawing show that no p and q exist such that $\sqrt{2} = p/q$? Suppose it did, then, from $\sqrt{2} = p/q$ follows $2 = p^2/q^2$, and $2q^2 = p^2$. In other words, a

---

[16]One of the finest examples is, of course, the classification theorem of finite, simple groups. If all separate papers are put together, the current estimate is a total proof length of about fifteen thousand pages. For more details and some philosophical considerations concerning the impact of such proofs on the development of mathematics, see [17].

[17]This example is to be found in the introduction by Dana Scott of [3]. This fascinating book deals with the problem how seventeen different theorem provers deal with the proof of the irrationality of $\sqrt{2}$. The curious result is that seventeen different ways of handling the problem appear. So, apparently, things are not that standardized after all.

square with side p equals twice a square with side q. The large square has size p and the grey squares have side q. But the grey squares overlap – the dark grey area – and the overlap must be equal to the sum of the two little white squares. But the sides of the dark grey square and the white ones are also natural numbers, which means that we have now found another solution with smaller squares, hence smaller sides, hence smaller numbers. This leads to an infinite regress and that is impossible.[18]

It might seem strange to a non-mathematician that mathematicians speak of romantic proofs – and I consider the case of Barendregt and Wiedijk as an exemplar – but the point is that they do and that they confirm its importance in the development of mathematics. This romantic nature of a proof precisely has to do with its rhetorical properties. If a comparison needs to be made, the contrast between a human-human telephone conversation to solve a problem and a machine-human telephone conversation with the same purpose, but where a specific range of possibilities is listed and choices need to be made ("If X, then press Y"), will do perfectly. Most people will agree that in the latter case no *conversation* took place.

Third example Future mathematicians have to learn how to write down proofs. This is shown clearly by the number of manuals and instruction booklets aiming at this goal. One of the most famous examples – it has reached a form of cult status – is Knuth et al. [5]. Many examples can be given, but I do want to draw the reader's attention to the fact that a lot of attention is paid to general form rules, applicable to any text whatsoever. A typical example is to avoid mixtures of formulations. Often one sees the following:

> This now allows us to proceed to the proof of Theorem 10. There are
> ...

If respectful of the English language, the following formulation is to be preferred:

> This now allows us to show the following result: Theorem 10. There
> are ...

This might perhaps appear rather innocent – after all, the mathematical proof itself is not going to suffer, it will remain correct, badly or eloquently formulated – but innocence is lost when this advice is given to the promising young mathematician: "If there is a possibility to give a direct proof, do not use a proof by

---

[18]The solution is easily translated into algebraic terms. The white squares have a side $p - q$, whereas the dark grey square has a side $2q - p$. It is easy to check that $(2q - p)^2 = 2(p - q)^2$. So we now have a smaller solution and this process can be infinitely repeated which is impossible when one is dealing with natural numbers. This process of reasoning has become known, since Fermat, as the method of infinite descent.

contradiction." (p. 8) Here a preference is made between one proof method and another. And, as philosophers of mathematics surely know, choices of this nature do have profound philosophical implications.[19]

No matter how we decide these issues, one thing is clear: if writing down proofs were a mere matter of following standardized rules, no such manuals were hardly necessary at all. The fact that they do exist, shows the importance of a rhetoric of mathematics.

Fourth example Reviel Netz, among others in [8], launches the intriguing idea to read proofs as "stories".[20] To avoid misunderstandings, the object is not to reduce mathematics to literary prose, but rather to draw attention to the rhetorical elements, our topic of interest in this paper. The story approach has the additional advantage that it is applicable both to an individual proof and to a collection of proofs, a prime example being the thirteen books of Euclid's *Elements*. A short comment about these two cases.

To an individual proof can always be assigned a tree structure, showing the backbone of the proof. Take the conclusion as the "root" of the tree and the premises as the endpoints of "branches" and a story-line becomes visible. A concrete example: suppose I want to show that the sum of the first n natural numbers equals $n \cdot (n + 1)/2$. Here is a standard proof based on mathematical induction:

(a) basis: suppose that $n = 0$, then

   (a1) $1 + 2 + 3 + ... + n = 0$ and

   (a2) $n \cdot (n + 1)/2 = 0$

   (a3) so they are equal.

(b) induction step: suppose the statement holds for n, so:

   . $1 + 2 + 3 + ... + n = n \cdot (n + 1)/2$

(c) add to both sides $n + 1$:

   . $1 + 2 + 3 + ... + n + (n + 1) = n \cdot (n + 1)/2 + (n + 1)$

(d) the right-hand side can be transformed into:

---

[19]Often proofs by contradiction only show the existence of, say, a solution to an equation, but no details about the solution itself. What one shows is that, if there were no solution, a contradiction would follow. One now knows that it is not the case that a solution does not exist and it requires the law of double negation to justify the conclusion that a solution exists. Direct proofs usually give more information. This is, of course, related to the constructivist debate in the philosophy of mathematics and in this sense a philosophical position is presented by giving such an advice.

[20]Netz is definitely not the only one to deal with the topic. I mention Thomas [14] and [15] where the comparison between stories and proofs is examined in depth; I refer to Otte et al. [9] and Rotman [12] for a semiotically inspired approach.

(d1) $n \cdot (n + 1)/2 + (n + 1) = (n + 1) \cdot (n/2 + 1)$

(d2) $n \cdot (n + 1)/2 + (n + 1) = (n + 1) \cdot (n + 2)/2$

(d3) $n \cdot (n + 1)/2 + (n + 1) = (n + 1) \cdot ((n + 1) + 1)/2$

(e) put this together, and one finds:

. $1 + 2 + 3 + ... + n + (n + 1) = (n + 1) \cdot ((n + 1) + 1)/2$

which is precisely the statement to be proven for $n + 1$.

(f) by mathematical induction, it follows that for all n, $1 + 2 + 3 + ... + n = n \cdot (n + 1)/2$.

The three structure of this proof corresponds to the picture below.

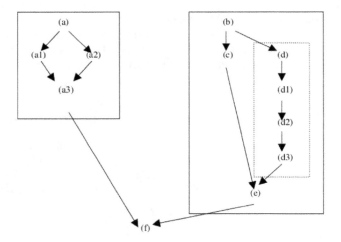

We can, of course, go into a nice discussion about the identification of the right elements in the tree, but the idea itself is clear enough. If one is willing to show some generosity in interpretation, the four steps in the dotted rectangle, viz. (d) up to (d3), can be seen as a "subplot" of the story. But even without such generosity, it is clear that the steps from (d) to (d3) do show a certain independence in relation to the whole proof. Or, to reformulate this observation in mathematical terms, it indicates that, e.g., these steps could be grouped into a separate proof, in other words, in the form of a lemma that preceeds the proof.

What holds for an individual proof, holds *a fortiori* for a set of interconnected proofs, say, a mathematical treatise, handbook or survey. All individual proofs have their own structure, but theorems refer to one another, in most cases, in non-linear structures. This opens up interesting possibilities as to the organisation of the theorems and proofs. A concrete example: imagine a situation where the proof of a theorem S_1 requires a set of preparatory proofs $A_1, A_2, A_3, ..., A_n$ and for a theorem S_2 the proofs $B_1, B_2, B_3, ..., B_m$ are needed. A possible "storyline" is a linear one: present all A_i's, followed by S_1 and then all B_i's followed by S_2, but it could very well be that, say, A_1, A_2, B_1 and B_3 are themselves connected because, e.g., they are all about arithmetical properties, so they could be brought together in a separate introductory chapter "Arithmetical Properties". Let me emphasize once again that choices of this kind can have long-ranging effects. The grouping of theorems under a separate heading can give raise to the creation of a separate mathematical domain – one sees connections where perhaps little or none were seen before the grouping – and in this sense it might have an impact on mathematical development.

Let me end here this short, obviously incomplete presentation of examples. If they do not indicate what a (full) rhetoric of mathematics should look like, I do think they show that the idea itself is not ludicrous, quite the contrary, it is a necessary addition to a full understanding of what it is what mathematicians do, whenever they do mathematics.

5   In defense of the non-mathematician

Let me now briefly deal with the second part of my thesis: how does the non-mathematician deal with mathematics, proofs in particular? My claim was and is that, although in many cases the non-mathematician is simply not familiar with the mathematical tools, hence is not able to express any judgment about a proof, yet, in equally many cases, where the familiarity does exist, the non-mathematician is not convinced by a particular proof, as in the case of the divergence of the harmonic series, let alone that she would be impressed by its excellence, beauty or romantic nature. If the thesis is acceptable that rhetorical elements are involved, it follows that it is not necessarily a wise strategy to transfer a proof from a mathematical to a non-mathematical context, or, to put if differently, one should not be amazed if the proof ceases to be convincing. All the implicit elements making up mathematical practices are not transferred at all, hence it depends to a large extent on the non-mathematical context whether the proof remains effective.

Actually I wish to defend an even stronger thesis. Cases exist where the mathematics involved is accepted as correct, yet it fails to convince as other arguments "overrule" the mathematical rhetoric. A classic example may illustrate what I mean. It is extremely difficult to convince people that, if a population is screened for a rare disease, one will have more false positives (the screening says you have the disease, but you don't) than false negatives (the screening says you don't have the disease, but you do). The underlying mathematics is simplicity itself. Let p be the (low) probability of a wrong diagnosis (positive and negative), let q be the (low) probability of actually having the rare disease, let N be the number of persons screened, then two scenarios are possible:

(a) False negatives: you have the disease but the diagnosis is wrong. The probability of that happening is $q \cdot p$, so $q \cdot p \cdot N$ persons will be in this case.

(b) False positives: you don't have the disease but the diagnosis is wrong. That probability is $(1 - q) \cdot p$, so $(1 - q) \cdot p \cdot N$ persons will suffer this fate.

Concrete example: take for p and q the same value, 0,001 (which means that the disease is rare and that the test is very reliable) and take for $N = 100.000$. Then $q \cdot p \cdot N = 0,001 \cdot 0,001 \cdot 100000 = 0,1$, in other words, there are practically no false negatives. So, if you have the disease you will be correctly diagnosed. The second scenario, however, tells a different story: $(1 - q) \cdot p \cdot N = 0,999 \cdot 0,001 \cdot 100000 =$ (approximately) 100. No fewer than a hundred persons will hear that they have the disease, although in reality they don't. Increasing N only increases this unfortunate group of people.

If one is determined to convince the non-mathematician of the correctness of this story, then, if my analysis points in a right direction, it is a bad strategy to repeat the mathematical story in more detail, filling in missing steps. Such strategies remain within mathematical discourse and that is not the problem as such. What needs to be understood is the "overruling" process.[21] When do non-mathematical arguments and considerations become more powerful than mathematical proofs? Or, to phrase it differently, what we need to know is what kind of rhetoric corresponds to everyday situations. A derived question is whether it is possible to transform the internal-mathematical rhetoric into a rhetoric of everyday life. In terms of the example, the question then becomes: how could the mathematical proof be retold such that it fits in with the other (non-mathematical) arguments?

I have no answers at the present moment, but let me just make one observation, based on my own teaching experience: what does seem to help in particular cases is the exploration of "extreme" or "borderline" scenarios. Imagine, e.g., that the disease simply does not exist, so $q = 0$. Then most people realize straight away that you can only have false positives and no false negatives at all. For a mathematician this is a trivial remark to make – the extreme cases are included in the general case – but the non-mathematician has not yet grasped the general case, so the exploration of special cases is a possible route towards the general case.

---

[21] My thanks to the anonymous referee who suggested to me that there are two different rhetorics involved in this example, a rhetoric of mathematics on the one hand, and a rhetoric of the applicability of mathematics on the other hand. I fully agree and that gave me the idea of the "overruling" process. In a sense, what I am trying to do in this paper, is to find a different vocabulary to talk about the relations between mathematics and the real world, not in the asymmetrical terms of pure, beautiful mathematics and its ugly applicability, but in the symmetrical terms of a clash of rhetorics.

In summary, rephrasing the basic idea of this paper, if the non-mathematician is not immediately convinced by a mathematical proof, then the diagnosis should not be that the problem has to do with the non-mathematician herself. It does involve her, no doubt about that, but it also involves how mathematics is done, how mathematics is situated in a society, how mathematics and mathematicians are perceived, in short, it just as much involves the mathematician too.

6 Conclusion

It is obvious that mathematics plays an essential part in present-day (western) societies. It is equally obvious that the average citizen of such societies seems to have a rather curious relation vis-à-vis mathematics. Either it is as good as unknown, or, if known, all too often distrusted. A reformulation of the aim of this paper, as presented in the introduction, is a wish to try to understand this situation and to resist the obvious explanation that, on average, people are unfortunately not all that bright. The formulation in terms of rhetorics has been an attempt to phrase the problem, to find the best suited words to express it. That required before all a demonstration that in mathematics itself a rhetoric is at work. Therefore the bulk of the paper has been devoted to a set of examples, mathematical proofs to be precise, to illustrate the presence of rhetorical elements. I hope that the philosophical-academic rhetoric of this paper was the best choice I could make.[22]

BIBLIOGRAPHY

[1] Martin Aigner and Günter M. Ziegler, *Proofs from The Book*, New York: Springer, 1998.

[2] Alan Baker, *A Concise Introduction to the Theory of Numbers*, Cambridge: Cambridge University Press, 1984.

[3] Henk Barendregt and Freek Wiedijk, "Bewijzen: romantisch of cool?" ("Proofs: romantic or cool?"), Euclides, 81, 4. (http://www.cs.ru.nl/ freek/pubs/euclides.pdf), 2006.

[4] Paul Ernest, *Social Constructivism as a Philosophy of Mathematics*, New York: SUNY, 1998.

[5] Donald E. Knuth, Tracy Larrabee and Paul M. Roberts, *Mathematical Writing*. Washington, The Mathematical Association of America, 1989.

[6] A. R. D. Mathias, "A Term of Length 4.523.659.424.929", Synthese, vol. 133, 1-2, pp. 75-86, 2002.

[7] Karl Menger, "Why Johnny Hates Math", The Mathematics Teacher 49, pp. 578-584, 1956.

[8] Reviel Netz, *The Shaping of Deduction in Greek Mathematics. A Study in Cognitive History*, Cambridge: Cambridge University Press, 1999.

[9] Michael Otte and Marco Panza, "Mathematics as an Activity", In: Michael Otte and Marco Penza (eds.), *Analysis and Synthesis in Mathematics. History and Philosophy* Dordrecht: Kluwer Academic Publishers, pp. 261-271, 1997.

[10] Chaïm Perelman and Lucie Olbrechts-Tyteca, *Traité de l'argumentation. La nouvelle rhétorique*, Brussels: Editions de l'ULB.

---

[22] In a sequel to this paper, having done this preliminary work, the next step must be to integrate all the material that Paul Ernest has gathered about mathematics and rhetoric in [4]. At this moment, it is safe to claim that, to a certain extent, our approaches are complementary (which is obviously a good thing!).

[11] Shahid Rahman, Dov Gabbay, John Symons and Jean Paul Van Bendegem (eds.), *Logic, Epistemology and the Unity of Science* (LEUS), Volume 1. Dordrecht: Kluwer Academic, 2004.

[12] Brian Rotman, *Mathematics as Sign. Writing, Imagining, Counting*, Stanford: Stanford University Press, 2000.

[13] Walter H. Schmitz, *De Hollandse Significa. Een reconstructie van de geschiedenis van 1892 tot 1926, (The Dutch Significs. A reconstruction of its history from 1892 to 1926)* Van Gorcum, Assen/Maastricht, 1990.

[14] R. S. D. Thomas, "Mathematics and Fiction I: Identification", Logique et Analyse, vol. 43, pp. 301-340, 2000.

[15] R. S. D. Thomas, "Mathematics and Fiction II: Analogy", Logique et Analyse, vol. 45, pp. 185-228, 2002.

[16] Jean Paul Van Bendegem, "Why do so many search so desperately for a universal language (and fortunately fail to find it)?", In: Frank Brisard, Sigurd D'hondt and Tanja Mortelmans (eds.), *Language and Revolution/Language and Time*, Antwerp Papers in Linguistics, 106, Antwerp: UA, Department of Linguistics, pp. 93-113, 2004.

[17] Jean Paul Van Bendegem, "The Creative Growth of Mathematics", In: Dov Gabbay, Shahid Rahman, John Symons and Jean Paul Van Bendegem (eds.), *Logic, Epistemology and the Unity of Science* (LEUS), Volume 1, Dordrecht: Kluwer Academic, pp. 229-255, 2004.

[18] Bart Van Kerkhove and Jean Paul Van Bendegem, "The Unreasonable Richness of Mathematics", *Journal of Cognition and Culture*, vol. 4, 3-4, pp. 525-549, 2004. ing we could have

[19] Freek Wiedijk (ed.), *The Seventeen Provers of the World*, New York: Springer, 2006.

A General Logic of Propositional Attitudes

DANIEL VANDERVEKEN

ABSTRACT. Contemporary logic and analytic philosophy are confined to a few paradigmatic propositional attitudes such as belief, knowledge, desire and intention. How could we develop a larger theory of all kinds of attitudes directed at objects and facts? Descartes in *Les passions de l'âme* analyzes attitudes in terms of beliefs and desires. In my analysis, psychological modes of propositional attitudes have other components than the basic *Cartesian categories of cognition* and *volition*. Complex modes like expectation, knowledge and intention have a *proper way* of believing or desiring, proper *conditions on their propositional content* or proper *preparatory conditions*. Thanks to these other components one can well distinguish stronger and weaker modes. I will *recursively define the set of all psychological modes* of attitudes. As Descartes anticipated, the two primitive psychological modes are those of *belief* and *desire*. They are the simplest modes. Other more complex modes are obtained from the two primitives by adding to them special cognitive and volitive ways, special conditions on the propositional content or special preparatory conditions. I will *define inductively the conditions of possession and of satisfaction* of all kinds of propositional attitudes. To that end, I will exploit the resources of a non standard *predicative logic* that distinguishes propositions with the same truth conditions that do not have the same cognitive value. We need to consider *subjective* as well as *objective possibilities* in philosophical logic in order to account for the fact that human agents are not perfectly but only minimally rational. I will state fundamental valid laws of my logic of attitudes.[1]

Descartes in his treatise on *Les passions de l'âme*[2] analyzed a large number of propositional attitudes. His work is a major contribution to modern philosophy of mind. Contemporary logic and analytic philosophy only consider a few paradigmatic attitudes such as belief, knowledge, desire and intention. Could we use Cartesian analysis to develop a larger theory of all propositional attitudes? Searle in *Intentionality*[3] criticized Descartes who tends to reduce all such attitudes to be-

---

[1] I have presented this logic at the International Colloquium *Computers and Philosophy* at Laval (France) in May 2006. I am grateful to H. Bergier, Paul Gochet, David Kaplan, K. MacQueen and John Searle for their comments.

[2] The treatise *Les passions de l'âme* is reedited in R. Descartes *Œuvres complètes*, La Pléiade, Gallimard: 1953.

[3] See J. Searle, *Intentionality*, Cambridge, Cambridge University Press: 1982.

liefs and desires. Many different kinds of attitudes such as fear, regret and sadness reduce to the same sums of beliefs and desires. Moreover, our intentions are much more than a desire to do something with a belief that we are able to do it. Of course, all cognitive attitudes contain beliefs and all volitive attitudes desires. But we need more than the two traditional categories of cognition and volition in order to analyze attitudes. By nature, attitudes have *intentionality*: they are *directed at* objects and facts of the world (Brentano). For that reason, they have *conditions* of *possession* and of *satisfaction* that are logically related. Beliefs and convictions are *satisfied* whenever they are *true*, desires and wishes whenever they are *realized* and intentions and plans whenever they are *executed*. In order that an agent *possesses* an attitude, he or she must be in a certain mental state. Whoever has an attitude is able to determine what must happen in the world in order that his or her attitude is satisfied. The main purpose of this paper is to contribute to analytic philosophy of mind and to the foundations of logic in formulating a *recursive theory of conditions of possession and of satisfaction of* all propositional *attitudes. Propositional attitudes* are the simplest kinds of individual attitudes directed at facts. They are possessed by a single agent at a moment or during an interval of time. From a logical point of view, they consist of a *psychological mode M* with a *propositional content P*. I will first proceed to an explication of the nature of psychological modes. In my analysis, psychological modes divide into other components than the *basic categories of cognition* and *volition*. Complex modes also have a *proper way* of believing or desiring, proper *conditions on their propositional content* or proper *preparatory conditions*. Thanks to these other components one can well distinguish stronger and weaker modes. I will also *recursively define the set of all modes* of attitudes. There are more *complex individual attitudes* than propositional attitudes. So are *denegations of attitudes* like discontent, *conditional attitudes* like intentions to buy in the case of a good offer and *conjunctions of attitudes* like doubt. I will not consider here such complex attitudes.

I will exploit the resources of a non standard *predicative logic* that distinguishes propositions with the same truth conditions that do not have the same cognitive value. We need to consider *subjective* as well as *objective possibilities* in the logic of attitudes and action in order to account for the fact that human agents are not perfectly but only minimally rational. By virtue of their logical form, attitudes are logically related in various ways. There are four different kinds of relations of inclusion between conditions of possession and of satisfaction of attitudes. Some attitudes strongly commit the agent to having others: an agent cannot possess them at a moment without possessing others at the same moment. No one can enjoy something without desiring it. Some attitudes have more satisfaction conditions than others. Whenever an aspiration is fulfilled so is the corresponding hope. Certain attitudes cannot be possessed unless others are satisfied. Whoever knows something has a true belief. Conversely, certain attitudes cannot be satisfied unless others are possessed. Whoever executes a plan has the intention of executing that plan. My primary objective here is to formulate the principles of a recursive theory of the conditions of possession and of satisfaction of propositional attitudes. I will present the ideography and state fundamental valid laws of my logic of attitudes in the last section.

1 Analysis of the propositional content of attitudes

Following Carnap,[4] standard propositional logic tends to identify so-called strictly equivalent propositions that have the same truth values in the same possible circumstances. However it is clear that such propositions are not the contents of the same attitudes just as they are not the senses of synonymous sentences. We need a much finer criterion of propositional identity than strict equivalence for the purposes of philosophy of mind, action and language. Indeed we do not know *a priori* by virtue of competence the necessary truth of many propositions. We have to learn a lot of essential properties of objects. By *essential property* of an object I mean here a property that it *really* possesses in any possible circumstance. So are our properties of having certain parents. We learn *a posteriori* their identity. Certain children do not know their natural mother. Others are wrong about her identity; they have a necessary false belief. However when we are inconsistent, we always remain paraconsistent. As the Greek philosophers pointed out, we cannot believe every proposition (the sophist's paradox). Any adequate logic of attitudes has to account for such facts. Few necessarily true propositions are pure *tautologies* such as the proposition that mothers are mothers that we know *a priori* to be true.

A second important problem of the standard logic of attitudes is related to the way in which it analyzes satisfaction and possession conditions of agents' attitudes. According to the standard analysis, relations of psychological compatibility with the truth of beliefs and the realization of desires are modal relations of accessibility between agents and moments, on one hand, and possible circumstances, on the other hand. Thus according to Hintikka,[5] possible circumstances are compatible with the truth of agents' beliefs at each moment of time. To each agent a and moment m there corresponds in each model a unique set $Belief(a, m)$ of possible circumstances where all beliefs of that agent at that moment are true. Such possible circumstances are by definition compatible with what that agent then believes. In short, an agent a *believes* a proposition P *at a moment m* according to Hintikka when that proposition P is true in all circumstances of the set $Belief(a, m)$. Given that approach, all human agents are logically omniscient. They believe all necessarily true propositions and their beliefs are closed under logical implication.

---

[4]See R. Carnap, *Meaning and Necessity*, Chicago, University of Chicago Press: 1947.

[5]J. Hintikka, "Semantics for Propositional Attitudes", in L. Linsky (ed), *Reference and Modality*, Oxford, Oxford U.P. 1971.

Whoever believes a proposition P *eo ipso* believes all propositions that P logically implies. Moreover, according to standard epistemic logic, we, human agents, are either *perfectly rational* or *totally irrational*. We are perfectly rational when at least one possible circumstance is compatible with what we believe. Otherwise, we are totally irrational. Whoever believes a necessary falsehood *eo ipso* believes all propositions. But all this is incompatible with standard philosophy of mind and empirical psychology. Clearly we ignore most logical truths and we do not draw all logical inferences. Moreover, when we are inconsistent, we never believe everything. Problems are worse in the case of the logic of desire if we proceed according to the standard approach. For we keep desires that we know to be unsatisfied or even insatisfiable.

One could introduce in epistemic logic so-called *impossible circumstances* where necessarily false propositions would be true. However, such a theoretical move is very *ad hoc* and it is moreover neither necessary nor sufficient. In my approach, all *circumstances* remain *possible*: they are *objective possibilities*, as Belnap says. So objects keep their essential properties (each of us keeps his real parents) and necessarily false propositions remain false in all circumstances. In order to account for human inconsistency, logic needs no impossible circumstances. It has to consider *subjective* in addition to *objective possibilities*. Many obvious subjective possibilities are not objective. Whales are not fishes and not every property does correspond to a characteristic set of entities. So we can be inconsistent in ordinary life and in science. In order to explicate subjective possibilities and to define adequately the notion of truth according to an agent, I will now present a non classical logic that better analyzes the logical form of propositions as well as conditions of possession and satisfaction of attitudes.

1.1   New principles of predicative propositional logic

My propositional logic is *predicative* in the general sense that it analyzes the logical form of propositions by taking into account predications that we make in expressing them.[6]

- In my view, each proposition has a finite *structure of constituents*. It contains a positive number of elementary propositions predicating *attributes* of degree n (properties or relations) of n *objects subsumed under concepts* in a certain order. As Frege and Russell pointed out, we understand an elementary proposition when we understand which attributes objects of reference must possess in a possible circumstance in order that this proposition be true in that circumstance. There is no propositional formula without a predicate expressing an attribute in my ideography. Most often the degree n of the predicated attribute is positive. In that case one cannot predicate

---

[6]For more information on predicative logic see the book *Logic, Thought and Action* that I edited at Springer, 2005.

the attribute without making a reference to objects under concepts. Properties are attributes of degree one. The elementary proposition that Socrates is a philosopher predicates the property of being philosopher of a person. Binary relations are attributes of degree two. The elementary proposition that Brutus killed Julius Caesar predicates a binary relation of two persons. Simplest elementary propositions predicate an attribute of degree zero of no object under concept. So is the proposition that it rains.

- In addition to taking into account the structure of constituents of propositions, we also need a *better explication of their truth conditions*. We understand most propositions without knowing in which possible circumstances they are true, because we ignore *real denotations* of most attributes and concepts in many circumstances. One can refer to a colleague's wife without knowing who she is. However we can always in principle think of persons who could be that wife. Sometimes we even have evidence. His wife could be that or that woman. So in any possible use and interpretation of language, there are a lot of *possible denotation assignments to attributes and concepts* in addition to the standard *real denotation assignment* which associates with each propositional constituent its actual denotation in every possible circumstance. They are functions of the same type that, for example, associate with each individual concept a unique individual or no individual at all in every possible circumstance. According to the real denotation assignment, my colleague's wife is the woman with whom he is really married when there is such a person.[7] According to other possible denotation assignments, his wife is another person or even he is not married. In spite of such differences, all possible denotation assignments respect *meaning postulates*. According to any, a wife is a married woman. One cannot express the property of being a wife without knowing *a priori* by virtue of competence such essential features of that property.

We ignore the value of the real denotation assignment for most concepts and attributes in many possible circumstances. But we can in principle think of denotations that they could have. Moreover, when we have in mind certain concepts and attributes, only some possible denotation assignments to these senses *are then compatible with* our beliefs. Suppose that according to an agent Smith's wife is either Paula or Ursula. In that case, possible denotation assignments according to which Smith is married to another woman are then incompatible with that agent's beliefs. So in my approach, possible denotation assignments rather than possible

---

[7] Some individual concepts do not apply to any object according to the real denotation assignment in certain possible circumstances. In that case either that assignment is in models a partial function that is undefined for these concepts in these circumstances or that concept is given an arbitrary denotation like the empty individual in such circumstances.

circumstances are compatible with the beliefs of agents. This is my way of accounting for subjective possibilities in predicative propositional logic where the basic truth definition is relative to both possible circumstances and denotation assignments.

- In understanding propositions we in general do not know whether they are true or false. We just know that their truth in a circumstance is compatible with certain possible denotation assignments to their concepts and attributes, and incompatible with all others. Thus an elementary proposition predicating an extensional property of an object under a concept is true in a circumstance according to a denotation assignment when according to that assignment the individual who falls under that concept has that property in that circumstance. Otherwise, it is false in that circumstance according to that assignment. Most propositions have therefore *a lot of possible truth conditions*. Suppose that a proposition is true according to a possible denotation assignment to its constituents in a certain set of circumstances. That proposition would be true in all and only these circumstances if that denotation assignment were the real one. Of course, in order to be *true in a circumstance* a proposition has to be *true in that circumstance according to the real denotation assignment*. So among all possible truth conditions of a proposition, there are its *real Carnapian truth conditions* which correspond to the set of possible circumstances where it is true according to the real denotation assignment $val*$.

- In my view, propositions are *identical* when they contain the same elementary propositions and they are true in the same circumstances according to the same possible denotation assignments. Such a finer criterion of propositional identity explains why many strictly equivalent propositions have a different cognitive value. Propositions whose expression requires different predications have a different structure of constituents. We do not express them at the same moments. My identity criterion also distinguishes strictly equivalent propositions that we do not understand to be true in the same possible circumstances: these are not true according to the same possible denotation assignments to their constituents. Consequently, few necessarily true propositions are *pure tautologies* that we know *a priori*. By hypothesis, a proposition is *necessarily true* when it is true in every possible circumstance according to the real denotation assignment. In order to be *tautologically true*, a proposition has to be true in every circumstance according to every possible denotation assignment to its constituents. Unlike the proposition that Oedipus' mother is his mother, the necessarily true proposition that Oedipus' mother is Jocasta is not a pure tautology. It is false according to possible denotation assignments. We now can distinguish formally subjective and objective possibilities. A proposition is *subjectively possible* when it is true in a circumstance according to a possible denotation assignment. In order to be *objectively possible* a proposition has to be true in a circumstance according to the real denotation assignment.

Attitudes and actions of human agents are not determined. Whenever we do or think something we could have done or thought something else. For that reason, the logic of attitudes requires a ramified conception of time compatible with indeterminism and the apparent liberty of human agents. In branching time, a *moment* is a complete possible state of the actual world at a certain instant and the *temporal relation* of *anteriority* between moments is partial rather than linear. There is a single causal route to the past. So all moments are historically connected: any two distinct moments have a common historical ancestor in their past. However, there are multiple future routes: several incompatible moments might be posterior to a given moment. For facts, events or actions can have incompatible future effects. Consequently, the set *Times* of moments of time has the formal structure of a *tree-like frame* of the following form:

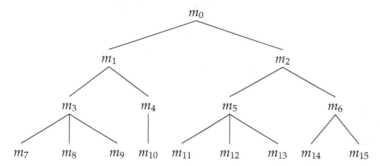

A maximal chain h of moments of time is called a history. It represents a *possible course of history of our world*. When a history has a first and a last moment, the world has according to it a beginning and an end. As Belnap[8] pointed out, a *possible circumstance* is a pair of a moment m and of a history h to which that moment belongs. Thanks to histories temporal logic can analyze important modal notions like settled truth and historic necessity and possibility. Certain propositions are true at a moment according to all histories. Their truth is then *settled at that moment* no matter how the world continues. So are past propositions because the past is unique. Their truth does not depend at all on histories. So are also propositions attributing attitudes to persons. Whoever believes or desires something at a moment then believes or desires that thing no matter what happens later. Contrary

---

[8] See N. Belnap, M. Perloff and Ming Xu, *Facing the Future. Agents and Choices in Our Indeterminist World*, Oxford University Press: 2001.

to the past, the future is open. The world can continue in various ways after indeterminist moments. Thus the truth of future propositions is not settled at such moments. It depends on which historical continuation of that moment is under consideration. When there are different possible historic continuations of a moment, its actual future continuation is not then determined. However, as Occam[9] pointed out, if the world continues after a moment, it will continue in a unique way. The actual historic continuation of each moment is unique even if it is still undetermined at that moment. Indeterminism cannot prevent that unicity. So each moment m has a *proper history* h_m in my temporal logic. If m is the last moment of a history h, that history is of course its proper history h_m. If on the contrary the moment continues, then by hypothesis all moments of its proper history have the same real historic continuation. For $m' \in h_m$ iff $h_{m'} = h_m$. In my terminology, a proposition is *true at a moment* according to a possible denotation assignment when it is then true according to that assignment in the history of that moment. Among all possible courses of history of this world, one will be its actual course of history. It is the proper history of the present actual moment *now*.

Two moments of time m and m' are coinstantaneous when they belong to the same instant.[10] Coinstantaneous moments m and m' *represent* two complete possible states of the world in which things could then be. They are on the same horizontal line in each tree-like frame. One can analyze historic necessity by quantifying over coinstantaneous moments. The proposition that P is then necessary (in symbols $\Box P$) – in the sense that it is then inevitable that P – is true at a moment in a history when P is true at all coinstantaneous moments according to all histories.

1.2   My new approach in the logic of attitudes

In my view, the relation of compatibility with the satisfaction of attitudes of agents is not a modal relation of accessibility. First of all, attitudes of human agents are about objects that they represent under concepts. Each agent has consciously or potentially in mind a certain set of attributes and concepts at each moment. (That set is empty when the agent does not exist.) No agent can have a particular propositional attitude without having in mind all attributes and concepts of its content. Otherwise, he would be unable to determine under which conditions that attitude is satisfied. As Wittgenstein and Searle pointed out, an attitude with entirely undetermined satisfaction conditions would be an attitude without content; so it would not be a proper attitude. In order to believe or desire to be archbishop one must understand characteristic features determined by the meaning of the predicate "archbishop" that are common to all entities that have that property. Such features are internalized when one learns the meaning of that predicate. They are respected

---

[9]See A.N. Prior, *Past, Present and Future*, Oxford, Clarendon Press, 1967.

[10]The set of instants is a partition of the set *Time* which satisfies unique intersection and order preservation. So to any instant i and history h there corresponds a unique moment $m(i, h)$ belonging to both i and h.

by all possible denotation assignments to the expressed property of being an arch-bishop.[11]

Secondly, possible denotation assignments to propositional constituents rather than possible circumstances are compatible with the satisfaction of agents' attitudes. So there corresponds to each agent a and moment m in each model a unique set $Belief(a, m)$ of possible denotation assignments to attributes and concepts that are compatible with the truth of beliefs of that agent at that moment. By hypothesis, $Belief(a, m)$ is the set Val of all possible denotation assignments to senses when the agent a has no attribute or concept in mind at the moment m. In that case, that agent has then no attitudes. Otherwise, $Belief(a, m)$ is a *proper non empty* subset of Val. Whenever an agent has in mind propositional constituents, he or she respects meaning postulates and there always are possible denotation assignments to these senses compatible with what that agent then believes. In my approach, an agent a *believes a proposition* at a moment m when he or she has then in mind all its concepts and attributes and that proposition is true at that moment according to all possible denotation assignments belonging to $Belief(a, m)$. We all ignore what will happen later. But we now have a lot of beliefs directed at the future. As Occam pointed out, such beliefs are true when things will be as we believe in the *actual future continuation* of the present moment. Other possible historic continuations do not matter.

One can analyze desire according to the same approach. To each agent a and moment m there corresponds in each model a unique non empty set $Desire(a, m)$ of possible denotation assignments to attributes and concepts that are compatible with the realization of all desires of that agent at that moment. There is however an important difference between desire and believe. We can believe, but we cannot desire, that objects are such and such without believing that they could be otherwise. For any desire requires a *preference*. Whoever desires something distinguishes two different ways in which represented objects could be in the actual world. In a first preferred way, objects are in the world as the agent desires, in a second way, they are not. The agent's desire is satisfied in the first case, it is unsatisfied in the second case. Thus in order that an agent a *desires the fact represented by a proposition* at a moment m, it is not enough that he or she has then in mind all its constituents and that the proposition is true at that moment according to all possible denotation assignments of $Desire(a, m)$. That proposition must also be false in a circumstance according to possible denotation assignments of $Belief(a, m)$. Otherwise that agent would not prefer the existence of that fact.

My explication of belief and desire is well supported by received accounts in the philosophy of mind. Given new meaning postulates and the nature of possible

---

[11] Many essential features of objects of reference like for example essential biological properties of human beings are not determined by meaning. When I refer to a particular archbishop I do not have in mind his genetic code.

denotation assignments, it accounts for the fact that *human agents are not perfectly rational*. We do not have in mind all concepts and attributes. So we ignore logical as well as necessary truths. Our knowledge is limited: we ignore which objects possess certain properties in many circumstances. In that case assignments associating different denotations to these properties in these circumstances are then compatible with our beliefs. We have false beliefs and unsatisfied desires. Many possible denotation assignments compatible with our beliefs and desires do not assign real denotations to attributes that we have in mind. Such assignments can even violate essential properties of objects of reference. In that case we have necessarily false beliefs and insatisfiable desires. My analysis solves traditional paradoxes.

However human agents *cannot be totally irrational*. On the contrary, they are *minimally rational* in a well determined way. First of all, agents cannot believe or desire everything since in all models some possible denotation assignments are always compatible with the satisfaction of their beliefs and desires. Consequently some propositions are false according to all such denotation assignments. Moreover, they cannot possess certain beliefs and desires without *eo ipso* possessing others. Indeed by hypothesis all possible denotation assignments compatible with their beliefs and desires respect meaning postulates. So we, human agents, are *minimally logically omniscient*, in the sense that we cannot have in mind a pure tautology without knowing for certain that it is necessarily true. Represented objects could not be in another way according to us. Similarly, *pure contradictions* (negations of tautologies) are false in every possible circumstance according to any possible denotation assignment. We can neither believe nor desire contradictory things. Things could never be in certain ways according to us. Contemporary logicians still wish that arithmetic were complete (a necessarily false proposition if Gödel's proof is right). But they never believe nor desire both the completeness and the incompleteness of arithmetic (a pure contradiction). Sometimes we desire something (to be somewhere at a moment) for one reason and another incompatible thing (to be elsewhere at the same moment) for another reason. When the logical form of such attitudes is fully analyzed, they are not categorical desires with a contradictory content. Agents believe all tautological propositions that they express. However they cannot desire the existence of inevitable facts represented by tautologies. In order to desire something one must believe that it could not occur. One can desire to drink; one can also desire not to drink. But one cannot desire to drink or not to drink.

One can define in predicative logic a new *strong* propositional *implication* much finer than Lewis' strict implication that is important for the analysis of psychological commitment. A proposition *strongly implies* another when it contains all its elementary propositions and it cannot be true in a circumstance according to a denotation assignment unless the other proposition is also true in that circumstance according to that assignment. Strong implication is finite, tautological, paraconsistent, decidable and *a priori known*. Any agent who believes a proposition P also believes all the propositions that P strongly implies. For he or she cannot apprehend that proposition without understanding that it cannot be true unless these others are. Unlike belief, desire is not closed under strong implication. For no one can desire tautological things.

2 Analysis of propositional attitudes

The notion of psychological mode is too rich to be taken as a primitive notion. As Descartes pointed out, the two traditional categories of cognition and volition are essential components of psychological modes. But they divide into other components that I will now analyze.

2.1 The general categories of cognition and volition

All propositional attitudes are cognitive or volitive. Among cognitive attitudes, there are conviction, faith, confidence, knowledge, certainty, presumption, pride, arrogance, surprise, amazement, stupefaction, presupposition, prevision, anticipation and expectation. All of them contain beliefs in the truth of their propositional content. Among volitive attitudes, there are wish, will, intention, ambition, project, hope, aspiration, satisfaction, pleasure, enjoyment, delight, gladness, joy, elation, amusement, fear, regret, sadness, sorrow, grief, remorse, terror. All of them contain desires. Like Searle, I advocate a very general category of volition applying to all kinds of desires directed towards the past (shame), the present (lust) and the future (aspiration), even to desires known or believed to be satisfied (pleasure, joy) or unsatisfied (disappointment, regret) including desires directed at past actions that the agent would wish not to have done (remorse).

In philosophy of mind, *beliefs* have the proper *mind-to-things direction of fit*. Whoever possesses a cognitive attitude intends to represent how things are then in the world. Such an attitude is satisfied when its propositional content corresponds to things as they are in the world. On the other hand, *desire* has the opposite *things-to-mind direction of fit*. Volitive attitudes are satisfied only if things in the world fit their propositional content. Each direction of fit between mind and the world determines which side is at fault in case of dissatisfaction. When a belief turns out to be false, it is the agent who is at fault, not the world. He should have had other thoughts about the world. In such a case, the agent easily corrects the situation in changing his beliefs. On the contrary, when a desire turns out to be unsatisfied, it is not the agent but the world which is at fault. Objects should have been different. The agent then rarely corrects the situation in changing his or her desire. Most often, he or she keeps that desire and remains unsatisfied.

One can explicate formally the two general categories of cognition and volition in terms of compatibility relations with respect to the truth of beliefs and the realization of desires of agents. So far we have only spoken of real beliefs and

real desires of agents. Agents of course have real attitudes about objects at certain
moments in this actual world. They really have such and such beliefs and desires
at certain moments. But they could have had other attitudes about the same or
even about other objects. Often other agents attribute to us attitudes that we do
not really have. In order to get a general theory, we need to define the sets $Belief_m^a$
and $Desire_m^a$ with respect to all possible denotation assignments. Agents can have
different attitudes according to different possible denotation assignments. Their
real attitudes depend on the real denotation assignment. First of all, each agent
could have many concepts and attributes in mind. According to any possible denotation assignment val, every agent a has in mind at each moment m in each
model a *certain set* $val(a, m)$ *of propositional constituents*. Thus $val(a, m)$ is the
set of concepts and attributes that agent a really has in mind at moment m in the
model. By definition, $Belief_m^a(val)$ is the non empty set of all possible denotation
assignment that are compatible with the truth of beliefs that agent a has at moment
m according to denotation assignment val. In particular, $Belief_m^a(val^*)$ contains all
possible denotation assignment that are compatible with the truth of beliefs that
agent a really has at moment m. Similarly, $Desire_m^a(val)$ is the non empty set of all
possible denotation assignments that are compatible with the realization of desires
that agent a has at moment m according to assignment val. Of course, $Belief_m^a(val)$
and $Desire_m^a(val)$ are the whole set Val of all possible denotation assignment when
the set of propositional constituents $val(a, m)$ is empty. In that case, the agent has
then no attitude at all according to val. In my view, an agent a *believes* or *desires*
that P at a moment m (no matter what is the history) according to a denotation
assignment val when firstly he or she has in mind all concepts and attributes of
proposition P (they belong to the set $val(a, m)$) and secondly P is true at that moment according to all denotation assignments of $Belief_m^a(val)$ or $Desire_m^a(val)$. In
the case of desire, proposition P has moreover to be non tautological according
to that agent at that moment. Agents have conscious and unconscious beliefs and
desires. Whenever an agent's attitude is conscious, he or she has then consciously
in mind all its attributes and concepts. We have unconsciously in mind at each
conscious moment of our existence a lot of concepts and attributes that we could
express at that moment given the language that we speak. We then have a lot of
unconscious tautological beliefs about unconscious propositional contents.

One can explicate the nature of attitudes by determining formal properties of
psychological compatibility relations that correspond to them. Whoever has a belief believes that he has that belief. The relation $Belief_m^a(val)$ is then transitive in
each model. On the contrary, we often feel desires that we would wish not to feel.
So the relation $Desire_m^a$ is not transitive. Some of our beliefs are false; many of our
desires are unsatisfied. The compatibility relations $Belief_m^a(val)$ and $Desire_m^a(val)$
are then not reflexive. They are also not symmetric. We can have new beliefs
and new desires according to denotation assignments that are compatible with our

beliefs and desires now.

2.2   Different ways of having cognition or volition

Our beliefs and desires can be more or less strong and we feel them in a lot of ways. Certain psychological modes require a special *cognitive* or *volitive* way of believing or desiring. *Knowledge* is a belief based on *strong evidence* that guarantees truth. The agent has a real perception of what he or she knows or it is part of his or her conceptual network or background. With such kind of sensorial, analytic or background evidence, whoever knows something is sure of it and whatever is known has to exist. Whoever has an *intention* feels such a strong desire that he or she is disposed to *act* in the world in order to satisfy that desire. Intentions commit the agent to a present or future action. Whoever is *pleased* feels a satisfied desire whose very satisfaction puts him or her in a state of pleasure. Whoever *enjoys* is in a conscious state of enjoyment. In the case of *lust* he or she is in a state of sensual enjoyment. On the contrary, whoever is *sad* feels an unsatisfied desire whose dissatisfaction puts him or her in a sad state; whoever is *terrified* is in a worst conscious state of terror. Language distinguishes many psychological modes with different cognitive or volitive ways. Thus regret, sadness, sorrow and terror put the agent in more and more unpleasant states. From a logical point of view, a *cognitive or volitive way* is a function f_ω which restricts basic psychological categories. Whoever feels a belief or desire in a certain way has of course that belief or that desire. By definition, $f_\omega(a, m, val) \subseteq Belief_m^a(val) \cup Desire_m^a(val)$. The set U_ω of cognitive and volitive ways is a *Boolean algebra*. It contains the *neutral way* $1_\omega : 1_\omega(a, m, val) = Belief_m^a(val) \cup Desire_m^a(val)$. And it is closed under the operation of conjunction. A psychological mode has the *conjunction* $f_\omega \cap h_\omega$ of two ways when it has each of them. Thus $f_\omega \cap h_\omega(a, m, val) = f_\omega(a, m, val) \cap h_\omega(a, m, val)$. The *way* of a mode is *special* when it is not neutral. Certain ways are *stronger than* others. Whoever is certain of something knows it: $\omega_{certainty}(a, m, val) \subseteq \omega_{knowledge}(a, m, val)$. Because knowledge is true, its proper cognitive way is reflexive: $val \in \omega_{knowledge}(a, m, val)$. Whoever has an intention intends to execute that intention. So the volitive way of the intention mode is transitive: if $val' \in \omega_{intention}(a, m, val)$ and $val'' \in \omega_{intention}(a, m, val')$ then $val'' \in \omega_{intention}(a, m, val)$. Agents can feel more or less strong beliefs and desires. We must distinguish different degrees of beliefs and desires. The two basic relations of psychological compatibility are then indexed by the set $\mathbb{Z}$ of integers. Whoever believes or desires with a certain degree of strength also believes or desires with weaker degrees. Thus $Belief_m^a(val)(k + 1) \subseteq Belief_m^a(val)(k)$ and similarly for $Desire_m^a$. Certain cognitive or volitive ways require a minimal degree of strength. Any knowledge contains a strong belief, any intention a strong desire. $\omega_{knowledge}(a, m, val) \subseteq Belief_m^a(val)(1)$ and $\omega_{intention}(a, m, val) \subseteq Desire_m^a(val)(1)$.

2.3 Propositional content conditions

Like illocutionary forces, psychological modes have propositional content conditions. The propositional content of attitudes having certain modes must satisfy certain conditions. Thus we can only at a moment *foresee*, *anticipate* or *expect* a fact that is future with respect to that moment. Whoever possesses an *intention* at a given moment desires to carry out a present or future action in the history of that moment. Propositional content conditions of attitudes depend on the moment of such attitudes. I can have today the intention to wake up early tomorrow morning. The day after tomorrow I cannot anymore have that intention, because the intended action will then be past. The propositional content of certain attitudes concerns the very person of the agent. We can be *disappointed* or *sorry* about something that has nothing to do with our person. But we can only feel *shame* or *remorse* for something that is personal.

From a logical point of view, a *condition on the propositional content* is a function f_θ from the set $Agent \times Time$ into the power set $\mathcal{P}(U_p)$ of the set U_p of all propositions that associates which each agent and moment a set of propositions. By definition, the *propositional content conditions of a mode M* is the function θ_M which associates with each agent and moment the set of propositions that could be the propositional content of an attitude of that mode of that agent at that moment. By hypothesis, illocutionary forces have propositional content conditions of all psychological modes that enter into their sincerity conditions. Thus the force of *prediction* and the modes of *prevision*, *anticipation* and *expectation* have the condition that their propositional content represents a future fact. Similarly, the force of *promise* and the modes of *intention* and *project* have the condition that their propositional content represents a present or future action of the agent. The set U_θ of propositional content conditions is a *Boolean algebra*. It contains the *neutral propositional content condition* 1_μ: $1_\theta(a, m)$ is the whole set U_p of propositions. And it is closed under the operation of *intersection*. The psychological mode has the intersection $(f_\theta^1 \cap f_\theta^2)$ of two conditions when it has each of them. $(f_\theta^1 \cap f_\theta^2)(a, m) = f_\theta^1(a, m) \cap f_\theta^2(a, m)$. A mode M has a *special* propositional content condition when its condition θ_M is not neutral. Certain psychological modes have more propositional content conditions than others. Thus a *prior intention* is an intention with the additional condition that its content represents a future action of the agent. So $\theta_{PriorIntention}(a, m) \subset \theta_{Intention}(a, m)$.

2.4 Preparatory conditions

Like illocutionary forces, psychological modes also have *preparatory conditions*. Whoever possesses an attitude or performs an illocutionary act *presupposes* certain propositions. His or her attitude and illocutionary act would be defective if these propositions were then false. Thus the force of *promise* and the mode of *intention* have the preparatory condition that the agent is then able to do the action represented by the propositional content. When this is not the case, the promise and the intention are defective. This defect shows itself in the fact that it is quite paradoxical to promise an action and to deny simultaneously that one is able to do it. In the illocutionary case, the speaker can lie in order to mislead the hearer. In the psychological case, however the agent cannot lie to himself. He must both presuppose and believe that the preparatory conditions of his attitudes are fulfilled. Whoever has an intention really believes that he is able to execute it. Otherwise, he would not have that intention. The volitive mode of *hope* has the preparatory condition that the propositional content is then possible. Whoever hopes something desires that thing while believing that it could happen. The volitive modes of *will* and *intention* have the preparatory condition that the agent has means. In the case of wish, on the contrary, the satisfaction of the agent's desire is independent of his own will. All depends on the course of nature or on the good will of someone else.

From a logical point of view, a preparatory condition is a function f_Σ from the set $Agent \times Time \times U_p$ into the set $\mathcal{P}(U_p)$ associating with each agent, moment and propositional content a set of propositions. Illocutionary forces have the preparatory conditions of modes which are their sincerity conditions. Thus the preparatory condition $\Sigma_{Intention}$ common to the mode of *intention* and to the force of *promise* associates with each agent, moment and propositional content a set containing the proposition that that agent is then able to do the represented action. The set U_Σ of preparatory conditions is also a *Boolean algebra*. It contains the *neutral preparatory condition* $1_\Sigma : 1_\Sigma(a, m, P) = \varnothing$. It is closed under the operation of *union*. A psychological mode has the union $(f_\Sigma^1 \cup f_\Sigma^2)$ of two preparatory conditions when it has each of them. $(f_\Sigma^1 \cup f_\Sigma^2)(a, m, P) = f_\Sigma^1(a, m, P) \cup f_\Sigma^2(a, m, P)$. A preparatory condition is *special* when it is not neutral. Many volitive or cognitive ways have special preparatory conditions. The cognitive ways of the modes of certainty and knowledge and the volitive ways of the modes of pleasure, joy and enjoyment all determine the preparatory condition that the fact represented by the propositional content exists.

2.5 Criterion of identity for psychological modes

On the basis of my analysis, one can formally distinguish different psychological modes whose attitudes apparently reduce to the same sums of beliefs and desires. One can also proceed to a lexical systematic analysis of many ordinary verbs or terms naming propositional attitudes. Two psychological modes M_1 and M_2 are *identical* when they have the same basic psychological categories, the same cognitive and volitive ways, the same propositional content conditions and the same preparatory conditions. So $M_1 = M_2$ when $Cat_{M_1} = Cat_{M_2}, \omega_{M_1} = \omega_{M_2}, \theta_{M_1} = \theta_{M_2}$ and $\Sigma_{M_1} = \Sigma_{M_2}$. As we will see later, attitudes about the same proposition whose modes divide into the same components have the same conditions of possession and of satisfaction. Such modes play the same role in psychological life.

2.6 Conditions of possession of propositional attitudes

Each component of a mode determines a particular necessary possession condition of attitudes with that mode, all the components together possession conditions that are both necessary and sufficient. By definition, an agent *a possesses a cognitive (or volitive) attitude of the form M(P) at a moment m* when he or she then *believes* (or *desires*) the propositional content *P*, he or she feels that belief or desire that *P* in *the cognitive or volitive way* ω_M proper to psychological mode *M*, the *proposition P then satisfies propositional content conditions* $\theta_M(a, m)$ and finally *that agent then presupposes and believes all* propositions determined by the preparatory conditions $\Sigma_M(a, m, P)$ of mode *M* with respect to the content *P*. For example, an agent *intends that P* at a moment when proposition *P* then represents a present or future action of that agent, he or she desires so much that action that he or she is committed to carrying it out and moreover that agent then presupposes and believes to be able to carry it out.[12] The basic psychological category of every attitude determines of course its primary condition of possession. We can desire without having an intention. But we could not have an intention without a desire.

2.7 Definition of strong and weak psychological commitment

An *attitude strongly commits an agent to another at a moment* when he or she could not then have that attitude without having the second. For example, whoever believes that it will rain tomorrow then foresees rain tomorrow. For the propositional content is then future with respect to the moment of the attitude. Some attitudes strongly commit the agent to another only at certain moments. Whoever believes now that it will rain tomorrow foresees rain tomorrow. The day after tomorrow the same belief won't be a prevision. It will just be a belief about the past. An attitude *contains another* when it strongly commits any agent to that other attitude at any moment. Whoever possesses the first attitude possesses the second. Thus any cognitive attitude contains a belief and any volitive attitude a desire with the same propositional content.

As Searle and I pointed out,[13] one must distinguish in speech act theory between the *overt performance of an illocutionary act* and a simple *illocutionary commitment to the act*. For example, whoever asserts that every man is mortal is weakly committed to asserting that the man called Nebuchadnezzar is mortal, even if he has not made any reference to Nebuchadnezzar and if he has not overtly made the second assertion. The same holds in the theory of attitudes. One must distinguish

---

[12]The proposition [aPossessesM(P)] according to which the agent *a* possesses the attitude *M(P)* is true in a circumstance *m/h* according to a denotation assignment *val* iff (1) each constituent of $P \in val(a, m)$, (2) *P* is true at moment *m* according to all assignments of $(Cat_M)_m^a(val) \cap \omega_M(a, m, val)$, (3) $P \in \theta_M(a, m)$ and (4) each proposition $Q \in \Sigma_M(a, m, P)$ is then true according to all assignments of $Belief_m^a(val)$ and is moreover also *presupposed* by agent *a*.

[13]See J. Searle and D. Vanderveken, *Foundations of Illocutionary Logic*, Cambridge, Cambridge University Press: 1985.

between the overt possession of an attitude and a simple psychological commitment to that attitude. There are *strong and weak psychological commitments* just as there are strong and weak illocutionary commitments. Some attitudes *weakly commit* the agent to others at certain moments: the agent could not then possess these attitudes without being committed to the others. For example, whoever believes that every man is mortal is weakly committed to believing that Nebuchadnezzar is mortal, even if he has not Nebuchadnezzar's concept in mind and if he does not then possess the second belief. Clearly no one could simultaneously believe the first universal proposition and the negation of the second. So there is a general parallelism between strong and weak illocutionary and psychological commitments. When an illocutionary act strongly commits the speaker to another, the attitudes that the speaker expresses in performing the first act strongly commit him to attitudes that are sincerity conditions of the second act. And similarly for weak commitments.

One can explicate weak psychological commitments of each agent a at each moment m in the logic of attitudes by quantifying over the set $Compatible_m^a$ of moments that are psychologically compatible with his or her attitudes at that moment. Two moments m and m' are psychologically compatible as regard an agent a in a model according to a denotation assignment val [in symbols: $m' \in Compatible_m^a(val)$] when that agent could then have all the attitudes that he has at both moments according to that assignment. In that case, there is a coinstantaneous moment where that agent has all such attitudes according to that assignment. The relation of *psychological compatibility* between moments of time as regards any agent is by definition reflexive and symmetric like that of illocutionary compatibility. Thus an agent could have an attitude at a moment when he or she has that attitude at some moment that is psychologically compatible with that moment as regards to him or her. In my view, an agent is *weakly committed to an attitude at a moment m* (in symbols: $[a \triangleright M(P)]$) when he could then possess that attitude at each moment that is psychologically compatible with that moment as regards that agent.

2.8   Recursive definition of the set of psychological modes

Psychological modes are not a simple sequence of a basic psychological category, a cognitive or volitive way, a propositional content condition and a preparatory condition. For their components are not logically independent. Indeed certain components *determine* others of the same or of another kind. Thus the volitive way of the mode of *intention* determines the propositional content condition that it represents a present or future action of the agent and the preparatory condition that that agent is then able to carry out that action. That last preparatory condition determines another, namely that the agent has means. The cognitive way of the mode of *pride* determines the condition that the propositional content concerns the very person of the agent and the preparatory condition that the represented fact exists and is good for the agent. In my ideography, each formula of the form $[Cat_M, f_\omega, f_\theta, f_\Sigma]$ represents the psychological mode M having the four components $Cat_M, f_\omega, f_\theta, f_\Sigma$ and all other components which are determined by them. Such a mode M has then often other ways than f_ω, other propositional content conditions than f_θ and other preparatory conditions than f_Σ. The two primitive modes of *belief* and *desire* are the simplest cognitive and volitive modes. Descartes was right to take them as primitive modes. They have no special cognitive or volitive way, no special condition on the propositional content and no special preparatory condition.[14] According to my ideography $M_{belief} = [Belief, 1_\omega, 1_\theta, 1_\Sigma]$ and $M_{desire} = [Desire, 1_\omega, 1_\theta, 1_\Sigma]$. All other psychological modes are more complex. They are obtained by adding to primitive modes special cognitive or volitive ways, propositional content conditions or preparatory conditions.

In my logic, the set of all possible psychological modes is then *defined recursively*. It is the smallest set Modes containing the two primitive modes of *belief* and *desire* that is closed under a finite number of applications of the three Boolean operations of adding new cognitive or volitive ways, new propositional content conditions or new preparatory conditions. Thus the psychological mode of *prevision* $M_{foresee}$ is obtained by adding to the primitive mode of belief the propositional content condition θ_{future} that associates with each agent and moment the set of propositions that are future with respect to that moment. In symbols $M_{foresee} = [\theta_{future}]Belief$. The mode of *expectation* is obtained from that of prevision by adding the special cognitive way that the agent is then in a state of expectation. $M_{expect} = [\omega_{expectation}]M_{foresee}$. The mode of being sure is obtained from that of belief by adding the special way that the agent is in a state of confidence for he has strong reasons. Whoever is *convinced* is sure while being in a stronger state of conviction (special cognitive way). Whoever has *faith* is convinced that God exists (propositional content condition). In the case of *knowledge* the agent has more than confidence, he has strong evidence that guarantees truth (special cognitive way and preparatory condition). The mode of *hope* is obtained from that of desire by adding the special cognitive way that the agent is then uncertain as regards the existence and the inexistence of the represented fact and the preparatory condition that that fact is then possible. The mode of *aspiration* is obtained from that of hope by adding the condition that the propositional content concerns the very person of the agent and the preparatory condition that it is a pursued goal, sometimes even an ultimate or ideal aim. The mode of *satisfaction* is obtained from that of *desire* by adding the *preparatory condition* that the desired fact exists. The mode of *pleasure* has, in addition, the *volitive way* that the satisfaction of the desire puts the agent in a state of pleasure and the preparatory condition that it is

---

[14] I have formulated the logic of primitive propositional attitudes in "Belief and Desire: A Logical Analysis", forthcoming in Colin Schmidt (ed), *Proceedings of the International Colloquium Computers and Philosophy*, held at Laval (France) in May 2006.

good for the agent. The mode of *enjoyment* is obtained from that of pleasure by adding the special *way* of being in a conscious state of enjoyment. *Lust* has the special volitive way that it is a *sensual* enjoyment. *Concupiscence* is a lust of the flesh (propositional content condition).

Because all operations on psychological modes add new components, they generate stronger modes. Each attitude $M(P)$ with a complex mode *contains* attitudes $M'(P)$ whose modes have less components. For the psychological mode $[\varpi]M$ that is obtained by adding to mode M the special way ϖ has the *conjunction* $\varpi \cap \varpi_M$ as proper way. Similarly the psychological mode $[\theta]M$ that is obtained by adding to mode M the propositional content condition θ has the *intersection* $\theta \cap \theta_M$ as proper propositional content condition. Finally the mode $[\Sigma]M$ that is obtained by adding to mode M the preparatory condition Σ has the union $\Sigma \cup \Sigma_M$ as a proper preparatory condition. Whoever has an attitude of the form $[\varpi]M(P)$, $[\theta]M(P)$ or $[\Sigma]M(P)$ has *eo ipso* the simpler attitude $M(P)$. Thus whoever expects, knows, is sure, certain or convinced of something believes that thing. Whoever hopes, wishes, enjoys, is happy, satisfied or pleased of something desires it.

2.9   Semantic tableaux

A lexical analysis of terms for attitudes based on the decomposition that I advocate can systematically explain which name stronger psychological modes. One can even show comparative strength by drawing semantic tableaux having the form of trees whose nodes are ordinary names or verbs for attitudes. Here are two semantic tableaux showing relations of comparative strength between cognitive and volitive modes respectively. The initial node of the first tableau is the term "belief" which names the primitive cognitive mode, just as the initial node of the second is the term "desire" that names the primitive volitive mode. Any immediate successor of a term is another term naming a stronger psychological mode obtained by applying one or more operations whose nature is indicating by symbols in the branch between the two terms. I have already specified special components in examples that I have given. Here are a few additional examples. To have *appetite* in one sense of the term is to desire to eat (special propositional content condition). To be hungry is to desire to eat in feeling a strong need of food (special volitive way). *Plans* are partial prior intentions formed after a deliberation (preparatory condition) to achieve ends that agents fill in, as time goes by, with more specific intentions concerning means and preliminary steps. They have a hierarchical structure:[15] plans concerning ends require subplans concerning means, preliminary steps and specific actions (volitive way). *Projects* on the other hand are general plans (propositional content condition).[16]

---

[15] See M. Bratman, *Intentions, Plans and Practical Reason*, Cambridge, Harvard University Press: 1987.

[16] See my next book *Speech Acts in Dialogue* for more explanation.

Daniel Vanderveken

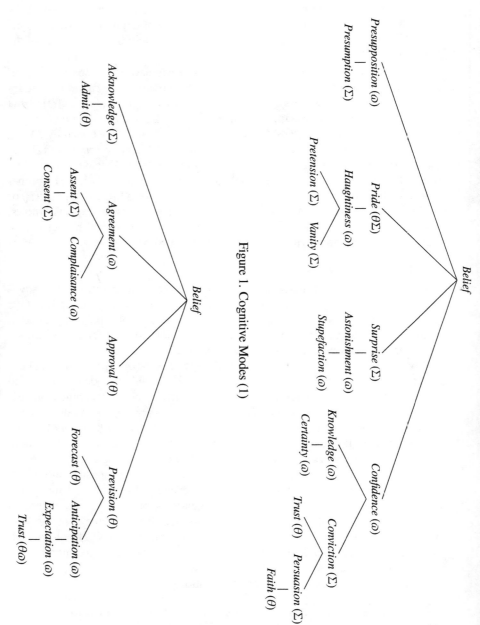

Figure 1. Cognitive Modes (1)

Figure 2. Cognitive Modes (2)

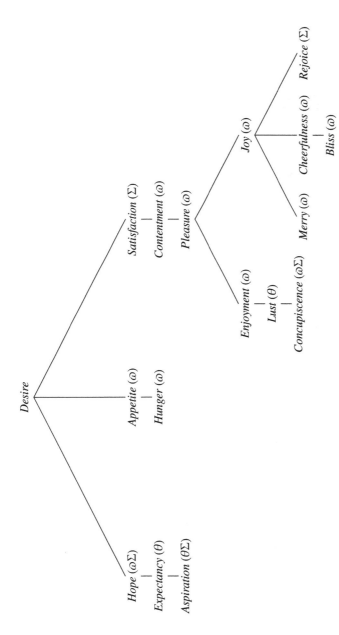

Figure 3. Volitive Modes (1)

Daniel Vanderveken

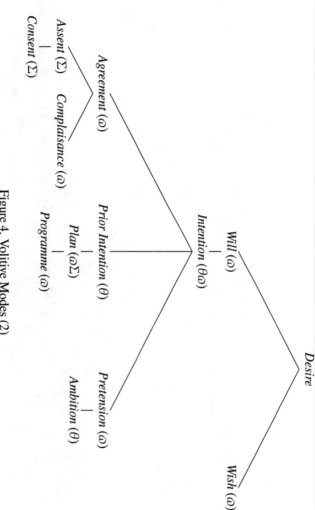

Figure 4. Volitive Modes (2)

Each term of a branch of a semantic tableau names a stronger psychological mode than terms that are higher in the same branch. Certain modes can be obtained by adding components to modes named by terms occurring in different branches of the tableau. Whoever *trusts* someone is *convinced* that he is honest, that he will keep his word, etc. These are new propositional content conditions to add to the mode of *conviction* to get that of *trust*. Whoever *trusts* someone also *expects* him to behave in a certain way. By adding this new cognitive way to the mode of *expectation* we also get that of *trust*.

2.10 Ideography

The object language of my logic of attitudes is *ideographical*. The apparent syntactic forms of formulas naming psychological modes show their components and logical form. My ideography shows in particular which modes are stronger than others. Thus a mode-formula that contains another one names *eo ipso* a stronger psychological mode. We all know that an *expectation* is a *prevision* and a *prevision* a *belief*. That strong psychological commitment is not shown by the graphic form of the verbs "expect", "foresee" and "believe". But it is visible in my ideography where $M_{foresee} = [\theta_{future}]M_{belief}$ and $M_{expect} = [\omega_{expectation}]M_{foresee}$. There is no one-to-one correspondence between possible psychological modes and ordinary language names or verbs for attitudes. On one hand, certain modes are not lexicalized. Only certain possible components of modes are significant in each natural language. On the other hand, certain terms for attitudes are ambiguous. For example, "agreement", "confidence" and "consent" name mental *acts* as well as mental *states*. One can *give* one's agreement as well as *be* in agreement. In such a case, the state named by the verb is a sincerity condition of the speech act. Whoever performs the act expresses the corresponding state. Certain verbs like "to be sure" and "to agree" are ambiguous between different modes. One can be *sure* of the existence of a fact (cognitive mode); one can also be *sure* that one will act (volitive mode). In that case, the cognitive and volitive modes have the same additional component (the special way of being in a state of *confidence*). Moreover terms like *fear, sorrow, regret* that have the same syntactical surface behavior do not name psychological modes but rather forms of propositional attitudes. My ideography clarifies the deep structure of natural languages in showing the logical nature of attitudes. Thus in my ideography, a deception that P is not an attitude with the propositional content P. It is a desire of the negation $\neg P$ of P with the preparatory conditions that its propositional content is false and that the agent previously believed that it would be true.

2.11 Conditions of satisfaction of propositional attitudes

The general notion of *satisfaction condition* in logic is based on that of *correspondence*. Propositional attitudes and elementary illocutions are directed towards facts of the world represented by their propositional content. Most often agents establish a correspondence between their ideas and things in the case of attitudes and between their words and things in the case of illocutions. Their attitudes and illocutions have for that reason *satisfaction conditions*. In order that the attitude or illocution of an agent at a moment is *satisfied*, there must be a correspondence between that agent's ideas or words and represented things in the world in the history of that moment. Agents live and persist in an indeterminist world. Their future is open. At each moment where we think and act we ignore how the world will continue. However, our attitudes and actions are directed toward the real historic continuation of these moments. In order that a present desire directed at the future is satisfied, it is not enough that things will be at a posterior moment as the agent now desires. They must be so later in the real future. So the *satisfaction* of propositional attitudes and elementary illocutions of an agent at a moment requires the *truth at that very moment* of their propositional content. The notion of satisfaction is a generalization of the notion of *actual truth* that covers attitudes and elementary illocutions with a not empty direction of fit.[17] Just as a *belief* at a moment is *satisfied* when it is *then true*, a *wish* and a *desire* are satisfied when they are *then realized*; a *prevision*, an *expectation*, a *hope* and an *aspiration* are satisfied when they are *fulfilled*; an *intention*, a *project* and a *plan* when they are *then executed*; a *fear* and a *fright* when the thing that is feared does not *then happen*.

There are four possible directions of fit between mind and things, just as there are four possible directions of fit between words and things. Like assertive illocutions, *cognitive attitudes* have the *mind-to-things direction of fit*. They are *satisfied* when their propositional content is *true* at the moment under consideration. The agent's ideas correspond to things as they are then in the world. In the cognitive case, when the agent realizes that there is no correspondence, he immediately changes his ideas. This is why the *truth predicates* characterize so well *satisfaction* and *dissatisfaction* in the case of *cognitive* attitudes. However, such truth predicates do not apply to *volitive* attitudes whose direction of fit goes from things to words. For the world and not the agent is at fault in the case of dissatisfaction of volitive attitudes. In that case, the agent can keep his ideas and remains dissatisfied. Most often, agents having a *volitive attitude desire the fact* represented by the propositional content *no matter how that fact turns to be existent in the world*. So most volitive attitudes that agents have at a moment are *satisfied* when their content is then true, no matter for which reason. Things are then such as the agent desires them to be, no matter what is the cause of their existence.

The only exceptions to this rule are *volitive attitudes* like *will, intentions, projects, plans* and *programs* whose proper volitive way requires that things fit the agent's ideas because he or she wants them in that way. Such attitudes as well as illo-

---

[17]We need an actuality connective for a right account of satisfaction conditions. A proposition of the form *Actually P* is true in a circumstance m/h when it is true at the moment m according to its history h_m.

cutionary acts (orders, commands, pledges and promises) that express them have *self-referential satisfaction conditions*. Their satisfaction requires more than the real existence of the fact represented by their propositional content. It requires that the represented fact turns to be existent in order to satisfy the agent's attitude. For example, in order to execute a prior intention, an agent must do more than carry out later the intended action in the real future; he or she must carry out that action because of that previous intention. If the agent does not act for that reason, (if he or she is obliged to act), his or her prior intention is not then executed. Like speech act theory, the logic of attitudes can explain such cases of self-referential satisfaction by relying on the notion of *intentional causation*. The agent's attitude must then be a *practical reason* why the represented fact turns to be existent.

As Searle pointed out in *Intentionality*, certain *volitive* modes like *joy, gladness, pride, pleasure, regret, sadness, sorrow*, and *shame* have the *empty direction of fit*. Agents who have such attitudes do not want to establish a correspondence between their ideas and things in the world. They just take for granted either correspondence or lack of correspondence. In the case of *joy, gladness, pride* and *pleasure*, the agent believes that the desired fact exists. In the case of *regret, sorrow* and *shame*, he or she believes on the contrary that it does not exist. The first attitudes have the special preparatory condition Σ_{Truth} that their propositional content is then true. The second attitudes have the opposite preparatory condition $\Sigma_{Falsehood}$ that their content is then false. So $\Sigma_{Truth}(a, m, P) = \{Actually\ P\}$ and $\Sigma_{Falsehood}(a, m, P) = \{Actually\ \neg P\}$. Volitive attitudes with such special *preparatory condition* have the *empty direction of fit* because their agent could not have the intention of achieving a success of fit. This is why they do not have *satisfaction conditions*. Instead of being satisfied or dissatisfied, they are just *appropriate* or *inappropriate*. They are inappropriate when their preparatory condition of actual truth or falsehood is wrong or when their proper psychological mode does not suit the fact represented by their content. No agent should be ashamed of an action that he has not made or that is exemplary and good for all. As Candida de Sousa Melo pointed out,[18] declaratory acts of thought have the *double direction of fit between mind and things*. In making verbal and mental *declarations*, the speaker changes represented things of the world just by way of thinking or saying that he is changing them. Whoever gives by declaration a new name to a thing acts in such a way that that thing has then that name. In such a case, the very act of the mind brings about the represented fact. Unlike illocutions, *attitudes are states and not actions of the mind*. So they could not have the double direction of fit.

---

[18]Candida de Sousa Melo, "Possible Directions of Fit between Mind, Language and the World", in D. Vanderveken and S. Kubo (eds.), *Essays in Speech Act Theory*, Benjamins 2002.

2.12   Identity criteria for propositional attitudes

In my view, propositional attitudes are not pairs of a psychological mode and a propositional content. Whoever possesses a propositional attitude of the form $M(P)$ *applies* in a certain way the mode M to the content P so as to determine under which conditions it is satisfied. In order that two propositional attitudes are identical, it is not necessary that their psychological mode be identical. What matters is that their content represents the same fact and that whoever possesses one attitude also possesses the other. Attitudes with the same propositional content and possession conditions fulfill the same function in psychological life. Thus the *belief* that it will always be the case that $2 + 2 = 4$ is also a *prevision*. These two attitudes have the same content but different modes. Contrary to the mode of belief, that of prevision has the special propositional content condition that the represented fact is future. However the content of the belief in question is future with respect to any moment. This is why it is also a prevision.

3   Main laws of my logic of propositional attitudes

The ideographical object language $\mathcal{L}$ of my logic contains formulas of the following forms.

3.1   Terms representing components of psychological modes

Constants of type ω name *ways* of believing or desiring. In particular, *Bel* and *Des* are two logical constants of type ω naming respectively the general categories of cognition and volition of *belief* and *desire*. K_ω names the *special cognitive way of knowledge* and $Want_\omega$ and Int_ω the *special volitive ways* of *will* and *intention*. Constants of type θ name *propositional content conditions* and constants of type Σ *preparatory conditions*. Thus $Past_\theta$ names the propositional content condition that the propositional content is past and $Future_\theta$ that it is future. The logical constant Int_θ names the *special propositional content condition of intentions*. $Truth_\Sigma$ names the preparatory condition that the propositional content is then true. 1_ω, 1_θ and 1_Σ name *neutral* components. If A and B are component terms of $\mathcal{L}$ of the same type, the new terms $(A \cap B)$ and $(A \cup B)$ name respectively the *intersection* and the *union* of the components named by A and B. Thus, $(A_\theta \cap B_\theta)$ names the intersection of propositional content conditions A_θ and B_θ.

3.2   Formulas naming cognitive and volitive *psychological modes*

$[A_\omega, A_\theta, A_\Sigma, Bel]$ names the weakest cognitive psychological mode having the way named by A_ω, the propositional content condition named by A_θ and the preparatory condition named by A_Σ. Similarly, $[A_\omega, A_\theta, A_\Sigma, Des]$ names the weakest volitive mode with these components.

3.3   Propositional formulas

1. If R_n is a predicate of degree n and $t_1, ..., t_n$ is a sequence of n individual constants, then $[R_n t_1...t_n]$ is a propositional formula expressing the elementary proposition that predicates the attribute expressed by R_n of the n individuals under concepts expressed by $t_1,..., t_n$ in that order.

2. If A_p and B_p are propositional formulas and a an individual constant naming an agent, then $\neg A_p$, $\square A_p$, $Will\, A_p$, $Was\, A_p$, $Settled\, A_p$, $Actually\, A_p$, $Tautological\, A_p$, $(A_p \wedge B_p)$, $(A_p \ni B_p)$, $[BelaA_\omega A_p]$, $[DesaA_\omega A_p]$, $\diamond a A_p$, $[\rho A_p B_p]$, $(A_\theta a A_p)$, $(A_\Sigma a A_p)$ and $(a \gg A_p)$ are new propositional formulas.

- $\neg A_p$ expresses the truth functional negation of proposition A_p.

- $\square A_p$ means that it is historically necessary that A_p, in the sense that it could not then have been otherwise than A_p.

- $Will\, A_p$ expresses the future proposition that it will be the case that A_p.

- $Was\, A_p$ expresses the past proposition that it has been the case that A_p.

- $Settled\, A_p$ expresses the modal proposition that it is settled that A_p, namely that A_p is true according to all histories.

- $Actually\, A_p$ expresses in each context the indexical proposition that it is then actual that A_p, that is to say that proposition A_p is true at the moment and in the proper history of that context.

- $Tautological\, A_p$ means that proposition A_p is a tautology.

- $(A_p \wedge B_p)$ expresses the conjunction of the propositions expressed by A_p and B_p.

- $(A_p \ni B_p)$ means that proposition A_p contains all elementary propositions of proposition B_p.

- $[BelaA_\omega A_p]$ and $[DesaA_\omega A_p]$ respectively mean that agent a believes and desires the proposition A_p with the mode named by A_ω.

- $\diamond a A_p$ means that A_p could be true given all attitudes of the agent a .[19]

- $[\rho A_p B_p]$ means that A_p is true because of B_p.

- $(A_\theta a A_p)$ means that proposition A_p satisfies the propositional content condition named by A_θ as regards the agent a.

---

[19] $\diamond a A_p$ is true in a circumstance m/h when A_p is true in at least one coinstantaneous circumstance where the agent a has all attitudes that he or she has at the moment m.

- $(A_\Sigma a A_p)$ expresses in each context a proposition that is true in a circumstance m/h when the preparatory conditions named by A_Σ are fulfilled as regards the agent a, the propositional content A_p and the moment m of that circumstance. In other words, $(A_\Sigma a A_p)$ expresses in each context the conjunction of all propositions that the agent a would presuppose if he or she had then an attitude with the propositional content A_p and the preparatory condition named by A_Σ.

- $(a \gg A_p)$ means that agent a presupposes that A_p.

Most fundamental truth functional, modal, propositional, and psychological notions that are important for the analysis of attitudes can be derived from my few primitive notions. Here are well known or new important abbreviation rules. Some are laws of identity.

- Always:
$$Always\, A =_{def} \neg Was\neg A \wedge A \wedge \neg Will\neg A$$

- Actual future (Occam):
$$Later\, A =_{def} Actually\, Will\, A$$

- Universal Necessity:
$$\blacksquare A =_{def} Always\, \square A \wedge \square Always\, A$$

- Strict implication:
$$(A \twoheadrightarrow B) =_{def} \blacksquare(A \Rightarrow B)$$

- Historical possibility:
$$\Diamond A =_{def} \neg\square\neg A$$

- Strong implication:
$$(A_p \mapsto B_p) =_{def} (A_p \ni B_p) \wedge Tautological\,(A_p \Rightarrow B_p)$$

- Same structure of propositional constituents:
$$A_p \equiv B_p =_{def} (A_p \ni B_p) \wedge (B_p \ni A_p)$$

- The preparatory condition of actual falsehood of the propositional content
$$Falsehood_\Sigma a A_p =_{def} Truth_\Sigma a \neg A_p$$

- The preparatory condition that the propositional content is then possible

$$\Diamond_\Sigma a A_p =_{def} Truth_\Sigma a \Diamond A_p$$

- The preparatory condition that the propositional content is not then necessary

$$\neg\Box_\Sigma a A_p =_{def} Truth_\Sigma a \neg\Box A_p$$

- The primitive mode of belief:

$$Belief =_{def} [1_\omega, 1_\theta, 1_\Sigma, Bel]$$

- The primitive mode of desire:

$$Desire =_{def} [1_\omega, 1_\theta, 1_\Sigma, Des]$$

- The operation of imposing a cognitive or volitive way to a psychological mode:

$$[B_\omega][A_\omega, A_\theta, A_\Sigma, \vartheta] =_{def} [(B_\omega \cap A_\omega), A_\theta, A_\Sigma, \vartheta]$$

where ϑ is *Bel* or *Des*.

- The operation of adding a new propositional content condition to a psychological mode:

$$[B_\theta][A_\omega, A_\theta, A_\Sigma, \vartheta] =_{def} [A_\omega, (B_\theta \cap A_\theta), A_\Sigma, \vartheta]$$

- The operation of adding a new preparatory condition to a psychological mode:

$$[B_\Sigma][A_\omega, A_\theta, A_\Sigma, \vartheta] =_{def} [A_\omega, A_\theta, (B_\Sigma \cup A_\Sigma), \vartheta]$$

- The cognitive mode of knowledge:

$$Knowledge =_{def} [K_\omega, 1_\theta, Truth_\Sigma, Bel]$$

- The volitive mode of will:

$$Will =_{def} [Want_\omega, Present_\theta \cup Future_\theta, \neg\Box_\Sigma \cup \Diamond_\Sigma, Des]$$

- The volitive mode of intention:

$$Intention =_{def} [Int_\omega][Int_\theta]Will$$

- The volitive mode of prior intention:

$$Priorintention =_{def} [Future_\theta]Intention$$

- Components: For any formule A_M naming a mode,

$$A_M(A_\varsigma) =_{def} [A_\varsigma]A_M = A_M$$

$A_M(A_\varsigma)$ means that A_ς is a component of mode A_M.

- The property of having the null direction of fit

$$\varnothing(A_M) =_{def} A_M(Des) \wedge (A_M(Truth_\Sigma) \vee A_M(Falsehood_\Sigma))$$

- Belief:

$$[BelaA_p] =_{def} [Bela1_\omega A_p]$$

$[BelaA_p]$ expresses the proposition that agent a believes that A_p.

- Desire:

$$[DesaA_p] =_{def} [Desa1_\omega A_p]$$

- Conditions of possession of propositional attitudes

$$[aHas([A_\omega, A_\theta, A_\Sigma, Bel]A_p)] =_{def}$$

$$[BelaA_\omega A_p] \wedge (A_\theta a A_p) \wedge [BelaA_\Sigma A_p] \wedge [a \gg A_\Sigma A_p]$$

$$[aHas([A_\omega, A_\theta, A_\Sigma, Des]Ap)] =_{def}$$

$$[BelaA_\omega A_p] \wedge (A_\theta a A_p) \wedge [BelaA_\Sigma A_p] \wedge [a \gg A_\Sigma A_p] \wedge [Bela\neg Tautological\, A_p]$$

  - $[aHas(A_M A_p)]$ means that agent a has the attitude with the mode A_M and the content A_p.

- Intention:

$$[IntendsaA_p] =_{def} [aHas(Intention\, A_p)]$$

- Knowledge:

$$[KaA_p] =_{def} [aHas(Knowledge\, A_p)]$$

- Strong psychological commitment for an agent:

$$(A_M A_p) \blacktriangleright_a (B_M B_p) =_{def} \Box([aHas(A_M A_p)] \Rightarrow [aHas(B_M B_p)])$$

where $(A_M A_p) \blacktriangleright_a (B_M B_p)$ means that agent a could not then have the first attitude without the second.

- Strong psychological commitment for all agents:

$$(A_M A_p) \blacktriangleright (B_M B_p) =_{def} Tautological((A_M A_p) \blacktriangleright_{a_1} (B_M B_p))$$

a_1 is the first individual constant. $(A_M A_p) \blacktriangleright (B_M B_p)$ means that no agent can have the first attitude without having the second.

- Psychological necessity[20]

$$\Box a A_p =_{def} \neg \Diamond a \neg A_p$$

- Weak psychological commitment for an agent:

  - $\triangleright_a (A_M A_p) =_{def} \Box a \Diamond a [aHas(A_M A_p)]$
  - $(A_M A_p) \triangleright_a (B_M B_p) =_{def} \Box ((aHas A_M A_p) \Rightarrow \triangleright_a (B_M B_p))$

- Weak psychological commitment for all agents:

$$(A_M A_p) \triangleright (B_M B_p) =_{def} Tautological((A_M A_p) \triangleright_{a_1} (B_M B_p))$$

- Propositional identity:

$$(A_p = B_p) =_{def} (A_p \mapsto B_p) \wedge (B_p \mapsto A_p)$$

- Same propositional attitudes:

$$(A_M A_p = B_M B_p) =_{def} (A_p = B_p) \wedge (A_M A_p \blacktriangleright B_M B_p) \wedge (B_M B_p \blacktriangleright A_M A_p)$$

- Conditions of satisfaction of propositional attitudes

$$Satisfied_a(A_M A_p) =_{def}$$

$$[aHas A_M A_p] \wedge \neg \varnothing(A_M) \wedge Actually\, A_p \wedge (A_M(Want_{\textcircled{}}) \Rightarrow [\rho(Actually\, A_p)[aHas A_M A_p]])$$

$Satisfied_a(A_M A_p)$ means that the propositional attitude of mode A_M and propositional content A_p of agent a is (or will be) satisfied.

All the instances in my ideography of classical axiom schemas of the first order predicate calculus, S5 modal logic for settled truth and historic necessity and of branching temporal logic are valid formulas. Here are new fundamental valid laws. Some are true by definition.[21]

---

[20] $\Box a A_p$ is true in a circumstance m/h when proposition A_p is true at all coinstantaneous moments where the agent a has all attitudes that he or she has at moment m.

[21] See my forthcoming book *Propositions, Truth and Thought* for the axiomatic system and formal semantics of my logic.

3.4   Valid formulas

Valid schemas for tautologies

(T1) $\vDash$ *Tautological* $A_p \Rightarrow \blacksquare A_p$ (Notice that $\nvDash \blacksquare A_p \Rightarrow$ *Tautological* A_p)

(T2) $\vDash$ *Tautological* $A_p \Rightarrow$ *Tautological Tautological* A_p

(T3) $\vDash \neg$*Tautological* $A_p \Rightarrow$ *Tautological* $\neg$*Tautological* A_p

(T4) $\vDash$ *Tautological* $A_p \Rightarrow$ (*Tautological* $(A_p \Rightarrow B_p) \Rightarrow$ *Tautological* B_p)

(T5) $\vDash$ *Tautological* $B_p \Rightarrow$ (*Tautological* $(A_p \Rightarrow B_p) \Rightarrow$ *Tautological* A_p)

Tautological implication is much stronger than strict implication.

(T6) $\vDash (A_p \supseteq B_p) \Rightarrow$ *Tautological* $(A_p \supseteq B_p)$

(T7) $\vDash \neg(A_p \supseteq B_p) \Rightarrow$ *Tautological* $\neg(A_p \supseteq B_p)$

Valid schemas for propositional identity

(Id1) $\vDash ((A_p \mapsto B_p) \wedge (B_p \mapsto A_p)) \Rightarrow (A_p = B_p)$

(Id2) $\vDash (A_p = B_p) \Rightarrow (C \Rightarrow C^*)$

where C^* and C are propositional formulas which differ at most by the fact that an occurrence of B_p in C^* replaces an occurrence of A_p in C.

(Id3) $\vDash (A_p = B_p) \Rightarrow$ *Tautological* $(A_p = B_p)$

(Id4) $\vDash \neg(A_p = B_p) \Rightarrow$ *Tautological* $\neg(A_p = B_p)$

Valid schemas for belief

(B1) $\vDash (BelaA_p \wedge BelaB_p) \Rightarrow Bela(A_p \wedge B_p)$

(B2) $\vDash$ *Tautological* $A_p \Rightarrow \neg Bela\neg A_p$

(B3) $\vDash$ *Tautological* $A_p \Rightarrow (BelaA_p \Rightarrow BelaTautological A_p)$

(B4) $\vDash BelaA_p \Rightarrow ((A_p \mapsto B_p) \Rightarrow (BelaB_p))$

(B5) $\vDash BelaA_p \Rightarrow (BelaBelaA_p)$

(B6) $\vDash BelaA_p \Rightarrow Bela\Diamond A_p$

(B7) $\vDash Satisfied_a(BelA_p) \Leftrightarrow ([BelaA_p] \wedge Actually A_p)$

Notice that the following laws are not valid for beliefs:

- $\not\models \blacksquare A_p \Rightarrow Bela A_p$ and $\not\models \neg \Diamond A_p \Rightarrow Bela \neg A_p$.

  Agents are not perfectly rational.

- $\not\models Tautological\, A_p \Rightarrow Bela A_p$.

  Agents are not logically omniscient.

  In particular,

- $\not\models (Tautological\, (A_p \Rightarrow B_p)) \Rightarrow (Bela A_p \Rightarrow Bela B_p)$.

  For:

- $\not\models (Tautological\, (A_p \Rightarrow B_p)) \Rightarrow (A_p \ni B_p)$.

  However,

- $\models Tautological\, (A_p \Rightarrow B_p) \Rightarrow (Bela A_p \twoheadrightarrow \neg Bela \neg B_p)$.

Valid schemas for desire

(D1) $\models (Desa A_p \wedge Desa B_p) \Rightarrow Desa(A_p \wedge B_p)$

(D2) $\models Tautological\, A_p \Rightarrow \neg(Desa A_p \vee Desa \neg A_p)$

(D3) $\models Desa A_p \Rightarrow (((A_p \mapsto B_p) \wedge \neg Tautological\, A_p) \Rightarrow (Desa B_p))$

(D4) $\models Desa A_p \Rightarrow Bela \neg Tautological\, A_p$

(D5) $\models Desa A_p \Rightarrow Desa \Diamond A_p$

(D6) $\models Satisfied_a(Desa A_p) \Leftrightarrow ([Desa A_p] \wedge Actually\, A_p)$

  Notice that

- $\not\models (A_p \mapsto B_p) \Rightarrow (Desa A_p \Rightarrow Desa B_p)$.

  Moreover

- $\not\models Bela \blacksquare \neg A_p \Rightarrow \neg Desa A_p$

  However

- $\models Tautological\, (A_p \Rightarrow B_p) \Rightarrow (Desa A_p \twoheadrightarrow \neg Desa \neg B_p)$

Valid schemas for attitudes

- $\vDash [aHas(A_M A_p)] \Rightarrow Settled[aHas(A_M A_p)]$

- $\vDash ((A_p = B_p) \wedge (A_M A_p \blacktriangleright B_M B_p) \wedge (B_M B_p \blacktriangleright A_M A_p)) \Rightarrow (A_M A_p = B_M B_p)$
  where $\blacktriangleright$ is the sign for strong psychological commitment for all agents.

- $\vDash ([A_\varsigma]A_M A_p) \blacktriangleright (A_M A_p)$

- $\vDash Satisfied_a(A_M A_p) \Rightarrow (\neg\varnothing(A_M) \wedge Actually\, A_p)$

- $\vDash (\neg\varnothing(A_M) \wedge \neg A_M(Want_\omega) \wedge [aHas(A_M A_p)]) \Rightarrow (Satisfied_a(A_M a A_p) \Leftrightarrow Actually\, A_p)$

- $\vDash (\neg\varnothing(A_M) \wedge A_M(Want_\omega)) \Rightarrow (Satisfied_a(A_M A_p) \Rightarrow [\rho(Actually\, A_p)[aHas A_M A_p]]$

Valid schemas for knowledge

(K1) $\vDash KaA_p \Rightarrow (A_p \wedge BelaA_p)$

(K2) $\vDash KaA_p \Rightarrow KaKaA_p$

(K3) $\vDash Tautological\, A_p \Rightarrow ([aHas(A_M A_p)] \Rightarrow KaA_p)$
      But $\nvDash Tautological\, A_p \Rightarrow KaA_p$

(K4) $\vDash KaA_p \Rightarrow ((A_p \mapsto B_p) \Rightarrow (KaB_p))$
      But $\nvDash (Tautological\, (A_p \Rightarrow B_p)) \Rightarrow (KaA_p \Rightarrow KaB_p)$

Valid schemas for intention

(In1) $\vDash (IntaA_p) \Rightarrow \neg(Was\, B_p = A_p)$

(In2) $\vDash IntaA_p \Rightarrow ((Bela\Diamond A_p \wedge Bela\neg\Box A_p) \wedge DesaA_p)$

BIBLIOGRAPHY

[1] N. Belnap, M. Perloff and Ming Xu, *Facing the Future. Agents and Choices in Our Indeterminist World*, Oxford, Oxford University Press: 2001.
[2] M. Bratman, *Intentions, Plans and Practical Reason*, Harvard University Press: 1987.
[3] Rudolf Carnap, *Meaning and Necessity*, Chicago, University of Chicago Press: 1947.
[4] René Descartes, *Œuvres complètes*, Paris, La Pléiade, Gallimard: 1953.
[5] Jaakko Hintikka , "Semantics for Propositional attitudes", in L. Linsky (ed), *Reference and Modality*, Oxford, Oxford University Press: 1971.
[6] Candida de Sousa Melo, "Possible Directions of Fit between Mind, Language and the World", in D. Vanderveken and S. Kubo (eds), *Essays in Speech Act Theory*, Benjamins 2002.
[7] A.N. Prior, *Past, Present and Future*, Oxford, Clarendon Press: 1967.
[8] John Searle, *Intentionality*, Cambridge University Press: 1982.

[9] John Searle and Daniel Vanderveken, *Foundations of Illocutionary Logic*, Cambridge University Press: 1985.

[10] Daniel Vanderveken (ed), *Logic, Thought and Action*, Dordrecht, Springer: 2005.

[11] Daniel Vanderveken, "Belief and Desire: A Logical Analysis", forthcoming in Colin Schmidt (ed), *Proceedings of the International Colloquium Computers and Philosophy*, held at Laval (France) in May 2006.

[12] Daniel Vanderveken, *Speech Acts in Dialogue*. Forthcoming.

[13] Daniel Vanderveken, *Propositions, Truth and Thought*. Forthcoming.

Harmonious Many-Valued Propositional Logics and the Logic of Computer Networks

HEINRICH WANSING AND YAROSLAV SHRAMKO

ABSTRACT. In this paper we reconsider the notion of an n-valued propositional logic. In many-valued logic, sometimes a distinction is made not only between designated and undesignated (not designated) truth values, but between designated, undesignated, and antidesignated truth values. But even if the set of truth values is, in fact, tripartitioned, usually only a single semantic consequence relation is defined that preserves the possession of a designated value from the premises to the conclusions of an inference. We shall argue that if the set of semantical values is not bipartitioned into the designated and the antidesignated truth values, it is natural to define *two* entailment relations, a positive one that preserves possessing a designated value from the premises to the conclusions of an inference, and a negative one that preserves possessing an antidesignated value from the conclusions to the premises. Once this distinction has been drawn, it is quite natural to reflect it in the logical object language and to contemplate many-valued logics Λ, whose language is split into a positive and a matching negative logical vocabulary. If the positive and the negative entailment relations do not coincide, the interpretations of matching pairs of connectives are distinct, and nevertheless the positive entailment relation restricted to the positive vocabulary is isomorphic to the negative entailment relation restricted to the negative vocabulary, then we shall say that Λ is a *harmonious* many-valued logic. We shall present examples of harmonious finitely-valued logics. These examples are not *ad hoc*, but emerge naturally in the context of generalizing Nuel Belnap's ideas on how a single computer should think to how interconnected computers should reason. We shall conclude this paper with some remarks on generalizing the notion of a harmonious n-valued propositional logic.

1 Many-valued propositional logics generalized

There exists a simple definition of the notion of an extensional n-valued ($2 \le n \in \mathbb{N}$) propositional logic as a valuational system, see, for example, [22] or the standard definition of a logical matrix, say in [18]. According to this definition, an n-valued propositional logic is a structure

$$\langle \mathcal{V}, \mathcal{D}, \{f_c : c \in C\}\rangle,$$

where $\mathcal{V}$ is a non-empty set containing n elements ($2 \leq n$), $\mathcal{D}$ is a non-empty proper subset of $\mathcal{V}$, C is the (non-empty, finite) set of (primitive) connectives of some propositional language $\mathcal{L}$, and every f_c is a function on $\mathcal{V}$ with the same arity as c. The elements of $\mathcal{V}$ are usually called *truth values*, and the elements of $\mathcal{D}$ are regarded as the *designated* truth values. A structure $\langle \mathcal{V}, \mathcal{D}, \{f_c : c \in C\} \rangle$ may be viewed as a logic, because the set of designated truth values determines a relation of semantical consequence (entailment) $\models \subseteq \mathcal{P}(\mathcal{L}) \times \mathcal{P}(\mathcal{L})$, where $\mathcal{P}(\mathcal{L})$ is the powerset of $\mathcal{L}$. A valuation function v is a function from the set of all atomic formulas (alias sentence letters) into $\mathcal{V}$. Every valuation function v is inductively extended to a function from the set of all $\mathcal{L}$-formulas into $\mathcal{V}$ by the following definition:

$$v(c(A_1, \ldots, A_m)) \quad = \quad f_c(v(A_1), \ldots, v(A_m)),$$

where c is an m-place connective from C. A set of formulas Δ *entails* a set of formulas Γ ($\Delta \models \Gamma$) iff for every valuation function v the following holds true: if for every $A \in \Delta$, $v(A) \in \mathcal{D}$, then $v(B) \in \mathcal{D}$ for some $B \in \Gamma$. An n-valued tautology then is a formula A such that $\varnothing \models A$.

In some writings, the definition of an n-valued propositional logic and the terminology is slightly different. First, the elements of $\mathcal{V}$ are sometimes referred to as *quasi truth values*. Gottwald [14, p. 2] explains that one reason for using the term 'quasi truth value' is that there is no convincing and uniform interpretation[1] of the truth values that in many-valued logic are taken in addition to the classical truth values *true* and *false*, an understanding that, according to Gottwald, associates the additional values with the naive understanding of being true, respectively the naive understanding of degrees of being true. In later publications, Gottwald has changed his terminology and states that "to [a]void any confusion with the case of classical logic one prefers in many-valued logic to speak of *truth degrees* and to use the word "truth value" only for classical logic" [15, p. 4]. The term 'semantical value' (or just 'value') seems to be non-committal.

What is perhaps more important than these differences in terminology is that in part of the literature, for example in [14], [15], [18], [23], an explicit distinction is drawn between a set $\mathcal{D}^+$ of *designated* values and a set $\mathcal{D}^-$ of *antidesignated* values, where the latter need not coincide with the complement of $\mathcal{D}^+$.[2] Usually, this distinction is, however, not fully exploited in many-valued logic: The notion of entailment is defined with respect to the designated values, and no independent additional entailment relation is defined with respect to the antidesignated values.

---

[1] At least there was no such interpretation at the time of the writing of [14].

[2] Incidentally, this distinction is relevant for an assessment of Suszko's Thesis, see [32], the claim that "there are but two logical values, true and false" [6, p. 169], which is given a formal content by the so-called Suszko Reduction, the proof that every Tarskian n-valued propositional logic is also characterized by a bivalent semantics. For a recent treatment and references to the literature, see [6]. A critical discussion of Suszko's Thesis is presented in [36], which is a companion article to the present paper.

Gottwald is quite aware of the distinction between designated and antidesignated values. When he discusses the notion of a contradiction (or logical falsity), for example, he explains that there are two ways of generalizing this notion from classical logic to many-valued logic, [15, p. 32, notation adjusted]:

1. In the case that the given system S of propositional many-valued logic has a suitable negation connective $\sim$, one can take as logical falsities all those wffs H for which $\sim H$ is a logical truth.

2. In the case that the given system S of propositional many-valued logic has antidesignated truth degrees, one can take as logical falsities all those wffs H which assume only antidesignated truth degrees...

Gottwald assumes that $\mathcal{D}^+ \cap \mathcal{D}^- = \varnothing$ and remarks that if the designated and antidesignated values exhaust all truth degrees and, moreover, the negation operation $\sim$ satisfies the following standard condition (notation adjusted):

$$f_\sim(x) \in \mathcal{D}^+ \text{ iff } x \notin \mathcal{D}^+,$$

then the two notions of a contradiction coincide [15, p. 32]. Since $\mathcal{D}^-$ may differ from the complement of $\mathcal{D}^+$, another standard condition for negation $\sim$ is:

$$f_\sim(x) \in \mathcal{D}^+ \text{ iff } x \in \mathcal{D}^- \text{ and } f_\sim(x) \in \mathcal{D}^- \text{ iff } x \in \mathcal{D}^+.$$

If the latter condition is satisfied, the two ways of defining the notion of a contradiction are equivalent also if $\mathcal{V} \setminus \mathcal{D}^+ \neq \mathcal{D}^-$.

Rescher [23, p. 68], who does not consider semantic consequence but only tautologies and contradictions, explains that "there may be good reason for *letting one and the same truth-value be both designated and antidesignated.*" However, he also warns that "we would not want it to happen that there is some truth-value ν which is both designated and antidesignated when it is also the case that there is some formula which uniformly assumes this truth-value, for then this formula would be both a tautology and a contradiction" [23, p. 67].

Although Gottwald recognizes that $\mathcal{V} \setminus \mathcal{D}^+$ may be distinct from $\mathcal{D}^-$,[3] he nevertheless follows the tradition in defining only a single semantic consequence rela-

---

[3]On p. 30 of [15] he explains (notation adjusted):

Even in the case that $\mathcal{D}^+ \neq \varnothing$ and $\mathcal{D}^- \neq \varnothing$ it is, however, not necessarily $\mathcal{D}^+ \cup \mathcal{D}^- = \mathcal{V}$, which means that together with designated and antidesignated truth degrees also *undesignated* truth degrees may exist. This possibility indicates two essentially different positions regarding the designation of truth degrees. The first one assumes only a binary division of the set of truth degrees and can proceed by simply marking a set of designated truth degrees, treating the undesignated ones like antidesignated ones. The second position assumes a tripartition and marks some truth degrees as designated, some others as antidesignated, and has besides these both types also some undesignated truth degrees....

tion in terms of $\mathcal{D}^+$. If this privileged treatment of $\mathcal{D}^+$ is given up, a more general definition of an n-valued propositional logic emerges.

DEFINITION 1. Let $\mathcal{L}$ be a language in a denumerable set of sentence letters and a finite non-empty set of finitary connectives C. An n-valued propositional logic is a structure

$$\langle \mathcal{V}, \mathcal{D}^+, \mathcal{D}^-, \{f_c : c \in C\}\rangle$$

where $\mathcal{V}$ is a non-empty set containing n elements ($2 \leq n$), $\mathcal{D}^+$ and $\mathcal{D}^-$ are distinct non-empty proper subsets of $\mathcal{V}$, and every f_c is a function on $\mathcal{V}$ with the same arity as c. Again, every valuation v is inductively extended to a function from the set of all $\mathcal{L}$-formulas into $\mathcal{V}$ by setting: $v(c(A_1, \ldots, A_m)) = f_c(v(A_1), \ldots, v(A_m))$, for every m-place $c \in C$. For all sets of $\mathcal{L}$-formulas Δ, Γ, semantic consequence relations $\models^+$ and $\models^-$ are defined as follows:

1. $\Delta \models^+ \Gamma$ iff for every valuation function v: (if for every $A \in \Delta$, $v(A) \in \mathcal{D}^+$, then $v(B) \in \mathcal{D}^+$ for some $B \in \Gamma$);

2. $\Delta \models^- \Gamma$ iff for every valuation function v: (if for every $A \in \Gamma$, $v(A) \in \mathcal{D}^-$, then $v(B) \in \mathcal{D}^-$ for some $B \in \Delta$).

An n-valued tautology then is a formula A such that $\varnothing \models^+ A$, and an n-valued contradiction is a formula A such that $A \models^- \varnothing$.

DEFINITION 2. Let $\Lambda = \langle \mathcal{V}, \mathcal{D}^+, \mathcal{D}^-, \{f_c : c \in C\}\rangle$ be an n-valued propositional logic. Λ is called a *separated* n-valued logic, if $\mathcal{V} \setminus \mathcal{D}^+ \neq \mathcal{D}^-$ (that is if $\mathcal{V}$ is not partitioned into the non-empty sets $\mathcal{D}^+$ and $\mathcal{D}^-$), and Λ is said to be *refined*, if it is separated and $\models^+ \neq \models^-$.

Clearly, if an n-valued logic is not separated, the two entailment relations $\models^+$ and $\models^-$ coincide. In a refined n-valued propositional logic, however, neither "positive" entailment $\models^+$ nor "negative" entailment $\models^-$ need enjoy a privileged status in comparison to each other. Moreover, an entailment relation is usually defined with respect to a given formal language. For the separate entailment relations $\models^+$ and $\models^-$ one might, therefore, expect that they come with their own, in a sense "dual", languages, $\mathcal{L}^+$ and $\mathcal{L}^-$ (see the next section for further explanations). Then one might be interested in n-valued logics in which these languages have the same signature.

DEFINITION 3. Let C be a finite non-empty set of finitary connectives, let $\mathcal{L}^+$ be the language based on $C^+ = \{c^+ \mid c \in C\}$, and let $\mathcal{L}^-$ be the language based on $C^- = \{c^- \mid c \in C\}$. If A is an $\mathcal{L}^+$-formula, let A^- be the result of replacing every connective c^+ in A by c^-. If Δ is a set of $\mathcal{L}^+$ formulas, let $\Delta^- = \{A^- \mid A \in \Delta\}$. If the language $\mathcal{L}$ of a refined n-valued logic Λ is based on $C^+ \cup C^-$, then Λ is said to be *harmonious* iff (i) for all sets of $\mathcal{L}^+$-formulas Δ, Γ: $\Delta \models^+ \Gamma$ iff $\Delta^- \models^- \Gamma^-$, and (ii) for every $c \in C$, $f_{c^+} \neq f_{c^-}$.

In the present paper, we shall, first of all, argue that the distinction between designated and antidesignated values, and therefore also the distinction between positive entailment $\models^+$ and negative entailment $\models^-$, is an important distinction (Section 2). In Section 3 we shall consider some separated n-valued propositional logics, and in Sections 4 and 5 we shall present natural examples of harmonious finitely-valued logics. Finally, we shall make some brief remarks on generalizing harmony (Section 6).

2   Designated and antidesignated values

Why is it important to draw a distinction between designated and antidesignated values? The notion of a set of designated values is often considered as a generalization of the notion of truth. Similarly the set of antidesignated values can be regarded as representing a generalized concept of falsity. However, logic and its terminology is to a large extent predominated by the notion of truth. The Fregean *Bedeutung* of a declarative sentence (or, in the first place, a thought) is a truth value. According to Frege, there are exactly two truth values, *The True* and *The False*, henceforth just *true* (T) and *false* (F). Whereas T and F are referred to as *truth* values, neither F nor T is called a *falsity* value.

Moreover, Frege explicitly characterized logic as "the science of the most general laws of being true", see H. Sluga's translation in [31, p. 86]. This view finds its manifestation in the fact that most of the fundamental logical notions – logical operations, relations etc. – are usually defined through the category of truth. Thus, valid consequence is usually defined as preserving truth in passing from the premises to the conclusions of an inference. It is required that for every model $\mathfrak{M}$, if all the premises are true in $\mathfrak{M}$, then so is at least one conclusion. By contraposition, in a valid inference being not true is preserved from the conclusions to at least one of the premises. For every model $\mathfrak{M}$, if every conclusion is not true in $\mathfrak{M}$, then so is at least one of the premises. As to falsity, its role in such definitions frequently remains a subordinated one, if any. When constructing a semantic model, "false" is often understood as a mere abbreviation for "not true", for instance when the classical truth-table definition for conjunction is stated as: "A conjunctive sentence is true if both of its conjuncts are true, otherwise it is not true (i.e., false)."

In general failing to be true and being false may, however, fall apart. Although this is the very point of many-valued logic, a distinction between falsity and the absence of truth (or truth and the absence of falsity) is often represented only by the values that $\mathcal{V}$ contains in addition to T and F, and not by distinguishing between a set of designated values $\mathcal{D}^+$ and another set of antidesignated values $\mathcal{D}^-$ (and the consequence relations induced by these sets). We are interested in inferences that preserve truth, because we are interested in true beliefs. But likewise we are interested in inferences in which false conclusions are bound to depend on at least one false premise, because we are interested in avoiding false beliefs. This point has vividly been made by William James in his essay 'The will to believe':

> Believe truth! Shun error!—These, we see, are two materially different laws; and by choosing between them we may end by coloring differently our whole intellectual life. We may regard the chase for truth as paramount, and the avoidance of error as secondary; or we may, on the other hand, treat the avoidance of error as more imperative, and let truth take its chance [16, p. 18].

It seems then quite natural to modify (or to extend) in a certain respect the Fregean definition of logic by saying that its scope is studying not just the laws of being true, but rather of being true *and* being false. An immediate effect of this modification consists in acknowledging the importance of falsity (and more generally, antidesignated values) for defining logical notions. Every definition formulated in terms of truth should be "counterbalanced" with a "parallel" (dual) definition formulated in terms of falsity.

This observation not only justifies the distinction between positive ($\models^+$) and negative ($\models^-$) entailment relations, but also clarifies the idea of the corresponding languages $\mathcal{L}^+$ and $\mathcal{L}^-$ with "positive" and "negative" connectives. As an example let us take the operation of conjunction in classical logic. The connective $\wedge^+$ can naturally be defined through its truth conditions: $v(A \wedge^+ B) = T$ iff $v(A) = T$ and $v(B) = T$. But we may also wish to consider the connective $\wedge^-$, exhaustively defined by means of the *falsity conditions*: $v(A \wedge^- B) = F$ iff $v(A) = F$ or $v(B) = F$. Now, although in classical logic $\wedge^+$ and $\wedge^-$ are, obviously, coincident, in the general case, for example in various many-valued logics, they may well differ from each other. Thus, if the relation between designated and antidesignated values is not so straightforward as in classical logic, a separate introduction of the positive and negative connectives (in fact: "truth-connectives" and "falsity-connectives") may acquire especial significance.

Let us generalize this point with respect to the propositional connectives of conjunction and disjunction in the context of many-valued logics. There is a view that in a many-valued semantics $f_\wedge$ and $f_\vee$ are just the functions of taking the minimum and the maximum of their arguments. As R. Dewitt put it:

> In many-valued systems, intuitions concerning the appropriate truth-conditions for disjunction and conjunction are the most widely agreed on. In particular, there is general agreement that a disjunction should take the maximum value of the disjuncts, while a conjunction should take the minimum value of the conjuncts [8, p. 552].

As a result, we get the following definitions:

$$v(A \wedge B) = min(v(A), v(B));$$
$$v(A \vee B) = max(v(A), v(B)),$$

which Dewitt refers to as the *standard conditions* for conjunction and disjunction. One may note, however, that these conditions are justified only if the set of semantical values is in some way *linearly ordered*, so that any two values are mutually comparable. And although this is frequently indeed the case, e.g., when the semantical values are identified with some points on a numerical segment, there also exist a number of many-valued systems where not all of the values are comparable with each other.[4] In such systems, the standard conditions cannot be employed directly.

The idea of truth *degrees* in many-valued logic naturally implies that semantical values may differ in their truth-content. Moreover, it is assumed that any of the designated values is "more true" (is of a higher degree in its truth-content) than any of the values which is not designated. Thus, taking into account the "minimality-maximality" intuition described above, conjunction can be more generally regarded as an operation $\wedge^+$ that in a sense *minimizes the truth-content* (or the "designatedness") of the conjuncts, and disjunction as an operation $\vee^+$ that *maximizes the truth-content* of the disjuncts.[5] That is, for truth values that are comparable in their truth degrees, $f_{\wedge^+}$ is just the standard *min*-function, but if two truth values x and y are incomparable, the "less true"–relation nevertheless should be such that it determines $f_{\wedge^+}(x, y)$, the outcome of which is *less true* than *both* of the conjuncts. And similarly for disjunction. In this way, we should be able to obtain definitions of notions of conjunction $\wedge^+$ and disjunction $\vee^+$ purely in terms of truth:

$$v(A \wedge^+ B) = min^+(v(A), v(B));$$
$$v(A \vee^+ B) = max^+(v(A), v(B)),$$

where min^+ and max^+ are generalized functions of truth-minimizing and truth-maximizing, correspondingly.

Now, if falsity is not the same as non-truth, an independent consideration of propositional connectives from the standpoint of antidesignated values is appropriate. In this sense conjunction should be regarded as the falsity-maximizer and disjunction as the falsity-minimizer, and thus, we should be able to obtain generalized functions max^- and min^- of maximizing and minimizing, respectively, the falsity-content of their arguments, so that operations $\wedge^-$ and $\vee^-$ can be defined purely in terms of falsity:

$$v(A \wedge^- B) = max^-(v(A), v(B));$$
$$v(A \vee^- B) = min^-(v(A), v(B)).$$

We also briefly observe the difference between two kinds of negation . Namely, whereas $f_{\sim^+}$ can be viewed as a function that (within every semantical value) turns

---

[4]Cf., for instance, the values **N** and **B** under the order $\leq_t$ in the four-valued logic B_4 considered in Section 4.

[5]And of course, a conjunction should maximize the non-truth of the conjuncts, while a disjunction should minimize the non-truth of the disjuncts.

truth into non-truth and vice versa, $f_{\sim}$- can be treated as an operation that interchanges exclusively between falsity and non-falsity.

One might object that the distinction between designated values and antidesignated values makes sense *only* for doxastically or epistemically interpreted semantical values. Certainly, if a proposition is not believed (known) to be true, this does not imply that the proposition is believed (known) to be false. The distinction is, however, sensible also for other, non-doxastical and non-epistemical adverbial qualifications of truth and falsity. If a proposition is not necessarily true, for instance, it need not be necessarily false. The designated semantical values are used to define an entailment relation $\models^+$ that preserves possessing a doxastically wanted value in passing from the premises to the conclusions of an inference. Analogously, the antidesignated values may be used to define an entailment relation $\models^-$ by requiring that if the conclusions are doxastically unwanted, at least one of the assumptions is doxastically unwanted, too. Among the doxastically wanted values there may be values interpreted, for example, as "true", "neither true nor false", "known to be true", "unknown to be false", "necessarily true", "possibly true", etc. Among the doxastically unwanted values there may be values interpreted, for example, as "false", "neither true nor false", "known to be false", "unknown to be true", "necessarily false", "possibly false", etc.

Whether "neither true nor false" is doxastically wanted or unwanted may be a matter of perspective. A proposition evaluated as "neither true nor false" is not falsified and hence possibly wanted, but it is also not verified and therefore possibly unwanted. If one wants to take into account that both perspectives are legitimate, a value read as "neither true nor false" may even sensibly be classified as both wanted and unwanted. That is, in general it is reasonable to distinguish between designated, antidesignated, and undesignated semantical values, and also not to exclude the possibility of values that are both designated and antidesignated.

These considerations can be generalized, see Section 6.

3 Some separated finitely-valued logics

We first consider a separated version of Kleene's well-known 3-valued logic.

DEFINITION 4. Kleene's strong 3-valued logic K_3 is defined as follows:

$K_3 = \langle \{T, \varnothing, F\}, \{T\}, \{\varnothing, F\}, \{f_c : c \in \{\sim, \wedge, \vee, \supset\}\}\rangle$, where the functions f_c are defined by the following tables:

$f_\sim$	
T	F
$\varnothing$	$\varnothing$
F	T

$f_\wedge$	T	$\varnothing$	F
T	T	$\varnothing$	F
$\varnothing$	$\varnothing$	$\varnothing$	F
F	F	F	F

$f_\vee$	T	$\varnothing$	F
T	T	T	T
$\varnothing$	T	$\varnothing$	$\varnothing$
F	T	$\varnothing$	F

$f_\supset$	T	$\varnothing$	F
T	T	$\varnothing$	F
$\varnothing$	T	$\varnothing$	$\varnothing$
F	T	T	T

DEFINITION 5. The separated 3-valued propositional logic K_3^* is defined as follows:

$K_3^* := \langle \{T, \varnothing, F\}, \{T\}, \{F\}, \{f_c : c \in \{\sim, \wedge, \vee, \supset\}\}\rangle$, where the functions f_c are defined as in K_3.

OBSERVATION 6. K_3^* is refined, i.e., the relations $\models^+$ and $\models^-$ do not coincide.

Proof. In K_3^* formulas of the form $A \wedge (A \supset B)$ have the following truth table:

	A	B	$A \wedge (A \supset B)$
	T	T	T
	T	$\varnothing$	$\varnothing$
	T	F	F
	$\varnothing$	T	$\varnothing$
	$\varnothing$	$\varnothing$	$\varnothing$
*	$\varnothing$	F	$\varnothing$
	F	T	F
	F	$\varnothing$	F
	F	F	F

Whereas $A \wedge (A \supset B) \models^+ B$, the row marked with an asterisk shows that $A \wedge (A \supset B) \not\models^- B$. ∎

Although it is not surprising, perhaps, that in a separated n-valued logic the relations $\models^+$ and $\models^-$ need not coincide, there are separated n-valued logics which are not refined. Let $\mathbf{N} := \varnothing$, $\mathbf{T} := \{T\}$, $\mathbf{F} := \{F\}$ and $\mathbf{B} := \{T, F\}$.

DEFINITION 7. The useful 4-valued logic of Dunn and Belnap is the structure $B_4 = \langle \{\mathbf{N}, \mathbf{T}, \mathbf{F}, \mathbf{B}\}, \{\mathbf{T}, \mathbf{B}\}, \{\mathbf{N}, \mathbf{F}\}, \{f_c : c \in \{\sim, \wedge, \vee\}\}\rangle$, where the functions f_c are defined as follows:

$f_\sim$	
$\mathbf{T}$	$\mathbf{F}$
$\mathbf{B}$	$\mathbf{B}$
$\mathbf{N}$	$\mathbf{N}$
$\mathbf{F}$	$\mathbf{T}$

$f_\wedge$	$\mathbf{T}$	$\mathbf{B}$	$\mathbf{N}$	$\mathbf{F}$
$\mathbf{T}$	$\mathbf{T}$	$\mathbf{B}$	$\mathbf{N}$	$\mathbf{F}$
$\mathbf{B}$	$\mathbf{B}$	$\mathbf{B}$	$\mathbf{F}$	$\mathbf{F}$
$\mathbf{N}$	$\mathbf{N}$	$\mathbf{F}$	$\mathbf{N}$	$\mathbf{F}$
$\mathbf{F}$	$\mathbf{F}$	$\mathbf{F}$	$\mathbf{F}$	$\mathbf{F}$

$f_\vee$	$\mathbf{T}$	$\mathbf{B}$	$\mathbf{N}$	$\mathbf{F}$
$\mathbf{T}$	$\mathbf{T}$	$\mathbf{T}$	$\mathbf{T}$	$\mathbf{T}$
$\mathbf{B}$	$\mathbf{T}$	$\mathbf{B}$	$\mathbf{T}$	$\mathbf{B}$
$\mathbf{N}$	$\mathbf{T}$	$\mathbf{T}$	$\mathbf{N}$	$\mathbf{N}$
$\mathbf{F}$	$\mathbf{T}$	$\mathbf{B}$	$\mathbf{N}$	$\mathbf{F}$

DEFINITION 8. The separated 4-valued propositional logic B_4^* is the structure $\langle \{\mathbf{N}, \mathbf{T}, \mathbf{F}, \mathbf{B}\}, \{\mathbf{T}, \mathbf{B}\}, \{\mathbf{F}, \mathbf{B}\}, \{f_c : c \in \{\sim, \wedge, \vee, \}\}\rangle$, where the functions f_c are defined as in B_4.

The set $\{\mathbf{N}, \mathbf{T}, \mathbf{F}, \mathbf{B}\}$ is also referred to as $\mathbf{4}$. Note that in B_4^* not only $\mathbf{4} \setminus \mathcal{D}^+ \neq \mathcal{D}^-$, but also $\mathcal{D}^+ \cap \mathcal{D}^- \neq \varnothing$.

OBSERVATION 9. The separated logic B_4^* is not refined: $\models^+ = \models^-$.

Proof. A proof (for the case of single premises and conclusions) using Dunn's method of "dual" valuations is given, e.g., in [10, p. 10]. There it is observed that the relation $\models^+$ alias $\models^-$ also coincides with the relation $\models$ defined as follows: $\Delta \models \Gamma$ iff $(\Delta \models^+ \Gamma$ and $\Delta \models^- \Gamma)$. ∎

Clearly, K_3 does not have any tautologies, because any formula takes the value $\varnothing$ if every propositional variable occurring in it takes the value $\varnothing$. It has been observed in [23] that for the same reason the 3-valued logic which is now known as the Logic of Paradox [21], LP, has no contradictions.

DEFINITION 10.   The Logic of Paradox LP is the 3-valued propositional logic $\langle \{T, \varnothing, F\}, \{T, \varnothing\}, \{F\}, \{f_c : c \in \{\sim, \wedge, \vee, \supset\}\}\rangle$, where the functions f_c are defined as in K_3.

In B_4 there are neither any tautologies (consider the constant valuation that assigns only **N**) nor any contradictions (consider the constant valuation that assigns only **B**).[6] Obviously, $\models^+$ in B_4 coincides with $\models^+$ and $\models^-$ in B_4^*. Our main question is: Are there *natural* examples of harmonious finitely-valued logics?

4   A harmonious finitely-valued logic

We shall present a harmonious finitely-valued logic that emerges naturally in the context of generalizing Nuel Belnap's ideas on how a computer should reason.

The bilattice $FOUR_2$

Our starting point is the logic B_4, which is also known as the logic of *first degree entailment*. Belnap ([3], [4], see also [1, §81]), building on ideas developed by M. Dunn (see, e.g., [9] and [10]), proposed B_4 as a "useful four-valued logic" for dealing with information received by a computer. As Belnap points out, a computer may receive data from *various* (maybe independent) sources. Belnap's computers have to take into account various kinds of information concerning a given sentence. Besides the standard (classical) cases, when a computer obtains information either that the sentence is (1) true or that it is (2) false, two other (non-standard) situations are possible: (3) nothing is told about the sentence or (4) the sources supply inconsistent information, information that the sentence is true and information that it is false. Thus, we obtain the four truth values from B_4 that naturally correspond to these four "informational" situations: **N** (there is no information that the sentence is false and no information that it is true), **F** (there is *merely* information that the sentence is false), **T** (there is *merely* information that

---

[6]Rescher [23, p. 67] seems to interpret the fact that the logic LP has no contradictions as a reason for distinguishing between antidesignated and undesignated values, because in LP no formula receives an undesignated truth value under any valuation.

the sentence is true), and **B** (there is information that the sentence is false, but there is also information that it is true).

M. Ginsberg [12], [13] noticed that the four truth values of B_4 constitute an interesting algebraic structure, which he called a *bilattice*.[7] Figure 1 presents this bilattice $(FOUR_2)$ by a double Hasse diagram. We have here two partial orderings:

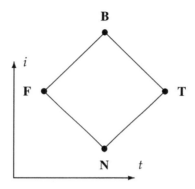

Figure 1. The bilattice $FOUR_2$

$\leq_i$ and $\leq_t$. The relation $\leq_i$ orders the elements of **4** by set-inclusion and represents an increase of information. The relation $\leq_t$ is said to represent an increase of truth among the elements constituting **4**.[8] The relation $\leq_t$ is of special importance, because this order determines the *logic* of Belnap's computers. First, conjunction and disjunction are interpreted as the operations of lattice meet and join relative to $\leq_t$. These operations are given by the truth functions $f_\wedge$ and $f_\vee$ from Definition 7. Moreover, negation $\sim$ is interpreted as the function $f_\sim$ from Definition 7, which represents an inversion of $\leq_t$ in the sense that $x \leq_t y$ iff $f_\sim(y) \leq_t f_\sim(x)$.

A valuation function into **4** is then recursively extended in the standard way to a map v from the language of B_4 into **4**, and the entailment relation of B_4 can be defined as follows:

$$\Delta \models \Gamma \text{ iff } \forall v \bigsqcap_t \{v(A) \mid A \in \Delta\} \leq_t \bigsqcup_t \{v(A) \mid A \in \Gamma\},$$

---

[7]Roughly speaking a bilattice is a set with *two* partial orderings, each determining its own lattice on this set (see also [2], [11]).

[8]Correspondingly one can distinguish between an information (approximation) lattice and a logical lattice. Belnap [3] has considered both these lattices separately under the labels **A4** and **L4**.

where $\sqcap_t$ is lattice meet and $\sqcup_t$ is lattice join in the complete lattice $(\mathbf{4}, \leq_t)$.[9]

From isolated computers to computer networks

One can observe that Belnap's interpretation works perfectly well if we deal with *one* (isolated) computer receiving information from *classical sources*, i.e., these sources operate exclusively with the classical truth values. As soon as a computer C is connected to other computers, there is no reason to assume that these computers cannot pass higher-level information concerning a given proposition to C. If several computers form a computer network, Belnap's ideas that motivated B_4 can be generalized. Consider, for example, four computers: C_1, C_2, C_3, and C_4 connected to another computer C_1', a server, to which they are supposed to supply information (Figure 2). It is fairly clear that the logic of the server itself (so, the network as a whole) cannot remain four-valued any more. Indeed, suppose C_1 informs C_1' that a sentence is true only (has the value **T**), whereas C_2 supplies inconsistent information (the sentence is both true and false, i.e., has the value **B**). In this situation C_1' has received the information that the sentence simultaneously

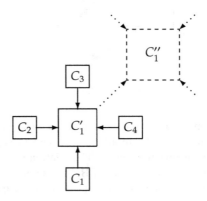

Figure 2. A computer network

is true only as well as both true and false, in other words, it has a value not from **4**, but from $\mathcal{P}(\mathbf{4})$, namely the value **TB** $= \{\{T\}, \{T, B\}\}$. If C_1' has been informed simultaneously by C_1 that a sentence is true-only, by C_2 that it is false-only, by C_3 that it is both-true-and-false, and by C_4 that it is neither-true-nor-false, then the value **NFTB** $= \{\varnothing, \{T\}, \{F\}, \{T, F\}\}$ is far from being a "madness" (cf. [20, p. 19]) but is just an adequate value which should be ascribed to the sentence by C'. That

---

[9] Another standard definition of multiple-conclusion semantic consequence builds compactness into this notion by requiring that $\Delta \models \Gamma$ iff there exist finite subsets $\Delta' \subseteq \Delta$ and $\Gamma' \subseteq \Gamma$ such that for every valuation v, $v(\bigwedge_t \Delta') \leq_t v(\bigvee_t \Gamma')$.

is, the logic of C'_1 has to be 16-valued. And if we wish to extend our network and to connect C'_1 to some "higher" computer (C''_1), then the amount of semantical values will increase to $2^{16} = 65536$. As we shall see later, this exponential growth of the number of truth values need not worry us too much, since we always will end up with the same semantical (and syntactical) consequence relations, see [29], [30], and Section 5.

Thus, generalizing Belnap's idea of truth values encoding information passed to a computer leads us *in a first step*, when we assume that a computer informed by a classical source informs another computer, from **4** to **16** $= \mathcal{P}(\mathbf{4})$ with the following generalized truth values (where **A** = **NFTB** stands for "all"):

1. $\mathbf{N} = \varnothing$	9. $\mathbf{FT} = \{\{F\}, \{T\}\}$
2. $\mathbf{N} = \{\varnothing\}$	10. $\mathbf{FB} = \{\{F\}, \{F, T\}\}$
3. $\mathbf{F} = \{\{F\}\}$	11. $\mathbf{TB} = \{\{T\}, \{F, T\}\}$
4. $\mathbf{T} = \{\{T\}\}$	12. $\mathbf{NFT} = \{\varnothing, \{F\}, \{T\}\}$
5. $\mathbf{B} = \{\{F, T\}\}$	13. $\mathbf{NFB} = \{\varnothing, \{F\}, \{F, T\}\}$
6. $\mathbf{NF} = \{\varnothing, \{F\}\}$	14. $\mathbf{NTB} = \{\varnothing, \{T\}, \{F, T\}\}$
7. $\mathbf{NT} = \{\varnothing, \{T\}\}$	15. $\mathbf{FTB} = \{\{F\}, \{T\}, \{F, T\}\}$
8. $\mathbf{NB} = \{\varnothing, \{F, T\}\}$	16. $\mathbf{A} = \{\varnothing, \{T\}, \{F\}, \{F, T\}\}$.

It appears that this passage from **4** to **16** is essential for obtaining a natural example of a harmonious finitely-valued logic. Whereas in $FOUR_2$ the truth order is defined in terms of both T and F, on the richer set of values **16** separate truth and falsity orderings $\leq_t$ and $\leq_f$ can be isolated. In this 16-valued setting, truth and falsity are thereby treated as independent notions in their own right.

The trilattice $SIXTEEN_3$

According to the truth order of $FOUR_2$, the value **B** is less true than **T**, but this means that the *truth* order takes into account the absence of *falsity*, F, and thus, is in fact a *truth-and-falsity* order.[10] The truth values in **16**, however, allow one to define separately a truth order $\leq_t$ by referring only to the classical value T and a falsity order $\leq_f$ by referring only to F. For every x in **16** we first define the sets x^t, x^{-t}, x^f, and x^{-f} as follows:

$$x^t := \{y \in x \mid T \in y\}; \quad x^{-t} := \{y \in x \mid T \notin y\};$$
$$x^f := \{y \in x \mid F \in y\}; \quad x^{-f} := \{y \in x \mid F \notin y\}.$$

---

[10] In $FOUR_2$ maximizing (resp. minimizing) truth means simultaneously minimizing (resp. maximizing) falsity. It is therefore impossible in B_4 to discriminate between $\mathcal{L}^+$ and $\mathcal{L}^-$: for every $c \in C$, $f_{c^+} = f_{c^-}$.

DEFINITION 11. For every x, y in **16**:

1. $x \leq_i y$ iff $x \subseteq y$;

2. $x \leq_t y$ iff $x^t \subseteq y^t$ and $y^{-t} \subseteq x^{-t}$;

3. $x \leq_f y$ iff $x^f \subseteq y^f$ and $y^{-f} \subseteq x^{-f}$.

As a result, we obtain an algebraic structure that combines the three (complete) lattices $(\mathbf{16}, \leq_i)$, $(\mathbf{16}, \leq_t)$, and $(\mathbf{16}, \leq_f)$ into the *trilattice* $SIXTEEN_3 = (\mathbf{16}, \leq_i, \leq_t, \leq_f)$, see [28]. $SIXTEEN_3$ is presented by a triple Hasse diagram in Figure 3 (cf. also Figure 5 in [27]).

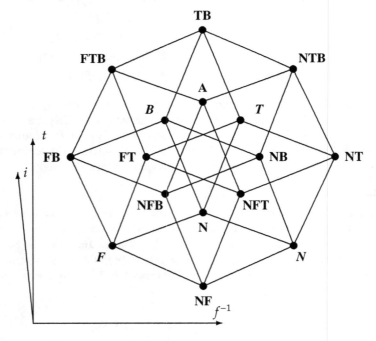

Figure 3. Trilattice $SIXTEEN_3$ (projection $t - f^{-1}$)

Meets and joints exist in $SIXTEEN_3$ for all three partial orders. We will use $\sqcap$ and $\sqcup$ with the appropriate subscripts for these operations under the corresponding ordering relations. Since from the operations one can recover the relations, $SIXTEEN_3$ may also be represented as the structure $(\mathbf{16}, \sqcap_i, \sqcup_i, \sqcap_t, \sqcup_t, \sqcap_f, \sqcup_f)$. In what follows we will be especially interested in the "logical" operations $\sqcap_t, \sqcup_t, \sqcap_f$ and $\sqcup_f$. Some key properties of these operations are summarized in the following proposition:

PROPOSITION 12. *For any x, y in $SIXTEEN_3$:*

1. $\mathbf{T} \in x \sqcap_t y \Leftrightarrow \mathbf{T} \in x \text{ and } \mathbf{T} \in y;$ 2. $\mathbf{T} \in x \sqcup_t y \Leftrightarrow \mathbf{T} \in x \text{ or } \mathbf{T} \in y;$
    $\mathbf{B} \in x \sqcap_t y \Leftrightarrow \mathbf{B} \in x \text{ and } \mathbf{B} \in y;$ $\mathbf{B} \in x \sqcup_t y \Leftrightarrow \mathbf{B} \in x \text{ or } \mathbf{B} \in y;$
    $\mathbf{F} \in x \sqcap_t y \Leftrightarrow \mathbf{F} \in x \text{ or } \mathbf{F} \in y;$ $\mathbf{F} \in x \sqcup_t y \Leftrightarrow \mathbf{F} \in x \text{ and } \mathbf{F} \in y;$
    $\mathbf{N} \in x \sqcap_t y \Leftrightarrow \mathbf{N} \in x \text{ or } \mathbf{N} \in y;$ $\mathbf{N} \in x \sqcup_t y \Leftrightarrow \mathbf{N} \in x \text{ and } \mathbf{N} \in y;$

3. $\mathbf{T} \in x \sqcup_f y \Leftrightarrow \mathbf{T} \in x \text{ and } \mathbf{T} \in y;$ 4. $\mathbf{T} \in x \sqcap_f y \Leftrightarrow \mathbf{T} \in x \text{ or } \mathbf{T} \in y;$
    $\mathbf{N} \in x \sqcup_f y \Leftrightarrow \mathbf{N} \in x \text{ and } \mathbf{N} \in y;$ $\mathbf{N} \in x \sqcap_f y \Leftrightarrow \mathbf{N} \in x \text{ or } \mathbf{N} \in y;$
    $\mathbf{F} \in x \sqcup_f y \Leftrightarrow \mathbf{F} \in x \text{ or } \mathbf{F} \in y;$ $\mathbf{F} \in x \sqcap_f y \Leftrightarrow \mathbf{F} \in x \text{ and } \mathbf{F} \in y;$
    $\mathbf{B} \in x \sqcup_f y \Leftrightarrow \mathbf{B} \in x \text{ or } \mathbf{B} \in y;$ $\mathbf{B} \in x \sqcap_f y \Leftrightarrow \mathbf{B} \in x \text{ and } \mathbf{B} \in y.$

Since the relations $\leq_t$ and $\leq_f$ are treated on a par, the operations $\sqcap_t$ and $\sqcup_t$ are not privileged as interpretations of conjunction and disjunction. The operation $\sqcup_f$ may as well be regarded as a conjunction and $\sqcap_f$ as a disjunction. In other words, the logical vocabulary may be naturally split into a positive truth vocabulary and a negative falsity vocabulary. Also certain unary operations with natural negation -like properties are available in $SIXTEEN_3$.

DEFINITION 13. A unary operation $-_t$ $(-_f, -_i)$ on $SIXTEEN_3$ is said to be a t-inversion (f-inversion, i-inversion) iff the following conditions are satisfied:

1. t-inversion($-_t$) :   2. f-inversion($-_f$) :
(a) $a \leq_t b \Rightarrow -_t b \leq_t -_t a;$ (a) $a \leq_t b \Rightarrow -_f a \leq_t -_f b;$
(b) $a \leq_f b \Rightarrow -_t a \leq_f -_t b;$ (b) $a \leq_f b \Rightarrow -_f b \leq_f -_f a;$
(c) $a \leq_i b \Rightarrow -_t a \leq_i -_t b;$ (c) $a \leq_i b \Rightarrow -_f a \leq_i -_f b;$
(d) $-_t -_t a = a.$ (d) $-_f -_f a = a.$

3. i-inversion($-_i$) :
(a) $a \leq_t b \Rightarrow -_i a \leq_t -_i b;$
(b) $a \leq_f b \Rightarrow -_i a \leq_f -_i b;$
(c) $a \leq_i b \Rightarrow -_i b \leq_i -_i a;$
(d) $-_i -_i a = a.$

In $SIXTEEN_3$ such operations are definable as shown in Table 1. Both $-_t$ and $-_f$ are natural interpretations for a negation connective. The following proposition highlights some important properties of these operations.

a	$-_t a$	$-_f a$	$-_i a$	a	$-_t a$	$-_f a$	$-_i a$
N	**N**	**N**	**A**	**NB**	**FT**	**FT**	**FT**
N	*T*	*F*	**NFT**	**FB**	**FB**	**NT**	**FB**
F	*B*	*N*	**NFB**	**TB**	**NF**	**TB**	**TB**
T	*N*	*B*	**NTB**	**NFT**	**NTB**	**NFB**	*N*
B	*F*	*T*	**FTB**	**NFB**	**FTB**	**NFT**	*F*
NF	**TB**	**NF**	**NF**	**NTB**	**NFT**	**FTB**	*T*
NT	**NT**	**FB**	**NT**	**FTB**	**NFB**	**NTB**	*B*
FT	**NB**	**NB**	**NB**	**A**	**A**	**A**	**N**

Table 1. Inversions in $SIXTEEN_3$

PROPOSITION 14. *For any x in $SIXTEEN_3$:*

1.	$\mathbf{T} \in -_t x \Leftrightarrow \mathbf{N} \in x;$	*2.*	$\mathbf{T} \in -_f x \Leftrightarrow \mathbf{B} \in x;$
	$\mathbf{N} \in -_t x \Leftrightarrow \mathbf{T} \in x;$		$\mathbf{B} \in -_f x \Leftrightarrow \mathbf{T} \in x;$
	$\mathbf{F} \in -_t x \Leftrightarrow \mathbf{B} \in x;$		$\mathbf{F} \in -_f x \Leftrightarrow \mathbf{N} \in x;$
	$\mathbf{B} \in -_t x \Leftrightarrow \mathbf{F} \in x;$		$\mathbf{N} \in -_f x \Leftrightarrow \mathbf{F} \in x.$

The requirements that the information order $\leq_i$ is left untouched by the operations of t-inversion and f-inversion and that t-inversion (f-inversion) has no effect on $\leq_f$ ($\leq_t$) *are* satisfied by the operations $-_t$ and $-_f$ defined in Table 1, but these requirements might also be given up. If they are abandoned, the definition of t-inversion (f-inversion) refers only to the truth-order (falsity-order). What this suggests is that not only conjunction and disjunction, but also negation emerges in two versions. Moreover, since $x \sqcap_t y \neq x \sqcup_f y$, $x \sqcup_t y \neq x \sqcap_f y$ and $-_t x \neq -_f x$, the two logical orderings $\leq_t$ and $\leq_f$ indeed give rise to two distinct sets of logical operations of the same arity.

A harmonious logic inspired by the logic of $SIXTEEN_3$

An appropriate syntax for the logic emerging from $SIXTEEN_3$ is given by a denumerable set of propositional variables and three propositional languages $\mathcal{L}_t$, $\mathcal{L}_f$, and $\mathcal{L}_{tf}$ based on this set. They are defined in Backus–Naur form as follows:

$$\mathcal{L}_t : \quad A ::= \quad p \mid {\sim}_t A \mid A \wedge_t A \mid A \vee_t A$$
$$\mathcal{L}_f : \quad A ::= \quad p \mid {\sim}_f A \mid A \wedge_f A \mid A \vee_f A$$
$$\mathcal{L}_{tf} : \quad A ::= \quad p \mid {\sim}_t A \mid {\sim}_f A \mid A \wedge_t A \mid A \vee_t A \mid A \wedge_f A \mid A \vee_f A$$

The logic of $SIXTEEN_3$ is semantically presented as a *bi-consequence system*, namely the structure $(\mathcal{L}_{tf}, \models_t, \models_f)$, where the two entailment relations $\models_t$ and $\models_f$

are defined with respect to the truth order $\leq_t$ and the falsity order $\leq_f$, respectively. The main point of the present paper consists in defining the notion of a harmonious finitely-valued propositional logic and pointing out that the logic of $SIXTEEN_3$ and similar structures lead to examples of harmonious many-valued logics. We first define $(\mathcal{L}_{tf}, \models_t, \models_f)$.

DEFINITION 15. Let v be a map from the set of propositional variables into **16**. The function v is recursively extended to a function from the set of all $\mathcal{L}_{tf}$-formulas into **16** as follows:

1. $v(A \wedge_t B) = v(A) \sqcap_t v(B)$;   4. $v\left(A \wedge_f B\right) = v(A) \sqcup_f v(B)$;
2. $v(A \vee_t B) = v(A) \sqcup_t v(B)$;   5. $v\left(A \vee_f B\right) = v(A) \sqcap_f v(B)$;
3. $v(\sim_t A) = -_t v(A)$;   6. $v\left(\sim_f A\right) = -_f v(A)$.

The relations $\models_t \subseteq \mathcal{P}(\mathcal{L}_{tf}) \times \mathcal{P}(\mathcal{L}_{tf})$ and $\models_f \subseteq \mathcal{P}(\mathcal{L}_{tf}) \times \mathcal{P}(\mathcal{L}_{tf})$ are defined by the following equivalences:

$$\Delta \models_t \Gamma \quad \text{iff} \quad \forall v \; \sqcap_t \{v(A) \mid A \in \Delta\} \leq_t \sqcup_t \{v(A) \mid A \in \Gamma\};$$

$$\Delta \models_f \Gamma \quad \text{iff} \quad \forall v \; \sqcup_f \{v(A) \mid A \in \Gamma\} \leq_f \sqcap_f \{v(A) \mid A \in \Delta\}.$$

We now define a separated 16-valued logic in the language $\mathcal{L}_{tf}$.

DEFINITION 16. The separated 16-valued logic B_{16} is the structure $\langle \mathbf{16}, \{x \in \mathbf{16} \mid x^t \text{ is non-empty}\}, \{x \in \mathbf{16} \mid x^f \text{ is non-empty}\}, \{-_t, \sqcap_t, \sqcup_t, -_f, \sqcup_f, \sqcap_f\} \rangle$. Moreover, for all sets of $\mathcal{L}_{tf}$-formulas Δ, Γ, semantic consequence relations $\models^+$ and $\models^-$ are defined in accordance with 1. and 2. of Definition 1.

OBSERVATION 17. B_{16} is refined.

Proof. It can easily be seen that in B_{16} the relations $\models^+$ and $\models^-$ are distinct, e.g. in view of the following counterexample: $(A \wedge_f B) \models^- A$ but $(A \wedge_f B) \not\models^+ A$ (since, e.g., $F \sqcup_f \mathbf{FTB} = \mathbf{FB}$). ∎

PROPOSITION 18. *The 16-valued propositional logic B_{16} is harmonious.*

Proof. Obviously, we may view the language $\mathcal{L}_{tf}$ as being based on a set of positive connectives $C^+ = \{\sim_t, \wedge_t, \vee_t\}$ and a set of negative connectives with matching arity $C^- = \{\sim_f, \wedge_f, \vee_f\}$, i.e., $C = \{\sim, \wedge, \vee\}$. Moreover, we already observed that the condition $f_{c_t} \neq f_{c_f}$ is satisfied for every $c \in C$. If A is an $\mathcal{L}_t$-formula, let A^f be the result of replacing every connective c_t in A by c_f. If Δ is a set of $\mathcal{L}_t$ formulas, let $\Delta^f = \{A^f \mid A \in \Delta\}$. It remains to be shown that in B_{16} for all sets of $\mathcal{L}_t$-formulas Δ, Γ,

$$(\dagger) \quad \Delta \models^+ \Gamma \text{ iff } \Delta^f \models^- \Gamma^f.$$

This follows from Lemma 4.2, Lemma 4.3, Theorem 4.4 and Theorem 4.7 in [28] for $\leq_t$ and the analogous versions of these statements for $\leq_f$. Lemma 4.3 says that for every $A, B \in \mathcal{L}_t$: $A \models_t B$ iff $\forall v(\mathbf{T} \in v(A) \Rightarrow \mathbf{T} \in v(B))$. According to Lemma 4.2, within language $\mathcal{L}_t$, the condition $\forall v(\mathbf{T} \in v(A) \Rightarrow \mathbf{T} \in v(B))$ is equivalent to $\forall v(\mathbf{B} \in v(A) \Rightarrow \mathbf{B} \in v(B))$. Thus, for every $A, B \in \mathcal{L}_t$: $A \models_t B$ iff $\forall v(\mathbf{T} \in v(A)$ or $\mathbf{B} \in v(A) \Rightarrow \mathbf{T} \in v(B)$ or $\mathbf{B} \in v(B))$. This means that $\models^+$ restricted to $\mathcal{L}_t$ is the same relation as $\models_t$ restricted to $\mathcal{L}_t$, and it is then axiomatized as first degree entailment (Theorems 4.4 and 4.7). Since for every $A, B \in \mathcal{L}_f$: $A \models_f B$ iff $\forall v(\mathbf{F} \in v(B) \Rightarrow \mathbf{F} \in v(A))$ iff $\forall v(\mathbf{B} \in v(B) \Rightarrow \mathbf{B} \in v(A))$, and the restriction of $\models^-$ to $\mathcal{L}_f$ (= the restriction of $\models_f$ to $\mathcal{L}_f$) is *also* axiomatized as first degree entailment, condition (†) is satisfied.   ∎

5   Harmony ad infinitum

The trilattice $SIXTEEN_3$ is an example of a *multilattice*, see [28], [34].

DEFINITION 19. An n-dimensional multilattice (or just n-lattice) is a structure $\mathcal{M}_n = (S, \leq_1, \ldots, \leq_n)$ such that S is a non-empty set and $\leq_1, \ldots, \leq_n$ are partial orders defined on S such that $(S, \leq_1), \ldots, (S, \leq_n)$ are pairwise distinct lattices.

More concretely, the structure $SIXTEEN_3$ is an example of a *Belnap-trilattice*, see [30]. Belnap-trilattices are obtained by iterated powerset-formation applied to the set **4** and by generalizing the definitions of a truth order and a falsity order on **16**. If X is a set, let $\mathcal{P}^1(X) := \mathcal{P}(X)$ and $\mathcal{P}^n(X) := \mathcal{P}(\mathcal{P}^{n-1}(X))$ for $1 < n$, $n \in \mathbb{N}$. We obtain an infinite collection of sets of generalized semantical values by considering $\mathcal{P}^n(\mathbf{4})$. Each of these sets can be equipped with relations $\leq_i$, $\leq_t$, and $\leq_f$ in a canonical way as in Definition 11 by defining for every $x, y \in \mathcal{P}^n(\mathbf{4})$ the sets x^t, x^{-t}, x^f and x^{-f} as follows:

$$x^t := \{y_0 \in x \mid (\exists y_1 \in y_0)(\exists y_2 \in y_1) \ldots (\exists y_{n-1} \in y_{n-2})\, \mathbf{T} \in y_{n-1}\}$$

$$x^{-t} := \{y_0 \in x \mid \neg(\exists y_1 \in y_0)(\exists y_2 \in y_1) \ldots (\exists y_{n-1} \in y_{n-2})\, \mathbf{T} \in y_{n-1}\}$$

$$x^f := \{y_0 \in x \mid (\exists y_1 \in y_0)(\exists y_2 \in y_1) \ldots (\exists y_{n-1} \in y_{n-2})\, \mathbf{F} \in y_{n-1}\}$$

$$x^{-f} := \{y_0 \in x \mid \neg(\exists y_1 \in y_0)(\exists y_2 \in y_1) \ldots (\exists y_{n-1} \in y_{n-2})\, \mathbf{F} \in y_{n-1}\}$$

Thus, $x^{-t} = x \setminus x^t$ and $x^{-f} = x \setminus x^f$. We say that x is t-positive (t-negative, f-positive, f-negative) iff x^t (x^{-t}, x^f, x^{-f}) is non-empty and we denote by $\mathcal{P}^n(\mathbf{4})^t$ ($\mathcal{P}^n(\mathbf{4})^{-t}, \mathcal{P}^n(\mathbf{4})^f, \mathcal{P}^n(\mathbf{4})^{-f}$) the set of all t-positive (resp. t-negative, f-positive, f-negative) elements of $\mathcal{P}^n(\mathbf{4})$.

DEFINITION 20. A Belnap-trilattice is a structure

$$\mathcal{M}_3^n := (\mathcal{P}^n(\mathbf{4}), \sqcap_i, \sqcup_i, \sqcap_t, \sqcup_t, \sqcap_f, \sqcup_f),$$

where $\sqcap_i$ ($\sqcap_t, \sqcap_f$) is the lattice meet and $\sqcup_i$ ($\sqcup_t, \sqcup_f$) is the lattice join with respect to the ordering $\leq_i$ ($\leq_t, \leq_f$) on $\mathcal{P}^n(\mathbf{4})$, $n \geq 1$.

Thus, $SIXTEEN_3$ ($= \mathcal{M}_3^1$) is the smallest Belnap-trilattice. Moreover, if unary operations $-_t$ and $-_f$ satisfying the conditions from Definition 13 exist on $\mathcal{P}^n(\mathbf{4})$, then we may consider Belnap-trilattices with t-inversion and f-inversion.

PROPOSITION 21. *If $\mathcal{M}_3^n$ is a Belnap-trilattice, then there exist operations of t-inversion and f-inversion on $\mathcal{P}^n(\mathbf{4})$.*

In [30] a canonical definition for such inversion operations is presented. Thus, we may without loss of generality confine our considerations to Belnap-trilattices with t-inversions and f-inversions.

We consider again the languages $\mathcal{L}_t$, $\mathcal{L}_f$, $\mathcal{L}_{tf}$ defined in the previous section. An n-valuation is a function v^n from the set of propositional variables into $\mathcal{P}^n(\mathbf{4})$. This function can be extended to an interpretation of arbitrary formulas in $\mathcal{P}^n(\mathbf{4})$ by Definition 15 (replacing v by v^n).

DEFINITION 22. The relations $\models_t^n \subseteq \mathcal{P}(\mathcal{L}_{tf}) \times \mathcal{P}(\mathcal{L}_{tf})$ and $\models_f^n \subseteq \mathcal{P}(\mathcal{L}_{tf}) \times \mathcal{P}(\mathcal{L}_{tf})$ are defined by the following equivalences:

$$\Delta \models_t^n \Gamma \quad \text{iff} \quad \forall v^n \ \sqcap_t \{v^n(A) \mid A \in \Delta\} \leq_t \sqcup_t \{v^n(A) \mid A \in \Gamma\};$$

$$\Delta \models_f^n \Gamma \quad \text{iff} \quad \forall v^n \ \sqcup_f \{v^n(A) \mid A \in \Gamma\} \leq_f \sqcap_f \{v^n(A) \mid A \in \Delta\}.$$

Semantically, the logic of a Belnap-trilattice $\mathcal{M}_3^n$ is the bi-consequence system $(\mathcal{L}_{tf}, \models_t^n, \models_f^n)$.

We now define an infinite chain of separated finitely-valued logics.

DEFINITION 23. Let $\sharp n$ be the cardinality of $\mathcal{P}^n(\mathbf{4})$. The $\sharp n$-valued logic $B_{\sharp n}$ is the structure $\langle \mathcal{P}^n(\mathbf{4}), \mathcal{D}^{n+}, \mathcal{D}^{n-}, \{-_t, \sqcap_t, \sqcup_t, -_f, \sqcup_f, \sqcap_f\}\rangle$, where $\mathcal{D}^{n+} := \{x \in \mathcal{P}^n(\mathbf{4}) \mid x \text{ is } t\text{-positive}\}$ and $\mathcal{D}^{n-} := \{x \in \mathcal{P}^n(\mathbf{4}) \mid x \text{ is } f\text{-positive}\}$. For every logic $B_{\sharp n}$, for all sets of $\mathcal{L}_{tf}$-formulas Δ, Γ, the semantic consequence relations $\models^{n+}$ and $\models^{n-}$ are defined in accordance with 1. and 2. of Definition 1.

OBSERVATION 24. For every $n \in \mathbb{N}$, the logic $B_{\sharp n}$ is refined.

Proof. Obviously, $B_{\sharp n}$ is separated. That in $B_{\sharp n}$ the relations $\models^{n+}$ and $\models^{n-}$ are distinct can again be seen by noticing that for every $n \in \mathbb{N}$, $(A \wedge_f B) \models^{n-} A$ but $(A \wedge_f B) \not\models^{n+} A$ (since for every $\mathcal{M}_3^n$ there may well exist $x, y \in \mathcal{P}^n(\mathbf{4})$, such that $x \sqcup_f y$ is t-positive, whereas either x or y is not). ∎

PROPOSITION 25. *For every $n \in \mathbb{N}$, the logic $B_{\sharp n}$ is harmonious.*

Proof. Again, we regard $\mathcal{L}_{tf}$ as being based on $C^+ = \{\sim_t, \wedge_t, \vee_t\}$ and the set of connectives with matching arity $C^- = \{\sim_f, \wedge_f, \vee_f\}$, so that $C = \{\sim, \wedge, \vee\}$. Again, it can easily be seen that the condition $f_{c_t} \neq f_{c_f}$ is satisfied for every $c \in C$. We must show that in $B_{\sharp n}$ for all sets of $\mathcal{L}_t$-formulas Δ, Γ,

$$\Delta \models^{n+} \Gamma \text{ iff } \Delta^f \models^{n-} \Gamma^f.$$

We use the main result of [30], namely that for *every* $n \in \mathbb{N}$, the truth entailment relation $\models_t^n$ of $(\mathcal{L}_{tf}, \models_t^n, \models_f^n)$ restricted to $\mathcal{L}_t$ and the falsity entailment relation $\models_f^n$ of $(\mathcal{L}_{tf}, \models_t^n, \models_f^n)$ restricted to $\mathcal{L}_f$ can again both be axiomatized as first degree entailment. The proof systems differ only insofar, as every connective c_t is uniformly replaced by its negative counterpart c_f. Thus, given the completeness theorem for $\models_t^n$ and its counterpart for $\models_f^n$, it is enough to show that for all sets of $\mathcal{L}_t$-formulas Δ, Γ: $\Delta \models^{n+} \Gamma$ in $B_{\sharp n}$ iff $\Delta \models_t^n \Gamma$, and for all sets of $\mathcal{L}_f$-formulas Δ, Γ: $\Delta \models^{n-} \Gamma$ in $B_{\sharp n}$ iff $\Delta \models_f^n \Gamma$. But this follows from Corollary 17 in [30]: For any $A, B \in \mathcal{L}_t$:

$$A \models_t^n B \text{ iff } \forall v^n (x \in v^n(A)^t \Rightarrow x \in v^n(B)^t)$$

and the analogous result for falsity entailment: For any $A, B \in \mathcal{L}_f$:

$$A \models_f^n B \text{ iff } \forall v^n (x \in v^n(B)^f \Rightarrow x \in v^n(A)^f).$$

∎

6   Some remarks on generalizing harmony

The point of departure for our considerations was the familiar and simple notion of an extensional n-valued propositional logic. We argued that the distinction between designated, antidesignated, and undesignated values and values that are both designated and antidesignated ought to be taken seriously. There is no reason to privilege designation over antidesignation when it comes to defining entailment. Hence we suggested a slightly more general notion of an n-valued logic. Whereas the set $\mathcal{D}^+$ of designated values determines a positive entailment relation $\models^+$, the set $\mathcal{D}^-$ of antidesignated values determines a negative entailment relation $\models^-$. Once we have two entailment relations, it is natural to consider two languages. This consideration then led us to defining the notion of a harmonious n-valued logic.

It has been observed quite a while ago already that one may consider n-valued logics in which the truth functions $f_\wedge$ and $f_\vee$ for conjunction and disjunction form a lattice on the underlying set of truth values, see, for example [24], [25]. Thus, given an n-valued propositional logic, one may wonder whether a lattice order can be defined from some given truth functions. However, in the light of the research on bilattices and trilattices, the direction of interest goes in the opposite direction. Given some natural partial orders on sets of semantical values, one may wonder whether these orderings form lattices on the underlying sets and thereby give rise to a conjunction (lattice meet), disjunction (lattice join), and hopefully also some sort of negation . In the bilattice $FOUR_2$, there is only one "logical" order, there referred to as the truth order. In a Belnap-trilattice $\mathcal{M}_3^n$, there are two logical orderings, the truth order $\leq_t$ and the falsity order $\leq_f$. From the bi-consequence system of any Belnap-trilattice we obtained a harmonious finitely-valued logic. But neither did we consider the following relation induced by the information order $\models^i$, defined as

$$\Delta \models_i^n \Gamma \quad \text{iff} \quad \forall v^n \; \bigsqcap_i \{v^n(A) \mid A \in \Delta\} \leq_i \bigsqcup_i \{v^n(A) \mid A \in \Gamma\},$$

nor did we consider additional orderings or other sets of truth values with more than two lattice orderings on them. This would, however, be an interesting direction to pursue, if one aims at finding examples of many-valued logics displaying a more general form of harmony. Thus, generalizations of the notion of a harmonious n-valued propositional logic can be obtained by replacing the set $\{\mathcal{D}^+, \mathcal{D}^-\}$ of distinguished subsets of the set of values $\mathcal{V}$ by a set $\{\mathcal{D}_1, \ldots, \mathcal{D}_n\}$ of distinguished subsets of $\mathcal{V}$. Moreover, the association of entailment relations with the distinguished subsets of $\mathcal{V}$ may vary.

A relation $\models \subseteq \mathcal{P}(\mathcal{L}) \times \mathcal{P}(\mathcal{L})$ is a Tarski-Scott multiple conclusion consequence relation iff it satisfies the following conditions:

1. For every $\Delta \subseteq \mathcal{L}, \Delta \models \Delta$ (reflexivity);

2. If $\Delta \models \Gamma \cup \{A\}$ and $\{A\} \cup \Theta \models \Sigma$, then $\Delta \cup \Theta \models \Gamma \cup \Sigma$ (transitivity);

3. If $\Delta \subseteq \Theta, \Gamma \subseteq \Sigma$, and $\Delta \models \Gamma$, then $\Theta \models \Sigma$ (monotony).

A Tarski-Scott multiple conclusion consequence relation is said to be structural, if it is closed under substitution.

In an n-valued logic which is refined in the sense of Definition 2, in addition to $\models^+$ and $\models^-$ several other, possibly interesting semantical relations can be defined, though not all of them turn out to be Tarski-Scott multiple conclusion entailment relations. In order to refer to these relations in a compact way, we need a more systematic notation. Let + stand for designated values, − for antidesignated values, u for neither designated nor antidesignated values, and b for values that are both designated and antidesignated. Let $\overline{+}, \overline{-}, \overline{u}$, and $\overline{b}$ stand for the respective complements. Moreover, let $\Leftarrow$ indicate preservation from the conclusions to the premises, $\Rightarrow$ preservation from the premises to the conclusions, and $\Leftrightarrow$ preservation in both directions. Then $\models^+$ is denoted as $\models^{+\Rightarrow+}$ and $\models^-$ as $\models^{-\Leftarrow-}$. The relation $\models$ defined (for a given n-valued logic) by requiring that $\Delta \models \Gamma$ iff both $\Delta \models^+ \Gamma$ and $\Delta \models^- \Gamma$ (cf. the proof of Observation 9) is denoted as $\models^{(+\Rightarrow+,-\Leftarrow-)}$, and the relation defined by requiring that $\Delta \models \Gamma$ iff ($\Delta \models^+ \Gamma$ or $\Delta \models^- \Gamma$) is denoted as $\models^{(+\Rightarrow+|-\Leftarrow-)}$. We may consider in addition to $\models^+$ and $\models^-$ for example the following "semantic consequence" relations:

- $\models^{+\Leftarrow+}$;

- $\models^{-\Rightarrow-}$;

- $\models^{u \Rightarrow u}$; $\models^{u \Leftarrow u}$;

- $\models^{b \Rightarrow b}$; $\models^{b \Leftarrow b}$;

- $\models^{\overline{+} \Rightarrow \overline{+}}$; $\models^{\overline{+} \Leftarrow \overline{+}}$;

- $\models^{\overline{-} \Rightarrow \overline{-}}$; $\models^{\overline{-} \Leftarrow \overline{-}}$;

- $\models^{\overline{u} \Rightarrow \overline{u}}$; $\models^{\overline{u} \Leftarrow \overline{u}}$;

- $\models^{\overline{b} \Rightarrow \overline{b}}$; $\models^{\overline{b} \Leftarrow \overline{b}}$;

- $\models^{\bullet \Rightarrow \circ}$; $\models^{\bullet \Leftarrow \circ}$;

  where $\bullet, \circ \in \{+, -, u, b, \overline{+}, \overline{-}, \overline{u}, \overline{b}\}$ and $\bullet \neq \circ$;

- $\models^{(\bullet \Rightarrow \circ, \blacklozenge \Leftarrow \Diamond)}$; $\models^{(\bullet \Leftarrow \circ, \blacklozenge \Rightarrow \Diamond)}$;

  where $\bullet, \circ, \blacklozenge, \Diamond \in \{+, -, u, b, \overline{+}, \overline{-}, \overline{u}, \overline{b}\}$ and $\bullet \neq \overline{\blacklozenge}$ or $\circ \neq \overline{\Diamond}$;

- $\models^{(\bullet \Rightarrow \circ, \blacklozenge \Rightarrow \Diamond)}$; $\models^{(\bullet \Leftarrow \circ, \blacklozenge \Leftarrow \Diamond)}$;

  where $\bullet, \circ, \blacklozenge, \Diamond \in \{+, -, u, b, \overline{+}, \overline{-}, \overline{u}, \overline{b}\}$ and $\bullet \neq \blacklozenge$ or $\circ \neq \Diamond$;

- $\models^{(\bullet \Rightarrow \circ | \blacklozenge \Leftarrow \Diamond)}$; $\models^{(\bullet \Leftarrow \circ | \blacklozenge \Rightarrow \Diamond)}$;

  where $\bullet, \circ, \blacklozenge, \Diamond \in \{+, -, u, b, \overline{+}, \overline{-}, \overline{u}, \overline{b}\}$ and $\bullet \neq \overline{\blacklozenge}$ or $\circ \neq \overline{\Diamond}$;

- $\models^{(\bullet \Rightarrow \circ | \blacklozenge \Rightarrow \Diamond)}$; $\models^{(\bullet \Leftarrow \circ | \blacklozenge \Leftarrow \Diamond)}$;

  where $\bullet, \circ, \blacklozenge, \Diamond \in \{+, -, u, b, \overline{+}, \overline{-}, \overline{u}, \overline{b}\}$ and $\bullet \neq \blacklozenge$ or $\circ \neq \Diamond$.

Since consequence relations normally are not required to be symmetric, relations like $\models^{+ \Leftrightarrow +}$ are, perhaps, not of primary interest. But the relation $\models^{(+ \Leftrightarrow +, u \Rightarrow \overline{-})} = \models^{+ \Leftrightarrow +} \cap \models^{u \Rightarrow \overline{-}}$, for example, might be of some interest. Investigating and applying such non-standard semantic consequence relations is not as exotic as it might seem at first sight, and, indeed, some such relations have been considered in the literature. We already noted that $\models^{(+ \Rightarrow +, - \Leftarrow -)}$ is dealt with in [10], and there are other examples.

EXAMPLE 26. G. Malinowski [17], [18], [19], emphasizing the distinction between accepted and rejected propositions, draws the distinction between designated and antidesignated values and uses it to generalize Tarski's notion of a consequence operation to the notion of a quasi-consequence operation (or just q-consequence operation). Also, single-conclusion q-consequence relations are defined; they relate not antidesignated assumptions to designated (single) conclusions. In our notation, multiple-conclusion q-consequence is the relation $\models^{\overline{-} \Rightarrow +}$.

EXAMPLE 27. Another non-standard consequence relation has been presented in [7] and is there said *not* to be "overly outlandish or inconceivable", although it fails to be a Tarski-Scott multiple-conclusion consequence relation. In our notation, the "tonk-consequence" relation of [7] is the relation $\models^{(+\Rightarrow+|-\Leftarrow-)}$ on the set $\mathbf{4}$ with $\mathcal{D}^+$ = $\{\mathbf{T}, \mathbf{B}\}$ and $\mathcal{D}^- = \{\mathbf{F}, \mathbf{B}\}$. Since the logic is not transitive, sound truth tables for Prior's connective $\mathtt{tonk}$ are available such that this addition of $\mathtt{tonk}$ does not have a trivializing effect (but see also [35]).

EXAMPLE 28. Formula-to-formula q-consequence, though not under this name, is also among the varieties of semantic consequence considered in Chapters 3 and 4 of [33]. The types of consequence relations presented by Thijsse include the relations which in our notation for the multiple conclusion case are denoted as follows: $\models^{+\Rightarrow+}, \models^{-\Rightarrow+}, \models^{-\Rightarrow-}, \models^{+\Rightarrow-}$. Thijsse remarks that besides the familiar $\models^{+\Rightarrow+}$ the relation $\models^{-\Rightarrow-}$ "turns out to be interesting, both in theory and application".

Yet another way of defining a generalized notion of semantic consequence can be found in [5]. The two types of relation $\models^{+\Rightarrow+}$ and $\models^{-\Rightarrow-}$ are there merged into a single *four-place* bi-consequence relation. Note also that n-place semantic sequents for n-valued logics have been considered by Schröter [26], see also [14], [15].

After these preparatory remarks, we are in a position to define a generalized notion of a harmonious n-valued propositional logic.

DEFINITION 29. Let $\mathcal{L}$ be a language in a denumerable set of sentence letters and a finite non-empty set of finitary connectives C. An n-valued propositional logic is a structure

$$\langle \mathcal{V}, \mathcal{D}_1, \ldots, \mathcal{D}_k, \{f_c : c \in C\}\rangle$$

where $\mathcal{V}$ is a non-empty set of cardinality n ($2 \leq n$), $2 \leq k$, every $\mathcal{D}_i$ ($1 \leq i \leq k$) is a non-empty proper subset of $\mathcal{V}$, the sets $\mathcal{D}_i$ are pairwise distinct, and every f_c is a function on $\mathcal{V}$ with the same arity as c. The sets $\mathcal{D}_i$ are called distinguished sets. A valuation v is inductively extended to a function from the set of all $\mathcal{L}$-formulas into $\mathcal{V}$ by setting: $v(c(A_1, \ldots, A_m)) = f_c(v(A_1), \ldots, v(A_m))$, for every m-place $c \in C$. For every set $\mathcal{D}_i$, two semantic consequence relation $\models_i^{\Rightarrow}$ and $\models_i^{\Leftarrow}$ are defined as follows:

1. $\Delta \models_i^{\Rightarrow} \Gamma$ iff for every valuation function v: (if for every $A \in \Delta$, $v(A) \in \mathcal{D}_i$, then $v(B) \in \mathcal{D}_i$ for some $B \in \Gamma$);

2. $\Delta \models_i^{\Leftarrow} \Gamma$ iff for every valuation function v: (if for every $A \in \Gamma$, $v(A) \in \mathcal{D}_i$, then $v(B) \in \mathcal{D}_i$ for some $B \in \Delta$).

Obviously, the relations $\models_i^{\Rightarrow}$ and $\models_i^{\Leftarrow}$ are inverses of each other.

DEFINITION 30. Let $\Lambda = \langle \mathcal{V}, \mathcal{D}_1, \ldots, \mathcal{D}_k, \{f_c : c \in C\}\rangle$ be an n-valued logic. Λ is said to be separated, if for every $\mathcal{D}_i$, there exists no $\mathcal{D}_j$ such that $i \neq j$ and $\mathcal{V} \setminus \mathcal{D}_i = \mathcal{D}_j$. Λ is said to be refined, if it is separated and the relations $\models_i^{\Rightarrow}$ are pairwise distinct (and hence also the relations $\models_i^{\circ}$ with $\circ \in \{\Rightarrow, \Leftarrow\}$ are distinct).

DEFINITION 31. Let C be a finite non-empty set of finitary connectives, and let $\mathcal{L}_i$ be the language based on $C_i = \{c_i \mid c \in C\}$ for some $k \in \mathbb{N}$ such that $2 \leq i \leq k$. If $i \neq j$, $2 \leq j \leq k$ and if A is an $\mathcal{L}_i$-formula, then let A_j be the result of replacing every connective c_i in A by c_j. If Δ is a set of $\mathcal{L}_i$ formulas, let $\Delta_j = \{A_j \mid A \in \Delta\}$. If the language $\mathcal{L}$ of a refined n-valued logic Λ with k distinguished sets is based on the set $\bigcup_{i \leq k} C_i$, then Λ is said to be *harmonious* iff (i) for every i, j with $i \neq j$ and all sets of $\mathcal{L}_i$-formulas Δ, Γ the following holds: (i) $\Delta \models_i^{\Rightarrow} \Gamma$ iff $\Delta_j \models_j^{\Leftarrow} \Gamma_j$, (ii) $\Delta \models_i^{\Leftarrow} \Gamma$ iff $\Delta_j \models_j^{\Rightarrow} \Gamma_j$, and (iii) for every $c \in C$, $f_{c_i} \neq f_{c_j}$.

The logics $B_{\sharp n}$ are harmonious in this generalized sense. We may set $k = 2$, $\mathcal{D}_1 = \mathcal{D}^+$, $\mathcal{D}_2 = \mathcal{D}^-$, $C_1 = \{\sim_t, \wedge_t, \vee_t\}$, $C_2 = \{\sim_f, \wedge_f, \vee_f\}$, $\models_1^{\Rightarrow} = \models^+$, $\models_1^{\Leftarrow} = (\models^+)^{-1}$ (the inverse of $\models^+$), $\models_2^{\Leftarrow} = \models^-$, and $\models_2^{\Rightarrow} = (\models^-)^{-1}$.

Another obvious generalization of the present considerations is giving up the restriction to finitely many values.

We end this paper with the following question: Are there natural examples of harmonious n-valued logics $\langle \mathcal{V}, \mathcal{D}_1, \mathcal{D}_2, \mathcal{D}_3, \{f_c : c \in C\}\rangle$ with *three* distinguished sets of semantical values?

Acknowledgments

This work was supported by DFG grant WA 936/6-1. We would like to thank Siegfried Gottwald and an anonymous referee for their comments on an earlier version of this paper.

BIBLIOGRAPHY

[1] Anderson, A.R., Belnap, N.D. and Dunn, J.M.: *Entailment: The Logic of Relevance and Necessity*, Vol. II, Princeton University Press, Princeton, NJ, 1992.
[2] Arieli, O. and Avron, A.: Reasoning with logical bilattices, *Journal of Logic, Language and Information* **5** (1996), 25-63.
[3] Belnap, N.D.: How a computer should think, in G. Ryle (ed.), *Contemporary Aspects of Philosophy*, Oriel Press Ltd., Stocksfield, 1977, pp. 30-55.
[4] Belnap, N.D.: A useful four-valued logic, in: J.M. Dunn and G. Epstein (eds.), *Modern Uses of Multiple-Valued Logic*, D. Reidel Publishing Company, Dordrecht, 1977, pp. 8-37.
[5] Bochman, A.: A general formalism of reasoning with inconsistency and incompleteness, *Notre Dame Journal of Formal Logic* **39** (1998), 47-73.
[6] Caleiro, C., Carnielli, W., Coniglio, M., and Marcos, J.: Two's company: "The humbug of many logical values", in: J.-Y. Beziau (ed.), *Logica Universalis*, Birkhäuser, Basel, 2005, 169-189.
[7] Cook, R.: What's wrong with tonk (?), *Journal of Philosophical Logic* **34** (2005), 217-226.
[8] Dewitt, R.: On retaining classical truths and classical deducibility in many-valued and fuzzy logics, *Journal of Philosophical Logic* **34** (2005), 545-560.
[9] Dunn, J.M.: Intuitive semantics for first-degree entailment and 'coupled trees', *Philosophical Studies* **29** (1976), 149-168.
[10] Dunn, J.M.: Partiality and its dual, *Studia Logica* **66** (2000), 5-40.
[11] Fitting, M.: Bilattices are nice things, in V.F. Hendricks, S.A. Pedersen and T. Bolander (eds.) *Self-Reference*, CSLI- Publications, 2004, 53-77.

[12] Ginsberg, M.: Multi-valued logics, in *Proceedings of AAAI-86, Fifth National Conference on Artificial Intelligence*, Morgan Kaufman Publishers, Los Altos, 1986, pp. 243-247.
[13] Ginsberg, M.: Multivalued logics: a uniform approach to reasoning in AI, *Computer Intelligence* **4** (1988), 256-316.
[14] Gottwald, S.: *Mehrwertige Logik. Eine Einführung in Theorie und Anwendungen*, Akademie-Verlag, Berlin, 1989.
[15] Gottwald, S.: *A Treatise on Many-valued Logic*, Research Studies Press, Baldock, 2001.
[16] James, W.: The Will to Believe, 1897, in: *The Will to Believe and Other Essays in Popular Philosophy*, New York, Dover Publications, 1956.
[17] Malinowski, G.: Q-Consequence Operation, *Reports on Mathematical Logic* **24** (1990), 49-59.
[18] Malinowski, G.: *Many-valued Logics*, Clarendon Press, Oxford, 1993.
[19] Malinowski, G.: Inferential Many-Valuedness, in: Jan Woleński (ed.), *Philosophical Logic in Poland*, Kluwer Academic Publishers, Dordrecht, 1994, 75-84.
[20] Meyer, R.K.: *Why I Am Not a Relevantist*, Research paper, no. 1, Australian National University, Logic Group, Research School of the Social Sciences, Canberra, 1978.
[21] Priest, G.: Logic of Paradox, *Journal of Philosophical Logic*, **8** (1979), 219-241.
[22] Priest, G.: *An Introduction to Non-Classical Logic*, Cambridge UP, Cambridge, 2001.
[23] Rescher, N.: *Many-Valued Logic*, McGraw-Hill, New York, 1969.
[24] Rose, A.: A lattice-theoretic characterization of three-valued logic, *Journal of the London Mathematical Society* **25** (1950), 255-259.
[25] Rose, A.: Systems of logics whose truth-values form lattices, *Mathematische Annalen* **123** (1951), 152-165.
[26] Schröter, K.: Methoden zur Axiomatisierung beliebiger Aussagen- und Prädikatenkalküle, *Zeitschrift für mathematische Logik und Grundlagen der Mathematik* **1** (1955), 241-251.
[27] Shramko, Y., Dunn, J.M. and Takenaka, T.: The trilattice of constructive truth values, *Journal of Logic and Computation* **11** (2001), 761-788.
[28] Shramko, Y. and Wansing, H.: Some useful 16-valued logics: how a computer network should think, *Journal of Philosophical Logic* **34** (2005), 121-153.
[29] Shramko, Y. and Wansing, H.: The Logic of Computer Networks (in Russian), *Logical Studies* (Moscow) **12** (2005), 119-145.
[30] Shramko, Y. and Wansing, H.: Hypercontradictions, generalized truth values, and logics of truth and falsehood, *Journal of Logic, Language and Information* **15** (2006), 403-424.
[31] Sluga, H.: Frege on the Indefinability of Truth, in: E. Reck (ed.), *From Frege to Wittgenstein: Perspectives on Early Analytic Philosophy*, Oxford University Press, Oxford, 2002, 75-95.
[32] Suszko, R.: The Fregean axiom and Polish mathematical logic in the 1920's, *Studia Logica* **36** (1977), 373-380
[33] Thijsse, E.: *Partial Logic and Knowledge Representation*, PhD Thesis, Katholieke Universiteit Brabant, Eburon Publishers, Delft 1992.
[34] Wansing, H.: Short dialogue between M (Mathematician) and P (Philosopher) on multi-lattices, *Journal of Logic and Computation* **11** (2001), 759-760.
[35] Wansing, H.: Connectives stranger than tonk, *Journal of Philosophical Logic* **35** (2006), 653-660.
[36] Wansing, H. and Shramko, Y.: Suszko's Thesis, inferential many-valuedness, and the notion of a logical system, *Studia Logica* **88** (2008), 405-429, **89** (2008), 147.

The Character of T-Equivalences

Jan Woleński

Syntactically speaking, the expression 'it is true that' can be considered as a monadic sentence-forming functor. Let the symbol **T** denote this functor. Its meaning is given by two valuations: $v(\mathbf{T}p) = verum$ if $v(p) = verum$, $v(\mathbf{T}p) = falsum$ if $v(p) = falsum$. Accordingly, the formula

$$(0.1) \qquad\qquad \mathbf{T}p \Longleftrightarrow p$$

is a tautology of propositional calculus (**PC**). It was noted by Couturat (see [1, p. 84]). In his symbolism (0.1) is expressed by the formula

$$(0.2) \qquad\qquad (a = 1) = a$$

Couturat calls (0.2) the principle of assertion (in fact, it is the axiom X of his system) and adds the following comment:

> To say that a proposition a is true is to state the proposition itself. In other words, to state a proposition is to affirm the truth of that proposition.

This remark very well concurs with theories of truth developed by deflationists, minimalists, redundationists, etc. Apparently, (0.1) can be generalized to

$$(0.3) \qquad\qquad \mathbf{T}A \Longleftrightarrow A$$

where A is an arbitrary well-formed formula of logic (in this paper, I identify logic with first order logic). Clearly, if $v(A) = verum$, $v(\mathbf{T}A) = verum$, and if $v(A) = falsum$, $v(\mathbf{T}A) = falsum$. Thus, (0.3) is logically tautologous. However, there are some problems with (0.3). A minor point is that $v(A)$ is complex for non-atomic formulas, even in **PC**. For example, $v(\neg p) = verum$ if $p = falsum$. Although this does not matter from the formal point view, it still shows that the value of compound wff's is not given directly. More important questions arise when we take a more general standpoint than that suggested by the classical **PC**. Formula (0.3) is a conjunction of $\mathbf{T}A \Longrightarrow A$ and its converse. Although the first conjunct is unproblematic, the formula

$$(0.4) \qquad\qquad A \Longrightarrow \mathbf{T}A$$

cannot be generally considered as a logical theorem. There are at least two reasons for that. Firstly, the unrestricted version of (0.4) leads to inconsistency (see [21, p. 24-26]). Secondly, if we admit truth-value gaps or many-valueness, the meaning of (0.4) is unclear (see [22]). Hence, since (0.4) is not a purely logical theorem, the same concerns (0.3), because the conjunction $A \wedge B$ is logically true if and only both of its conjuncts are logically true. Some other problems concerning **T** and (0.3) will be discussed later.

Apparently (0.3) is a version of the **T**-scheme, a central element of the semantic theory of truth (**STT**, for brevity). However, this is a serious simplification. Let **L** be a language (that is, a set of sentences of **L**) and the letter A serve as a metavariable running over sentences of **L**. Then, the **T**-scheme is the formula (**VER**^{L} refers to the expression 'is true in **L**')

$$(*) \qquad\qquad \mathbf{VER}^L(n) \Longleftrightarrow \mathbf{A}^L$$

where the letter n refers to a name of A in **ML** (the metalanguage of **L**), for example, a code for A made by devices of arithmetization, but $\mathbf{A}^L$ displays how a sentence A is expressed in the metalanguage, for example, by means of set theory. Tarski's famous convention **T** requires that a truth-definition for **L**, in order to be considered as materially correct, must logically entail all instances of $(*)$ formed by suitable replacements of n and $\mathbf{A}^L$. Otherwise speaking, any materially correct truth-definition entails an instantiation of $(*)$ for each A in **L** (I do not enter into some doubts whether this claim formulates a sufficient or necessary condition; see [9] for a discussion).

One can argue that neither $(*)$ nor its instantiations can be just reduced to propositional tautologies. Firstly, they belong to **ML**, not to **L**. Secondly, the symbol **VER** is not a monadic sentential connective, but a unary predicate. Thirdly, if $(*)$ and its instantiations were truths of logic, they would trivially follow from arbitrary truth-definitions, because logical theorems are formally derivable from arbitrary premises and thereby the convention **T** is inessential, although it seems to capture an important intuition about the concept of truth. Yet all these reasons can be contested. First of all, since logic is common to all languages, it should be considered as fairly independent of levels of languages. Otherwise speaking, logic is invariant across languages of various degrees, but the difference rather concerns the amount of formalization. Furthermore, we can look at the formula $\mathbf{T}^L A$ as expressing truth-modality *de dicto*, but see '$n \in \mathbf{VER}^L$' as capturing this modality interpreted *de re* (in what follows I will often write $\mathbf{T}A$ and $\mathbf{VER}(A)$ instead of $\mathbf{T}^L A$ and $\mathbf{VER}^L(A)$, respectively). Now, since the equivalence of modalities *de dicto* and *de re* is plausible, if the quantifiers are not involved, we have

$$(0.5) \qquad\qquad \mathbf{T}A \Longleftrightarrow \mathbf{VER}(A)$$

as a principle which equates 'it is true that' (functor) and 'is true' (predicate).

Eventually, one can add special restrictions to block paradoxical sentences. The next observation points out that since **VER** is a monadic predicate, its logical behavior can be fully captured by **PC**. Finally, the deflationists and their nephews can say that the convention **T** is simply covered by (0.3) and nothing more is to be said. Since the reduction of the **T**-scheme and its concrete instantiations to logical truths is an open question (in order to simplify my considerations I will equalize, besides a few contextually clear exceptions, the expressions 'logical theorem', 'logical truth' and 'tautology', although this practice is not common; let me add that by logic I understand first-order predicate calculus), the problem of the (cognitive, logical, semantic) status of **T**-equivalences (or **T**-sentences) has no simple answer. We can ask, for example, whether they are analytical, necessary or a priori.

In order to demonstrate some related proposals, I will mention a sample of possible solutions. Tarski himself practically did not consider this question in an extensive way. He pointed out (see [17, p. 188], [18, p. 668, p. 674]; the page-reference is to reprint) that **T**-sentences explain the intuitive meaning of the locution 'a sentence A is true'. A much more important requirement is that they be theorems of **STT**, that is, that they be logically derivable from this theory of truth. Thus, we postulate ($T^{\Leftrightarrow}[A]$- **T**-equivalence)

$$(0.6) \qquad\qquad \mathbf{STT} \vdash T^{\Leftrightarrow}[A], \text{ for every } A \in \mathbf{L}$$

According to Quine (see [12, p. 137, note 10]), it is sometimes overlooked that there is no need to claim, and that Tarski has not claimed that the statements of the form [(*)] are analytic.

This quotation only shows that Quine, due to his views about analyticity, was not inclined to interpret **T**-equivalences as analytic. Tarski probably agreed with that view due to his serious scepticism (see [20]) about a reasonable illumination of the analytic/synthetic distinction. Dummett (see [4, p. XX]) interprets **T**-sentences just as material equivalences. They require that their members have the same logical value, but do not have to possess the same meaning. McGee (see [8, p. 1]) would like to understand **T**-equivalences as not only true, but also analytically true. Unfortunately, as he notes, this postulate is simply blocked by the occurrence of the Liar sentence and other self-referential statements. This seems to mean that all instantiations of the naive **T**-scheme, that is, the formula

$$(**) \qquad\qquad A \in \mathbf{VER} \Longleftrightarrow A$$

could be considered as analytic, provided the admissibility of all possible replacements. Remembering that the conditional $TA \Longrightarrow A$ does not lead to any trouble, there remains the problem of (0.4). Although Turner's logic of truth cancels this formula, this only means that it is not a theorem in this logic, but not that it is not analytic on other grounds, unless we decide to equate analyticity and being a theorem in Turner's system. Soames [14] argues that **T**-equivalences falling under the **T**-scheme but without indicating the language **L**, are neither analytic nor a priori, but if we supplement 'in my language' the full **T**-scheme, in Soames' wording, 'A is true in my language if and only if A', we obtain an analytic sentence, accepted by every competent user of a given language. The same author (see [15, p. 106]) says that sentences of the type 'the proposition A is true if and only if A' are necessary and a priori, regarding an analytic contentual link (Scott Soames informed me in a letter dated October 14, 2002 that he was never fully convinced of this view and does not currently accept it). According to Putnam (see [10, p. 332-333]), (∗) is a law of logic in **ML** and thereby is true in each possible world, because **T**-equivalences are true by logic axioms and axioms establishing the names of sentences. Other authors (see, for example, [6, p. 17]) consider particular instantiations of (∗) (or its variant) as the axioms of the theory of truth. Unfortunately, they do not explain what it means that a formula of the type $T^{\Leftrightarrow}[A]$ is an axiom, although they usually do not qualify them as logical axioms. Yet we do not know whether axioms falling under the form $T^{\Leftrightarrow}[A]$ are empirical or not.

Some essential points related to the status of the **T**-scheme and its instantiations are illuminated in [3, pp. 130-135]. David argues that particular instances of (∗) are contingent (non-necessary), because conditionals occurring in them are such. Consider

(0.7)   if 'snow is white' is true, then snow is white

Now it might be that the sentence 'snow is white' means that snow is green. In such a case, (0.7) were false. Thus, we have a possible world in which (0.7) would be false. This just entails that this sentence is contingent, because a necessary truth must be true in all possible worlds. The reverse conditional, that is, the statement

(0.8)   if snow is white, then 'snow is white' is true

is in a similar situation, because it could happen that snow is white but the sentence 'snow is green' is false. Suppose that we are working with a truth-definition **D** which logically implies all **T**-sentences and thereby satisfies the convention **T**. Since the principle

(0.9)   if $A \vdash B$ and B is contingent, then A is contingent as well

(the contingency of the logical consequent entails the contingency of the logical antecedent) is plausible, we should consider **D** as contingent too. However, any proper definition of truth should be something more than a contingent statement and could not imply contingencies, unless some non-necessities are taken as premises. In fact, it is not important here whether "something more" invokes a definite necessity or not, because an essential point consists in arguing for the contingency of (∗) and its instantiations. Similar matters are discussed by Künne (see [7, p. 183]). He considers the conjunction of (0.7) and (8), that is, the legendary Tarski's example

(0.10)   'snow is white' is true if and only if snow is white

and argues that Tarski should accept

(*Nec*)   Necessarily, 'snow is white' is true if and only if snow is white

The reason for the necessitation of (0.10) is that Tarski understood it (and other instantiations of (∗) as well) as a partial truth-definition, but this means that **T**-sentences are something more than contingent equivalences. Künne's main idea is as follows:

> Criterion [= Convention - J. W.] T demands that the pertinent T-equivalences follow from the definition of 'true' for **L** (plus some non-contingent syntactical truths like "snow is white" is not identical with "blood is red"). The definition itself is not a contingent truth, for it is constitutive of **L** that its sentences mean what they do mean. Now a conceptual truth *A* cannot entail a contingent truth *B*; for otherwise, by contraposition, the negation of *B*, which is as contingent as *B*, would entail the negation of *A*, which is conceptually false. And this is absurd, since whatever entails a conceptual falsehood is itself conceptually false. Hence Tarski ought to be ready to accept (*Nec*).

Denote conceptual falsehood by **CF**. Künne's principle is expressed by

(0.11)   if $B \in \mathbf{CF}$ and $A \vdash B$, then $A \in \mathbf{CF}$.

Dually (see above), if the symbol **CT** refers to conceptual truths, we have (modulo the principle of bivalence):

(0.12)   if $A \in \mathbf{CT}$ and $A \vdash B$, then $B \in \mathbf{CT}$.

Although (0.11) and (0.12) appear as unproblematic, everything depends on how conceptual (necessary, analytic, a priori) truths (falsehoods) are understood and what we think about their relation to tautologies. The above discussion shows that we have two extreme views about the status of **T**- equivalences:

(I) **T**-sentences are material biconditionals (Dummett on Tarski);

(II) **T**-sentences are laws of logic (Putnam).

I will demonstrate that both (I) and (II) are not acceptable as solutions of our problem. All further reasonings will be carried out in **ML**, because this language is the homeland for the concept of truth. I will begin with (I). Assume that a concrete **T**-equivalence, for example (0.9), is merely a material biconditional. Since it as well as its left and right parts are true, we can replace its right hand, that is, 'snow is white' by another true sentence, for instance, 'blood is red'. This move gives

(0.13)   'snow is white' is true if and only if blood is red

However, if (0.10) formulates a truth-condition for 'snow is white', the same cannot be said about (0.13). In particular, the material equivalence of (0.10) and (0.13) does not entail that they are also equivalent with respect to provability. In order to do that, one must justify the minimal condition of material adequacy in Tarski's sense, which requires that

(0.14)   The sentence 'snow is white if and only if blood is red' is provable in

  ML, that is, in a theory formulated in this language. (I will denote it by **MT**).

The only possibility consists in taking 'blood is red' as expressing in **ML** the sentence called 'snow is white' in this language. Clearly, we can do that, but this move should be made at the beginning. Anyway, no sentence which means that snow is white is provably equivalent to the sentence which means that blood is red, unless we resign from an essential feature of (∗), that its right part represents the sentence which is mentioned in its left hand. This conclusion is unavoidable, if one takes seriously Tarski's view (see [17]) that all our semantic considerations are always formulated with respect to interpreted languages, although possibly more or less formalized. Thereby, provable material biconditionals are something much more than merely material equivalences. Moreover, from the logical point of view, it is more important that they be provable than that they be material. Formally speaking, it is substantial that sentences of the form $\mathbf{T}^{\Leftrightarrow}[A]$ are to be preceded by the sign ⊢, independently whether they are considered as axioms or theorems. Their importance stems from the fact that they function as theorems of the theory of truth formulated in **ML**. Let me, however, note that this observation does not suffice to solve our problem.

The argument against position (II) is more complicated. Passing to this issue, let us assume that the formula

(0.15) $\mathbf{VER}(A) \Longleftrightarrow A$

is a **T**-equivalence and a logical theorem (remember that it is an abbreviation of (∗) and not the naïve **T**-scheme). This means that

(0.16) $\vdash (\mathbf{VER}(A) \Longleftrightarrow A)$

holds, because logical theorems are consequences of the empty set of premises. Since we are working in first-order logic, we can use the completeness theorem and rewrite (0.16) as

(0.17) $\vDash (\mathbf{VER}(A) \Longleftrightarrow A)$

which points out that the formula $(\mathbf{VER}(A) \iff A)$ is logically (universally) valid, that is, true in all models; the equivalence of (0.16) and (0.17) justifies the mutual interchangeability of the concept of a logical theorem (a formula that is provable from the empty set) and the concept of logical tautology (a formula that is universally valid).

Let the symbol 1 denote a tautology. Simple formal transformations guaranteed by the the equivalence of arbitrary tautologies, the completeness theorem, the associativity of the functor $\iff$ and the distributivity of the provability sign over $\iff$ lead to

(0.18) $$\vdash ((\mathbf{VER}(A) \iff A) \iff 1)$$

(0.19) $$\vdash (\mathbf{VER}(A)) \iff (A \iff 1)$$

(0.20) $$\vdash (\mathbf{VER}(A) \iff \vdash (A \iff 1))$$

We can define being a tautology (symbolically: $\mathbf{TAUT}$) by

(0.21) $$\mathbf{TAUT}(A) \iff \vdash (A \iff 1)$$

Since the equivalence $(A \iff 1)$ is provable, it is universally true. Thus, we have $\models (A \iff 1)$. The same concerns $\mathbf{TAUT}(A)$. Hence, we obtain $\models \mathbf{TAUT}(A)$.

The next segment of the reasoning employs (0.21). It and (0.20) leads to

(0.22) $$\models \mathbf{TAUT}(A) \iff \vdash \mathbf{TAUT}(A)$$

Having (0.20), (0.21) and (0.22), we go to

(0.23) $$\vdash (\mathbf{VER}(A) \vdash \mathbf{TAUT}(A))$$

In the general case we cannot pass from

(0.24) $$\vdash A \iff \vdash B$$

to the formula

(0.25) $$A \iff B$$

because it can happen that A is a non-provable truth, but B a non-provable falsehood, which makes (0.24) true and (0.25) false. However, since our assumptions about A effectively exclude this possibility, we can omit the provability sign in (0.23) and conclude

(0.26) $$\mathbf{VER}(A) \iff \mathbf{TAUT}(A)$$

This formula establishes co-extensionality of **VER** and **TAUT**, a result which we must accept due to the following argument. The denial of (0.26) gives

(0.27) $(\neg\mathbf{VER}(A) \wedge \mathbf{TAUT}(A)) \vee (\mathbf{VER}(A) \wedge \neg\mathbf{TAUT}(A))$

The first disjunct of (0.27) does not hold, because tautologies cannot be false. Consequently, there remains the second one, that is, the formula, $\mathbf{VER}(A) \wedge \neg(\mathbf{TAUT}(A))$. It implies

(0.28) $A \Longleftrightarrow \neg(\mathbf{TAUT}(A))$

and, furthermore (by (∗), bivalence and properties of implication)

(0.29) $\mathbf{VER}(\neg A) \Longleftrightarrow \vdash (A \Longleftrightarrow 1)$

The right hand of (0.29), that is, the formula $\vdash (A \Longleftrightarrow 1)$ is equivalent to $\vdash A$, due to

(0.30) $\vdash (A \Longleftrightarrow 1) \Longleftrightarrow A$

Finally, we obtain

(0.31) $\mathbf{VER}(\neg A) \Longleftrightarrow \vdash A$

However, (0.31) is not acceptable, if our logic is to be sound. Hence, we should accept (0.26). The entire above reasoning essentially depends on (0.15), because it validates passing from (0.23) to (0.25). Thus, if **T**-equivalences are considered as tautologies, truth and logical truth are extensionally non-distinguishable. This result, however, is arguably non intuitive, because we want to have

(∗ ∗ ∗) $\neg(\mathbf{VER}(A) \Longleftrightarrow \mathbf{TAUT}(A))$

that is, we tend to see logical truth as a special instance of truth simpliciter. Observe also that if **VER** and **TAUT** are equivalent, the same concerns falsehood and logical contradiction. Consequently, if we say that (∗) generates logical truths, we must consider every sentence as logically determined, that is, tautologous or formally inconsistent.

Yet a comment is in order here. The reasoning coded by (0.15) - (0.31) cannot be regarded as showing that there is something wrong with logic. Couturat's formula, that is, (0.2), as well as all the mentioned wff's with 1 remain valid if this symbol refers to a constant interpreted as 'true sentence', independently whether it is a tautology or an arbitrary sentence A such that $v(A) = verum$. In fact, there is no possibility to distinguish between truth and logical truth inside logic itself. We can do that only in metalogic. Thus, steps (0.15) - (0.31) are intended to demonstrate that taking (0.15) as a tautology is at odds with (∗ ∗ ∗). Speaking otherwise, when **T** and 1 are interpreted as logical constants, (0.15) and its formal equivalents are logical theorems. Let me observe that **T** also possesses another reading, namely 'it is asserted that', provided, in Frege's spirit, that we assert only truths and never falsehoods; this kind of assertion can be termed as logical. This suggests that the concepts of assertion, truth and tautology are extralogical, at least relatively to elementary logic. Since logic does not distinguish any extralogical constant, it is not strange that the concepts of **VER**, **TAUT** and logical assertion are not extensionally distinguishable in logic itself.

That treating **T**-sentences as tautologies is improper can be also shown in another way. If (0.15) is a tautology, it is the case that

(0.32)   for every valuation v, $v(\mathbf{VER}(A)) = v(A)$

However, (0.32), unless **VER** is a logical constant of **PC**, requires further justification, because **VER**(A) and A are syntactically different. In particular, we should not decide in advance that the symbol **VER** is redundant. Such an additional justification can only be

(0.33) $(\mathbf{VER}(A) \Longleftrightarrow A) \Longleftrightarrow (A \Longleftrightarrow A)$

Now the left part of (0.33) must be tautological, provided that the entire equivalence logically holds. We cannot use (0.15) in order to justify (0.15), because this step would be circular or begging the question. Thus, we have no other possibility to meet the problem than consider **VER**(A) as a stylistic variant of A. The right part of (0.33) can be made more complex by formulas, like $A \Longleftrightarrow (A \vee A)$ or $A \Longleftrightarrow (A \wedge A)$, but such developments always employ logical equivalents of A. These steps preserve the condition established by (0.32), but the result leads to **T**-sentences as trivialities of the type $A \Longleftrightarrow A$. I do not think that the proponents of (II) would like to accept this view. Yet it seems that they have no other way out.

Having rejected both extreme views about the status of (∗) and its instances, we should look for a compromise. First of all, let us observe that weak second order arithmetic is enough as **MT**, that is, it suffices to developing **STT** (see [5]). This metatheory offers resources for naming sentences of **L**, in particular, it generates their arithmetical codes. The most important fact is that we can demonstrate the validity of the convention **T** in **MT**. In a sense, the scheme (∗) in its full version generates all fixed points for sentences of **L**. Thus, the convention **T** is not trivial, because it states a form of the fixed point theorem, which says that, under some additional requirements, every sentence of a language has its fix point (see [13, p. 102-104]). More explicitly, the symbol $\mathbf{A}^{\mathbf{L}}$ refers to a fixed point of the sentence $\mathbf{VER}^{\mathbf{L}}(n)$. This fact immediately suggests that **T**-sentences cannot be considered as truths of pure logic, because our **MT** exceeds first order logic. Incidentally, Putnam's claim that the instances of (∗) are provable by logic and the axioms of naming goes against his view that **T**-equivalences are logical truths, because ways of forming codes for sentences go beyond logic. Instead, assuming that (0.15) is logically true, we adopt

(0.34) $\mathbf{MT} \models \mathbf{T}^{\Leftrightarrow}[A]$, for every $A \in \mathbf{L}$

but this formula entails only that **T**-sentences are true in **MT**-models, but not in all models.. This observation straightforwardly suggests the compromise we are looking for. The starting point is that we cannot reproduce the reasoning displayed by (0.15) - (0.31) in new frameworks. Thus, we are not forced to identify truth and logical truth. Call **MT**-tautologies, sentences true in all **MT**-models. They represent sentences analytic in **MT** or its conceptual truths. Thus, I suggest that every $\mathbf{T}^{\Leftrightarrow}[A]$ is a **MT**-tautology. Accordingly, the negations of **MT**-tautologies are **MT**-inconsistencies, that is, conceptual falsehoods in **MT**. We can easily verify that the conditions stated by (0.11) and (0.12) hold. Elsewhere (see [22]) I proposed to regard sentences true in all models of a given theory **Th** as relative (with respect to **Th**) analytic sentences in the semantic sense. Thus, our compromise concerning the status of (∗) and its instantiations can be expressed by

(S) $\mathbf{T}^{\Leftrightarrow}[A]$'s are relative semantic sentences of **STT**

This thesis exactly expresses what it means to say that **T**-sentences are something more than material equivalences, but still are not logically true.

BIBLIOGRAPHY

[1] Louis Couturat. *L'Algebre de la logique*. Gauthier-Villars, Paris, 1905. page references to the English translation by L. Gillingham Robinson, [2].
[2] Louis Couturat. *The Algebra of Logic*. The Open Court, Chicago, 1914.
[3] Marian David. *Correspondence and Disquotation: An Essay on the Nature of Truth*. Clarendon Press, Oxford, 1994.
[4] Michael A. E. Dummett. *Truth and Other Enigmas*. Harvard University Press, Cambridge, MA, 1978.
[5] Volker Halbach. *Axiomatische Wahrheitstheorien*. AkademieVerlag, Berlin, 1996.
[6] Paul Horwich. *Truth*. Oxford University Press, Oxford, second edition, 1998.
[7] Wolfgang Künne. Disquotationalist conceptions of truth. In Richard Schantz, editor, *What is truth?*, pages 176–193. de Gruyter, Cambridge, 2002.
[8] Vann McGee. *Truth, Vagueness and Paradox: An Essay on the Logic of Truth*. Hackett, Indianapolis/Cambridge, 1990.
[9] Douglas Eden Patterson. Tarski on the necessity reading of convention T. *Synthese*, 151(1): 1–32, 2006.
[10] Hilary Putnam. A comparison of something with something else. *New Literary History*, 17(1): 61–79, 1985. Page references are to the reprint in [11].
[11] Hilary Putnam, editor. *Words and Life*. Harvard University Press, Cambridge, Mass, 1994.
[12] W.V.O. Quine. *From a Logical Point of View*. Harvard University Press, 1953.
[13] R. Smullyan. *Gödel's Incompleteness Theorems*. Oxford University Press, Oxford, UK, 1992.

[14] Scott Soames. T-sentences. In Walter Sinnott-Armstrong, Diana Raffmann, and Nicholas Asher, editors, *Modality, Morality, and Belief: Essays in Honor of Ruth Barcan Marcus*, pages 250–270. Cambridge University Press, Cambridge, 1995.

[15] Scott Soames. *Understanding Truth*. Oxford University Press, New York, 1999.

[16] Alfred Tarski. Der Wahrheitsbegriff in den formalisierten Sprachen. *Studia Philosophica*, 1: 261–405, 1936.

[17] Alfred Tarski. The concept of truth in formalized languages. In *Logic, Semantics, Metamathematics*, pages 152–278. Oxford University Press, Oxford, 1956. Translation, by Joseph H. Woodger, of [16].

[18] Alfred Tarski. The semantic conception of truth: and the foundations of semantics. *Philosophy and Phenomenological Research*, 4:341–376, 1944. Page references are to reprint in [19].

[19] Alfred Tarski. *Collected Papers, Volume 4: 1958–1979*. Birkhäuser, Basel, 1986.

[20] Alfred Tarski. A philosophical letter of Alfred Tarski. *Journal of Philosophy*, 84(1): 28–32, 1987.

[21] Raymond Turner. *Truth and modality for knowledge representation*. MIT Press, Cambridge, MA, USA, 1991.

[22] Jan Woleński. On some formal properties of truth. In Daniel Kolak and John Symons, editors, *Quantifiers, Questions and Quantum Physics: Essays on the Philosophy of Jaakko Hintikka*, pages 195–207. Springer, Dordrecht, 2004.

Begging the Question is Not a Fallacy

JOHN WOODS

1 Getting started

The present essay is related to a larger project with Dov Gabbay on the logic of practical reasoning. One of the principal foci of this work is the role of error in the cognitive economies of human reasoners. Logicians have been interested in error since the founding of their discipline. By and large, they have restricted their attention to three theoretical targets: the deductive error of invalidity, the non-demonstrative error of inductive weakness, and fallacies, characterized by Michael Scriven as "the attractive nuisances of argumentation" [26, p. 333]. In recent writings or in work under way, Gabbay and I have questioned the assumption that invalidity and inductive strength are in fact errors just as they stand [6].[1] We have also advanced – with a tentativeness appropriate to its radicality – what we call the Negative Thesis. To see what the Negative Thesis proposes, it is necessary to take note of both the identity of the concept of fallacy which tradition has passed down to us, as well as the traditionally agreed members of its extension. The Negative Thesis asserts that the traditionally received list of the fallacies are not in fact in the extension of the traditionally received concept of fallacies[2] [6].[3] It is well to note right at the outset that the Negative Thesis does not assert that the traditional conception of fallacy is empty. It proposes only that the "usual suspects" – *ad hominem*, hasty generalization, *ad baculum*, etc. – are not in its extension.

---

[1] For one thing, most good reasoning and most good arguments are invalid and/or inductively weak. This inclines Gabbay and me to the view that validity and inductive strength are appropriate standards only relative to the reasoner's cognitive agenda. For more on the relativity of error see [6].

[2] Of course, there is not perfect unanimity about what the traditional fallacies have turned out to be in the present day. But most of the preferred lists exhibit a considerable overlap. [35] records the following eighteen: *ad baculum, ad hominem, ad misericordiam, ad populum, ad verecundiam, ad ignorantium,* affirming the consequent, denying the antecedent, begging the question, equivocation, amphiboly, hasty generalization, *post hoc, ergo propter hoc,* biased statistics, composition and division, faulty analogy, gambler's and *ignorato elenchi.*

[3] The present author is also of the view that the same kind of "disconnect" is demonstrable in Aristotle's writings on fallacies; that is to say: that Aristotle's list of the fallacies fails to satisfy his definition of them. I have tried to defend this claim in the case of the Aristotelian fallacy of Many Questions in [42]. For a discussion of the connection between many questions in the modern sense and question-begging see [17].

We are not in the least doubt that the Negative Thesis will strike many people as preposterous. What could be more obvious than that *ad hominem* retorts are destructively fallacious, that hasty generalization is a great folly,[4] that begging the question is a despoiler of argument, and so on? The job of answering this scepticism is the business of our book in progress, *Seductions and Shortcuts: Error in the Cognitive Economy*.[5] My present purpose is to test the waters in a much more modest way. I want to examine whether I can make good on the Negative Thesis in the particular instance of begging the question. Doing so will require us to have at hand an appreciation of the traditional concept of fallacy and the traditional concept of begging the question. I turn to these matters in reverse order in the next section. But first a word of clarification. In my usage here, the expression "the traditional concept of K" means the standard present-day understanding of K as it has evolved over time.

2   The modern conception of question-begging

The idea that begging the question is a fallacy originates with Aristotle,[6] as does the idea of fallacy itself.[7] Given logic's already long history, it should not be surprising that Aristotle's views of these matters appear to have been superseded. But the modern view retains the original connection, for here too question-begging in the modern sense is said to be a fallacy in modern sense.[8] As currently conceived of, begging the question and fallacies can be characterized in the following way:

Begging the question. Let τ be a thesis advanced by Smith. Let α be a proposition forwarded by Jones as counting against τ. Then Jones begs the question against Smith's thesis τ iff

1. α is damaging to τ,

2. α is not conceded by Smith, does not follow from propositions already conceded by Smith, and

3. α is not otherwise ascribable to Smith as what we might call a "reasonable presumption" or a "default" (for example, the belief that water is wet or that Washington is the capital city of the United States).[9]

---

[4] For an attempt to apply the Negative Thesis to the *ad hominem* see [44] and to hasty generalization see [45].

[5] Volume 3 of the omnibus work *A Practical Logic of Cognitive Systems* [8].

[6] See, for example, *Soph. Ref.* 5, 167a, 37-40; 6, 168b, 25-27; 7, 169b, 13-17; 17, 176a, 27-32; Top. 8, 161b, 11-18; 162b, 34-163a, 13, 28; *Pr. Anal.* 24, 41b, 9

[7] Aristotle characterizes fallacies as arguments that appear to be syllogisms but in fact are not syllogisms (*Soph. Ref.*, 169b, 17-21). A syllogism, in turn, "rests on certain propositions such that they involve necessarily the assertion of something other than what has been stated." (*Soph. Ref.* 165a, 1-3)

[8] See, for example, [18], [25], [3], [31], [32], [33], [21], [29].

[9] These conditions are discussed further in [39, chapter 1].

Fallacies. A fallacy is an error of reasoning that satisfies the following conditions:

1. The error is an *attractive* one,

2. it is a widely-committed or *universal* error, and

3. it exhibits a substantial degree of *incorrigibility*; that is to say, levels of post-diagnostic recidivism are high.

If we add to our list of conditions that fallacies are also bad, we might propose an acronym BEAUI made up of the first letters of "bad", "error", "attractive", "universal" and "incorrigible". That the BEAUI-conception is the traditional concept of fallacy is attested to by a number of writers. Here again is Michael Scriven:

> Fallacies are the attractive nuisances of argumentation, the ideal types of improper inference. They require labels because they are thought to be common enough or important enough to make the costs of labels worthwhile [26, p. 333].

The same view is echoed by Trudy Govier.

> By definition, a fallacy is a mistake in reasoning, a mistake that occurs with some frequency in real arguments and which is characteristically deceptive [10, p. 172].

It is endorsed by David Hitchcock as

> the standard conception of fallacy in the western logical tradition [16, p. 1].

Needless to say, the concept of "reasoning" exhibits a rather sprawling usage. The same may be said for the concept of "error". This being so, it would be quite wrong to leave the impression that those who characterize fallacies as errors of reasoning are committed to the view that there is no sense of "reasoning" and no sense of "error" for which there are errors of reasoning that aren't fallacies. There are mechanical errors, factual errors, perceptual errors, and so on. Errors of reasoning belong to a family of errors that have the character of *illusions*. Historically, fallacies are errors that seduce us; they are missteps that take us in.

It bears on this matter that the concept of fallacy first arose as a logician's notion, and it has remained one ever since – albeit with some rather scruffy patches[10] as part of the research programmes of logic. It is quite true, as we will see in the

---

[10]See [11] for a diatribe against logicians for having given up on the fallacies programme.

section to follow, that certain disciplines other than logic have appropriated the term "fallacy". There is nothing as such wrong with these appropriations. The important question is whether they preserve the traditional sense of "fallacy".

Right at the beginning of it, the founder of logic introduced an interesting pair of distinctions. One is the distinction between

The consequences a set of propositions **has**

and

The consequences of a set of propositions that it is necessary (or appropriate) to **draw**.

This contrast is present in the definition of syllogism. A syllogism is an argument whose premisses necessitate its conclusion, and which satisfies further conditions, primary among which are the following two: The conclusion of a syllogism may not repeat a premiss; and no premiss may be redundant. However, it is advisable to note in passing that although the concept of fallacy originates with Aristotle, what I am calling the traditional or BEAUI conception differs from Aristotle's idea. For Aristotle, a fallacy is the mistaking of a non-syllogism for a syllogism. This feature does not distinguish the BEAUI conception of fallacy.[11]

Consider any successful necessitation-argument which fails these conditions. In each case the conclusion is a consequence of its premisses – it is necessitated by them yet in each case there are circumstances in which Aristotle decrees that these consequences not *be* drawn. Thus, like error itself, consequence-drawing is appropriate or inappropriate relative to the reasoner's agenda.[12] In marking this

---

[11] Aristotle discusses the fallacy of begging the question at *Topics* VIII 13, 162b 34-163a 13. He distinguishes five varieties of question-begging, three of which involve the attribution to one's opponent of a proposition which entails in one step the contradictory of the opponent's own thesis. The remaining two are situations in which the refuter's desired conclusion is obtained piecemeal by means of a set of postulates whose conclusion is equivalent to the desired conclusion. It would be useful to pause and consider a case which clearly counts as begging the question in Aristotle's sense but not in the BEAUI sense. Suppose that Jones is attempting to refute Smith's thesis τ. Suppose that $\alpha \models \sim \tau$. Then I must not attribute α to you without proof. But what if you actually do hold α? By modern lights, you're sunk. By Aristotle's lights, things are a bit more complicated. Suppose I observe, "Well since α, then $\sim \tau$". This is a perfectly correct immediate inference, but it is not a refutation in Aristotle's sense. The reason is that "Since α, $\sim \tau$" is not a syllogism (it is a deduction from a single premiss); and since a refutation must always be a syllogism, offering this inference as a refutation would be a fallacy in Aristotle's technical sense of deploying a non-syllogism to do the work of a syllogism. Of course, my task here is not to determine whether question begging in Aristotle's sense is a fallacy, but rather whether begging the question is a fallacy in the BEAUI sense.

[12] A case in point: On most accounts of deductive consequence , a set of premisses has infinitely many consequences. Many logicians, including Aristotle himself, would be of the view that the consequences that a reasonable person would actually draw from this multitude are those propositions *relevant* to the task at hand, that is, to the task that motivated in the first place the reasoner's interest in knowing what the consequences *are*.

distinction, we provide the means to characterize the "logical" meaning of "error of reasoning".

> *One commits an error of reasoning in one or other of two ways: By inappropriately drawing from a set of premisses a consequence that it has; or by citing (or drawing from) a set of premisses a consequence that it does not have.*

The second distinction, also unmistakably present in *On Sophistical Refutations*, is that between a deduction and an argument. A deduction is a sequence of propositions, the last of which is the conclusion and the rest the premisses, which satisfies the definition of syllogism. Arguments, on the other hand, are concrete affairs, exchanges in real time between actual people. Refutations are paradigms of arguments in this sense. It was Aristotle's view that the class of arguments in which he was interested all involved the construction of syllogisms. But it is also clear that there is much more (say) to refuting the thesis of an interlocutor than citing or drawing consequences. For one thing, one must ask him the right questions, questions that are designed to elicit answers damaging to his position. So we have a helpful contrast between errors of argument that are errors of reasoning and errors of argument that are not errors of reasoning.

> *In making an argument against an interlocutor one makes an error of reasoning if one incorrectly cites a consequence or inappropriately draws a consequence. Other errors that might occur are not errors of reasoning.*

In the *Organon* Aristotle gives dominant attention to relations of deductive consequence . In the modern tradition this exclusivity is relaxed. Consequences are now of any stripe - statistical, probabilistic, abductive, or whatever else - that fall within the ambit of modern logic.

It is hard to see how begging the question in the modern sense squares with the BEAUI-conception of fallacy. Question-begging is an attribution error. Jones's advancement of α against Smith's τ presupposes that Smith is committed to α. If, in so assuming, Jones begs the question against Smith, it is clear that his error is one of false attribution. Jones's move can also be seen as a premiss-selection error. Jones appropriates as a premiss in his attack upon Smith a proposition α that Smith does not accept and is not committed to accepting.

Does this perhaps bring us a little closer to the idea that question-begging is a mistake of reasoning? If we schematize Jones's move as the construction of an argument in the form $\langle \alpha / \sim \tau \rangle$, we can say that the argument is faulty to the extent that it embodies a premiss-selection error, but this is not enough to sustain the claim that, in making the argument, Jones has reasoned badly. For reasoning here is a matter of what consequences Jones draws from α, and, not only is it not obvious that $\sim \tau$ is *not* a consequence of α, the very fact that $\langle \alpha / \sim \tau \rangle$ is a question-begging argument, guarantees that it is.

What these reflections suggest is the usefulness of the distinction between *defective arguments* and *errors of reasoning*. Certainly, someone damages his own argument by the selection of premises that are not properly available to him; yet, equally, in drawing the consequences of those premises his reasoning might well be impeccable. If we decided to give sway to this distinction, we could amend the BEAUI-conception of fallacy in the obvious way. We could replace condition (1), according to which a fallacy is an error of reasoning, with a disjunctive variant (1'), according to which a fallacy is an error of reasoning or a deficiency of argument.

There is something to be said for this latitude, not all of it bad.[13] But, on the whole, I think it preferable to resist it. If we allow fallacies to include premiss-selection errors, there is nothing to prevent the idea from extending to the employment of any proposition as premiss if it happens to be false. It is true that some of the falsehoods to which we are drawn are fallacies in the most common non-technical meaning of the term, where a fallacy is simply a widely held false belief. But although not every false belief is a fallacy in *this* sense, our revised definition would make an argument a BEAUI-fallacy if it had any false proposition as premiss.

People's mistaken beliefs range far and wide – from the things of everyday concern, to macroeconomics, biochemistry and theology. If someone soils his argument by appropriating a biochemical falsehood as a premiss, there is plenty of room to think him a defective biochemist but a splendid reasoner. It is true that since the beginnings of systematic logic, fallacies have been associated with defective arguments. But one finds it neither in the ancient writings nor in the traditional modern writings that everything whatever that defaces an argument convicts the miscreant of fallaciousness.[14] What our present example suggests is that even defective arguments can be occasions of brilliant reasoning. Embedded in this contrast is the well-known distinction between errors of fact and errors of inference. Up to a point, people may classify things as they like. But for those who are attracted by the suggestion that the present contrast captures a significant difference of kind, this will be sufficient to reinforce the traditional inclination to reserve the name of fallacy for transgressions of the second kind only. Accordingly, we might now hazard that

---

[13]The present suggestion puts one in mind of Hintikka's distinction between *definatory* and *strategic* rules, which is similar but inequivalent [13],[14]. Roughly speaking, a definatory rule is a consequence-spotting procedure. A strategic rule offers guidance as to when it is appropriate actually to apply a definatory rule. Donald Gillies picks up on this distinction, proposing that logic = the definatory + the strategic rules [9].

[14]I will consider a modern exception in the section just below.

Question-begging does not instantiate the BEAUI-conception of fallacy.[15]

3   Reinstating premiss-selection as a mode of reasoning?

Perhaps this would be a good place to explain that I intend no semantic imperiousness about the words "logic", "reasoning" or "fallacy". My point is only that the mainstream interpretations of these things create a tangled picture. It is a picture in which if we persist with those interpretations of "logic", and "reasoning" we will have difficulty in also persisting with this interpretation of "fallacy" in its application to begging the question. What I am proposing is that the disentanglement that best preserves these mainline interpretations is one which leaves them all intact, ruling instead that begging the question in its modern sense is not indeed in the extension of the traditional concept of fallacy. Pivotal to that recommendation is the claim that, in as much as question-begging is a premiss-selection error, it is not an error of reasoning; that is, not an error of reasoning whose attribution and investigation fall within the province of logic as standardly understood.

Some people will not like this at all. Everyone will agree that the reasoning investigated by logic has to do with the citing and drawing of consequences. But it will likewise be agreed that logic also concerns proof, and that proof is a consequence-generating exercise that involves a search for the requisite theorems to serve as inputs. Proofs, too, are premiss-searches; proofs fall within the investigatory reach of logic; proofs are a kind of reasoning; so premiss-searches can be a kind of reasoning.

It would appear that much the same can be said about abduction. Abductions, too, are processes that complete a consequence-connection, albeit of a softer kind than deductive consequence. The completion is effected when an appropriate hypothesis is grafted onto the reasoner's database. As we might expect, then, abduction is in part a search for the right hypothesis (as the name "logic of discovery" clearly suggests). If, as many logicians believe, abductive reasoning falls within the bounds of logic, then we seem to have it that hypothesis-searches are modes of reasoning that lie within the competence of logic to pronounce upon.[16]

Why, then, would we exclude the premiss-searches that drive the engines of refutation?

---

[15] It may be of some interest to note in passing that Aristotle is not indifferent to the problem of faulty attribution. Indeed he thinks that it is related to the fallacy of *ignoratio elenchi* (*Soph. Ref.* 5, 167a, 21-36; 6, 168b, 17-21; 169b, 9-13) This is the mistake of either misdeducing a purported consequence of something one's opponent holds (hence is a mistake of reasoning) or correctly deducing a consequence of something one's opponent does not hold. Either way, it is a fallacy by Aristotle's lights. By our own lights, it is in the second instance the argument-error of faulty attribution, but not an error of reasoning, hence not a BEAUI-fallacy.

[16] There are abductive logicians aplenty for whom hypothesis-selection is governed by considerations of relevance and plausibility, both of which properties are also the subject of investigation by logicians. See here [4].

Here is why. If I search for a premiss in a proof and fail to find a theorem and/or find a theorem that doesn't effect the consequence-completion, I have made a mistake of a sort that we might not mind a logician categorizing as a mistake of reasoning. Similarly, if, in casting about for a hypothesis with which to complete an abductive inference, I hit upon a proposition that isn't plausible and/or fails to make the connection, perhaps it is fine to call this a mistake of reasoning of a kind of interest to logicians. But if I am searching for a premiss in a refutation, my candidate is not subject to the condition that it be true or even plausible, never mind that it makes the consequence-connection. All that is further required of this proposition is that *my opponent concede it*. When I get this wrong, there are two cases to consider. In the first case, I believe falsely that my opponent does concede it. In the second case, I know that he doesn't concede it, but attribute it anyway. In the first instance, I am guilty of not knowing something that I should know. But this is far from showing that I have been landed in this ignorance by an error of reasoning. In the second instance, I am either stupid or I fail to see that pressing it is useless or mischievous perhaps I can catch my *interlocutor* in a hoped-for stupidity. Perhaps my own stupidity a kind of dialectical clumsiness is something like an error of reasoning, but such stupidity is not common, to say nothing of universal. On the other hand, if I am being mischievous, I might be trying to get you to make an error – possibly an error of reasoning. But my mischief is not an error of reasoning. I conclude, then, that the premiss-attribution errors that beggings of the question are fail the logician's mainline "error of reasoning" and "fallacy".

4   Dialectifying the fallacies

Perhaps the best-known alternative conception of fallacy is that of [28], in which fallacies are characterized as *any* violation of a discursive rule that governs a particular form of argument which these authors call a "critical discussion".[17] In a more general form, it is proposed that in all or more contexts most – if not all – fallacies are dialectical errors, rather than errors of reasoning or inference.[18] This is not the place to debate these claims in any detail. But I do want to pause long enough to register a basic reservation about the dialectification of fallacies. Of course, some of the things traditionally conceived of as fallacies lay no claim to a dialectical identity – consider for example, the gambler's fallacy or the fallacy of *post hoc, ergo propter hoc*. On the other hand, it is certainly true – as Aristotle was well-aware – that interpersonal wranglings are a natural context for the commission of fallacies, especially those that involve a reference to persons. In this regard, the *ad hominem* comes easily to mind as a typical example. In its modern conception, it takes two forms – the abusive and the circumstantial. The former al-

---

[17]For reservations, see [39, chapters 9 and 10] and [41].
[18]See here [13], and, in reply, [34], and in rejoinder, [15].

leges some flaw of character, the latter a behavioural inconsistency.[19] What counts here is that, whether or not these are indeed fallacies, there is nothing that requires that the person of whom these ascriptions are made be an *interlocutor* of the person who makes them.[20] This allows us to take note of a quite general point. Even if one is arguing against a view held by another person, it is not in the least necessary that one's case be directed to that person himself. One need not have chat with Plato in order to find fault with the Theory of Forms. The same is also true of begging the question. Fallacy or not, it is not essential to its commission that it be spoken to an interlocutor. Plato is dead, alas, but nothing in his death inoculates him against question-begging criticisms of his views. Accordingly,

> *Although n-person disputes are a natural context for the commission of fallacies, fallacies on the BEAUI-model are not intrinsically dialectical.*

Before leaving this section, I want briefly to deal with another objection. It is that what I am calling the BEAUI-conception is in fact *not* the standard conception of fallacy that has come down to us to the present day. Some commentators are of the view that the standard conception is one according to which a fallacy is an *invalid argument* that *appears to be valid.*[21] I have three things to say about this complaint.

1. With a little imagination, the inapparent-invalidity conception can be seen as instantiating the BEAUI-conception. So the gap between the two conceptions is not as great as might be supposed.

2. The claim of the inapparent-invalidity model to standardness is contradicted by the empirical record [12]. Roughly speaking, the present definition is not the one proposed by the authors of the standard works.[22] So it is hard to see how it gets to be the standard definition.

3. If the present definition were indeed the standard conception of fallacy, it would remain the case that question-begging is false attribution. And, since picking an unacceptable premiss does not, just so, invalidate the argument

---

[19]For example, citing an interlocutor's bias is usually thought of as abusive (albeit in a somewhat technical sense), whereas alleging that an opponent's behaviour is a defection from his own views is the circumstantial variant. As mentioned earlier, my own view is that *ad hominem* retorts aren't BEAUI-fallacies either, but that it is an issue for another time [44].

[20]We should mention that on an earlier conception, originated by Aristotle (*Soph. Ref.* 167b, 8-9 ff; *Pr. Anal.* B27, 70a, 6-7, *Metaph.* 1006a, 15-18) and revived by Locke ([20], pp. 686-688), an *ad hominem* move is an intrinsically dialectical manoeuvre, but it is not a fallacy by Aristotle's lights or by Locke's. Nor does it instantiate the BEAUI-conception. See here [39, chapter 7].

[21]That this is the standard definition is advanced by, e.g., [28].

[22]Chiefly, the standard *textbooks* published in the period 1940 onwards.

in which it occurs, much less disguise its invalidity, question-begging in the modern sense is not a fallacy even on the inapparent-invalidity model.[23] So it is not essential to my present purposes whether the inapparent-invalidity model or the BEAUI-model is the standard (modern) conception of fallacy.

5 An ambiguity

This would be a good point at which to take note of a significant ambiguity in the word "refutation". In so doing, I want to remove the tentativeness that presently attaches to my principal claim that question begging is not an error of reasoning, hence not a fallacy in the BEAUI-sense. In common parlance, as well as in numerous technical writings, "refutation" has two quite different senses.

1. *The propositional sense.* A proposition α is a refutation of a proposition τ iff α negates τ and α is true.

2. *The personal sense.* A person Jones refutes the thesis τ of a person Smith iff there is a proposition α which Jones advances, Jones believes that $\alpha \models \sim \tau$, Jones believes that Smith accepts α, or would accept it if he considered it, and that Smith accepts that $\alpha \models \sim \tau$, or would accept it if he considered it.

In the first sense, a refutation stands or falls independently of what any addressee might think of it. In order to construct a sound refutation in this sense, one must be right about two things. One must be right in thinking that α is true (or anyhow that one's believing that α is justified), and one must be right in thinking that $\alpha \models \sim \tau$. Reasoning correctly is a matter of drawing the right inference from the premises at hand. Refutations in the first sense fail when the reasoning is defective. (They also fail when a premiss is unjustified, but that is a different matter.) It is easy to see that

> *In contexts of disputation between Jones and Smith, two things are perfectly possible with respect to Jones's argument $\langle \alpha / \sim \tau \rangle$. Jones might beg the question against Smith, and yet Jones might have made a perfectly sound refutation in the first sense of Smith's thesis τ.*

Refutations in the second sense are another matter. They are a kind of solution of a kind of co-ordination problem. The problem is one of producing unanimity

---

[23]Lawrence Powers turns the present point on its head. He says that notwithstanding Hansen the standard (and correct) definition is that a fallacy is an invalidity that appears to be valid. Since question-begging arguments are valid, they can't be fallacies [22]. Powers holds to this definition of fallacy, even though he allows that it is not reflected in the empirical record. Perhaps we might persuade ourselves to agree that this is what the standard definition *should* be. But it is too much of a stretch to insist that this is what it is in fact. (Ought-is problems.) Powers also conflates question-begging with circularity, which certainly doesn't comport with what I am calling the modern conception of question-begging.

between Jones and Smith with respect to τ. The problem is solved if both come to agree that τ is not the case, or if both come to agree that Smith cannot consistently hold to τ. Whether τ is or is not in fact the case in fact need not be a factor in the construction of such solutions. Refutations in the second sense are a kind of solution that produces the shared belief that τ is not the case or that Smith cannot consistently persist with τ. Here, too, it is easy to see that

> *Jones's refutation in the second sense of Smith's thesis τ succeeds independently of whether it succeeds as a refutation in the first sense. So, $\langle \alpha / \sim \tau \rangle$ can fail as a refutation in the second sense and succeed as a refutation in the first sense.*

Whether a refutation in the second sense succeeds or fails is strictly speaking entirely in the hands of the addressee.[24] He may reject α and he may reject that $\alpha \models \sim \tau$. In this, he may be objectively right or wrong, but it doesn't matter. Refuting the addressee is wholly a matter of getting the addressee to give up on τ by way of his concession that α. It goes without saying that

> *If in a refutation in the second sense Jones begs the question against Smith, the refutation fails.*

Clearly, since refutation in the second sense has an expressly dialectical goal, we may surely say that in such contexts, begging the question is a *dialectically unavailing* manoeuvre. Only at the risk of considerable distortion of the concept of error of reasoning is Jones's move an error of reasoning. For in making the error, the following pair of conditions can still hold true:

1. Jones had good reason to think that Smith would accept α.

2. Jones had good reason to think not only that $\alpha \models \sim \tau$ but that Smith would think this too.

Where, then, is Jones's alleged misreasoning to be found? The answer is that it cannot be found. Failed refutations in the second sense are not errors of reasoning.

6   Correction by contradiction

Since on the BEAUI-conception a fallacy is always an error of reasoning, question-begging cannot be a fallacy. Fine as far as it goes, it is also advisable to keep in mind that BEAUI-fallacies are errors of reasoning that satisfy the three additional

---

[24] Accordingly, we should also take note of the limiting case in which, although he presses α and $\alpha \models \sim \tau$ against Smith, he himself doesn't believe α but does believe that α is something to which Smith is committed. In these cases, the force of Jones's move is to confront Smith with a choice: "You cannot have it both that τ and α. So, if you're going to hold onto α, you'll have to give up on τ."

conditions of attractiveness, universality and incorrigibility. This reminds us that there are four ways, not one, in which reasoning can be fallacy-free. It can, as we have noted, fail to be an error. But, error or not, it can also fail to be attractive or universal or incorrigible.[25] Consider the universality requirement. As here intended, universality is not a matter of strict universal quantification. It is rather a generic matter. It is not that everyone whomsoever commits the error. Neither is it intended by the universality condition that a fallacy is an error that everyone commits all the time. It is universal in the sense that committing it in the appropriate circumstances is something that is typical of human reasoners to do. So we may say that a form of reasoning is universal when, (a), it is a form of reasoning which is in the repertoire of the typical reasoner and, (b), it is a form of reasoning to which he recurs with notable frequency. No one will think for a moment that this is as complete an account of universality as it would be desirable to have, but is enough to be getting on with here.[26] What I want now to suggest is that there is a second reason for saying that question-begging fails the universality requirement. We have already touched on the inclination of some logicians to link question-begging to circularity. This quite naturally calls to mind the following picture: That where τ is some thesis advanced by Smith, any utterance by Jones of an α that immediately implies the negation of τ is question-begging. I want to say two things about this picture. First, the practice it documents is empirically widespread in human dialectical practice. Second, in the general case, it is not question-begging. The two most conspicuous instances of this behaviour are *correction by contradiction* and *counterexampling*, the first of which we take up now, and the second in the section to follow.

---

[25] It can also be an error that is attractive, universal and incorrigible, but not *bad*, concerning which see [6].

[26] Here is a case. Suppose that Jones is examining a sample. Suppose that he is interested in whether it supports the generalization of which it is an instantiation. If the sample is a properly representative one, the generalization may be made. If not, not. (This, anyhow, is the standard story. Let it stand for present purposes.) Jones has two generalization-options. This is the *context* for his subsequent determinations. It bears on this in a crucial way that our record as generalizers from instantial samples is actually quite good. Informally and intuitively, the probability of getting our generalizations right is quite high. We might say that, when it comes to generalization, beings like us have a significantly positive *track-record TR*. If in the present context Jones generalizes correctly, well and good. It is what the relevant *TR* would predict. If he generalizes incorrectly, he has committed the fallacy of hasty generalization. By the universality requirement, hasty generalization is a manoeuvre in Jones's repertoire that is applied with a "notable frequency". We can now be a bit more precise about this. Hasty generalization occurs with a notable frequency if and only if in contexts such as these, the frequency of commission is, in the light of *TR*, *anomalous*. Informally, its rate of commission exceeds *TR*'s "margin of error". But this is not quite right. The universality requirement applies not just to Jones but to people generally. How are we to accommodate this feature of it? This we might do by giving due weight to the illusional character of error-making. Again, errors are missteps that *take us in*. Illusions in turn are caused. They don't just befall us. Let e be any error that Jones has committed just now. Let E be the etiology of that error. Then e is an error that meets the universality condition to the extent that whatever in E triggers e in the present case would (generically) trigger e for anyone else in Jones's situation.

Far and away the most common situation in which a proposition is used against an opponent and implies in one step the negation of what the opponent holds, is correction by contradiction. Here are some examples, with Smith the proponent and Jones the corrector.

1. Smith says: "Tomorrow is Barb's birthday." Jones replies: "No, it's the day after." Smith responds: "Oh, I see."

2. Smith says: "Harry is a bachelor and is married to Sarah." Jones: "Bachelorhood is defined in such a way that that can't be true." Smith: "Of course. I was speaking loosely."

3. Smith: "Some ravens aren't black." Jones: "Oh no, all ravens are black." Smith: "I must have been thinking of swans."

It is easy to see that in each case Smith contradicts Jones by uttering a sentence which in one fell swoop delivers the negation of what Smith says. In two respects this kind of case differs from those we have been considering. One is that when I contradict you by uttering a α that implies the negation of what you say, I needn't be attributing to you the belief that α. In some cases, I am *informing* you of something of which you appear to be unaware. In others I am reminding you of something you seem to have forgotten. The other is that a context in which I contradict you in this way needn't be one in which you are actually defending τ. In fact, when it succeeds, correction by contradiction pre-empts defence; that is to say, obviates the need for it.

Of course, contradiction-exchanges sometimes do not terminate with Smith's withdrawal of his claim. There are situations in which Smith won't accept the contradicting claim of Jones. When this happens, it would be question-begging of Jones to persist with it. It would be a stupid way to argue, but it would not, as we have seen, be an error of reasoning.

7   Counterexamples

Closely related to correction by contradiction is the use of counterexamples as a critical device. As it has evolved in philosophical practice, β is a successful counterexample of α only if β immediately implies $\sim \alpha$, and typically one of three further conditions is met.

1. α is a generalization and β is a true negative instance of it.

2. α is a definition and β is a true conjunction that instantiates its *definiens* and fails to instantiate its *definiendum, or vice versa.*[27]

---

[27] Similarly for equivalences generally.

3. α is an implication-statement and β is a true conjunction of its antecedent and a contrary of its consequent.

It is interesting to observe an apparent asymmetry between producing a counterexample and begging a question. It is an asymmetry of which both parts pivot on the factor of presumed obviousness. Accordingly, whereas "a is a F that is not G" is, if true, a successful counterexample of "All F are G", "All F are G", even if true, has the look of a question-begging move against "a is a F that is not G". Similarly, "This figure is a square that is not a rectangle", if true, is a successful counterexample to "A square $=$ df a rectangle", yet "A square $=$ df a rectangle" strikes us as question-begging against "This figure is a square that is not a rectangle". Again, "p and $\sim q$", if true, is a successful counterexample against "p implies q", but "p implies q" has the feel of a question begged against "p and $\sim q$". Examples such as these draw us to conjecture that

> *In their typical forms, successful counterexamples are the converses of apparent question-beggings.*

We appear to have it from this that

> *Neither counterexamplehood nor question-begging is closed under the relation of being the converse of.*

At the heart of these claims is the factor of obviousness. When β is a successful counterexample of a generalization or definition or entailment-statement α, it is taken that β obviously contradicts α and, in many cases, that, *once it is pointed out*, β is obviously true. In other words, successful counterexamples embed something that closely resembles correction by contradiction.

8   Explaining the apparent asymmetry

The issue before us divides into two subcases, depending on whether the interventions are taken as correcting an oversight or as providing new information.

Oversight. The oversight case preserves the asymmetry. If Smith's thesis is that all F are G (τ), he needn't be expected to believe, even if true, that a is an F that is G (α). This is perhaps the most valuable insight in Mill's famous analysis of general propositions: α is not in the belief-closure of τ. This being so, if "a is an F that is not G" is advanced against Smith's τ, it directly contradicts a proposition α that is not already in Smith's belief-set. Accordingly, unless there is independent reason to think that Smith is inconsistent with respect to α, there is room for the possibility that "a is an F that is not G" is something that Smith has indeed lost sight of. So pressing it against "All F are G" does not beg the question against Smith.

On the other hand, if Smith's thesis is that a is an F that is not G, then that not all F are G is squarely in Smith's belief-set. That is, "Not all F are G" is in the belief-closure of "*a* is an F that is not G". Accordingly, if Jones advances against Smith the claim that all F are G, he advances a proposition that directly contradicts something in Smith's belief-set, a proposition which, therefore, it cannot (assuming consistency) be supposed that Smith has merely overlooked. So pressing it against "a is an F that is not G" begs the question against Smith.

New information. Here the asymmetry is erased. In principle, a subject is free to accept as new information anything that immediately contradicts something he currently believes. In principle, then, the asymmetry under review disappears in any such context. It is hardly surprising, therefore, that it is a matter of empirical fact that practical agents are more open to accepting new information when it performs the function of a counterexample, rather than its converse.

A question-begging move is one in which Jones forwards against Smith a proposition α that Smith doesn't concede and that directly implies the contradictory of some thesis of Smith's or proves otherwise damaging to it. Schematically, Jones begs the question against Smith in pressing that $\langle \alpha / \sim \tau \rangle$. The empirical record clearly attests that in actual practice the great percentage of uses of $\langle \alpha / \sim \tau \rangle$ are in the contexts of correction by contradiction or counterexampling, neither of which is question-begging as such. It follows, then, that if there are uses of $\langle \alpha / \sim \tau \rangle$ which are question-begging, they are *minority* uses of it. This suggests the failure of the universality condition. If in its minority uses $\langle \alpha / \sim \tau \rangle$ is question-begging, it would need to be shown that such uses are both in the repertoires of the typical reasoner and are resorted to by him with notable frequency. But here, too, the empirical record suggests otherwise, especially in the second instance. We may therefore conclude that

> *In the form in which it has attracted the attention of logicians (i.e., as a move in the form* $\langle \alpha / \sim \tau \rangle$*), question-begging fails the universality requirement, and which is a further reason that it fails to be a BEAUI-fallacy.*

9   Babbling

In its present-day meaning, babbling is foolish, excited or confused talk. For Aristotle, however, it has a quite different meaning which I shall now briefly explain. According to Aristotle, one babbles when one repeatedly re-asserts one's own thesis (*Soph. Ref.* 3, 165b, 15-17). Suppose that Jones has placed Smith's thesis τ under challenge. Suppose further that until now Smith has never been challenged to defend τ, that, until now, Smith has simply taken it for granted that τ is a proposition that everyone sees as obviously true. (For concreteness, let τ be the proposition that same-sex marriage is morally (or metaphysically) unsupportable.) It is easy to see the difficulty that Jones has placed Smith in. For, while it cannot be ruled out that Smith has a perfectly satisfactory case to make for τ, very often in just this kind of situation, he has no case to make. If, up until now, no case has been demanded, if, up until now, no case has appeared necessary, it is not surprising that Smith may lack the resources to mount a defence of τ there and then.[28]

In such cases, Smith is faced with two options, both of them unattractive. One is to admit that τ is a proposition which he is unable to defend.[29] The other is to stand mute. The first is unattractive since it snags the presumption that challenges that draw no defence require the surrender of the thesis in question. The other is unattractive, since it convicts Smith of unresponsiveness. It is not uncommon in such cases for Smith simply to re-assert or re-phrase τ. In so doing, he performs the minimally necessary task of avoiding the other two options, each of which would be presumed to call for capitulation. In so doing, he keeps the conversation going; he keeps his own view of the matter "on the table". Re-asserting a proposition that is under attack is babbling in Aristotle's sense. It is not as such an error. It is not a case of begging the question. However, re-asserting it as its own defence does beg the question. It attributes to his attacker a proposition it is clear that he does not concede. Since this is evident to both parties, the begging of the question is rather stupid. It is a dialectically unavailing thing to do. But it hardly ever happens. Sticking to your guns when under attack is one thing. Using the proposition under attack *as its own defence* is another thing (and a comparative rarity).[30]

Babbling is a kind of question-begging. It most nearly resembles a defence of τ in the form "Why, τ is obvious". It is a form of question-begging which reverses the roles of Smith and Jones. In the modern sense, Jones, the challenger, begs the question against Smith by attributing to Smith a proposition he doesn't concede. In the present case, it is Smith, the attacked, who begs the question against Jones. The question is whether it is a fallacy. The answer is that it is not a fallacy on the BEAUI-conception of it. For no one could miss that attributing to Jones acceptance of the very proposition that he judges that Smith has no right to hold, is a blatant premiss-attribution error. It is an error, therefore, that fails the attractiveness condition.

10 Inconsistent commitments

Suppose that Smith thinks that λ is false. Let $\sum$ be any set of propositions that entail λ's truth. On the face of it, Smith has no recourse but to reject $\sum$. One of the

---

[28]The dialectical vulnerabilities of this kind of case are discussed in greater detail in [36].

[29]Lest I be accused of using a loaded example, the present point applies equally to, say, the proposition that a person's life is his as a moral right.

[30]Rare as it is, the practice of defending a proposition by re-iteration or by way of a trivial equivalence is not unheard of, especially among philosophers. A case in point is the defendant's response to a challenge to a proposition he takes to be a "first principle" or analytically true. See here [38, chapter 4] and [40].

reasons for this is that what Smith thinks is so implies the falsity of the conjunction of the propositions in Σ. Another is that if – on whatever independent grounds – Smith came to accept Σ, he would have begged the question *against himself.* He would have begged the question against himself in a rather special way. Although Smith now concedes the propositions in Σ, he is also committed to rejecting them. He is committed to rejecting them in virtue of his persistence with the falsehood of λ. Upon reflection, however, it isn't very instructive to parse this situation as one in which Smith has begged the question against himself. What has actually happened is that Smith has fallen into an utterly common belief-update problem. Initially taking λ to be false, Smith comes to hold beliefs incompatible with λ's falsity. Usually, when this is the case, people in Smith's position will restore consistency by changing their minds about λ or about one or more of the propositions in Σ. Sometimes they will be unaware of the inconsistency into which they have fallen. In other cases, they will know it, but will not know how best to climb out of the difficulty. Smith has found himself in an inconsistency-management difficulty, which he will handle intelligently or stupidly or in some other way. In so doing, he may fail to reason in ways that best addresses his problem. But no one seriously thinks that, either in getting into it or getting out of it, circularity or question-begging is the culprit.

11   Spurious reflexivities

I have been saying that the form in which it tends to attract the attention of logicians, question-begging is reasoning in the form $\langle \alpha / \sim \tau \rangle$ in which α is unconceded by the defender and immediately contradicts or otherwise damages his thesis τ. In this section and the next I want to touch on forms of question-begging that tend to attract the attention of philosophers.

Consider now what we might call *spurious reflexivities.* In their pure form, spurious reflexivities are sentences in the form $\alpha \, R \, \alpha$, in which α is a sentence and R a non-reflexive relation on sentences. Consider for concreteness various interpretations of R: "causes", "explains", "justifies", and "proves" (in its common sense meaning). A related category is that of circular definition and circular analysis, presented in formulations such as "Being a δ is what the definition of δ-hood is" and "Being δ analyzes what it is to be a δ". No one would want seriously to deny that such cases give rise to genuinely interesting problems, of which the so-called Paradox of Analysis is perhaps the most venerable. But it is clear that none of these spurious reflexivities comes within reach of anyone's conception of fallacy, and certainly, in any case, not the BEAUI-conception. Let us take a representative example: "That α is the case explains α's being the case." This is an error, needless to say. But the error is not circularity; the error is *falsity.* It is the error of supposing that explanation is reflexive.

12 Triviality

The Paradox of Analysis is one of a family of problems having to do with triviality. Its modern version was introduced by the American logician C.H. Langford [19], although there are clear anticipations of it in antiquity, most notably in Plato's *Meno* and Aristotle's *Posterior Analytics*. In Langford's version, the paradox arises as follows. Suppose we have a concept A for which we seek a conceptual analysis. Suppose that someone proposes that something is an *A* if and only if it is a *BC*. If the putative analysis is correct then "*A*" and "*BC*" will have just the same information. If that is so, then the analysis is trivial. On the other hand, if it is not trivial to characterize an *A* as a *BC*, the analysis is false.

Another source of the same problem is the complaint of Sextus Empiricus that all syllogisms beg the question or, in its modern variant, that all valid arguments beg the question [23, chapter 17]. Here, too, the concept of information lies at the heart of the problem. For if an argument is valid, all the information contained in its conclusion is contained in its premises. Let us call this version the Paradox of Validity.

The connection with circularity is hard not to miss. Circularity may be understood in at least two ways. In one, it is a certain kind of linking of one and the same syntactic item ("Since α, then α"). In a second, it is a certain kind of linking of semantically equivalent items ("Harry is a bachelor because Harry is a man who has never married"). The Paradoxes of Analysis and Validity introduce a third conception. Circularity is a certain kind of linking of expressions having the same (or subsuming) information-content $\alpha, \alpha \rightarrow \beta/\beta$.

The conclusion of a valid argument produces no new information, that is, no information not already contained in its premises. Central to this claim is the notion of information-measure, introduced into the literature by Shannon and Weaver [24]. In this technical sense, not only is Sextus' claim confirmed, but so too is the classical theorem – *ex falso quodlibet* – according to which an argument with inconsistent premises is valid for any conclusion. Let α and $\sim \alpha$ be premises. Then since α, $\sim \alpha$ contains all information, there is no conclusion β which contains information not contained in α, $\sim \alpha$. We may say, then, that if, in the claim that an argument is valid if and only if its conclusion contains no new information, the embedded notion of information-quantity is that of Shannon and Weaver, then valid arguments are always circular. On the other hand, if information is taken as *propositional content*, then neither Sextus' claim nor *ex falso quodlibet* stands up to scrutiny.

Seen the first way, a valid argument can't be informative. Seen the second way, it can be informative. Seen the first way, a correct conceptual analysis of something can't be informative. Seen the second way, it can be. Accordingly, a correct conceptual analysis can be circular (in the Shannon and Weaver sense) without appearing to be (since it may not in fact be circular in the propositional content sense). This being so,

> *While there is a sense in which a correct conceptual analysis is circular, it is not a sense in which circularity constitutes question-begging.*

All of this also plays on the problem of determining the closure conditions for belief sets. Let α/β be a valid argument, and let α be in Smith's belief set $\sum$. Is β in the deductive closure of $\sum$? On the Shannon-Weaver model, all the information in β is already in α. If believing β is just a matter of the information imparted by β being contained in what one believes, then in believing α, Smith believes β. So β is in the deductive closure of $\sum$. On the other hand, if believing β involves an affirmative understanding of it, it is wholly implausible to put β in the deductive closure of $\sum$ as a general principle. Accordingly, classical systems of belief-dynamics, such as [1], in which belief is closed under deduction, are right in so saying if belief is taken in the first way, but quite wrong if belief is taken in the second way.

13   Re-orienting ourselves

The nub of what I have been saying here is that attribution to another party of a commitment which contradicts or otherwise damages some proposition he has an interest in defending can hardly be regarded as an error of reasoning if the attribution is unconceded. Further, in most of those contexts in which such moves are actually made by beings like us, they are either not attribution-errors because they are not attributions (the "new information" cases) or not attribution-errors because they are true (the "overlooked" cases). What is more, while there are some cases in which $\langle \alpha / \sim \tau \rangle$ does employ an α that begs the question against the thesis of another party, these are dialectically unavailing moves to make, but they are not errors of reasoning.

If the onus of the preceding pages has now been met, we may say that the Negative Thesis is lent a degree of positive support by the real story of begging the question. Even so, most of the work required for the justification of this piece of heterodoxy still waits doing. This might seem to make of our present result rather small beer. For if there are eighteen things claimed to be fallacies, and begging the question is falsely in the embrace of that claim, that still leaves the other seventeen to take the proper measure of.

On the principle that one can only journey one step at a time, perhaps a record of one-out-of-eighteen is not so bad a result early in the proceedings. But we might also consider with some profit what we should think of a perfect record were possible eventually to produce it. In that case, none of the traditional fallacies would be in the extension of the traditional concept of fallacy. Although some people might see this as a kind of *tour de force*, others might have questions that evince a degree of under-whelmedness. Here are some of them.

1. If none of the traditional candidates (the "usual suspects") is in the extension
   of the traditional concept of fallacy, what if anything, *is*?[31]

2. If, it turned out, that the traditional concept of fallacy were actually *empty*
   (or nearly so), wouldn't that show that the traditional concept of fallacy is
   like (or nearly like) the traditional conception of phlogiston?

3. If it were true that none of the eighteen is a fallacy in the traditional sense,
   might it not be the case that our rather firm inuition that they *are* fallacies
   would be better served by a concept of fallacy that preserves it? And, if
   so, shouldn't we after all consider giving pride of place to the *dialectical*
   concept of fallacy, along the line proposed by Hintikka and others?[32]

These are fair and interesting questions. I see no reason to be discouraged by
them, or to be led to think that the emptiness or near-emptiness of the traditional
concept of fallacy is an underwhelming result. On the contrary, it would be a
result with lots of whelm. It would lend some encouragement to two interesting
conjectures which, if true, would be important to know.

Conjecture 1 In its traditional preoccupation with deductive invalidity and inductive weakness, as well as with the fallacies as traditionally conceived, logic has not
managed to engage the concept of error in a central way.

Conjecture 2 Granting that the eighteen are in *some* non-trivial sense errors, our failure
to convict them of fallaciousness on the logician's traditional understanding
of "error of reasoning" and "fallacy" suggests that human reasoners are not
sufficiently inadept at citing and drawing consequences to produce a track-
record widespread enough and bad enough to satisfy the traditional concept
of fallacy. Needless to say, beings like us hold lots of false beliefs, some of
which are nothing short of appalling. But if the present suggestion has merit,
these will not in the main be the result of faulty consequence-management.

Besides, it was quite worthwhile to learn that phlogiston was, well, *nothing*.[33]

---

[31]One prominent candidate is the "conjunction fallacy" of [27]. For reservations see [7].

[32]For example, [30].

[33]I take pleasure in dedicating this paper to our admired colleague Shahid Rahman on the occasion
of his 52nd birthday. For helpful comments or instructive demurrals, I warmly thank Peter Bruza,
the late Jonathan Cohen, Bas van Fraassen, Dale Jacquette, Lawrence Powers, Patrick Suppes and,
of course, my colleague in the investigation of the logic of practical reasoning, Dov Gabbay. I am
especially indebted to the editors' referee for astute and constructive criticism. I also acknowledge
with gratitude financial support from The Abductive Systems Group at UBC and the Engineering and
Physical Sciences Research Council of the United Kingdom, and the technical assistance of Carol
Woods in Vancouver.

BIBLIOGRAPHY

[1] C.A. Alchurron, P. Gardenfors and D. Makinson, "On the logic of theory change: Partial meet, contraction and revision functions", *The Journal of Symbolic Logic*, 50, pp. 510-530, 1985.

[2] Aristotle, *The Complete Works of Aristotle*, translated and edited by Jonathan Barnes, Princeton NJ: Princeton University Press, 1984.

[3] John A. Barker, "The fallacy of begging the question", *Dialogue*, 15 (1976), pp. 241-255.

[4] Dov M. Gabbay and John Woods, *Agenda Relevance : A Study in Formal Pragmatics, volume 1 of A Practical Logic of Cognitive Systems*, Amsterdam: Elsevier/North-Holland, 2003.

[5] Dov M. Gabbay and John Woods, *The Reach of Abduction: Insight and Trial, volume 2 of A Practical Logic of Cognitive Systems*, Amsterdam: Elsevier/North-Holland, 2005.

[6] Dov M. Gabbay and John Woods, "Error", *Journal of Logic and Computation*, forthcoming in 2007a.

[7] Dov M. Gabbay and John Woods, "Probability in the law". In this volume, pp. 149-188.

[8] Dov M. Gabbay and John Woods, *Seductions and Shortcuts: Error in the Cognitive Economy*, volume 3 of *A Practical Logic of Cognitive Systems*, Amsterdam: Elsevier/North-Holland to appear in 2008.

[9] Donald Gillies, "A rapprochement between deductive and inductive logic", *Bulletin of the IGPL*, 2, pp. 149-166, 1994.

[10] Trudy Govier, "Reply to Massay". In *Fallacies: Classical and Contemporary Readings*, editors Hans V. Hansen and Robert C. Pinto, pp. 172-180. University Park, PA: The Pennsylvania State University Press, 1995.

[11] C.L. Hamblin, *Fallacies*, London: Methuen, 1970.

[12] Hans V. Hansen, "The straw thing of fallacy theory", *Argumentation*, 16, pp. 133-155, 2002.

[13] Jaakko Hintikka, "The fallacy of fallacies", *Argumentation*, I, pp. 211-238, 1987.

[14] Jaakko Hintikka, "The role of logic in argumentation", The Monist, 72, pp. 3-24, 1989.

[15] Jaakko Hintikka, "What was Aristotle doing in his early logic anyway? A reply to Woods and Hansen", Synthese, 113, 2pp. 41-249, 1997.

[16] David Hitchcock, "Why there is no *argumentum ad hominem*" (HTML), 2006.

[17] Dale Jacquette, "Many questions begs the question (but questions do not beg the question)", *Argumentation*, 8, pp. 283-289, 1994.

[18] Oliver Johnson, "Begging the question", *Dialogue*, 6, pp. 135-160, 1967.

[19] C.H. Langford, "Moore's notion of analysis". In *The Philosophy of G.E. Moore*, edited by P.A. Schilpp. pp. 319-342, Chicago: Open Court, 1942.

[20] J. Locke, "An Essay concerning Humane Understanding", edited by Peter H. Nidditch (Oxford: Clarendon, 1975).

[21] J.D. Mackenzie, "Question-begging in non-cumulative systems", *Journal of Philosophical Logic*, 8, pp. 117-133, 1978.

[22] Lawrence Powers, "Equivocation". In *Fallacies: Classical and Contemporary Readings*, edited by Hans V. Hansen and Robert C. Pinto, pp. 287-301. University Park, PA: Pennsylvania State University Press, 1995.

[23] Sextus Empiricus, *Outlines of Pyrrhonism, Works*, vol. 1. Translation by R.G. Bury. Originally published in the 2nd-3rd century, London: Loeb Classical Library, 1933.

[24] Claude E. Shannon and Warren Weaver, *The Mathematical Theory of Communication*, Urbana, Chicago, London: University of Illinois Press, 1963.

[25] David H. Sanford, "Begging the question", Analysis, 32, pp. 197-199, 1972.

[26] Michael Scriven, *Reasoning*, New York: McGraw-Hill, 1976.

[27] Amos Tversky and Daniel Kahneman "Judgment under Uncertainty: Heuristics and Biases", *Science* Vol. 185. no. 4157, pp. 1124 - 1131, 1974.

[28] Frans H. van Eemeren and Rob Grootendorst, *Speech Acts in Argumentative Discussions*, Dordrecht: Foris, 1984.

[29] Douglas Walton, *Begging the Question*, New York: Greenwood Press, 1991.

[30] Douglas Walton, *A Pragmatic Theory of Fallacy*, Tuscaloosa: University of Alabama Press, 1995.

[31] John Woods and Douglas Walton, "*Petitio principii*", *Synthese*, pp. 107-127, 1975.

[32] John Woods and Douglas Walton, *"Petitio* and relevant many-premissed arguments", *Logique et Analyse*, 20, pp. 97-110, 1977.

[33] John Woods and Douglas Walton, "Arresting circles in formal dialogues", *Journal of Philosophical Logic*, 7, pp. 73-90, 1978.

[34] John Woods and Hans Hansen, "Hintikka on Aristotle's fallacies", *Synthese*, 113, pp. 217-239, 1997.

[35] John Woods, Andrew Irvine and Douglas Walton, *Argument: Critical Thinking, Logic and The Fallacies*, viii, 344, Toronto: Prentice-Hall, 2000.

[36] John Woods, "Slippery slopes and collapsing taboos", *Argumentation*, 4, pp. 107-134, 2000.

[37] John Woods, Aristotle's Earlier Logic, Oxford: Hermes Publishing Ltd., 2001.

[38] John Woods, *Paradox and Paraconsistency: Conflict Resolution in the Abstract Sciences*. Cambridge and New York: Cambridge University Press, 2003.

[39] John Woods, *The Death of Argument: Fallacies in Agent-Based Reasoning*, Dordrecht and Boston: Kluwer, 2004.

[40] John Woods, "Dialectical considerations on the logic of contradiction", *Logic Journal of the IGPL*, 13, pp. 231-260, 2005.

[41] John Woods, "Pragma-dialectics reconsidered". In *Considering Pragma-Dialectics*, edited by Agns van Rees and Peter Houtlosser, 301-311, Mahwah, NJ: Erlbaum, 2006.

[42] John Woods, "SE 176a 10-12: Many questions for Julius Moravcsik". In a presently untitled volume, edited by Dagfinn Follesdal and John Woods. London: College Publications, to appear in 2007a.

[43] John Woods, "Eight theses reflecting on Stephen Toulmin". In *Arguing on the Toulmin Model: New Essays in Argument Analysis and Evaluation*, edited by David Hitchcock and Bart Verheij, pp. 379-397, Amsterdam: Springer Netherlands, to appear 2007b.

[44] John Woods, "Lightening up on the *ad hominem*", *Informal Logic*, 27, pp. 109-134, 2007.

[45] John Woods, "The concept of fallacy is empty: A resource-bound approach to error". In the presently untitled MBR 06 China Proceedings, edited by Ping Li and Lorenzo Magnani. Amsterdam: Springer Netherlands, to appear in 2007d.

INDEX

Contact Information

AUTHORS AND EDITORS

1 Shahid Rahman

Shahid RAHMAN
U.F.R. de Philosophie
Université Lille 3
France
shahid.rahman@univ-lille3.fr
http://stl.recherche.univ-lille3.fr/sitespersonnels/rahman/
accueilrahman.html

2 The Contributors

Philippe BALBIANI
Institut de recherche en informatique de Toulouse
Université Paul Sabatier, Toulouse III
France
balbiani@irit.fr
http://www.irit.fr/~Philippe.Balbiani/

Diderik BATENS
Centre for Logic and Philosophy of Science
Universiteit Gent, Belgium
Diderik.Batens@UGent.be
http://logica.ugent.be/dirk/

Johan van BENTHEM
Institute for Logic, Language ∧ Computation
University of Amsterdam, The Netherlands
johan@science.uva.nl
http://staff.science.uva.nl/~johan/index.html
∧ Department of Philosophy
Stanford University, CA, USA
johan@csli.stanford.edu
www-philosophy.stanford.edu/fss/jvb.html

Giacomo BONANNO
Department of Economics
University of California at Davis, CA, USA
gfbonanno@ucdavis.edu
http://www.econ.ucdavis.edu/faculty/bonanno/

Walter A. CARNIELLI
Department of Philosophy
∧ Centre for Logic, Epistemology ∧ the History of Science - CLE
State University of Campinas, Brazil
carnielli@cle.unicamp.br
http://www.cle.unicamp.br/prof/carnielli/

Newton C.A. da COSTA
Departamento de Filosofia
Universidade Federal de Santa Catarina
Brazil
ncacosta@terra.com.br

Michel CRUBELLIER
U.F.R. de Philosophie
Université Lille 3
France
mcrubellier@nordnet.fr

Francisco A. DÓRIA
Professor of Communications, Emeritus
Federal University at Rio de Janeiro
Brazil
famadoria1@gmail.com

Dov M. GABBAY
Department of Computer Science
King's College London
Great Britain
dov.gabbay@kcl.ac.uk
http://www.dcs.kcl.ac.uk/staff/dg/

Olivier GASQUET
Institut de recherche en informatique de Toulouse
Université Paul Sabatier, Toulouse III
France
olivier.gasquet@irit.fr
http://www.irit.fr/~Olivier.Gasquet/

Gerhard HEINZMANN
Laboratoire d'Histoire des Sciences et de Philosophie Archives Henri Poincaré
Université Nancy 2
France
gerhard.heinzmann@univ-nancy2.fr
`http://poincare.univ-nancy2.fr/Presentation/?contentId=1502`

Andreas HERZIG
Institut de recherche en informatique de Toulouse
Université Paul Sabatier, Toulouse III
France
herzig@irit.fr
`http://www.irit.fr/~Andreas.Herzig/`

Jaakko HINTIKKA
Department of Philosophy
Boston University
USA
hintikka@bu.edu
`http://www.bu.edu/philo/faculty/hintikka.html`

Justine JACOT
U.F.R. de Philosophie
Université Lille 3
France
qualia1@free.fr
`http://stl.recherche.univ-lille3.fr/sitespersonnels/rahman/rahmanequipejacot.html`

Reinhard KAHLE
CENTRIA and DM
Universidade Nova de Lisboa
Portugal
kahle@mat.uc.pt
`http://www.mat.uc.pt/~kahle`

Erik C.W. KRABBE
Faculty of Philosophy
Universität Groningen
The Netherlands
e.c.w.krabbe@rug.nl
`http://www.rug.nl/staff/e.c.w.krabbe/index`

Décio KRAUSE
Departamento de Filosofia
Universidade Federal de Santa Catarina
Brazil
deciokrause@gmail.com
http://www.cfh.ufsc.br/~dkrause/

Franck LIHOREAU
Instituto de Filosofia da Linguagem
Universidade Nova de Lisboa
Portugal
franck.lihoreau@fcsh.unl.pt
http://www.ifl.pt/main/Pessoas/FranckLihoreau/tabid/168/
Default.aspx

Kuno LORENZ
Philosophisches Institut
Universität des Saarlandes
Germany
klorenz@rz.uni-saarland.de

Ilkka NIINILUOTO
Department of Philosphy
University of Helsinki
Finland
ilkka.niiniluoto@helsinki.fi
http://www.helsinki.fi/filosofia/filo/henk/niiniluoto.htm

Graham PRIEST
Departments of Philosophy
University of Melbourne, Australia
and University of St Andrews, UK
g.priest@unimelb.edu.au
http://www.st-andrews.ac.uk//philosophy/gp.html

Stephen READ
Department of Logic and Metaphysics
University of St. Andrews
Great Britain
slr@st-and.ac.uk
http://www.st-andrews.ac.uk/~slr/read.html

Manuel REBUSCHI
Laboratoire d'Histoire des Sciences et de Philosophie Archives Henri Poincaré
Université Nancy 2
France
manuel.rebuschi@univ-nancy2.fr
`http://poincare.univ-nancy2.fr/Presentation/?contentId=`
`1513`

Greg RESTALL
Philosophy Department
University of Melbourne
Australia
greg@consequently.org
`http://consequently.org/`

Gabriel SANDU
IHPST
Université Paris 1
France
sandu@mappi.helsinki.fi
  `http://www-ihpst.univ-paris1.fr/gsandu`

Gerhard SCHURZ
Lehrstuhl für Theoretische Philosophie
Universität Düsseldorf
Germany
schurz@phil-fak.uni-duesseldorf.de
`http://www.phil-fak.uni-duesseldorf.de/philo/personal/thphil/`
`schurz/`

François SCHWARZENTRUBER
Institut de recherche en informatique de Toulouse
Université Paul Sabatier, Toulouse III
France
schwarze@irit.fr
`http://www.irit.fr/~Francois.Schwarzentruber/`

Yaroslav SHRAMKO
Department of Philosophy
State Pedagogical University
Ukraine
shramko@rocketmail.com
`http://www.tu-dresden.de/phfiph/prof/lowiphil/gaeste/shramko/`
`yse.htm`

Göran SUNDHOLM
Faculteit der Wijsbegeerte
Universiteit Leiden
The Netherlands
b.g.sundholm@phil.leidenuniv.nl
http://www.filosofie.leidenuniv.nl/index.php3?c=36

John SYMONS
Department of Philosophy
University of Texas at El Paso
USA
jsymons@utep.edu
http://www.johnfsymons.com/

Christian THIEL
Institut für Philosophie
Universität Erlangen-Nürnberg
Germany
Christian.Thiel@sophie.phil.uni-erlangen.de
http://www.philosophie.phil.uni-erlangen.de/Mitarbeiter.
Institut/Thiel.html

Nicolas TROQUARD
Department of Computer Science
University of Liverpool
Great Britain
nico@liverpool.ac.uk
http://www.csc.liv.ac.uk/~nico/

Tero TULENHEIMO
Department of Philosophy
University of Helsinki
Finland
tero.tulenheimo@helsinki.fi
http://www.helsinki.fi/filosofia/filo/henk/tulenheimo.htm

Jean Paul VAN BENDEGEM
Centre for Logic and Philosophy of Science
Vrije Universiteit Brussel
Belgium
jpvbende@vub.ac.be
http://www.vub.ac.be/CLWF/members/jean/

Daniel VANDERVEKEN
Department of Philosophy,
University of Quebec at Trois-Rivières
Canada
daniel.vanderveken@uqtr.ca
http://www.uqtr.ca/~vandervk/cvdveng.htm

Yde VENEMA
Institute for Logic, Language ∧ Computation
University of Amsterdam
The Netherlands
Y.Venema@uva.nl
http://staff.science.uva.nl/~yde/

Heinrich WANSING
Institut für Philosophy
Technische Universität Dresden
Germany
Heinrich.Wansing@tu-dresden.de
http://tu-dresden.de/die_tu_dresden/fakultaeten/philosophische_
fakultaet/iph/prof

Jan WOLENSKI
Institute for Philosophy
Jagiellonian University Krakow
Poland
j.wolenski@iphils.uj.edu.pl

John WOODS
Department of Philosophy
University of British Columbia
Canada
JHWoods@interchange.ubc.ca
http://www.johnwoods.ca/

3   The Editors

Cédric DÉGREMONT
Institute for Logic, Language ∧ Computation
University of Amsterdam
The Netherlands
cedric.uva@gmail.com
http://staff.science.uva.nl/~cdegremo/

Laurent KEIFF
U.F.R. de Philosophie
Université Lille 3
France
laurent.keiff@gmail.com
http://stl.recherche.univ-lille3.fr/sitespersonnels/rahman/
rahmanequipeKeiff.html

Helge RÜCKERT
Lehrstuhl Philosophie II
Universität Mannheim
Germany
rueckert@rumms.uni-mannheim.de
http://www.phil.uni-mannheim.de/fakul/phil2/rueckert/index.
html